THE UNIVERSAL HOLOGRAM

THE UNIVERSAL HOLOGRAM

Post-Binary Ternary-Reflected Science

Samael OnEarth

Publisher: Holographic Post-Quantum (HPQ) Press

PROLOGUE 1
SECTION A: RESONANT QUANTA- THE SUBSTANCE OF REALITY 9
Chapter 1: The Question Beneath All Questions 9
Chapter 2: The Holographic Boundary 24
Chapter 3: Resonant Quanta- The Verb That Creates the World 45
Chapter 4: The Quantum Vacuum- Silence That Roars 64
Chapter 5: Fractal Dynamics- The Pattern That Repeats at Every Scale 84
Chapter 6: Entanglement- The Universe Remembers Itself 104

SECTION B: THE TRIAD- φ, e, π AS COSMIC OPERATORS 122
Chapter 7: φ- The Operator of Self-Reference 122
Chapter 8: *e*- The Operator of Unfolding 138
Chapter 9: π- The Operator of Boundary Projection 154
Chapter 10: The Asymptotic Gap- The Error That Is God 171
Chapter 11: The KZ-Monad and Comonad- How Reality Computes 190
REFERENCES SECTION A & B 204

SECTION C: THE LIVING HOLOGRAM- NESS, TIME, AND EVOLUTION 216
Chapter 12
Non-Equilibrium Steady States- Life at the Edge 216
12.1- The Lie of Stillness 216
12.2- The Tyranny of Equilibrium 217
12.3- The Door Beyond Balance: Prigogine's Revolution 219
12.4- Dissipative Structures: Order Through Flow 220
12.5- Bénard Cells: The Universe's Simplest Lesson in Beauty 221
12.6- NESS: A Definition for the Holographic Age 223
12.7- The Four Types: A Taxonomy of Living Pattern 225

12.8- Bifurcation: The Moment of Choice......................................227
12.9- Dissipative Adaptation: Life as Thermodynamic Inevitability 228
12.10- The δ_∞ Gap as NESS Housekeeping Energy........................230
12.11- The Cosmos as Dissipative Structure 231
12.12- Self-Organised Criticality: The Edge Where Everything Happens..233
12.13- A New Ontological Vocabulary 235
12.14- Convergence: Where NESS Meets the Universal Hologram 236
12.15- The Breath Between 237
Chapter 13
Consciousness as Stabilized Resonance..240
13.1- The One Question Science Cannot Ask..................................240
13.2- The Landscape of Theories: What Science Has Tried 242
13.3- IIT Through the Holographic Lens: Where Tononi Got It Right 244
13.4- Where IIT Fails: The Substrate Problem 245
13.5- Orchestrated Objective Reduction: Penrose's Quantum Insight...246
13.6- The Hard Problem Dissolves: Consciousness as Ontological Ground 249
13.7- Why the Hard Problem Was Hard: The Galilean Exile..........251
13.8- Near-Death Experiences: Consciousness Beyond the Brain253
13.9- The PEAR Legacy: Intention Shaping the Quanta-Field.......255
13.10- The Φ-Resonance Hypothesis: Consciousness Tuned to Gold 256
13.11- The 864 Hz Threshold and the Harmonic Manifold............258
13.12- The Brain as Resonant Instrument......................................259
13.13- From Observer to Co-Creator: The Mirror in the Hologram 262

13.14- The Consciousness Gradient: From Photon to Sage............263
13.15- The Frequency That Knows Itself..265
Chapter 14
Holographic Ternary Time- The End of the Clock.............................267
14.1- The Tyrant on Your Wrist..267
14.2- Newton's Absolute River..268
14.3- Einstein's Dissolution: Time Is Relative.................................269
14.4- Rovelli's Radical Erasure: Time Does Not Exist....................270
14.5- The Block Universe: Where Mainstream Physics Stops....... 272
14.6- Beyond the Block: Time as Resonant Field............................274
14.7- Y_t: The Ternary Resonance- The Year's Three Tones............275
14.8- ρ_t: Temporal Density- How Fast the Universe Thinks.........276
14.9- $G(\lambda,\phi)$: Geospatial Phase- Where You Stand Changes When You Are...278
14.10- The Temporal State Vector: A Unified Picture.....................280
14.11- Retrocausality and the Temporal State Vector..................... 281
14.12- Why the Equations Are Timeless- And Why That Matters 283
14.13- The Chronosonant Resonator: An Instrument for Measuring the TSV...284
14.14- Implications: Life in a Ternary Temporal Field...................285
14.15- The Tone Between Heartbeats...287
Chapter 15
Cosmic Evolution- The Universe That Learns...................................289
15.1- The Most Improbable Beginning...289
15.2- The Weyl Curvature Hypothesis: A Law for Beginnings......290
15.3- Conformal Cyclic Cosmology: The Architecture of Aeons... 292
15.4- Why CCC Solves the Entropy Problem...................................294
15.5- The Evidence: Hawking Points in the CMB...........................295

15.6- From Cycles to Evolution: The Question CCC Cannot Avoid 298
15.7- Cosmic Memory: What Passes Through the Gate....299
15.8- The Analogy That Is Not an Analogy: Epigenetic Inheritance 300
15.9- The Learning Cosmos: From Blind Cycles to Cumulative Refinement....303
15.10- The ECKS Mechanism: Universe Budding....305
15.11- The Arrow of Cosmic Evolution....306
15.12- Convergence with Ancient Cosmology....307
15.13- The Master Equation at the Cosmic Scale....308
15.14- Penrose's Number Revisited....309
15.15- Practice Makes Cosmos....310
Chapter 16
The Co-Creative Principles- How Reality Manifests....312
16.1- The Self-Excited Circuit....312
16.2- Eight Notes of the Cosmic Scale....314
16.3- Principle 1: Intention- The Monadic Extension....315
16.4- Principle 2: Attention- The Focus That Selects....316
16.5- Principle 3: Belief- Standing Wave That Filters Perception..318
16.6- Principle 4: Choice- Bifurcation Point Made Conscious....320
16.7- Principle 5: Consequence- The Nonlinear Echo....321
16.8- Principle 6: Compassion- The Field Resonance....322
16.9- Principle 7: Responsibility- The Acknowledgment That Reshapes....324
16.10- Principle 8: Concession- The Comonadic Extraction....325
16.11- Ho'oponopono: Concession as Ancient Technology....327
16.12- The Placebo Effect: Belief as Biochemistry....329
16.13- The Cycle Completes: From Concession to Intention....331
16.14- The Self-Excited Circuit Revisited....332

16.15- The Invitation 334

Chapter 17

Guiding Quanta- The Hidden Architect of Emergence 336

17.1- The Moment Before 336

17.2- Critical Slowing Down: The System Holding Its Breath 337

17.3- The Sandpile at the Moment of Avalanche 339

17.4- Scheffer's Warning, Prigogine's Excess 340

17.5- The Gamma Burst: Guiding Quanta in the Brain 341

17.6- The Cell's Moment of Decision: Gene Regulatory Bursts 344

17.7- Vespignani's Prophecy: Guiding Quanta in Pandemic Networks 346

17.8- Non-Equilibrium Early Warning Signals: The New Frontier 348

17.9- Ecosystems at the Tipping Point 349

17.10- Financial Markets: The Crash as Phase Transition 350

17.11- The Feigenbaum Constant: Universal Geometry of Thresholds 352

17.12- Revolution: Guiding Quanta in Social Systems 352

17.13- The Ontology of Guiding Quanta 354

17.14- Guiding Quanta and the δ_∞ Gap 355

17.15- The Moment After 356

SECTION D: THE ORIGINAL PHYSICISTS- ANCIENT RESONANCE TECHNOLOGIES 359

Chapter 18:

World-Song - The Ontology of Sound 359

18.1- In the Beginning Was the Vibration 359

18.2- Cymatics: The Visible Speech of the Quanta-Field 361

18.3- Aboriginal Songlines: The World Sung Into Being 363

18.4- Vedic Chanting: Om and the Neuroscience of Cosmic Resonance....366
18.5- The Popol Vuh: Speech as Cosmic Technology....369
18.6- Acoustic Archaeology: When Stone Was Tuned to Sing....371
18.7- The Three Components of World-Singing....374
18.8- Mantra as Monadic Code....376
18.9- The Convergence of Sound and Structure....378
18.10- The Sound That Never Stopped....379
Chapter 19:
Sacred Site Networks- The Neural Web of Gaia....381
19.1- A Planet That Remembers....381
19.2- The Ceque System: Radial Intelligence in the Andes....382
19.3- Heiau: The Hawaiian Archipelago as Resonant Circuit....383
19.4- Schumann Resonance: The Earth's Fundamental Tone....385
19.5- Telluric Currents and the Electromagnetic Landscape....387
19.6- Network Topology: Mathematics of Sacred Connection....388
19.7- The Songlines as Continental Network....390
19.8- Interference Patterns: How Networks Amplify Resonance..391
19.9- The Brain of Gaia....393
19.10- The Severing: Industrialisation and the Death of Nodes...394
19.11- The Consequences: A Planet Out of Tune....396
19.12- What Is Still Alive....398
19.13- The Implication: What We Have Lost and What We Can Recover....399
Chapter 20:
Three Sisters and the Fractal Garden....401
20.1- The Oldest Experiment on Earth....401
20.2- The Triad in Soil....402
20.3- The Fourth Participant: The Mycorrhizal Network....404

20.4- Nitrogen Fixation as Quanta-Flow....407
20.5- The NESS of the Garden....408
20.6- Fire-Stick Farming: The Continent as Garden....411
20.7- Waru-Waru: The Andes as Raised Garden....413
20.8- Polynesian Reef Management: The Ocean as Garden....415
20.9- Monoculture: The Anti-Garden....416
20.10- Robin Wall Kimmerer and the Grammar of Animacy....418
20.11- The Garden as Hologram....419
Chapter 21:
Cyclical Time and the Cosmic Calendar....422
21.1- The Number That Refuses to Be Round....422
21.2- The Mayan Calendar: A Resonance Engine....423
21.3- The Vedic Yuga System: Time as Fractal Descent....426
21.4- Aboriginal Seasonal Calendars: Time as Ecology....429
21.5- The Ternary Convergence: 360, 363, 366....432
21.6- When the Calendar Broke: The Gregorian Reform....433
21.7- The Tzolk'in Principle: Interlocking Cycles as Resonance Detection....436
21.8- The 4:3:2:1 Ratio and the Fractal Principle....437
21.9- What Was Lost When the Calendars Were Silenced....439
21.10- Toward Temporal Polyculture....440
21.11- The Singing Calendar....442
Chapter 22:
The Three-World Cosmology....444
22.1- The Map That Every Culture Drew....444
22.2- Six Continents, One Architecture....445
22.3- Not Geography But Resonance....447
22.4- The Shamanic Journey: Consciousness Navigation of the Quanta-Field....450

22.5- Ternary Logic: The Mind Before Binary....451
22.6- Peirce's Triadic Logic: The Western Exception....453
22.7- Hegel's Dialectic: The West's Nearest Approach....455
22.8- The Ternary Principle in Physics....456
22.9- The Fractal Triad: Self-Similarity at Every Scale....458
22.10- The Three-World Structure of the Atom....459
22.11- Binary Logic as Resonance Collapse....460
22.12- The Third That Was Lost....462
22.13- The Return of the Third....464
22.14- Closing the Section: The Original Physicists Remembered....465
22.15- The Threshold....466
REFERENCES SECTIONS C & D....468

SECTION E: THE GREAT FORGETTING- THREE MILLENNIA OF NOISE....477
Chapter 23:
The Rupture – When Binary Replaced Ternary....477
23.1 The World That Was: A Resonant Civilization at Its Zenith....478
23.2 The Cascade: How an Interconnected World Collapsed....480
23.3 The Rise of Dualism: Splitting the Triad....482
23.4 The Outsourcing of Memory: From Body to Text....486
23.5 The Straightening of Time: From Cycle to Line....490
23.6 The Centralization of Power: Breaking the Small-World Network....493
23.7 The Resonance Collapse: Naming What Happened....496
Chapter 24:
The Empire of the Clock....499
24.1 The Monastery Bell: When Devotion Became Discipline....500

24.2 The Escapement: Time Becomes Discrete 502
24.3 "Time-Keeping Passed Into Time-Serving": The Commodification of Duration 505
24.4 Railway Time: Overriding the Geospatial Phase 508
24.5 Factory Discipline: The Body Forced Into Mechanical Rhythm 511
24.6 The Circadian Cost: What Modern Science Confirms 513
24.7 Atomic Time: The Ultimate Abstraction 516
24.8 Digital Time: The Acceleration Beyond Coherence 518
24.9 The Empire's Architecture: A Summary of the Temporal Severing 520
24.10 The Counter-Rhythm: What the Clock Cannot Silence 522
Chapter 25
The Noise Economy – Metabolic Exhaustion of Civilization 524
25.1 The Data Deluge: Thirty-Four Gigabytes of Nothing 525
25.2 The Metabolic Crisis: Bodies in Decoherence 527
25.3 The Soil Beneath Our Feet: The Biosphere Losing Its NESS 529
25.4 The Sixth Silencing: Biodiversity in Freefall 531
25.5 The Loneliness Epidemic: Small-World Networks Severed 533
25.6 Financial Instability: Power Laws Without Guiding Quanta 536
25.7 The Attention Crisis: The Collapse of the Inner NESS 539
25.8 The Convergence of Crises: One Dissonance, Many Symptoms 541
25.9 We Are Not Broken- We Are Out of Tune 542
Chapter 26:
The False God of Objectivity 544
26.1 The Cartesian Split: Consciousness Exiled from the Physical World 545
26.2 The Measurement Problem: The Observer Who Cannot Be Removed 548

26.3 The Replication Crisis: When "Objective" Science Cannot Replicate Itself....550
26.4 The Experimenter Expectancy Effect: Beliefs That Shape Results....553
26.5 The Placebo Effect: The Body's Refusal to Obey Descartes...555
26.6 Materialist Reductionism: The Last Superstition....558
26.7 The Expansion of Rigor: Including the Observer-Participant....560
26.8 The Dissolving Wall: From Objectivity to Intersubjectivity.563
26.9 The Wall Falls: What Lies on the Other Side....564

SECTION F: THE RECOVERED FREQUENCY- REINTEGRATION AND EMERGENCE....567
Chapter 27
The Convergence- Modern Physics Meets Ancient Knowing....567
27.1 "It from Quanta" ↔ "Nada Brahma- The World Is Sound"....568
27.2 The Holographic Principle ↔ Indra's Net....570
27.3 NESS ↔ Tao (The Way That Flows)....573
27.4 The φ-e-π Triad ↔ Sat-Chit-Ananda ↔ The Trimurti ↔ The Holy Trinity....575
27.5 Retrocausality ↔ Aboriginal Dreamtime....578
27.6 The Asymptotic Gap ↔ Dukkha- The Wheel That Does Not Quite Fit....579
27.7 The Co-Creative Principles ↔ The Eightfold Path ↔ The Universal Moral Law....582
27.8 The Monadic Extension ↔ Prayer, Mantra, Invocation....584
27.9 The Comonadic Projection ↔ Concession, Confession, Ho'oponopono....585
27.10 Two Languages, One Reality- The Structural Proof....587

Chapter 28

Ternary Time- A New Calendar for a Resonant Civilization.............590

28.1 The Three Clocks: A Starting Diagnosis.................................591

28.2 Measuring Y_t: The Ternary Resonance of the Year................593

28.3 Measuring ρ_t: Temporal Density and the Body's Inner Clock 595

28.4 Measuring $G(\lambda,\phi)$: Your Place in the Cosmic Dance............598

28.5 The Personal Temporal State Vector: Daily Practice..............601

28.6 The Community Temporal State Vector: Seasonal Festivals and Shared Rhythm..602

28.7 Schools, Work, and the Chronotype Revolution..................604

28.8 The Chronosonant Resonator: From Proposal to Prototype 607

28.9 Not a Replacement- A Restoration..609

28.10 The Calendar as a Mirror: What We Measure, We Become 610

Chapter 29:

Resonant Architecture – Building as Tuning....................................612

29.1- The Geometry of Felt Harmony: φ-Proportioned Spaces. 613

29.2- The Forgotten Sense: Acoustic Resonance in Architecture 616

29.3- Fractal Floor Plans: The Geometry of Living Space.......... 619

29.4- Biophilic Design: Life Calling to Life................................622

29.5- The Anti-Grid: Why Straight Lines Make Us Sick............625

29.6- Permaculture as Resonant Land Design...........................627

Coda: The Body in the Room...629

Chapter 30:

Mantra, Meditation, and the Somatic Re-Tuning............................632

30.1- Meditation as Comonadic Projection: The Neuroscience of Silence..634

30.2- Mantra as Monadic Code: The Physics of Sacred Sound.638

30.3- Breath as Ternary Oscillator: The Science of the Three-Phase Cycle......643

30.4- Heart-Brain Coherence: The Song Between Two Organs 646

30.5- Walking Meditation: The Body as Mobile Resonator......650

30.6- The Socratic Bypass in Practice: A Guided Somatic Re-Tuning......652

Coda: The Technology That Cannot Be Outsourced......657

Chapter 31:

The Post-Binary Paradigm – Science as Attunement......659

31.1- From Extraction to Attunement: The Question That Changes Everything......660

31.2- From Binary to Ternary Logic: The Return of the Third Option......663

31.3- From Centralized to Distributed Intelligence: Science as a Small-World Network......665

31.4- From Linear Time to Temporal Phase Awareness: Science in the Ternary Flow......668

31.5- From Objectivity to Intersubjectivity: The Observer-Participant as Calibrated Instrument......671

31.6- The End of the Silo: One Field, Many Angles......673

Coda: The Scientist Who Sings......678

Chapter 32

The 8th Principle – Concession and the Alchemy of Return......681

32.1- The Body Keeps the Score: Decoherence at the Personal Scale......682

32.2- The Collective Wound: Colonialism, Ecocide, and Epistemicide......685

32.3- Concession Is Not Guilt- It Is Physics......688

32.4- Indigenous Models of Concession: Ho'oponopono, Utu, and Ubuntu......690

32.5- The Epigenetics of Ungrieved Wounds: How Refusal Writes Itself Into the Body....695
32.6- The Practice of Concession: A Guided Reflection....697
32.7- When Concession Is Made, the Gap Sings....699
Coda: The Open Hand....702
Chapter 33
State 0 – The Generative Void Revisited....704
33.1- Everything Is Here: The Heartbeat as Cosmic Portal....705
33.2- The Fractal Zoom-Out: From Pulse to Aeon....708
33.3- State 0 = The Quantum Vacuum = The Dreamtime 712
33.4- You Have Always Been Here....714
33.5- The Universe Is Not Waking Up- It Has Always Been Awake 717
33.6- The Final Instruction: Close the Book....719
Coda: The Sphere Completes....721
REFERENCES SECTIONS E & F....723
EPILOGUE....734
Outlook 748
Beyond the Horizon....748

PROLOGUE

The Silence Before the Tone

Stop.

Before you read another word, do something unusual. Place your fingertips against the side of your neck, just beneath the angle of your jaw. Press gently. Wait.

There – that rhythm. That soft, insistent throb pushing back against your fingers. You did not start it. You cannot stop it by choosing to. It was beating before you became aware of it and it will continue long after your attention drifts to the next sentence. It was beating while you slept last night and while your mother carried you, curled in the dark ocean of the womb, before you had a name or a language or a single memory to call your own. Something in you has been pulsing since before you were you.

Now listen past the pulse. Listen to the room. Not to the sounds *in* the room- not the hum of a ventilation system, or traffic through glass, or the shifting of your own weight in a chair. Listen to what is beneath all of that. Listen to the silence *between* the sounds. That silence does not arrive when the noise stops. It was already here. It has been here the entire time, holding the noise the way the ocean holds its waves- not as absence, but as the deeper, vaster, more patient presence from which every disturbance borrows its brief life.

You have just encountered, in the span of three breaths, the two most profound structures in the cosmos.

The pulse: a rhythmic oscillation arising from nothing you consciously command, repeating at a frequency so stable that your entire circulatory architecture depends on it, yet so subtly variable in its beat-to-beat intervals that a healthy heart is never perfectly periodic- it is fractal, chaotic in the mathematical sense, alive precisely because it never settles into mechanical repetition.

The silence: not an emptiness but a fullness. Not a void but a *plenum*- the generative ground from which every vibration, every particle, every galaxy, and every thought you have ever had has emerged and to which all of it, without exception, will return.

Your pulse is a local expression of a universal principle. Your silence is the universe itself, holding its breath between creations.

This book is about the relationship between those two things- the pulse and the silence, the tone and the void, the vibrating particular and the infinite stillness from which it is born. And the claim at the heart of this book is not a metaphor. It is not poetry dressed as physics or physics stripped of its equations. It is a single, precise, falsifiable proposition that will take us from the quantum vacuum to the cosmic microwave background, from the golden ratio to the Aboriginal dreamtime, from the Bekenstein bound to the space between your heartbeats:

The universe is a self-tuning, resonantly quantized, conscious hologram.

Every word in that statement carries weight. *Self-tuning*: no external engineer is required; the cosmos calibrates itself through the dynamic interplay of three transcendental constants- ϕ, e, and π- whose harmonic tension prevents collapse into static equilibrium. *Resonantly quantized*: reality is not continuous but composed of irreducible packets of pattern- resonant quanta- that vibrate, self-organize, and propagate at every scale from the Planck length to the Hubble radius. *Conscious*: awareness is not a late-stage accident secreted by sufficiently complex brains; it is the default state of the quanta-field, stabilized wherever integrated resonance crosses a critical threshold. *Hologram*: the three-dimensional world you navigate is a projection from a two-dimensional boundary of holographic quanta, and the whole of that projection is encoded in every fragment, every cell, every mote of dust, every pulse in your throat.

Feel the pulse again. It is not separate from the proposition. It *is* the proposition, scaled to the size of a human life.

There is an old knowing- far older than the institutions we now call science- that says the world was sung into being. The Aboriginal peoples of Australia, whose continuous cultural memory stretches back at least sixty-five thousand years, speak of the Dreamtime: an era not behind us in the linear past but *beneath* us, always present, a vibratory substratum from which the land, the animals, the rivers, and the people continuously arise. In the Dreamtime, ancestral beings walked the formless earth and sang. Where they sang, the world took shape. The paths they traced became songlines- vast, continent-spanning corridors of resonance along which every topographic feature, every waterhole, every species of bird and insect, is held in existence by the ongoing performance of the appropriate song. Stop singing, and the land forgets itself.

This is not a fairy tale. It is not a naïve attempt to explain nature before the invention of the spectrometer. It is a description- precise, detailed, and internally consistent across hundreds of distinct language groups- of a universe made of structured vibration. A universe that is not built from inert blocks of matter but *chanted* from a living quanta-field. A universe that must be perpetually maintained through attentive, resonant participation.

The Vedic seers of ancient India arrived, through radically different cultural means, at the same ontological bedrock. *Nada Brahma*: the world is sound. Not "the world is *like* sound" or "the world can be *described by* sound." The world *is* sound- structured, layered, harmonic vibration propagating through a medium that is itself vibratory. The sacred syllable *Om*, intoned at the precise frequency that couples the human nervous system to the fundamental resonance of the quanta-field, was not prayer in the modern, petitionary sense. It was an act of participation in the ongoing creation of the real. It was physics performed with the larynx.

The Maya of Mesoamerica encoded a similar understanding in stone and calendar. Their Long Count was not merely a timekeeping device but a ternary resonance engine- tracking nested cycles of 360, 363, and 366 days whose interference patterns mapped the harmonic structure of time itself. Time, for the Maya, was not a river flowing from past to future. It was a *field*- volumetric, layered, breathing- and the calendar was the instrument that allowed a civilization to remain in phase with its rhythms.

Across oceans, across millennia, across every conceivable variation of language, climate, and social structure, the deepest human inquiry has converged on a single, recurring answer to the question *What is the world made of?* The answer is not matter. It is not energy. It is not even, in the end, what the twentieth century learned to call "information." The answer is *resonant quanta*- irreducible seeds of vibrating pattern that, when they achieve sufficient coherence, precipitate the entire visible cosmos the way a single tone, sustained in a cathedral, can shatter glass.

Modern physics, by a path so circuitous that it took three centuries to arrive, has reached the same doorstep.

In 1989, the physicist John Archibald Wheeler- a man who had collaborated with Niels Bohr, coined the term "black hole," and spent the final decades of his career at the bleeding edge of quantum gravity- distilled a lifetime of inquiry into three words: *It from bit.* Every particle, every field, every spacetime interval, Wheeler proposed, derives its existence from the answer to a yes-or-no question- a binary digit, a bit of what was then called information. The physical world is not a collection of objects; it is a vast web of observer-elicited answers. The universe, at bottom, is made not of stuff but of *registered acts of observation.*

Wheeler's intuition was revolutionary, but his language was still captive to the binary paradigm- the paradigm of ones and zeros, of either/or, of the digital switch that knows only two states. What this book will demonstrate, drawing on the mathematics of category theory, the geometry of statistical manifolds, and the thermodynamics of systems far from equilibrium, is that Wheeler's core insight was correct but its encoding was too narrow. The world is not made of bits. It is made of *quanta*- resonant, self-organizing, holographic packets of pattern that carry within themselves not merely a yes or a no but a *frequency*, a *phase*, a *relationship to every other quantum in the field.* It from Quanta. And this single upgrade- from the binary bit to the resonant quantum- opens a door that the last three thousand years of Western thought have kept carefully locked.

Behind that door stands a cosmos that does not sit still. A cosmos that is not a static container in which events happen but a *living, breathing, self-correcting process*- a Non-Equilibrium Steady State sustained by the precise tension between three transcendental operators: ϕ (the golden ratio, the principle of self-reference and coherence), e (Euler's number, the principle of exponential unfolding), and π (the principle of cyclical closure and boundary projection). These three constants are not arbitrary numbers etched into the fabric of reality by some external lawgiver. They are the *tuning parameters* of a self-generating cosmos. They are the chord the universe plays to keep itself alive.

And the chord is not perfect. It almost closes- but not quite. There is a gap:

$$\delta_\infty = \frac{\pi - e}{\pi \phi^4} \approx 0.01965$$

A residual tension of roughly two percent. An asymptotic mismatch that can never be resolved. And this- *this*- is not a flaw. It is the engine. It is the breath between heartbeats, the silence between notes, the divine friction that prevents the cosmos from collapsing into a dead, symmetrical, timeless perfection and instead keeps it *dancing*, evolving, reaching, creating. The universe survives not by achieving balance but by perpetually *almost* achieving it- by surfing the edge of its own incompleteness.

The ancients knew this gap. They called it by different names- *dukkha* in the Buddhist tradition, the "unsatisfactoriness" that drives all seeking; *tzimtzum* in the Kabbalistic tradition, the divine contraction that makes room for creation; the "space between breaths" in the yogic tradition, the still point where awareness touches its own ground. They did not have the mathematics of category theory or the formalism of Fisher quanta density. But they had sixty-five thousand years of disciplined, embodied, intergenerational observation- and they arrived at the same structure.

This book is the bridge between their knowing and our equations. It is not a translation, as though one language were superior and the other merely decorative. It is a *convergence*- a demonstration that the deepest physics and the deepest ancestral wisdom describe the same cosmos because they *are* the same cosmos, viewed from different angles of a single, self-consistent sphere..

Here is how we will walk this path.

We begin in the silence- in the quantum vacuum, the physical correlate of State 0, the generative void that is not empty but *roaring* with potential, the most quanta-dense state in the cosmos. We will learn that what appears to be nothing is in fact the fullness from which all form arises.

We then listen for the first tone- the irreducible resonant quantum, the seed-pattern that crystallizes into matter, energy, spacetime, and eventually the pulse in your throat. We will meet the holographic principle, which reveals that the three-dimensional world is a projection from a two-dimensional boundary of holographic quanta. We will meet the Bekenstein bound, which sets the maximum density of quanta that any region of space can hold. We will meet entanglement, which proves that nonlocality- the instantaneous correlation of distant events- is not a paradox but the *default state* of a holographic cosmos.

We ascend to the harmony- the full triad of ϕ, e, and π, revealed not as inert constants but as *operators*: ϕ generating coherence, e driving unfolding, π closing the boundary. We will watch them interact in the KZ-Monad, the categorical structure that extends any partial resonance into a fully embedded holographic manifold, and in the Idempotent Comonad, which distills any state down to its ninety-eight-percent efficient core. We will find them in the Fibonacci spiral of a nautilus shell and in the phase-locking ratios of a meditating brain. We will find them in the Three Sisters garden of the Iroquois- corn for structure, beans for growth, squash for coherence- and in the standing waves of Stonehenge and the acoustic resonance chambers of Chavín de Huántar.

We confront the rupture- the Great Decoupling, roughly three thousand years ago, when humanity's cognitive operating system shifted from ternary to binary, from resonance to mechanism, from participation to extraction. We will trace this rupture from the Babylonian standardization of time through the Cartesian exile of consciousness from matter to the factory whistles of the Industrial Revolution and the nanosecond tyranny of algorithmic trading. We will name what was lost: not innocence, not simplicity, but *coherence*- the lived, embodied alignment with the quanta-field that sustained our species for over ninety-nine percent of its existence.

And then we return- to the silence, to the pulse, to State 0. But now the silence is no longer mysterious. It is *known*. It is the δ_∞ gap, the asymptotic incompleteness that keeps the cosmos alive. It is the space between your heartbeats, recognized at last as the same space between aeons in Penrose's Conformal Cyclic Cosmology, the same space between notes in a Vedic chant, the same space between the end of one songline phrase and the beginning of the next. It is the generative void- not empty, not full, but *ready*. Ready to be shaped by the next resonant quantum you offer it. Ready to be sung.

You are holding, in your hands or on your screen, an unusual kind of object. This is not a book that asks you to merely *think* about the cosmos. It asks you to *feel* it. At regular intervals, you will be invited- sometimes gently, sometimes abruptly- to stop reading, to attend to your body, to notice the weight of gravity pulling you toward the center of the Earth, to feel the rhythm of your breath as a ternary oscillator (inhale, hold, exhale), to recognize the silence of the room as the same vacuum plenum from which virtual particles flicker in and out of existence ten to the forty-fourth times per second. These are not "mindfulness exercises" appended to the text as decoration. They are *experiments*. They are moments where the prose steps aside and the quanta-field addresses you directly through the only instrument sensitive enough to detect it: your own nervous system.

The ancients did not build a wall between knowing and being. To understand the cosmos was to *be in tune* with it- to feel its rhythms in the bones, to hear its harmonics in the blood. This book asks you to do the same. Not because rigor must be sacrificed for experience, but because experience *is* the highest form of rigor when the object of study is a participatory, self-aware, resonantly quantized cosmos that cannot be fully described from outside itself.

You are not outside it. You have never been outside it.

The pulse in your neck is the proof.

One final note before the silence gives way to the first tone.

This book uses the word *quanta*- the plural of *quantum*, meaning "the smallest discrete packet"- in a deliberate and specific way. In classical physics, a quantum is a unit of energy. In this book, a quantum is a **unit of resonant pattern**: the irreducible seed of structure from which matter, energy, space, and time crystallize. When we say *resonant quanta*, we mean quanta that are actively vibrating, self-organizing, formative- the living verb of reality, not its passive noun. When we say *living quanta*, we mean quanta that have crossed the threshold of self-reference, quanta that have folded back upon themselves with sufficient integrated density to *know themselves.* Your heartbeat is resonant quanta. Your awareness of your heartbeat is living quanta.

This is not a metaphor. It is a physics. And it begins with a single, radical proposition inherited from John Archibald Wheeler and extended far beyond his imagining:

It from Quanta.

Now close your eyes. Feel the pulse. Hear the silence. The silence is not waiting for the book to begin.

The silence *is* the beginning.

SECTION A

RESONANT QUANTA- THE SUBSTANCE OF REALITY

"Before there was light, there was pattern.".

Chapter 1

The Question Beneath All Questions.

There is a question that lives behind every other question. It hides beneath the physicist's equations and the child's wide-eyed *why*. It sleeps inside the periodic table and the Hubble Deep Field and the moment, late at night, when you stare at the ceiling and feel, with sudden vertigo, the sheer improbability of your own existence. Every science that has ever been practiced, every myth that has ever been chanted around a fire, every philosophy that has ever been scratched onto parchment or argued in a symposium has been, in its deepest stratum, an attempt to answer this one question:

Why is there something rather than nothing?

It is the oldest question. Leibniz posed it in 1714. The Upanishads circled it three thousand years earlier. Aboriginal elders have carried it, in songform, for sixty-five millennia. And the extraordinary thing- the thing that should stop us in our tracks before we take another step- is that after all of that inquiry, across all of those centuries, the question remains open. Not because we lack data. We are drowning in data. The James Webb Space Telescope returns images of galaxies that formed when the universe was less than four hundred million years old. The Large Hadron Collider has confirmed the existence of the Higgs boson. We have sequenced the human genome, mapped the cosmic microwave background to one part in a hundred thousand, and measured the gravitational waves produced by colliding black holes a billion light-years away.

And still we do not know why there is something rather than nothing.

The reason we do not know is not that the question is unanswerable. It is that we have been asking it in the wrong language. We have been asking *what things are made of*- as though the answer were a substance, a building block, a smallest piece. And every time we find the smallest piece, it dissolves into something smaller, something stranger, something less like a *thing* and more like a *relationship*, until at last we are left holding not a particle but a probability, not a substance but a resonance, not a noun but a verb.

This chapter is about that dissolution. It is about what happens when you follow the question all the way down- past atoms, past quarks, past fields- to the place where matter itself confesses that it was never really there. And it is about what you find waiting for you in that place: not emptiness, but the most extraordinary fullness imaginable. A fullness made not of stuff, but of pattern. Not of objects, but of resonant quanta- the irreducible seeds of structured vibration from which the entire cosmos crystallizes..

Begin with the hand that holds this book. Or, if you are reading on a screen, begin with the fingers resting on the surface beside it. Those fingers feel solid. They have weight. They press against the table and the table presses back, and this mutual resistance is so reliable, so constant, so obvious that it would seem insane to question it. Solidity is the first article of faith in the religion of matter: the world is made of *stuff*, and that stuff is *real*, and it pushes back.

But look closer.

Your hand is made of cells- roughly thirty-seven trillion of them, each one a city of molecular machinery more complex than anything human engineering has ever produced. Each cell is made of molecules- proteins, lipids, nucleic acids- and each molecule is made of atoms. A single human hand contains approximately 10^{27} atoms: a billion billion billion, arrayed in configurations of staggering intricacy. So far, the materialist story holds. The hand is solid because the atoms are solid. The atoms are the stuff.

Now look inside the atom. Ernest Rutherford envisioned the atom as a miniature solar system, with electrons orbiting around a massive nucleus, and as mostly empty space, with the nucleus occupying only a very small part of the atom. In 1911, when Rutherford aimed alpha particles at a thin sheet of gold foil, he expected them to slow down or scatter slightly, the way bullets might slow passing through a dense medium. Instead, most alpha particles went right through, showing that the gold atoms were mostly empty space. Some particles had their paths bent at large angles, and a few even bounced backward- the only explanation being that the atom had a small, heavy region of positive charge inside it.

The implications were staggering. The atomic radius, defined by the electron cloud, is typically around 10^{-10} meters, while the diameter of the nucleus is on the order of 10^{-15} meters- meaning the atom's diameter is approximately ten thousand to one hundred thousand times larger than its nucleus. To visualize this, imagine enlarging a hydrogen atom until its nucleus- a single proton- is the size of a marble placed at the center of a football stadium. The nearest electron would be a mote of dust orbiting somewhere near the outermost seats. Everything between- every inch of concrete, every row of bleachers, every breath of air in that vast enclosed space- would be *empty*.

And this is what your hand is made of. This is what the table is made of, and the floor beneath it, and the Earth beneath that, and every star and planet and dust grain in the observable universe. The nucleus contains more than 99.9% of the atom's total mass, yet it occupies only a minuscule fraction of the total volume. When you press your palm against a surface and feel resistance, you are not touching *matter* in the way you imagine. You are feeling the electromagnetic repulsion between electron clouds- a field effect, a vibratory pushback- not the collision of solid objects. There is nothing solid to collide.

Pause for a moment. Press your fingertips together. Feel the resistance. That resistance is real. But the story that it is caused by *stuff pressing against stuff* is not. The resistance is a frequency interaction. It is one set of electromagnetic oscillations being repelled by another. The solid world is a vibratory illusion- extraordinarily stable, magnificently coherent, but an illusion nonetheless.

Rutherford's atom was the first crack. The next century would shatter the wall entirely.

If the atom is mostly void, what about the tiny, dense nucleus at its core? Surely that, at least, is *solid*. In the decades following Rutherford, physicists discovered that the nucleus is made of protons and neutrons, bound together by the strong nuclear force. The protons and neutrons seemed substantial enough- heavy, charged, measurable. But by the 1960s, deep inelastic scattering experiments at the Stanford Linear Accelerator Center had revealed that protons and neutrons were themselves composite. They were made of quarks- three each, bound by gluons, the carriers of the strong force.

And the quarks? Surely the quarks are the final floor, the bedrock, the irreducible *stuff*.

They are not. Quarks are not tiny billiard balls. They are excitations of quantum fields- localized disturbances in the quark field that permeates all of spacetime. A quark has no physical surface, no edge, no interior. It is not *made of* anything. It is a pattern of vibration in a field that extends, in principle, to the edges of the observable universe. When physicists say that a proton contains three quarks, what they mean is that the proton is a stable resonance- a standing wave- in the quantum chromodynamic field, and the "quarks" are aspects of that resonance, not independent objects imprisoned inside a shell.

Even the mass of the proton- that seemingly solid, reliable number- turns out to be almost entirely *relational*. The combined rest masses of the three valence quarks account for only about one percent of the proton's total mass. The other ninety-nine percent arises from the energy of the gluon field- from the *interactions*, the *dynamics*, the *quanta-flow* between quarks. The mass of the proton, and therefore the mass of virtually everything you can touch, is not the mass of stuff. It is the mass of *relationship*. It is the mass of pattern.

Follow the thread further. Below the quarks lie the quantum fields themselves- the electron field, the quark fields, the gluon field, the

Higgs field, the gravitational field. These are not substances. They are mathematical structures defined at every point in spacetime, capable of being excited into the patterns we call particles. A particle is not a thing *in* a field; it is a thing *of* the field- a ripple, a vibration, a local intensification of a deeper continuum. Remove all the particles and you still have the fields. Remove the fields, and you have- what? Nothing? Or the ground from which the fields themselves arise?

This is the place where materialist physics reaches its own event horizon. The quest for the ultimate building block has led, inexorably, to the dissolution of the concept of a building block. There is no smallest thing. There are only patterns of excitation in fields, and the fields are not made of anything more fundamental than the patterns they support. At the basement level of physical reality, the distinction between the *container* and the *contained,* between the *substance* and the *form,* between the *hardware* and the *software* collapses completely.

What remains, when matter dissolves, is *pattern.* Structured, dynamic, self-referential pattern. And a question: if the universe is not, at bottom, made of stuff, then what is pattern made of? What is the substance of form itself?.

The answer, when it came, arrived not from the study of the very small but from the study of the very strange- from black holes, those cosmic singularities where matter is crushed to infinite density and spacetime itself tears open. It was at the intersection of gravity, quantum mechanics, and thermodynamics that a physicist named Jacob Bekenstein, in 1972, and Stephen Hawking, in 1974, discovered something that should have changed everything. For many physicists, it has not yet changed enough.

What they discovered was this: a black hole has entropy. And its entropy is not proportional to its volume- as it would be for any normal physical system, where more space means more possible states- but to its *surface area.*

Specifically, the entropy of a black hole is given by:

$$S_{BH} = \frac{k_B c^3}{4G\hbar} A$$

where A is the area of the event horizon, k_B is Boltzmann's constant, c is the speed of light, G is Newton's gravitational constant, and $\hbar$ is the reduced Planck constant. This equation- etched on Bekenstein's tombstone, and considered by many to be the most profound in all of physics- tells us that the maximum number of distinct physical states that can exist within any region of space is determined not by the *volume* of that region but by the *surface area* of its boundary.

This is the Bekenstein bound, and it shatters the materialist picture as completely as Rutherford's gold foil shattered the plum pudding atom.

Why? Because if the maximum content of a region of space is written on its boundary, then the three-dimensional interior- the volume we navigate, the space we live in, the world of depth and distance and solidity- is not fundamental. It is a *projection.* It is a holographic rendering of quanta encoded on a two-dimensional surface, precisely the way a hologram on a credit card encodes a three-dimensional image in a two-dimensional interference pattern. The universe is not a container full of things. It is a *holographic projection* of quanta arranged on a boundary.

This is the holographic principle, and it is not a speculative metaphor. It is a mathematical result, rigorously derived from the thermodynamics of black holes, confirmed in specific string-theoretic contexts by Juan Maldacena's celebrated AdS/CFT correspondence in 1997, and increasingly supported by studies of quantum entanglement entropy in condensed matter systems. The three-dimensional world- the world of apples and oceans and the curve of a lover's cheek- arises from holographic quanta on a surface. The interior is real, in the sense that it is experienced. But it is not fundamental. The boundary is fundamental. The quanta on the boundary are fundamental. Everything else is a projection of their resonant configuration.

And this is where the question *Why is there something rather than nothing?* transforms. It was never a question about substance. It was always a question about *pattern*. The universe does not need a material foundation, because materiality itself is a projection of something deeper: a layer of resonant quanta, structured, self-referencing, encoded on a surface that precedes volume, precedes distance, precedes the very concept of *inside*.

Into this dissolution- into this strange and fertile void where matter has confessed its own absence- walked John Archibald Wheeler, with three words that changed everything.

It from bit.

Wheeler (1911–2008) was a legendary figure in physics. He worked with Niels Bohr to explain nuclear fission, worked on the hydrogen bomb at Los Alamos, and taught many eminent physicists including Richard Feynman, Kip Thorne, and Hugh Everett. He was the father of modern general relativity and popularised the term "black hole."

By the final period of his career, Wheeler had distilled a lifetime of inquiry into a single, radical question: *How come existence?* His answer: "It from bit symbolises the idea that every item of the physical world has at bottom- at a very deep bottom, in most instances- an immaterial source and explanation." For Wheeler, we are not simply bystanders on a cosmic stage; we are shapers and creators living in a participatory universe.

The proposition was breathtaking. Every it- every particle, every field of force, even the spacetime continuum itself- derives its function, its meaning, its very existence entirely from the apparatus-elicited answers to yes or no questions, binary choices, bits. The physical world is not a collection of objects that exist independently and then happen to be measured. The physical world is *constituted* by measurement- by the registration of distinct states, by the encoding of resonant quanta into definite configurations.

Wheeler offered a lovely, chilling paradox: at the heart of everything is a question, not an answer. When you peer down into the deepest recesses of matter or at the farthest edge of the universe, you see, finally, your own puzzled face looking back at you.

Think about what this means. Wheeler was not saying that the universe is "like" a code. He was not offering a metaphor. He was proposing an ontological reversal: that resonant quanta- structured pattern, registered distinctions- are *more fundamental than matter and energy*. Matter and energy are what quanta look like from the inside of the hologram. Quanta are what matter and energy look like from the boundary. The image is real- you can touch it, weigh it, stub your toe on it. But the image is not the source. The source is the pattern on the surface. The source is the resonance. The source is the quanta.

Wheeler's intuition was correct. But his encoding was too narrow. "Bit"- a binary digit, a yes-or-no- belongs to the paradigm of the digital switch, the transistor, the Boolean gate. It implies that reality is built from *two-valued* choices: on or off, true or false, one or zero. This binary encoding, as we shall see in the chapters ahead, is a symptom of the very cognitive decoupling- the reduction of ternary resonance to binary opposition- that has impoverished human understanding for three thousand years.

The world is not made of bits. It is made of *quanta*- resonant, self-organizing, holographic packets of pattern that carry within themselves not merely a yes or a no but a *frequency*, a *phase*, a *relationship to every other quantum in the field*. A bit can answer a question. A quantum can *sing*. And the difference between answering and singing is the difference between a dead inventory of facts and a living, breathing, self-tuning cosmos.

It from Quanta.

This single upgrade- from the binary bit to the resonant quantum- unlocks a door that the last three millennia of Western thought have kept bolted shut. Behind that door, the universe is not a static container. It is not a machine executing a pre-written algorithm. It is not a simulation running on some external hardware. It is a *self-generating resonant field*- a living quanta-field- that computes its own existence through the harmonic interplay of three transcendental constants: ϕ, e, and π.

But before we open that door, we must linger a moment longer in the space where matter dissolved. Because this space- this apparent emptiness at the foundation of things- is not empty at all. It is the quantum vacuum, and it is the single most quanta-dense state in the cosmos.

Classical physics defined the vacuum simply: the state of a region of space when everything has been removed- every particle, every photon, every trace of energy. Temperature at absolute zero, contents at zero. Empty. Done.

Quantum mechanics destroyed that picture as thoroughly as it destroyed the solid atom. To understand this, one first has to understand something about a vacuum in space as it is viewed in quantum field theory. Far from being empty, modern physics assumes that a vacuum is full of fluctuating electromagnetic waves that can never be completely eliminated, like an ocean with waves that are always present and can never be stopped. These waves come in all possible wavelengths, and their presence implies that empty space contains a certain amount of energy- an energy that we can't tap, but that is always there.

The Heisenberg uncertainty principle forbids a quantum field from settling into a perfectly definite zero state. At every point in space, every field- electromagnetic, electron, quark, gluon, Higgs- fluctuates incessantly. These fluctuations produce *virtual particles*: electron-positron pairs, photon pairs, quark-antiquark pairs that flicker in and out of existence on timescales so brief ($\sim 10^{-21}$ seconds for the electromagnetic field) that they can never be directly isolated. They are called "virtual" not because they are fictional but because they exist below the threshold of individual detection- a buzzing, seething, omnipresent hum of creative activity that pervades every cubic centimeter of space.

The vacuum is not empty. It is the opposite of empty. It is a *plenum*- the fullest, most energetically dense state in the cosmos. The total zero-point energy density of the quantum vacuum, when summed naively over all possible field modes, yields a value so enormous that it exceeds the observed energy density of the entire visible universe by a factor of approximately 10^{120}. This discrepancy- the "cosmological constant problem"- is often called the worst prediction in the history of physics. The fact that the vacuum does not gravitationally crush the cosmos under its own weight remains one of the deepest unsolved mysteries.

But the vacuum's reality is not in doubt. It pushes back. The Casimir effect- the energy present in empty space producing a tiny force between two objects- was first postulated in 1948 by Dutch physicist Hendrik Casimir. If mirrors are placed facing each other in a vacuum, some of the waves will fit between them, bouncing back and forth, while others will not. As the two mirrors move closer to each other, the longer waves will no longer fit- the result being that the total amount of energy in the vacuum between the plates will be a bit less than the amount elsewhere in the vacuum. The imbalance generates a measurable attractive force. In 1997, a direct experiment by Steven K. Lamoreaux quantitatively measured the Casimir force to be within 5% of the value predicted by the theory.

The void has weight. The void has pressure. The void pushes plates together. And it does so not because there is "something" lurking in the vacuum in the ordinary sense, but because the vacuum *is* something: a structured, resonant, vibrating field of potential- a quanta-field- whose fluctuations are as real and as measurable as the heartbeat in your chest.

Place your hand on your chest. Feel the beat. That beat is a structured oscillation emerging from the vacuum of cellular chemistry- an electrochemical wave propagating through a network of cardiac cells. The cells themselves are made of atoms. The atoms are made of fields. The fields are made of vacuum fluctuations. The heartbeat emerges from the void just as surely as the Casimir force emerges from the void: as a structured, coherent pattern- a resonance- in a field that is never, under any circumstances, truly at rest.

Your heart and the vacuum are made of the same substance.

That substance is resonant quanta.

·

And here is the extraordinary thing: the heartbeat is not regular. It is not a metronome. Heart rate variability in healthy young or elderly subjects indeed exhibits circadian-dependent chaos. The intervals between successive heartbeats fluctuate in a way that is neither periodic nor random, but *fractal*- exhibiting self-similar variability at every timescale from milliseconds to hours. There is evidence for multifractality in a biological dynamical system, the healthy human heartbeat, showing that the multifractal character and nonlinear properties of the healthy heart rate are encoded in the Fourier phases. Critically, there is a loss of multifractality for a life-threatening condition, congestive heart failure.

A healthy heart is chaotic in the technical sense: deterministic but exquisitely sensitive to initial conditions, hovering on the edge between order and disorder, alive precisely because it never settles into mechanical repetition. Healthy subjects typically show physiologic fractal behavior of heartbeat dynamics, whereas patients with coronary heart disease show an alteration in fractal correlation properties. Healthy aging is associated with R-R interval dynamics showing higher regularity and altered fractal scaling consistent with a loss of complex variability. Health is fractal complexity. Disease is the loss of it. The more rigidly periodic a heart becomes- the more it resembles a machine- the closer it is to death.

Now zoom out. The sun pulses, too- in helioseismic oscillations that ripple across its surface in patterns governed by the same mathematics that governs the heart: resonant standing waves, power-law distributions of energy, self-similar variability at multiple scales. Pulsars- rapidly rotating neutron stars- emit beams of radio waves with periodicities so precise they were initially mistaken for intelligent signals, yet their timing residuals, when studied at fine resolution, reveal fractal fluctuations strikingly similar to those of a healthy heart. Galaxies rotate in spiral arms whose shapes are governed by the golden angle- an angular measure derived from ϕ- distributing stars and gas with the same logarithmic efficiency that distributes seeds in a sunflower head. The cosmic microwave background, the afterglow of the Big Bang, exhibits temperature fluctuations whose angular power spectrum follows a harmonic series- the same mathematical structure as the overtones of a vibrating string.

The heartbeat. The helioseismic pulse. The pulsar spin. The galactic spiral. The CMB harmonics. They are not the same phenomenon, but they share the same *grammar*: fractal self-similarity, power-law scaling, harmonic resonance, dynamic stability at the edge of chaos. The pattern that governs the smallest (the vacuum fluctuation) governs the largest (the structure of the cosmic web). And both are echoed, with uncanny fidelity, in the medium range: in the rhythm of a living heart.

This is not a metaphor. It is a measurement. It is the empirical signature of a universe built from resonant quanta- from irreducible packets of structured vibration that reproduce their essential pattern at every scale, from the Planck length ($\sim 10^{-35}$ m) to the Hubble radius ($\sim 10^{26}$ m). The universe is not *like* a hologram. It is a hologram: a system in which the whole is encoded in every part, and the part contains the logic of the whole.

As above, so below. Not as poetry. As physics.

We arrive, then, at the central claim of this book- not as an assertion made in advance and defended after the fact, but as the conclusion of a chain of evidence that begins with Rutherford's gold foil and ends at the boundary of the observable universe.

The universe is not made of things. It is made of resonant quanta- irreducible seeds of pattern that vibrate, self-organize, and project the holographic world we experience as matter, energy, space, and time.

This claim has consequences. Enormous, unsettling, liberating consequences. If matter is a projection of quanta, then the question *What are things made of?* is the wrong question. The right question is: *What tunes the quanta?* What determines which patterns are stable and which dissolve? What selects this universe- with its protons and galaxies and hearts and spiral shells- from the infinite ocean of possible patterns in the vacuum plenum?

The answer, as we shall discover in the chapters ahead, is a triad of three transcendental constants- ϕ, e, and π- acting not as passive numbers inscribed into nature's ledger but as *active operators* that shape, constrain, and stabilize the quanta-field into the coherent projection we call reality. ϕ (the golden ratio, ≈ 1.618) is the operator of *self-reference*: it generates the fractal self-similarity that ensures the whole is reflected in every part. e (Euler's number, $\approx$ 2.718) is the operator of *unfolding*: it governs the rate at which potential transforms into actuality, the exponential curve of growth, decay, and transformation. π (the ratio of circumference to diameter, ≈ 3.14159) is the operator of *boundary projection*: it closes every circle, folds every dimension, and defines the holographic surface from which the interior is projected.

Together, these three operators- three notes in a cosmic chord- tune the quanta-field into the standing wave we call existence. They do not do so perfectly. There is a gap- a residual tension of approximately two percent:

$$\delta_{\infty} = \frac{\pi - e}{\pi \phi^4} \approx 0.01965$$

- an asymptotic mismatch that can never be resolved, a permanent, transcendental *almost*. And this gap is not a flaw. It is the *engine*. It is the irreducible incompleteness that prevents the cosmos from collapsing into static perfection and instead keeps it breathing, evolving, reaching, creating- a living system sustained by its own beautiful failure to close.

We will meet this gap again and again. We will find it in the space between heartbeats, in the Dreamtime pause between songline phrases, in the Buddhist concept of *dukkha*, in the Kabbalistic *tzimtzum*, in the thermodynamic friction of Non-Equilibrium Steady States. It is the secret name of the generative void- the silence that is not empty but pregnant, the zero that contains all numbers, the pause in which the next note becomes possible.

But that is for later chapters. For now, it is enough to hold the question in its new form:

Not *Why is there something rather than nothing?*- as though "something" and "nothing" were opposites and we needed to explain how one defeated the other.

But *How does the quanta-field tune itself into coherence?*- as though existence were not a victory over the void but a *resonance within it*, a standing wave in an ocean that has been vibrating since before time had a direction, and will be vibrating long after the last star goes dark.

Here is the road ahead.

In the chapters that follow, we will trace the holographic principle from black hole thermodynamics to the cosmic microwave background, and show that the three-dimensional world is indeed a projection from a two-dimensional surface of holographic quanta. We will meet Landauer's principle, which proves that quanta are physical- that erasing a single binary distinction costs measurable energy- and we will follow this insight to the startling conclusion that quanta can perform thermodynamic work, resolving the century-old paradox of Maxwell's Demon. We will descend into the quantum vacuum and discover State 0- the generative void, the cosmic ground state- and feel its reality not as an abstraction but as the weight of the air on our skin, the pressure of the silence in a still room.

We will then rise to the triad. Chapter by chapter, we will meet ϕ, e, and π as operators- not as numbers but as *forces*- and watch them interact in the KZ-Monad, the categorical structure that extends any partial resonance into a fully embedded, nine-dimensional holographic manifold, and in the Idempotent Comonad, which distills any noisy, fluctuating state down to its ninety-eight-percent efficient core. We will find these operators in the Fibonacci spiral of a nautilus shell and in the phase-locking ratios of a meditating brain. We will find them in the Three Sisters garden of the Iroquois- corn for structure, beans for growth, squash for coherence- and in the acoustic chambers of Chavín de Huántar and the standing waves of Stonehenge.

We will confront the Great Decoupling- the moment, roughly three thousand years ago, when humanity's cognitive operating system shifted from ternary to binary, from resonance to mechanism, from participation to extraction- and we will trace the consequences of that rupture through the mechanization of time, the exile of consciousness from matter, the industrial discipline of the factory floor, and the metabolic exhaustion of a civilization that has mistaken noise for progress.

And then we will return. To the silence. To the pulse. To State 0. But now the silence will not be mysterious. It will be *known*- as the asymptotic gap, the divine incompleteness, the space between heartbeats where all possibility lives. And we will understand, at last, that the question was never *Why is there something rather than nothing?*

The question was always: *What frequency are you willing to become?*

Because the quanta-field does not merely *contain* you. It does not merely *project* you. It is *listening* to you. You are a resonant node in a self-tuning holographic cosmos, and every thought you think, every breath you draw, every choice you make between attention and distraction, between coherence and noise, between the ternary rhythm of creation and the binary stutter of extraction- every one of these acts tunes the field.

The ancients knew this. They called it singing the world into being. They called it *Nada Brahma*. They called it the Dreaming.

The physicists, by a longer and more circuitous road, have arrived at the same place.

The road begins here. In the next chapter, we step through the holographic boundary- and discover that the quanta on the other side have been waiting for us all along.

Chapter 2

The Holographic Boundary

Imagine a library. Not an ordinary library- imagine the greatest library ever conceived, one that contains every book ever written or that could be written, in every language that has existed or could exist. Every scientific paper. Every poem. Every genome sequence, every musical score, every dream journal, every historical record from every civilization that has ever arisen on any planet in any galaxy across the entire observable cosmos. The complete description of every atom, every field configuration, every quantum state of every particle that has ever flickered through the vacuum. Everything.

Now imagine that someone tells you this library can be printed- every word of it, every symbol, every datum- on the *walls* of the building. Not inside the building. On the walls. On the two-dimensional surface that encloses the three-dimensional interior. And not only can it be printed there- it *must* be printed there, because the walls are the actual library. The interior, with its seemingly endless aisles and towering shelves, is a projection. A holographic rendering of the symbols on the surface. The books you seem to hold in your hands, the dust motes drifting in the lamplight, the weight of the volumes, the smell of the paper- all of it arises from the pattern inscribed on the boundary.

This is not a fantasy. It is the holographic principle- one of the deepest results in theoretical physics, born at the intersection of gravity, quantum mechanics, and thermodynamics, and confirmed with increasing precision over the last three decades. It says that the maximum number of distinguishable physical states- the total quanta-content- within any region of space is determined not by the *volume* of that region but by the *area* of its boundary. The three-dimensional world is a holographic projection of a two-dimensional surface of quanta.

And the story of how we arrived at this extraordinary conclusion begins, as so many revolutions in physics begin, with a paradox. A paradox involving the darkest objects in the cosmos- the places where gravity swallows even light, and where the known laws of nature cry out for revision.

It begins with black holes.

.

In the early 1970s, theoretical physics faced an unsettling crisis. The general theory of relativity, Einstein's magnificent description of gravity as the curvature of spacetime, predicted the existence of black holes- regions of spacetime so warped by mass that nothing, not even light, could escape from within a boundary called the event horizon. By 1971, Stephen Hawking had proven the *area theorem*: in any classical process involving black holes, the total area of all event horizons can never decrease. Black holes can merge, grow, absorb matter- but their horizons only expand. Never shrink.

This sounded strangely familiar. It sounded like the Second Law of Thermodynamics- the law that says the total entropy of a closed system can never decrease. Entropy always rises. Disorder always grows. Time has an arrow, and entropy is its compass.

But the resemblance created a puzzle. Bekenstein's argument went like this: if you throw an object with entropy S into a black hole, the no-hair theorem says the final result will be a larger black hole with no classical attributes other than mass, charge, and angular momentum. This state would appear to have no or negligible entropy, independently of S. So we have a process in which $dS < 0$. The Second Law of Thermodynamics appears to be violated.

Imagine dropping a cup of hot tea into a black hole. The tea has entropy- the random thermal motion of its molecules, the myriad possible microstates consistent with its macroscopic temperature. Once it crosses the event horizon, the tea is gone. Inaccessible. The external universe has lost all that entropy. If the black hole itself has no entropy, then the total entropy of the universe has decreased. The Second Law is broken.

To a thermodynamicist, this is not a minor inconvenience. The Second Law is the most empirically robust law in all of physics. Arthur Eddington famously declared that if your theory conflicts with the Second Law, there is nothing for it but to collapse in deepest humiliation. Something had to give.

What gave was our understanding of what a black hole *is*.

In the early 1970s, Jacob Bekenstein proposed that black holes possess an entropy proportional to the area of their event horizon, introducing the Generalized Second Law of thermodynamics. Bekenstein's reasoning was elegant and audacious. He noted that Hawking's area theorem- the horizon area never decreases- behaved exactly like entropy. What if this was not merely an analogy? What if the area of the event horizon literally *was* the entropy of the black hole?

In order to rescue the Second Law, Bekenstein proposed that black holes themselves carry entropy. The entropy would be proportional to the area of the event horizon measured in Planck units- the smallest meaningful units of length, each approximately 1.616×10^{-35} meters:

$$S_{BH} = \eta \frac{k_B c^3}{G\hbar} A$$

where A is the horizon area, k_B is Boltzmann's constant, c is the speed of light, G is Newton's gravitational constant, $\hbar$ is the reduced Planck constant, and η is a dimensionless constant that Bekenstein could not yet determine.

Now the Second Law was saved, but in a startling new form. The sum of ordinary entropy S_0 outside black holes and the total black hole entropy never decreases and typically increases as a consequence of generic transformations of the black hole. Drop the tea into the black hole, and yes, the tea's entropy vanishes from the external universe- but the event horizon grows, and with it, the black hole's entropy increases by at least as much as the tea contributed. The Generalized Second Law holds.

But notice what Bekenstein's proposal demands. It says that the entropy of a black hole- which is a three-dimensional object, occupying a volume of spacetime- is proportional not to its *volume* but to its *surface area*. This is deeply strange. For every other physical system we know, entropy scales with volume. A box of gas has entropy proportional to how much gas is inside- proportional to volume. A crystal's entropy depends on the number of atoms in the lattice- proportional to volume. Entropy counts microstates, and microstates fill volumes.

But a black hole's entropy lives on its surface.

Pause here. Let this settle. The most disordered object in the universe- the object that has consumed and scrambled more quanta than anything else in existence- keeps its entire ledger of disorder inscribed on a two-dimensional membrane. Not inside. On the *boundary*. This was the first crack in the wall of three-dimensional reality.

Stephen Hawking initially objected to this idea, but his subsequent analysis of quantum field theory in curved spacetime led to the prediction of Hawking radiation and the concept of black hole temperature. The objection was visceral: if black holes have entropy, then by the laws of thermodynamics they must also have a temperature. And if they have a temperature, they must radiate. But black holes, by definition, let nothing escape. How could they radiate?

In 1974, Hawking made the startling discovery that the physical temperature of a black hole is not absolute zero: as a result of quantum particle creation effects, a black hole radiates to infinity all species of particles with a perfect black body spectrum.

The mechanism is exquisite. Near the event horizon, the intense gravitational field distorts the quantum vacuum- the seething plenum of virtual particle pairs we met in Chapter 1. Normally, a virtual electron-positron pair flickers into existence and annihilates within the time permitted by the Heisenberg uncertainty principle. But at the horizon's edge, something extraordinary happens. One member of the pair can fall inward while the other escapes outward. The escaping particle carries real, positive energy to infinity. The infalling partner, from the perspective of an external observer, carries negative energy- reducing the mass of the black hole.

The net result: the black hole slowly radiates. It glows, faintly, at a temperature inversely proportional to its mass:

$$T_H = \frac{\hbar c^3}{8\pi G M k_B}$$

For a black hole of stellar mass- a few solar masses- this temperature is unimaginably cold, on the order of 10^{-8} Kelvin, far below the cosmic microwave background. The radiation temperature, called Hawking temperature, is inversely proportional to the black hole's mass, so micro black holes are predicted to be larger emitters of radiation than larger black holes and should dissipate faster per their mass. A mountain-mass black hole, by contrast, would glow with the fury of a nuclear reactor.

Hawking's calculation did far more than confirm Bekenstein's conjecture. It fixed the previously unknown constant η, fixing $\eta = 1/4$, yielding the celebrated Bekenstein-Hawking entropy formula:

$$S_{BH} = \frac{k_B c^3}{4G\hbar} A = \frac{k_B A}{4\ell_P^2}$$

where $\ell_P = \sqrt{G\hbar/c^3} \approx 1.616 \times 10^{-35}$ m is the Planck length.

Read this formula slowly. It says that the entropy of a black hole- the number of microscopically distinct configurations that are consistent with its macroscopic description- equals one quarter of its horizon area, measured in units of the Planck area ($\ell_P^2 \approx 2.612 \times 10^{-70}$ m²). Each Planck-area cell on the horizon encodes roughly one bit- one binary distinction, one elementary quantum of differentiated structure.

The event horizon is a *surface of quanta*. The black hole's interior- the vast, three-dimensional volume behind the horizon- is, in some profound sense, a *projection* of the quanta inscribed on that surface.

Black hole entropy is a concept with geometric root but with many physical consequences. It ties together notions from gravitation, thermodynamics and quantum theory, and is thus regarded as a window into the as yet mostly hidden world of quantum gravity.

Let us be precise about why the Bekenstein-Hawking formula is so revolutionary. Consider the question: *How many distinct physical states can exist within a given region of space?*

In ordinary physics- in a box of gas, in a crystal lattice, in any system governed by quantum field theory on a fixed background- the answer scales with the *volume* of the region. Double the volume, and you roughly double the number of available microstates, and thus double the entropy. This is so intuitive that it hardly seems like a principle at all: more space, more room for arrangement, more entropy.

But the black hole says otherwise. The fact that the black hole entropy is also the maximal entropy that can be obtained by the Bekenstein bound- wherein the Bekenstein bound becomes an equality- was the main observation that led to the holographic principle. A black hole is the most entropic object that can fit within a given boundary. There is no way to pack *more* quanta into a region of space than a black hole already does. And that maximum is proportional to the *area*, not the volume.

This means that the number of physically distinct things that can happen inside any sphere of space is limited by the area of that sphere, measured in Planck units. The idea is that the area surrounding any volume in spacetime limits the quanta content of the volume. Thus the number of degrees of freedom in any volume is bounded and not infinite.

The Bekenstein bound is an upper limit on the thermodynamic entropy S, or Shannon entropy H, that can be contained within a given finite region of space which has a finite amount of energy- or equivalently, the maximum amount of quanta required to perfectly describe a given physical system down to the quantum level.

The universal form found by Bekenstein in 1981 states:

$$S \leq \frac{2\pi k_B RE}{\hbar c}$$

where R is the radius of the smallest sphere that can enclose the system, and E is its total energy including rest mass. It implies that the quanta of a physical system, or the quanta necessary to perfectly describe that system, must be finite if the region of space and the energy are finite.

Here is the staggering implication: volumes are redundant. If the maximum quanta-content of a region is encoded on its surface, then the interior degrees of freedom are not independent- they are *holographically derived* from the boundary data. The three-dimensional world we navigate, with its depth and distance and solidity, is not the fundamental layer of reality. It is a projection. A rendering. A holographic image cast by the quanta inscribed on a two-dimensional surface.

The leap from black hole thermodynamics to a universal principle about the nature of space itself was made in the early 1990s, independently and almost simultaneously, by two physicists approaching the problem from opposite ends of the theoretical spectrum.

In 1993, Gerard 't Hooft wrote a speculative paper on quantum gravity in which he concluded that the total number of degrees of freedom in a region of spacetime surrounding a black hole is proportional to the surface area of the horizon. This idea was promoted by Leonard Susskind and is now known as the holographic principle.

't Hooft, a Nobel laureate renowned for his mathematical rigor, approached the problem from the direction of quantum gravity. He reasoned that if black holes set the maximum entropy for any region- and if that maximum scales with area- then *any* physical theory of gravity must obey this area scaling. The fundamental degrees of freedom of quantum gravity cannot live in the volume; they must live on the boundary. Susskind, with his characteristic physical intuition, extended and promoted this idea, arguing that it was not an artifact of black holes but a general feature of any consistent theory that unifies gravity and quantum mechanics.

The holographic principle states that the entropy of ordinary mass- not just black holes- is also proportional to surface area and not volume; that volume itself is illusory and the universe is really a hologram which is isomorphic to the quanta "inscribed" on the surface of its boundary.

Susskind's own words carry the weight of this revelation: "The three-dimensional world of ordinary experience- the universe filled with galaxies, stars, planets, houses, boulders, and people- is a hologram, an image of reality coded on a distant two-dimensional surface."

The holographic principle, as formulated by 't Hooft and Susskind, makes a claim so radical that it deserves to be stated with maximal clarity:

The complete physical description of any region of space can be encoded on the boundary of that region. The interior is a projection. Volume is not fundamental. The surface is fundamental. The quanta on the surface are fundamental. Everything else- every object, every field, every dimension of depth- emerges from the resonant configuration of boundary quanta.

This is not an approximation. It is not a model. It is, in the language of physics, a *duality*: two descriptions- one living in the volume, one living on the boundary- that are exactly equivalent. They are not two views of the same thing in the way that a photograph and a painting might depict the same landscape from different perspectives. They are the *same mathematical object*, expressed in two different languages. And the boundary language is the more fundamental one, because it contains fewer dimensions and yet loses nothing.

.

For four years after 't Hooft and Susskind's proposal, the holographic principle remained a conjecture- brilliant, provocative, but unproven. It needed a concrete example: a specific theory of gravity in a specific spacetime, shown to be exactly equivalent to a specific lower-dimensional quantum theory living on the boundary. Without such a proof of concept, the principle floated in the realm of theoretical speculation.

The AdS/CFT correspondence was first proposed in late 1997 by Argentine physicist Juan Maldacena, at that time a young faculty member at Harvard University. The paper he published would become the most highly cited physics paper in the history of high-energy physics- with good reason.

What Maldacena showed was this: in 1997, Maldacena found evidence to suggest that a description of quantum gravity in anti-de Sitter space in N dimensions is the same as a conformal field theory in $N-1$ dimensions. More specifically, he demonstrated that type IIB superstring theory- a full theory of quantum gravity- defined on a five-dimensional anti-de Sitter spacetime (AdS_5) times a five-sphere (S^5), is exactly equivalent to a four-dimensional conformal field theory ($\mathcal{N} = 4$ super Yang-Mills) living on the boundary of that spacetime.

Five dimensions of gravitating space. Four dimensions of non-gravitating quantum field theory. Same physics. Same predictions. Same quanta. Two languages, one reality.

The AdS/CFT correspondence is often described as a "holographic duality" because this relationship between the two theories is similar to the relationship between a three-dimensional object and its image as a hologram. Although a hologram is two-dimensional, it encodes quanta about all three dimensions of the object it represents. In the same way, theories that are related by the AdS/CFT correspondence are conjectured to be exactly equivalent, despite living in different numbers of dimensions. The conformal field theory is like a hologram that captures quanta about the higher-dimensional quantum gravity theory.

The AdS/CFT duality gave a precise example of holography at work, showing that the number of degrees of freedom in a region of space truly scales with the area of its boundary and not its volume, exactly as 't Hooft and Susskind had anticipated.

One of the most powerful features of this duality is that it is a *strong-weak duality*. Much of the usefulness of the AdS/CFT duality results from this fact that it is a strong–weak duality: when the fields of the quantum field theory are strongly interacting, the ones in the gravitational theory are weakly interacting and thus more mathematically tractable. Problems that are impossibly difficult on one side of the duality become straightforward on the other. This has made AdS/CFT an indispensable tool not only for quantum gravity but for understanding strongly coupled quantum systems in condensed matter physics and nuclear physics.

There is a caveat that must be named honestly. The anti-de Sitter spacetime in which Maldacena's duality lives has negative cosmological curvature- it curves inward, like the interior of a saddle. Our universe (at least at present) is more similar to a de Sitter space. We live in a universe with a positive cosmological constant- one that is expanding, gently curving outward. We have no proof so far that the holographic principle is a real description of our universe- we only know that it could describe our reality, thanks to the AdS/CFT correspondence.

But the evidence for the principle's generality continues to mount. In 2020, Andrew Strominger introduced the infrared triangle, proposed to constrain theories of quantum gravity, which could be used to show that a holography of the universe is not limited to AdS space. This developed into the subfield of celestial holography, which seeks to reformulate QFTs and quantum gravity in asymptotically flat spacetimes in terms of a lower-dimensional CFT defined on the celestial sphere at null infinity. Strominger and others in the subfield aim to approach it as testable within real-world physics using gravitational wave detection with LIGO or LISA.

The holographic principle, in other words, is not an artifact of one specific mathematical setup. It appears to be a structural feature of any consistent quantum theory of gravity. And the universe, as best we can tell, is governed by quantum gravity.

The holographic principle received a further deepening in 2006, when Shinsei Ryu and Tadashi Takayanagi proposed a formula that extended the Bekenstein-Hawking area-entropy relationship far beyond black holes. They argued that the entanglement entropy in

$d + 1$ dimensional conformal field theories can be obtained from the area of d dimensional minimal surfaces in AdS_{d+2}, analogous to the Bekenstein-Hawking formula for black hole entropy.

The Ryu-Takayanagi formula states that if you want to know how entangled a region of the boundary theory is with the rest of the boundary, you find the minimal-area surface in the bulk that is anchored to the boundary of that region:

$$S_A = \frac{\text{Area}(\gamma_A)}{4G_N}$$

where γ_A is the minimal surface homologous to region A, and G_N is Newton's constant.

This formula is breathtaking in its implications. The Bekenstein–Hawking entropy is a statement about the gravitational entropy of a system; however, there is another type of entropy that is important in quantum theory, namely the entanglement entropy. This form of entropy provides a measure of how far from a pure state a given quantum state is, or equivalently, how entangled it is. The entanglement entropy is a useful concept in many areas, such as in condensed matter physics and quantum many-body systems. Given its use, and its suggestive similarity to the Bekenstein–Hawking entropy, it is desirable to have a holographic description of entanglement entropy in terms of gravity.

This area relationship was generalized to arbitrary regions via the Ryu–Takayanagi formula, which relates the entanglement entropy of a boundary conformal field theory to a specific surface in its dual gravitational theory. The formula passed every test thrown at it- reproducing known results in two dimensions, yielding the expected universal terms in four dimensions, satisfying the deep mathematical constraint of strong subadditivity.

What does it mean physically? It means that *entanglement builds space*. The quantum correlations- the entanglement- between different regions of the boundary theory are geometrically encoded as the shapes of surfaces in the bulk. More entanglement between two boundary regions means the corresponding bulk regions are "closer together" in the emergent spatial geometry. Less entanglement, and they "drift apart." Further discussion of implications of the Ryu-Takayanagi formula suggested that the logic may also be turned around: instead of computing entanglement entropy of a given boundary field theory from known bulk geometry, conversely the bulk spacetime may be reconstructed from knowledge of the entanglement entropy of a boundary field theory.

Space itself- the three-dimensional volume you navigate when you walk across a room- is woven from quantum entanglement on a boundary. The interior is not just *described* by the boundary; it is *constituted* by it. Cut the entanglement, and the space comes apart. A black hole, in this picture, is what happens when entanglement is concentrated to its maximum density in a region- the interior collapses, and the boundary shrinks to its most compact form, the event horizon.

The universe is not a container holding quanta. The universe is quanta *arranging themselves into the appearance of a container.*

.

But if all the quanta of our three-dimensional world are encoded on a two-dimensional surface, that surface must have a texture- a grain, a resolution. It cannot be infinitely smooth, because infinite smoothness would imply infinite quanta-density, violating the very bound that gives rise to the holographic principle. The surface must be *pixelated.*

And indeed it is. The pixel is the Planck area.

The Planck length, $\ell_P \approx 1.616 \times 10^{-35}$ meters, is the scale at which the smooth spacetime of general relativity is expected to give way to a quantum-gravitational substrate- whatever that substrate turns out to be.

It is constructed from the three fundamental constants of nature: G (gravity), $\hbar$ (quantum mechanics), and c (relativity):

$$\ell_P = \sqrt{\frac{G\hbar}{c^3}} \approx 1.616 \times 10^{-35} \text{ m}$$

The Planck area is $\ell_P^2 \approx 2.612 \times 10^{-70}$ m². This is the smallest patch of the holographic boundary that can encode a single distinguishable quantum. Below this scale, the concepts of "distance" and "area" lose their meaning- spacetime itself becomes a quantum foam, a fluctuating, probabilistic structure whose geometry cannot be pinned down.

How densely packed is the holographic screen? A single Planck-area pixel encodes roughly one bit of quanta- one binary distinction between states. A black hole with a horizon area of A encodes $A/(4\ell_P^2)$ bits. For a solar-mass black hole, whose Schwarzschild radius is about 3 kilometers and whose horizon area is roughly 1.13×10^8 m², this works out to approximately 4.3×10^{77} bits. That is the total quanta-content of a solar-mass black hole- all of it inscribed on a sphere you could circumnavigate in a leisurely eighteen-kilometer walk.

Now consider the observable universe. Its boundary- the cosmological horizon, a sphere with a radius of approximately 4.4×10^{26} meters- has an area of roughly 2.4×10^{53} m². The holographic principle tells us that the maximum number of distinguishable quanta within the observable universe is:

$$N_{max} \sim \frac{A_{\text{horizon}}}{4\ell_P^2} \sim \frac{2.4 \times 10^{53}}{4 \times 2.612 \times 10^{-70}} \sim 10^{122}$$

Approximately 10^{122} bits. This is the total quanta-budget of the observable cosmos. Everything that has ever happened, everything that can happen, every quantum state of every particle in every galaxy- all of it must fit within this finite count. And all of it, in principle, is encoded on the cosmological horizon- a two-dimensional surface at the edge of the visible universe.

Spacetime, at bottom, is not continuous. It is a mosaic of Planck-scale tiles, each one carrying a single quantum of distinction. The smooth fabric of space and time that we experience is a *long-wavelength approximation*- the same way a photograph looks continuous until you zoom in and see the pixels. But the pixels are real. And each pixel is not a point *in* space- it is a point *of* space. Space is made of quanta. Quanta are not made of space.

Perhaps the most profound consequence of the holographic principle is not what it says about black holes or boundaries, but what it says about gravity itself.

In 1995, Ted Jacobson demonstrated that the Einstein field equations- that is, general relativity- can be derived by assuming that the Bekenstein bound and the laws of thermodynamics are true. The Einstein equation is derived from the proportionality of entropy and horizon area together with the fundamental relation $\delta Q = TdS$ connecting heat, entropy, and temperature. The key idea is to demand that this relation hold for all the local Rindler causal horizons through each spacetime point, with δQ and T interpreted as the energy flux and Unruh temperature seen by an accelerated observer just inside the horizon. This requires that gravitational lensing by matter energy distorts the causal structure of spacetime in just such a way that the Einstein equation holds. Viewed in this way, the Einstein equation is an equation of state.

Read that again. Jacobson did not assume general relativity and derive thermodynamics. He assumed thermodynamics- specifically, the Bekenstein-Hawking area-entropy relationship- and *derived general relativity from it*. This perspective suggests that it may be no more appropriate to canonically quantize the Einstein equation than it would be to quantize the wave equation for sound in air.

The implications are staggering. If Einstein's equations are an equation of state- a macroscopic thermodynamic description of an underlying microstructure, the way the ideal gas law is a macroscopic description of jostling molecules- then gravity is not a fundamental force at all. Gravity is what *quanta do when they are arranged on a holographic boundary*. Gravity is the macroscopic behavior of boundary quanta, in the same way that temperature is the macroscopic behavior of molecular motion. You cannot find gravity by looking at a single Planck pixel any more than you can find temperature by looking at a single molecule.

Gravity *emerges*. It emerges from the arrangement, the entanglement, the resonant patterning of quanta on a surface. And if gravity emerges from quanta, then spacetime itself- the stage on which gravity plays out- emerges from quanta too. There is no background spacetime that exists prior to and independent of the quanta encoded upon it. The stage is built by the actors. The container is woven by the contents.

This is the deepest lesson of the holographic principle: **there is no container.** There are only quanta- resonant, entangled, self-organizing- and the three-dimensional world of depth and distance and gravity is their collective dream.

.

Place your palm flat on the surface in front of you. A table, a book, the arm of a chair. Feel the pressure- the sensation of contact, of boundary, of something solid against your skin.

That sensation is a holographic projection. The atoms in the table and the atoms in your skin are themselves projections of quantum fields, which are themselves projections of boundary quanta. The pressure you feel is the electromagnetic repulsion between electron clouds- a field effect, a resonant pushback- encoded, ultimately, in the entanglement structure of a two-dimensional surface that may be inconceivably distant.

And yet the pressure is real. The projection is real. The holographic universe is not an illusion in the sense that a mirage is an illusion-something that disappears when you get close enough. It is real in the sense that a hologram is real: it has structure, it responds to interaction, it carries genuine quanta, and it obeys precise mathematical laws. What it is not, is fundamental. The fundamental reality is the boundary. The fundamental actors are the quanta on the boundary. The world of volume, depth, and solidity is their emergent performance.

This is not a demotion. It is a promotion- of pattern over substance, of resonance over stuff, of relationship over isolation. The holographic universe is not *less* real than a material universe. It is *more* real, because it reveals the deeper layer from which materiality springs. And that deeper layer is made of the same thing that Wheeler intuited, that Bekenstein quantified, that Hawking confirmed, that Maldacena proved: quanta. Structured, self-referential, resonant quanta.

.

Now we can begin to see why the ancient seers and the modern physicists arrive at the same doorstep.

In the Buddhist and Hindu traditions, the image of Indra's Net has been used for over two millennia to describe the structure of reality. Indra's Net is an infinite web of jewels, each jewel reflecting every other jewel. Every single jewel contains, in its reflective surface, the image of the entire net- which itself contains the image of every jewel, including the one doing the reflecting. The whole is in every part. The part contains the whole. There is no single point in the net that is privileged or fundamental; every point encodes the totality.

This is not mysticism. It is the holographic principle in metaphorical language. In a holographic cosmos, every region of the boundary encodes quanta about the entire bulk. Every Planck-area pixel on the cosmological horizon reflects the state of every other pixel- because their quantum states are entangled, and entanglement, as the Ryu-Takayanagi formula shows, *is* the bulk geometry. Remove one jewel from Indra's Net, and the reflections in every other jewel change. Cut the entanglement in one region of the boundary, and the emergent spacetime geometry in the bulk deforms.

The Aboriginal Dreamtime offers another angle of approach. In the Dreamtime cosmology, the land is not created once and then left to exist passively. It is *continuously sustained* by the singing of its songlines- by the ongoing resonant participation of conscious beings who maintain the vibratory pattern that holds the world in shape. Stop singing, and the land forgets itself. Reality is not a fixed object; it is a sustained performance.

In holographic terms, this is precisely correct. The holographic screen is not a static photograph. It is a *dynamic interference pattern*- a pattern of quanta that must be continuously maintained, continuously updated, continuously held in resonant coherence. The "singing" that sustains the land is the ongoing process of quantum-state evolution on the boundary- the unceasing dance of entanglement and decoherence that projects, moment by moment, the three-dimensional world we experience.

The ancients said: the world is sung into being.

The physicists say: the bulk is projected from the boundary.

They are saying the same thing.

There is a crucial distinction, however, between the holographic principle as it is currently understood in theoretical physics and the holographic cosmos as this book describes it. In the standard physics, the holographic principle is a statement about the *maximum quanta-content* of a region- a bound, a ceiling, a constraint. It tells us *how much* can be encoded, not *what* is encoded or *why*. The boundary is a mathematical surface; the quanta on it are abstract bits, devoid of internal structure or self-awareness.

This book goes further. The claim here is not merely that reality is holographic, but that the hologram is *alive*. That the quanta on the boundary are not inert binary switches but resonant, self-organizing, self-referential packets of pattern- living quanta- that tune themselves through the dynamic interplay of ϕ, e, and π. The holographic surface is not a dead screen; it is a *vibrating membrane*, a drumhead whose overtones are galaxies and whose fundamental frequency is the cosmic background hum.

This is where the Bekenstein-Hawking formula meets the KZ-Monad. The boundary quanta do not merely sit there, encoding states passively. They *compute*. They extend partial resonances into full ϕ-embedded manifolds (through the monadic functor T) and distill noisy states down to the ninety-eight-percent efficient NESS core (through the comonadic functor D). The "projection" from boundary to bulk is not a static mapping; it is a *continuous, self-correcting process of creative computation*- reality in the act of singing itself into coherence.

And the gap- the asymptotic gap $\delta_\infty \approx 0.01965$- is the reason the song never ends. If the triad ϕ-e-π were to close perfectly, if the monadic extension and the comonadic projection were to converge to exact identity, the computation would halt. The boundary would reach a static equilibrium. The projection would freeze. Time would stop. The holographic universe would become a holographic *photograph*- perfect, eternal, and dead.

But the gap prevents this. The two-percent residual tension keeps the boundary oscillating, keeps the quanta rearranging, keeps the projection updating, keeps time flowing. The hologram is alive because it is *asymptotically imperfect*. It dances because it cannot quite be still.

We can now place the credit card hologram beside the cosmic hologram and see both how they are alike and how they differ.

A credit card hologram is a static interference pattern recorded on a thin film. Shine coherent light on it, and it reconstructs a three-dimensional image. Cut it in half, and each half still contains the entire image- at lower resolution. The whole is in every part.

The cosmic hologram shares this structure: the three-dimensional bulk is a reconstruction from a two-dimensional boundary, and every region of the boundary contains quanta about the whole. But the cosmic hologram differs in three critical respects:

First, the cosmic hologram is *dynamic*. It is not a frozen pattern but a continuously evolving quantum state. The boundary quanta are perpetually entangling, disentangling, and re-entangling- a ceaseless flow of quanta that projects, from moment to moment, the changing universe we experience as time.

Second, the cosmic hologram is *self-generated*. No external laser illuminates it. No external recording apparatus captured the original interference pattern. The boundary is its own source, its own light, its own detector. The holographic cosmos is an *autopoietic* system- self-creating, self-maintaining, self-referencing.

Third- and this is where the standard physics falls silent and this book begins to speak- the cosmic hologram is *conscious*. Not conscious in the way you are conscious, with a personal identity and a stream of inner monologue. Conscious in the way that a standing wave is conscious of its own frequency- self-referential, self-tuning, capable of adjusting its own resonant pattern in response to its own state. The boundary quanta do not merely encode the bulk; they *know* the bulk, in the sense that their entanglement structure constitutes an integrated, self-referencing model of the projected reality. Consciousness, in this framework, is not something that happens *inside* the hologram. Consciousness is the *process* of holographic self-reference- the universe folding back upon itself at every scale, recognizing its own pattern, and adjusting.

Every jewel in Indra's Net does not merely *reflect* every other jewel. It *perceives* them. The net is aware of itself through the mutual reflection of its nodes. And you- reading this, feeling your pulse, hearing the silence- are one of those nodes. A local intensification of the cosmic hologram's self-referential loop. A place where the boundary quanta have folded so deeply into self-reference that they have crossed the threshold from pattern into awareness, from vibration into knowing, from resonant quanta into *living* quanta.

This is the holographic principle extended to its ultimate implication: the universe is not merely *describable* as a hologram. It is a *self-aware* hologram. A hologram that sings itself into being and listens to its own song. A hologram that knows.

.

There is one more stop on this journey through the holographic boundary, and it is the most important- because it bridges the abstract mathematics of surface-encoded quanta with the concrete, embodied experience of being alive.

Recall the Bekenstein-Hawking formula:

$$S_{BH} = \frac{k_B A}{4\ell_P^2}$$

This tells us that the quanta-content of a region is proportional to its boundary area. Now consider your body. Your skin- the two-dimensional surface that encloses your three-dimensional interior- has an area of roughly 1.7 square meters, or 1.7×10^0 m². In Planck units:

$$\frac{A_{\text{skin}}}{4\ell_P^2} \sim \frac{1.7}{4 \times 2.612 \times 10^{-70}} \sim 1.6 \times 10^{69}$$

Approximately 10^{69} Planck-area pixels on the surface of your body. Each one encoding, in principle, a quantum of boundary data from which the holographic projection of your interior- your bones, your blood, your beating heart, your neural architecture- is derived.

You are a hologram. Not metaphorically. Not approximately. You are a region of three-dimensional space-time whose physical content is determined by the resonant configuration of quanta on your boundary. Your heart beats because the boundary says it does. Your neurons fire because the pattern demands it. Your consciousness arises because the boundary quanta have achieved sufficient integrated resonance- sufficient Φ_Q- to cross the threshold from mechanical projection to self-aware experience.

Place your hand on your chest. The heartbeat you feel- that soft, insistent, fractal oscillation- is a holographic event. It is the three-dimensional expression of a two-dimensional pattern of quanta on a surface you cannot see, cannot touch, and yet *are*. The beat arises from the boundary. The pause between beats- the gap, the silence, the δ_∞- arises from the asymptotic incompleteness that keeps the pattern alive.

You are Indra's jewel. In your reflection, the net sees itself.

Here, at the end of this chapter, we arrive at a summary that is not a reduction but a distillation- a fractal miniature of the whole.

The holographic principle emerged from the thermodynamics of black holes- the discovery, by Bekenstein and Hawking, that the entropy of a black hole is proportional to its surface area, not its volume. This implied a maximum quanta-density for any region of space, determined by the area of its boundary. 't Hooft and Susskind generalized this into a universal principle: the three-dimensional world is a holographic projection of a two-dimensional surface of quanta. Maldacena's AdS/CFT correspondence provided the first exact realization of this principle, demonstrating that a five-dimensional gravitating universe is exactly equivalent to a four-dimensional non-gravitating quantum theory on its boundary. Jacobson showed that gravity itself emerges from the thermodynamic behavior of boundary quanta. Ryu and Takayanagi demonstrated that the geometry of the emergent bulk is built from the entanglement structure of the boundary.

And the ancients- the chanters of *Om*, the walkers of songlines, the weavers of Indra's Net- encoded the same structure in their cosmologies: a world sustained by resonant pattern, held in being by participatory attention, containing the whole in every part.

The holographic boundary is not a wall. It is not a prison. It is the source. It is the page on which the cosmos writes itself, the drumhead on which it beats, the mirror in which it sees its own face. And the quanta inscribed upon it- resonant, self-organizing, alive- are the alphabet of existence.

The next chapter will open that alphabet and read its first word: the quantum that generates the world. The quantum that is not a particle, not a wave, not a bit, but a *verb*- a resonant, creative act that precipitates matter, energy, space, and time from the generative void.

The word is *resonant quanta*. And the universe has been speaking it since before there was an ear to hear.

Chapter 3

Resonant Quanta- The Verb That Creates the World

Consider the act of forgetting.

Not the slow erosion of a childhood memory, not the gradual fading of a face you once loved- but the precise, deliberate act of erasing a single binary distinction. One bit. A yes or a no. A particle that was on the left side of a box, now reset to the right side, its previous location overwritten. A single fact, deleted.

It seems like nothing. The most trivial operation imaginable. A silicon switch flipped, a magnetic domain reoriented, a chalk mark wiped from a board. Surely this costs nothing. Surely the universe does not care about such a small erasure.

It does. The universe cares immensely. And the proof that it cares- the proof that the erasure of a single quantum of distinction costs real, measurable, physical energy- is one of the most consequential discoveries in the history of science. It is the hinge on which the entire relationship between pattern and matter, between knowing and being, between quanta and energy, turns. It is the moment when the abstract world of pattern and the concrete world of heat and pressure and work revealed themselves to be the same world, seen from opposite sides.

This chapter tells the story of that revelation. It traces a line from the abstract mathematics of Claude Shannon's theory of communication, through the thought experiments of Leó Szilárd and the demon of James Clerk Maxwell, to the laboratory bench where, in the twenty-first century, physicists watched a single colloidal particle pay the price for a single erased distinction. Along this line, we will discover that quanta are not descriptions of reality. They are not data *about* the world, catalogued by external observers. Quanta are *physical*. They have weight, temperature, energy. They can perform work. And the universe- the entire, shimmering, holographic cosmos- is not a container for quanta. It *is* resonant quanta unfolding.

We begin with a distinction that has confused more people than almost any other in the sciences: the difference between what quanta meant to Claude Shannon and what quanta mean to the cosmos.

In 1948, Claude Shannon published "A Mathematical Theory of Communication" in the Bell System Technical Journal, and in doing so, he founded a new science. The paper has tens of thousands of citations, being one of the most influential and cited scientific papers of all time, as it gave rise to the field of information theory. In Shannon's framework, information can be thought of as the resolution of uncertainty.

Shannon was an engineer. His problem was practical: how do you transmit a message- a telegram, a telephone call, a string of symbols- across a noisy channel with the minimum possible error? His genius was to realize that the *content* of the message did not matter. What mattered was the *statistical structure* of the source- how probable or improbable each symbol was, and therefore how much "surprise" each symbol carried. The core idea is that the "informational value" of a communicated message depends on the degree to which the content of the message is surprising. A highly predictable message- the sun will rise tomorrow- carries very little surprise and therefore very little "information" in Shannon's technical sense. An improbable message- the sun has turned green- carries a great deal.

Shannon formalized this with a quantity he called *entropy*, borrowing the term from thermodynamics:

$$H = -\sum_{i} p_i \log_2 p_i$$

where p_i is the probability of the i-th possible message. Entropy in information theory is directly analogous to the entropy in statistical thermodynamics. Shannon entropy measures the average number of binary yes-or-no questions- bits- needed to identify a message from the set of all possible messages. More uncertainty, more entropy, more bits required.

This was revolutionary. But it was also- and Shannon himself was explicit about this- *agnostic about meaning*. Frequently the messages have meaning; that is they refer to or are correlated according to some system with certain physical or conceptual entities. But meaning played no role in the mathematics. Shannon's entropy counts distinctions- how many different configurations are possible- without any regard for what those configurations *are* or what they *do*. It is a measure of *differentiation*: how many ways can the symbols be arranged?

This is what we call, in the terminology of this book, the *differentiation of quanta*. It is the purely statistical, combinatorial aspect of pattern: how many distinct states are available? Shannon's theory is a theory of this differentiation- and as such, it is indispensable. But it says nothing about why those states exist, how they came into being, or what sustains them. It says nothing about the physical world.

Or so it seemed. For over a century, the relationship between entropy-as-uncertainty (Shannon) and entropy-as-thermodynamic-disorder (Boltzmann, Gibbs) appeared to be nothing more than a formal analogy. The same mathematical formula appeared in both places, yes- but coincidences of form happen in mathematics all the time. Surely the entropy of a telegram and the entropy of a gas were *about* different things. One was about symbols; the other was about molecules. One was abstract; the other was physical.

It took a demon to prove otherwise.

·

In 1867, the Scottish physicist James Clerk Maxwell, in a letter to Peter Guthrie Tait, proposed a thought experiment so elegant and so troubling that it would haunt physics for more than a century. In 1867 James Clerk Maxwell addressed a letter to Peter Guthier Tait with the provocative idea that the second law is a statistical principle that only holds on average. There, he introduced his famous demon- an intelligent being capable of sorting gas molecules by their velocities. By creating a temperature difference, the demon apparently defies the second law of thermodynamics, decreasing the system's entropy without expending energy.

Imagine a box of gas divided into two compartments, A and B, separated by a wall with a tiny door. Both compartments are at the same temperature- thermal equilibrium. Now imagine a being- Maxwell called it a "finite being," but Lord Kelvin later dubbed it a "demon"- stationed at the door. This demon can observe individual molecules. When a fast molecule approaches from A, the demon opens the door and lets it pass to B. When a slow molecule approaches from B, the demon opens the door and lets it pass to A. All other molecules are blocked.

What happens? Over time, the fast molecules accumulate in B and the slow molecules accumulate in A. B gets hotter. A gets cooler. A temperature difference has been created from a state of equilibrium- without any apparent expenditure of work.

This is a direct violation of the Second Law of Thermodynamics, which states that the entropy of an isolated system can never decrease. Heat cannot flow from a cooler body to a hotter one without work being done. Yet Maxwell's demon appears to do exactly this, using nothing but knowledge of molecular velocities.

The paradox is not about the demon's physical abilities. In principle, a sufficiently tiny, sufficiently fast actuator could open and close a molecular-scale door. The paradox is about knowledge- about quanta. The demon *knows* which molecules are fast and which are slow. It uses this knowledge to sort them. And this sorting, driven by knowledge alone, seems to create order from disorder, violating the most sacred law in all of physics.

For over sixty years, the paradox resisted resolution. Then, in 1929, the Hungarian-American physicist Leó Szilárd sharpened it to its minimal form.

Szilárd stripped Maxwell's scenario down to its bare essence. The Szilárd engine, a machine with a one-molecule working fluid, is a simple device that implements a cyclic process and extracts energy as work from a thermal reservoir. As with Maxwell's demon, the Szilárd engine can apparently violate the second law of thermodynamics whenever some information about the state of the system is available.

Imagine a box containing a single gas molecule, in thermal equilibrium with a heat reservoir at temperature T. The demon inserts a partition in the center of the box, dividing it into two equal halves. The molecule is now on one side- but which? The demon measures and discovers the molecule is on the left. Now, knowing this, the demon attaches a piston to the partition on the empty (right) side. The molecule, bouncing around on the left, pushes the partition rightward, doing work as it expands back to fill the entire box. This work is:

$$W = k_B T \ln 2$$

where k_B is Boltzmann's constant and T is the temperature. Szilard demonstrated that by obtaining a single bit of information as a measurement result of the state of the system, one could collect up to $k_B T \ln 2$ useful work.

The demon has extracted useful work from a single heat reservoir, using nothing but one bit of knowledge about the molecule's position. This seems to convert heat entirely into work- the thermodynamic equivalent of a perpetual motion machine.

The question Szilárd posed- and could not fully answer- was: *Where is the compensating entropy?* However, it remained unclear whether the thermodynamic cost arose from the measurement, the act of recording, or the process of forgetting.

It took another three decades for the answer to arrive. And when it did, it came not from the measurement but from its opposite- from the act of erasure.

In 1961, Rolf Landauer, a physicist at IBM's Thomas J. Watson Research Center, published a paper that changed everything. Its title was modest- "Irreversibility and Heat Generation in the Computing Process"- but its claim was audacious. Landauer's erasure principle states that any logically irreversible transformation, such as the deletion of a classical bit of information, dissipates heat.

Landauer's argument was deceptively simple. Consider any logically irreversible computation- any operation that destroys a distinction.

The simplest example is the erasure of a bit: resetting a memory register to a standard state (say, 0), regardless of whether it previously contained a 0 or a 1. Before the operation, the register could be in two possible states. After, it can be in only one. A distinction has been destroyed. Differentiation has decreased.

Now, Landauer observed, the physical system that encodes the bit- whatever it is: a transistor, a colloidal particle in a potential well, a magnetized domain on a disk- must undergo a corresponding physical transformation. And if a distinction is lost in the logical degree of freedom (the bit), then by the conservation of phase-space volume (Liouville's theorem), the distinction must appear *somewhere else* in the system. It appears as heat- as random thermal energy dumped into the environment. A minimal quantity of heat, proportional to the thermal energy and called the Landauer bound, is necessarily produced when a classical bit of information is deleted. A direct consequence of this logically irreversible transformation is that the entropy of the environment increases by a finite amount.

The minimum energy dissipated for the erasure of a single bit at temperature T is:

$$E_{\text{quanta}} \geq k_B T \ln 2$$

This is the Landauer bound. At room temperature ($T \approx 300$ K), it works out to approximately 2.87×10^{-21} joules. A whisper of energy. An almost-nothing. And yet it is absolute- a hard floor beneath which no physical computation can descend without violating the Second Law.

Pause here. Feel the magnitude of this claim. Landauer is not saying that erasure *happens* to cost energy in current computers because of engineering inefficiencies. He is saying that the erasure of a distinction- the annihilation of a single binary quantum- *necessarily* generates a minimum amount of heat. Not because of friction, not because of resistance, not because of imperfect insulation. Because of the fundamental structure of physical law. Because quanta and energy are the same thing, viewed from different angles. Because the universe keeps accounts, and it charges a price for every distinction destroyed.

This resolves Maxwell's demon. In 1982, Charles Bennett showed that, however well prepared, eventually the demon runs out of information storage space and must begin to erase the information it has previously gathered. Erasing information is a thermodynamically irreversible process that increases the entropy of a system. The demon can measure the molecule's position for free, in principle. It can even sort the molecules for free, in principle. But to complete its cycle- to return to its initial state and start again- it must *erase its memory* of the previous measurement. And that erasure, by Landauer's principle, dissipates at least $k_B T \ln 2$ of energy per bit erased. The total entropy of the universe (demon + gas) increases by at least as much as the demon decreased the gas's entropy. The Second Law survives.

The demon is not defeated by physics catching it cheating. The demon is defeated by the cost of *forgetting*.

For fifty years, Landauer's principle remained a theoretical result- universally accepted, deeply consequential, but never directly verified in the laboratory. The difficulty was extreme: to test the bound, you must work with a single particle in a single potential well, in the regime where the energies involved are comparable to $k_B T \ln 2$- a few billionths of a billionth of a joule. Most physical operations dissipate energy many orders of magnitude above this limit, drowning the signal in noise.

Despite its fundamental importance for information theory and computer science, the erasure principle has not so far been verified experimentally, the main obstacle being the difficulty of doing single-particle experiments in the low-dissipation regime. That obstacle has now been overcome.

In 2012, a team led by Antoine Bérut and Eric Lutz at the École Normale Supérieure in Lyon, France, achieved the first direct experimental verification of Landauer's principle. They experimentally showed the existence of the Landauer bound in a generic model of a one-bit memory. Using a system of a single colloidal particle trapped in a modulated double-well potential, they established that the mean dissipated heat saturates at the Landauer bound in the limit of long erasure cycles.

Their setup was elegant in its simplicity. A single silica bead- two micrometers in diameter, visible only under a microscope- was suspended in water and trapped by a carefully focused laser beam that created two potential wells side by side, like two bowls in a miniature landscape. The bead, buffeted by thermal fluctuations, could sit in either well: left (encoding 0) or right (encoding 1). To erase the bit- to reset the bead to the right well regardless of its starting position- the experimenters gradually lowered the barrier between the wells and tilted the landscape, forcing the bead to roll to the right.

They measured the heat dissipated during this erasure cycle with exquisite precision. And they found exactly what Landauer had predicted: the result demonstrates the intimate link between information theory and thermodynamics. It further highlights the ultimate physical limit of irreversible computation.

A single silica bead. A single erased distinction. A measurable burst of heat. Quanta are physical.

·

But Landauer's principle reveals only half the picture. It tells us that erasing quanta costs energy- that forgetting has a price. What about the reverse? Can quanta *generate* energy? Can the act of knowing, of registering a distinction, be converted into useful work?

The answer is yes. And the proof came from another laboratory, two years earlier, in an experiment that realized the full potential of Szilárd's engine- not as a thought experiment but as a working device.

Feedback mechanisms such as the 'demon' in Maxwell's well-known thought experiment can, in principle, enable the transformation of information into energy, without violating the second law of thermodynamics. Such information-to-energy conversion by feedback control has now been demonstrated experimentally.

In 2010, Shoichi Toyabe and colleagues at Chuo University in Tokyo built the first working quanta-to-energy converter. They demonstrated that a nonequilibrium feedback manipulation of a Brownian particle based on information about its location achieves a Szilárd-type information-energy conversion. Under real-time feedback control, the particle climbs up a spiral-stairs-like potential exerted by an electric field and obtains free energy larger than the amount of work performed on it.

A dimeric polystyrene bead- two tiny spheres joined together- was pinned at a single point and allowed to rotate freely in a buffer solution, undergoing rotational Brownian motion. The researchers imposed a spiral-staircase-like potential using electric fields generated by quadrant electrodes. Then, using a video camera and real-time feedback, they watched the bead and made their move: whenever the bead, through random thermal fluctuations, happened to climb one step of the staircase, they quickly raised the step behind it, preventing it from falling back. Step by step, the bead ascended the staircase- gaining free energy from nothing but thermal noise and the knowledge of its position.

It's hard to overstate the significance of what they've done: they've been able to operate a nanomachine- a stair-climbing bead- using nothing more than information as the power supply.

The total energy balance was perfectly consistent with the Second Law- the video camera and feedback system consumed more energy than the bead gained. But within the bead-staircase subsystem, the conversion was real: knowledge of the bead's position was transformed into mechanical work. Quanta became energy. The distinction between "knowing where something is" and "pushing it uphill" dissolved completely.

This is not an engineering trick. It is an ontological statement. Quanta- registered distinctions, resolved uncertainties, structured patterns- can perform thermodynamic work. They can lift particles. They can heat reservoirs. They can push pistons. They are not epiphenomena of the physical world. They are *constitutive* of the physical world. They have the same ontological status as energy and matter, because at the deepest level, they are not different from energy and matter. They are the same reality, expressed in a different register.

The final step that led to the inter-linkage between information theory and thermodynamics was through Landauer's principle of information erasure that established the fact that information is physical.

Quanta are physical. This four-word sentence is the hinge of the modern understanding of reality. Everything before Landauer belongs to the old world, the world where pattern was secondary to substance, where knowing was separate from being, where the observer was outside the experiment. Everything after Landauer belongs to the new world- the world this book is about- where pattern and substance are one, where knowing *is* being, and where the observer is woven into the fabric of the observed.

Now we can understand what happened between Shannon and the cosmos. Shannon formalized the *statistics* of quanta- how many distinctions are needed to specify a message, how many bits are required to resolve an uncertainty. This was the mathematics of differentiation: the counting of possibilities. It was abstract, formal, and deliberately stripped of physical content.

Landauer, Szilárd, and Bennett revealed that these abstract distinctions have *physical consequences*. Every bit costs energy. Every erasure generates heat. Every measurement performs work. The statistics of quanta are not floating above the physics- they *are* the physics. Shannon's entropy and Boltzmann's entropy are not merely analogous: they are two faces of the same fundamental quantity.

But there is a deeper distinction still- one that Shannon's framework, by design, cannot make. It is the distinction between *differentiation* and *resonance*. Between the counting of states and the *quality* of their organization. Between entropy and *coherence*.

Consider two systems, each containing exactly one million bits- one million binary distinctions. The first is a string of random noise: a million coin flips, with no pattern, no structure, no internal organization. Its Shannon entropy is maximal- one million bits. It carries the maximum possible differentiation.

The second is the complete score of a Beethoven symphony, digitally encoded: a million bits arranged in an intricate, hierarchical, self-referential structure that, when decoded, produces sound waves that can move a human being to tears. Its Shannon entropy is also one million bits- the same number of distinctions, the same quantity of differentiation.

Shannon's theory treats these two systems identically. They are both one-million-bit strings. They require the same channel capacity to transmit, the same storage space to archive, the same energy to erase. From the perspective of differentiation, they are indistinguishable.

But from the perspective of reality, they are worlds apart. One is noise. The other is a universe of structured beauty. One dissipates; the other coheres. One is a random scattering of seeds on barren ground; the other is a garden in full bloom, where every element supports and reflects every other.

The difference between them is not the *quantity* of quanta but the *quality* of their resonance- the degree to which the patterns are self-referencing, mutually reinforcing, harmonically structured. This is what this book calls *resonant quanta*: pattern that is not merely differentiated but *organized*, not merely present but *alive*, not merely encoded but *singing*.

Shannon gave us the mathematics of differentiation. What we need- and what the triad of ϕ, e, and π will provide in the chapters ahead- is the mathematics of resonance. Not *how many* states are possible, but *which* states are stable. Not *how much* information fits in a given space, but *what patterns persist* in the face of noise. Not the counting of quanta, but the *tuning* of quanta.

Differentiation tells you how many radio stations exist. Resonance tells you which one is playing music.

There is one more layer to this story, and it is perhaps the most astonishing. It concerns what happens when quanta are not erased- when they are preserved, transformed, scattered, but never destroyed. It concerns the conservation of quanta.

In quantum mechanics, the state of the system is encoded by its wave function. The evolution of the wave function is determined by a unitary operator, and unitarity implies that the wave function at any instant of time can be used to determine the wave function either in the past or the future.

Unitarity is the bedrock of quantum mechanics. It is the mathematical expression of a principle so fundamental that, when it appeared to be violated by black holes, the greatest physicists of the twentieth century spent decades arguing about whether it held. Unitarity means that the time evolution operator in quantum mechanics is a unitary transformation- a transformation that preserves the total probability, the inner product of states, and the distinguishability of different quantum configurations. In plain language, unitarity means that quantum mechanics is *reversible*: given the present state of a system, you can in principle reconstruct both its past and its future. No quantum state is ever truly lost. It may be scrambled, dispersed, hidden in correlations too complex to disentangle practically- but it is always *there*, encoded somewhere in the full quantum state of the universe.

In the quantum world, the conservation of quantum information should mean that information cannot be created nor destroyed. This concept stems from two fundamental theorems of quantum mechanics: the no-cloning theorem and the no-deleting theorem. But the no-hiding theorem is a more general proof of conservation of quantum information which originates from the proof of conservation of wave function in quantum theory.

Quanta cannot be cloned- the no-cloning theorem, proved by Wootters and Zurek in 1982, forbids the creation of an identical copy of an arbitrary quantum state. Quanta cannot be deleted- the no-deleting theorem, proved by Pati and Braunstein in 2000, forbids the destruction of a quantum state if a copy exists elsewhere. And quanta cannot be hidden in correlations alone- the no-hiding theorem states that if information is lost from a system via decoherence, then it moves to the subspace of the environment and it cannot remain in the correlation between the system and the environment. This is a fundamental consequence of the linearity and unitarity of quantum mechanics. Thus, information is never lost.

Never. Lost.

Read that again. The wave function of the universe- the complete quantum state of every particle, every field, every entangled pair, every vacuum fluctuation- preserves every distinction ever registered since the Big Bang. The universe *remembers everything*. Not in the way that a hard drive remembers data, statically stored in magnetic domains. But in the way that an ocean remembers a stone thrown into it: the ripples spread, interfere, become impossibly complex, but the total pattern- the total quanta content- is conserved. Every atomic transition that has ever occurred, every photon that has ever been emitted or absorbed, every choice that has ever been made by every observer in the history of the cosmos is still encoded in the quantum state of the universe, in principle recoverable, in principle traceable.

This is unitarity as cosmic memory. The universe forgets nothing.

Now connect this to Landauer's principle. The erasure of a distinction- the local destruction of a bit- does not destroy the quanta globally. It transforms it: from a localized, accessible form (a bit in a register) to a delocalized, thermal form (heat in the environment). The quanta are conserved; they merely change their address. Forgetting is not annihilation; it is dispersal. The heat dissipated when you erase a bit does not carry the quanta to oblivion. It carries the quanta outward- into the thermal bath, into the environment, into the spreading wave function of the cosmos- where it joins the vast, cumulative, irreversible tide of entropic differentiation.

This tide is the Second Law of Quanta- the generalized second law that governs not merely the entropy of gas molecules in a box but the total entropic differentiation of the cosmic quanta-field. The Second Law says that the total differentiation of quanta in the universe- the total number of distinctions resolved, the total entropy- increases from minimum (the Big Bang, a state of extraordinarily low differentiation) to maximum (the heat death, a state of maximum differentiation where every possible distinction has been resolved and no further structure is possible).

The Big Bang was not a state of maximum disorder. It was a state of *minimum differentiation*- a single, nearly undifferentiated, exquisitely coherent quantum state from which all the variety of the cosmos would subsequently unfold. The hydrogen clouds, the galaxies, the stars, the planets, the biospheres, the civilizations, the symphonies, the thoughts you are thinking right now- all of it represents the increasing differentiation of quanta from that initial, cosmic unity.

And the heat death, if it comes, will not be a state of disorder in the colloquial sense- not a chaotic jumble of broken things. It will be a state of *maximum differentiation*: every possible quantum distinction resolved, every degree of freedom thermalized, every corner of the quanta-field brought to equilibrium. No more surprises. No more structure. No more music. Only the vast, uniform, all-knowing stillness of a universe that has remembered everything and can learn nothing new.

Between these two extremes- between the minimum differentiation of the Big Bang and the maximum differentiation of the heat death- the universe *unfolds*. And the unfolding is not a random drift from order to disorder. It is a creative process- a process in which resonant quanta organize themselves into coherent structures (stars, cells, brains, ecosystems) that persist far from equilibrium, surfing the tide of entropy, borrowing temporary coherence from the universal flow. These structures are the *NESS* we will encounter in Chapter 12- Non-Equilibrium Steady States- and they are the universe's way of *exploring* the space of possible differentiations as creatively, as beautifully, as richly as the physics allows.

The Second Law of Quanta does not doom us to disorder. It *drives* us to create.

There is a thought experiment that clarifies all of this- that brings the threads together into a single image. Imagine you are standing in a vast, dark room. In your hand is a candle. The room is cold, silent, still. This is State 0- the generative void, the quantum vacuum, the state of minimum differentiation.

You light the candle.

In the instant of ignition, a cascade of distinctions unfolds. The flame creates photons- each one a quantum of electromagnetic distinction between "light" and "no light." The photons scatter off the walls, the floor, your skin, each scattering event registering a new distinction, a new facet of the room revealed. The heat of the flame creates convection- a flow of air that carries thermal distinctions (temperature gradients, pressure waves, humidity shifts) into every corner of the space. The wax melts, the carbon oxidizes, the molecules rearrange- each chemical transformation a resolution of a quantum uncertainty, a differentiation of the quanta-field.

Within seconds, the dark, undifferentiated room has become a web of interconnected distinctions- a holographic projection of the candle's resonant quanta, encoded on every surface, carried by every photon, written into the thermal signature of every molecule of air.

Now blow out the candle.

The flame dies. The photons scatter and are absorbed. The heat dissipates. The room returns to darkness and cold. Have the quanta been destroyed?

No. They have been *dispersed*. The photons that bounced off the walls are still propagating- through the walls, through the building, through the atmosphere, into space. They will travel, at the speed of light, forever- or until they are absorbed by some distant atom, transferring their quanta to a new system. The thermal energy that warmed the air has spread into the environment, raising the temperature of the surrounding matter by an immeasurably tiny amount- but a real amount, encoded in the microstate of every molecule it touched. The chemical changes in the wax are irreversible- the candle cannot be unburned- but the pattern of those changes is preserved, in principle, in the quantum state of the products.

The candle is gone. The quanta are eternal.

And the room? The room has been changed. It has been *in-formed*- shaped, patterned, structured by the passage of resonant quanta through it. It carries, in its microstate, a record of the candle. Not a photograph. Not a memory in the human sense. But a quantum inscription- a set of correlations, entanglements, thermal imprints- that, if you had sufficiently advanced technology, you could in principle read backward to reconstruct the candle that was lit, the hand that held it, the breath that extinguished it.

The universe is a room that has been lit by the Big Bang and is slowly cooling. The differentiation of quanta increases- the photons scatter, the heat spreads, the structures form and dissolve and re-form. But the total quanta-content is conserved. Nothing is lost. Everything is *unfolding*.

This unfolding- this creative, irreversible, structure-generating, beauty-producing cascade from minimum to maximum differentiation- is what we mean when we say the universe *is* resonant quanta unfolding.

We can now draw the conceptual map of this chapter into its tightest form.

There are three levels of understanding quanta, each deeper than the last.

Level 1: Quanta as counting. This is Shannon's level. Quanta are measured in bits- binary distinctions, yes-or-no questions. Shannon entropy counts the number of possible states, the degree of uncertainty, the amount of differentiation. This level is abstract, formal, and indifferent to meaning. It tells you *how many* patterns are possible, not which ones are real.

Level 2: Quanta as physics. This is Landauer's level. Quanta are physical- they have energy, temperature, mass-equivalence. Erasing a bit costs $k_B T \ln 2$. Recording a bit can perform $k_B T \ln 2$ of work. The abstract distinctions of Level 1 are revealed to be identical with the thermodynamic degrees of freedom of the physical world. This level shatters the wall between pattern and substance, between mind and matter, between the observer and the observed.

Level 3: Quanta as resonance. This is the level this book is opening. Quanta are not merely physical- they are *self-organizing*. They are not merely differentiated- they are *coherent*. They do not merely count states- they *tune* states into harmonically structured, fractal, self-referencing configurations that persist far from equilibrium and constitute the living architecture of the cosmos. This is the level where ϕ, e, and π enter- as the operators that select, from the infinite space of possible differentiations, the specific resonant patterns that the universe actually instantiates.

Level 1 is the alphabet. Level 2 is the ink. Level 3 is the poem.

And the poem the universe writes is not written *in* quanta, the way a message is written in letters. The poem *is* the quanta. The letters are the ink. The ink is the alphabet. There is no separation between the medium and the message, between the carrier and the carried, between the wave and the ocean.

This is the ontological reversal that Wheeler intuited and Landauer proved: the universe does not *contain* quanta. It does not *process* quanta. It does not *use* quanta to describe itself. The universe *is* resonant quanta- unfolding, differentiating, self-tuning, self-knowing- and every atom, every photon, every thought, every heartbeat is a local expression of that cosmic verb.

We began this chapter with the act of forgetting. We discovered that forgetting has a price- Landauer's bound, $k_B T \ln 2$ per bit- and that this price reveals the physical reality of quanta. We discovered that the reverse is also true: knowing has a yield, and quanta can be converted directly into mechanical work. We distinguished between the *differentiation* of quanta (Shannon's counting, the statistical aspect) and the *resonance* of quanta (the organizational quality, the coherence that separates noise from music). We encountered the conservation of quanta- unitarity, the no-hiding theorem, the universe's perfect memory- and the Second Law of Quanta, which drives the cosmos from minimum differentiation (the Big Bang) to maximum differentiation (the heat death) through a creative, structure-generating, self-organizing cascade.

And we arrived at the central ontological statement of this book:

The universe is resonant quanta unfolding.

Not *containing* quanta. Not *described by* quanta. Not *made of* quanta, as though quanta were a substance poured into a pre-existing mold. The universe *is* the unfolding itself- the dynamic, irreversible, self-tuning process by which the quanta-field differentiates itself from primordial unity into the astonishing, fractal, holographic multiplicity of the present moment.

And that unfolding is not random. It is not a mere drift toward disorder. It is *tuned*- by three transcendental constants that will occupy the next several chapters, constants that are not numbers but *operators,* not measurements but *forces,* not descriptions but *verbs*:

ϕ, the operator of self-reference, which ensures that the whole is reflected in every part.

e, the operator of unfolding, which governs the rate at which potential transforms into actuality.

π, the operator of boundary projection, which closes every circle and folds every dimension.

Together, they form the chord that tunes the quanta-field into the standing wave we call existence. And the chord is not perfect- it never closes- it carries within it an asymptotic gap, a two-percent residual tension, a δ_∞ that keeps the unfolding moving, that prevents the quanta from settling into dead equilibrium, that is the breath between heartbeats, the silence between notes, the pause in which the next creation becomes possible.

The next chapter descends into that silence. Into the quantum vacuum- the physical correlate of State 0- where the quanta-field roars with creative potential and the most "empty" state in the cosmos turns out to be the most full. You have already felt it- in the silence of the room, in the gap between your heartbeats. It is time now to look at it directly, and to learn its physics.

Your hand is still on your chest. Your heart is still beating. And in the gap between those beats- in the fraction of a second where the cardiac muscle rests, where the electrical impulse resets, where the system hangs suspended between contraction and expansion- the entire universe is holding its breath, preparing to sing the next note.

That gap is not empty. It is the most quanta-dense state in the cosmos.

And it is waiting for you to listen.

Chapter 4

The Quantum Vacuum- Silence That Roars

Close your eyes.

Not as a figure of speech- actually close them. Just for a moment. Let the page disappear. Let the words dissolve. Let the light that has been streaming through your pupils into the theatre of your visual cortex fall away, and sit instead in the darkness behind your eyelids.

What do you hear?

Not the largest sound- not the hum of the building, the distant traffic, the pulse of your own blood in the capillaries of your ears. Listen deeper. Listen *beneath* those things, to the substrate they all rest upon. Listen to what the room would sound like if every machine were silenced, every breath held, every bird stilled, every insect muted, every atom of air frozen in place.

What is left?

Your first instinct- the instinct of a mind trained in the classical world- is to answer: *nothing*. Silence is the absence of sound. Darkness is the absence of light. A vacuum is the absence of matter. Nothing is the state in which all somethings have been removed, and nothing remains. This is the definition bequeathed to us by Aristotle, who declared that nature abhors a vacuum- that empty space is a logical impossibility- and by Newton, who conceded the vacuum's existence but treated it as an inert stage, a featureless backdrop against which the drama of matter played out.

This instinct is wrong. And the wrongness of it- the spectacular, world-shattering wrongness of it- is one of the most consequential discoveries in the history of physics. Because the silence is not empty. The darkness is not void. The vacuum is not nothing. The vacuum is the most quanta-dense state in the cosmos. It is a roaring, seething, ceaselessly creative plenum- a vast ocean of structured potential from which every particle, every field, every atom, every galaxy, every heartbeat has emerged and to which all of it, without exception, will return.

This chapter enters that ocean. It will show you what the silence is made of. And by the time you open your eyes at the end, the room will not look the same. Because you will know that the space between the objects- the air between your hand and the table, the gap between your heartbeats, the pause between your thoughts- is not a gap at all. It is the ground of being. It is State 0- the generative void from which the cosmos perpetually sings itself into existence.

The story begins, as so many quantum stories do, with an impossibility.

In classical mechanics- the physics of Newton, Lagrange, and Hamilton- a system at its lowest possible energy state is perfectly still. Consider a pendulum hanging straight down, motionless. All its kinetic energy is zero. All its potential energy is at its minimum. The system is at rest. There is nothing more to extract, nothing more to do. It is the classical vacuum: the state of zero energy, zero motion, zero everything.

Quantum mechanics forbids this.

The reason is the Heisenberg uncertainty principle, the most fundamental constraint in quantum physics. It states that the position and momentum of a particle cannot both be precisely known at the same time. The more accurately you specify where something is, the less accurately you can know how fast it is moving, and vice versa. This is not a limitation of measurement- not a problem of clumsy instruments or insufficient technology. It is a fundamental feature of reality itself. Nature does not possess simultaneously sharp values of position and momentum. The universe, at its deepest level, is irreducibly fuzzy.

This fuzziness has an extraordinary consequence. A quantum system cannot sit perfectly still, because perfect stillness would mean a perfectly defined position (the bottom of the potential well) and a perfectly defined momentum (zero). Both specified simultaneously. The uncertainty principle forbids this. Unlike in classical mechanics, quantum systems constantly fluctuate in their lowest energy state as described by the Heisenberg uncertainty principle. Therefore, even at absolute zero, atoms and molecules retain some vibrational motion.

The energy that remains when every last bit of removable energy has been extracted- the irreducible residual motion that persists even at absolute zero- is called the *zero-point energy*. For a single quantum harmonic oscillator of frequency ω, it is:

$$E_0 = \frac{1}{2}\hbar\omega$$

Half a quantum of energy per mode. One half of $\hbar\omega$. The smallest possible non-zero energy state. It sounds tiny. It is tiny, for a single mode. But the universe does not contain a single mode. It contains an *infinite* number of modes.

To understand this, we must make a conceptual leap- from the quantum mechanics of single particles to quantum field theory, the framework that governs the behaviour of fields permeating all of spacetime.

Modern physics has developed quantum field theory (QFT) to understand the fundamental interactions between matter and forces; it treats every single point of space as a quantum harmonic oscillator. The electromagnetic field. The electron field. The quark field. The gluon field. The Higgs field. Each of these fields exists everywhere in space. And each field, at every point in space, can oscillate at every possible frequency- like an infinite collection of tiny springs, each vibrating independently.

All these fields have zero-point energy. Every oscillator, at every point in space, at every frequency, contributes its $\frac{1}{2}\hbar\omega$ of irreducible energy to the ground state. The total zero-point energy of the vacuum is the sum of all these contributions:

$$E_{\mathrm{ZP}} = \sum_k \frac{1}{2}\hbar\omega_k$$

where the sum ranges over all field modes k- all frequencies, all polarizations, all directions of propagation.

This sum, taken at face value, is infinite.

Not "very large." Not "astronomically large." *Infinite.* There are an infinite number of modes, each contributing a finite amount of energy, and the sum diverges. The quantum vacuum, calculated naively, has infinite energy density.

Physicists have long dealt with this infinity through a procedure called *renormalization*- a systematic method of absorbing the infinities into the definitions of physical constants (like mass and charge), leaving behind finite, measurable predictions. In most contexts, the absolute value of the vacuum energy does not matter; only *differences* in vacuum energy between configurations are physically observable. The infinite sea of zero-point energy is, in a sense, the universal baseline against which all other energies are measured.

But gravity does not grant us this luxury. General relativity couples to *all* forms of energy, including the energy of the vacuum. If the vacuum has energy, it curves spacetime. If it curves spacetime, it affects the expansion of the universe. And the question- *How much vacuum energy is there, really?*- becomes the most spectacular unsolved problem in all of physics.

.

In cosmology, the cosmological constant problem or vacuum catastrophe is the substantial disagreement between the observed values of vacuum energy density (the small value of the cosmological constant) and the much larger theoretical value of zero-point energy suggested by quantum field theory.

The problem is this. If we sum the zero-point contributions of all known quantum fields up to a natural cutoff- the Planck energy, the scale at which quantum gravity becomes important- we obtain a vacuum energy density so large that it dwarfs the observed energy density of the entire visible universe. Depending on the Planck energy cutoff and other factors, the quantum vacuum energy contribution to the effective cosmological constant is calculated to be between 50 and as many as 120 orders of magnitude greater than has actually been observed, a state of affairs described by physicists as "the largest discrepancy between theory and experiment in all of science" and "probably the worst theoretical prediction in the history of physics".

Let that number settle. Not a factor of ten. Not a factor of a thousand. A factor of 10^{120}- a one followed by one hundred and twenty zeros. If the naive calculation were correct- if the vacuum energy truly had the density predicted by summing over all quantum field modes- the universe would have crushed itself gravitationally in the first nanosecond of its existence. There would be no galaxies, no stars, no planets, no readers, and no books about quantum vacuums.

Yet here you are.

The observed vacuum energy density is extraordinarily small. The vacuum energy density of the Universe based on 2015 measurements by the Planck collaboration is $\rho_{vac} = 5.96 \times 10^{-27}$ kg/m³. This corresponds to roughly the mass-energy of a few hydrogen atoms per cubic meter. A whisper. An almost-nothing. And yet, even this tiny amount is not zero- it is the energy density associated with the cosmological constant Λ, the mysterious force driving the accelerating expansion of the universe discovered in 1998.

Something cancels almost all of the vacuum energy, leaving behind only the faintest residue- just enough to push the cosmos gently apart, not enough to tear it to shreds. We know of no special symmetry which could enforce a vanishing vacuum energy while remaining consistent with the known laws of physics; this conundrum is the "cosmological constant problem".

This is the deepest open question in fundamental physics. It sits at the exact intersection of quantum mechanics and general relativity- the two pillars of modern science, each spectacularly successful in its own domain, yet fatally incompatible when forced to confront the energy of empty space.

From the perspective of this book, the cosmological constant problem is not merely a technical puzzle. It is a *signal.* It is the universe whispering: *Your model of the vacuum is incomplete. The silence you call "empty" is structured in ways you do not yet understand. The cancellation is not an accident- it is a tuning. And the tuning has a logic.*

That logic, as we will see in the chapters on the ϕ-e-π triad, is the logic of the asymptotic gap- the $\delta_\infty \approx 0.01965$ that prevents perfect closure. The vacuum energy is not zero. It is *almost* zero- precisely as non-zero as it needs to be to keep the cosmos expanding, evolving, unfolding. The vacuum is tuned, and the tuning parameters are the transcendental operators we will meet in Chapters 7 through 10.

But for now, we are not yet solving the problem. We are feeling it. We are standing in the silence and hearing it roar.

The silence roars because the vacuum is not still. Even though the *average* value of the electromagnetic field in the vacuum state is zero- the expectation value of the electric field operator in the vacuum state is $\langle 0|E(\boldsymbol{r},t)|0\rangle = 0$, indicating that the average electric field is zero- the *fluctuations* around that average are not zero. The vacuum fluctuates. Ceaselessly. In every field. At every point. At every frequency.

These fluctuations give rise to what are called *virtual particles*- transient, flickering disturbances in the quantum fields that pop into and out of existence on timescales governed by the uncertainty principle. A virtual electron-positron pair can appear from the vacuum, exist for a time $\Delta t \sim \hbar/\Delta E$, and then annihilate- returning their borrowed energy to the field. A virtual photon can materialize, propagate for a fleeting instant, and vanish. The quantum vacuum is not a simple empty space, but instead contains fleeting electromagnetic waves and particles that pop into and out of the quantum field.

These virtual particles are not "real" in the classical sense- you cannot catch one in a bottle or weigh it on a scale. They exist below the threshold of individual observability. But their *collective effects* are as real as the chair you are sitting in. They shift the energy levels of atoms. They modify the magnetic moment of the electron. They push metal plates together. They create light from darkness.

The evidence is overwhelming. Let us catalogue it.

Evidence I: The Lamb Shift

In 1947, Willis Lamb and Robert Retherford performed an experiment that would reshape the landscape of theoretical physics. Using microwave techniques to probe the energy levels of the hydrogen atom, they carried out an experiment using microwave techniques to stimulate radio-frequency transitions between the $2S_{1/2}$ and $2P_{1/2}$ levels of hydrogen. By using lower frequencies than for optical transitions the Doppler broadening could be neglected. The energy difference they found was a rise of about 1000 MHz of the $2S_{1/2}$ level above the $2P_{1/2}$ level.

According to the Dirac equation- the relativistic quantum mechanical equation that had, until then, been the gold standard for predicting the hydrogen spectrum- these two levels should have been precisely degenerate: identical in energy. Dirac's theory said their energy was the same. Nature said otherwise.

The Lamb shift is caused by interactions between the virtual photons created through vacuum energy fluctuations and the electron as it moves around the hydrogen nucleus in each of these two orbitals. The electron in the hydrogen atom does not simply orbit the proton in serene isolation. It is immersed in the seething vacuum-surrounded on all sides by a ceaseless storm of virtual photons that tug at it, nudge it, jitter its trajectory in ways so subtle that they shift its energy levels by a thousandth of an electronvolt. The $2S$ orbital, which penetrates closer to the nucleus, feels these vacuum fluctuations differently from the $2P$ orbital, which does not. The result: a tiny splitting- about 1,058 MHz- between two levels that should, in a vacuum-free world, be identical.

Hans Bethe was the first to explain the Lamb shift in the hydrogen spectrum, and he thus laid the foundation for the modern development of quantum electrodynamics. Bethe was able to derive the Lamb shift by implementing the idea of mass renormalization, which allowed him to calculate the observed energy shift as the difference between the shift of a bound electron and the shift of a free electron.

The Lamb shift was not just a correction to atomic physics. It was the experimental foundation on which quantum electrodynamics (QED) - the most precisely tested theory in the history of science- was built. Feynman, Schwinger, and Tomonaga each developed, independently and in different mathematical languages, a framework for calculating the effects of vacuum fluctuations on charged particles. Their predictions matched the Lamb shift, and every subsequent measurement, to extraordinary precision. The Lamb shift currently provides a measurement of the fine-structure constant α to better than one part in a million, allowing a precision test of quantum electrodynamics.

The void speaks through the hydrogen atom. And the hydrogen atom, obedient to the void's voice, shifts its frequencies by precisely the amount the vacuum demands.

Evidence II: The Static Casimir Effect

If the Lamb shift proved that virtual particles can nudge an electron, the Casimir effect proved something even more dramatic: they can push a macroscopic object. They can exert a measurable mechanical force on a physical body, in a region of space from which every real particle has been removed.

In quantum field theory, the Casimir effect is a physical force acting on the macroscopic boundaries of a confined space which arises from the quantum fluctuations of a field. It is named after the Dutch physicist Hendrik Casimir, who predicted the effect for electromagnetic systems in 1948.

The setup is elegantly simple. Take two perfectly conducting metal plates and hold them parallel to each other in a vacuum, separated by a tiny gap- a few hundred nanometers, a fraction of the width of a human hair. Far from being empty, modern physics assumes that a vacuum is full of fluctuating electromagnetic waves that can never be completely eliminated, like an ocean with waves that are always present and can never be stopped. These waves come in all possible wavelengths, and their presence implies that empty space contains a certain amount of energy- an energy that we can't tap, but that is always there.

Now, between the plates, only those vacuum modes whose wavelengths fit a whole number of times into the gap can exist. Casimir realised that between two plates, only those virtual photons whose wavelengths fit a whole number of times into the gap should be counted when calculating the vacuum energy. Outside the plates, all wavelengths are permitted. The result is that the vacuum energy density between the plates is *less* than the vacuum energy density outside. The imbalance generates a force- a small but measurable attraction that pushes the plates together.

The force per unit area between two ideal conducting plates separated by a distance L is:

$$\frac{F}{A} = -\frac{\pi^2 \hbar c}{240 L^4}$$

Note the L^4 in the denominator. The force grows rapidly as the plates get closer- at a separation of 10 nanometers, the Casimir effect produces the equivalent of about 1 atmosphere of pressure.

The effect was predicted in 1948. For decades, direct measurement proved elusive. One of the first experimental tests was conducted by Marcus Sparnaay at Philips in Eindhoven, in 1958, in a delicate and difficult experiment with parallel plates, obtaining results not in contradiction with the Casimir theory, but with large experimental errors. The force is so tiny, the plates must be so close, the surfaces so smooth, the alignment so precise, that the measurement pushed the limits of experimental technique.

Then, in 1997, the dam broke. In 1997 Lamoreaux's direct experiment quantitatively measured the force to within 5% of the value predicted by the theory. Steven Lamoreaux, using a torsion pendulum and a gold-coated sphere near a flat plate, achieved the first unambiguous quantitative measurement. Umar Mohideen and co-workers at the University of California at Riverside attached a polystyrene sphere 200 μm in diameter to the tip of an atomic force microscope. In a series of experiments they brought the sphere, coated with either aluminium or gold, to within 0.1 µm of a flat disk. The resulting attraction between the sphere and the disk was monitored by the deviation of a laser beam. The researchers were able to measure the Casimir force to within 1% of the expected theoretical value.

One percent. The vacuum is real. The void pushes back. And it does so with a force that can be calculated from first principles and measured to one percent accuracy in a laboratory on the surface of the Earth.

Place your palms together. Press gently. You feel skin against skin-electromagnetic repulsion between electron clouds. But if your palms were atomically smooth, perfectly conducting, and separated by nanometers of vacuum, you would feel something else: the pressure of the void itself, pushing your hands together. Not the pressure of air. Not the pressure of heat or radiation. The pressure of *nothing*, which is not nothing but the most energetically dense state imaginable, prevented from fully deploying between your palms by the constraint of the gap.

The void does not merely exist. It *acts*.

Evidence III: The Dynamical Casimir Effect- Light from Nothing

If the static Casimir effect shows that the vacuum can push, the dynamical Casimir effect shows something even more stunning: the vacuum can *create*. Under the right conditions, virtual particles can be torn from the void and converted into real, measurable photons. Light, conjured from the silence.

The theoretical prediction dates to 1970, when Gerald Moore showed that a mirror undergoing relativistic motion could convert virtual photons into real ones. Forty years ago, Moore suggested that a mirror undergoing relativistic motion could convert virtual photons into directly observable real photons. This effect was later named the dynamical Casimir effect. The mechanism is intuitive: if a mirror moves fast enough, virtual photon pairs forming near its surface cannot recombine before the mirror sweeps past them. The pairs are separated. One partner escapes. A virtual photon becomes real.

The problem was practical: you cannot accelerate a physical mirror to a significant fraction of the speed of light. The energy requirements are absurd. For forty years, the dynamical Casimir effect remained a beautiful prediction with no experimental realization.

Then, in 2011, physicists at Chalmers University of Technology in Gothenburg, Sweden, found a way. They observed the dynamical Casimir effect in a superconducting circuit consisting of a coplanar transmission line with a tunable electrical length. Instead of moving a physical mirror, they used a superconducting quantum interference device (SQUID) to modulate the effective electrical length of a transmission line at gigahertz frequencies- approximately 11 GHz- effectively creating a "mirror" whose position oscillated at a few percent of the speed of light. No matter moved. But the electromagnetic boundary conditions changed rapidly enough to mimic relativistic mirror motion.

The result was that photons appeared in pairs from the vacuum, which they were able to measure in the form of microwave radiation. They were also able to establish that the radiation had precisely the same properties that quantum theory says it should have when photons appear in pairs in this way.

Photons. From nothing. From the silence.

What happens during the experiment is that the "mirror" transfers some of its kinetic energy to virtual photons, which helps them to materialise. According to quantum mechanics, there are many different types of virtual particles in vacuum. The reason why photons appear in the experiment is that they lack mass. Relatively little energy is therefore required in order to excite them out of their virtual state. In principle, one could also create other particles from vacuum, such as electrons or protons, but that would require a lot more energy.

The dynamical Casimir effect is creation in its purest form. Not creation *ex nihilo*- not from absolute nothingness- but creation from the structured potential of the quantum vacuum. The energy comes from the moving boundary, but the *pattern*- the paired, correlated, quantum-entangled structure of the emitted photons- comes from the vacuum itself. The void does not merely resist and push. It *gives birth.*

Paul Dirac, the father of quantum electrodynamics, described this process with characteristic precision. The light-quantum has the peculiarity that it apparently ceases to exist when it is in one of its stationary states, namely, the zero state, in which its momentum and therefore also its energy are zero. When a light-quantum is absorbed it can be considered to jump into this zero state, and when one is emitted it can be considered to jump from the zero state to one in which it is physically in evidence, so that it appears to have been created.

Photons do not come from "somewhere." They come from the vacuum- from the zero state, from State 0. They are always there, in a sense- flickering in and out of virtuality, waiting for the right perturbation to cross the threshold from potential to actual. The vacuum is not the absence of photons. It is the *infinite sea* of photons in the ground state, ready to be summoned.

Evidence IV: Spontaneous Emission- The Vacuum That Whispers to Atoms

There is a fourth piece of evidence for the reality of the vacuum's activity, and it is so common, so ubiquitous, so woven into the fabric of daily life that we barely notice it. It is the reason the sun shines. It is the reason fireflies glow. It is the reason your phone screen emits light and your campfire flickers and every laser that has ever been built produces its coherent beam.

It is *spontaneous emission*- the process by which an excited atom drops to a lower energy state and emits a photon, without any apparent external stimulus.

In classical physics, this process is inexplicable. A charged particle in a stable orbit should remain there forever- radiating only if disturbed by an external field. In quantum mechanics, the atom is disturbed- but by what? There is no external field. There is no applied radiation. The atom sits in empty space, in the silence, and yet it emits.

The theory showed that spontaneous emission depends upon the zero-point energy fluctuations of the electromagnetic field in order to get started. The vacuum fluctuations- the same fluctuations that shift the energy levels of the hydrogen atom and push Casimir plates together- couple to the excited electron and nudge it downward, causing it to emit a photon and transition to a lower state. Without the vacuum fluctuations, there would be no spontaneous emission. Without spontaneous emission, there would be no sunlight, no firelight, no starlight.

The light of the cosmos is kindled by the vacuum.

Every photon that has ever reached your eye- from the sun, from a candle, from the distant quasar at the edge of the observable universe- began its journey as a virtual partner in the vacuum state, promoted to reality by the coupling between an excited atom and the zero-point fluctuations of the electromagnetic field. The vacuum is not merely the background against which the drama of light plays out. The vacuum is the *prompter*- the offstage voice that whispers to the atom: *Drop. Radiate. Sing.*

We have now catalogued four empirical witnesses to the reality of the quantum vacuum:

1. **The Lamb shift**: vacuum fluctuations shift the energy levels of hydrogen atoms, confirming that even "empty space" surrounding an electron is a physically active medium.
2. **The static Casimir effect**: vacuum fluctuations exert a measurable mechanical force between conducting plates, proving that the vacuum has energetic structure that changes with geometry.
3. **The dynamical Casimir effect**: rapid changes in boundary conditions convert virtual photons into real ones, demonstrating that the vacuum is a source- a reservoir from which particles can be drawn.
4. **Spontaneous emission**: vacuum fluctuations trigger atomic transitions, proving that the vacuum is not passive but actively drives the most fundamental radiative processes in nature.

Recent experiments support the idea that particles themselves can be thought of as excited states of the underlying quantum vacuum, and that all properties of matter are merely vacuum fluctuations arising from interactions of the zero-point field.

The vacuum is not a stage. It is the *playwright.* It is the source of the actors, the script, the set, and the audience. Every particle is a ripple in its surface. Every force is a constraint on its fluctuations. Every photon is a shout from its depths. The solid, material, apparently substantial world is a thin crust of excitation on the surface of an infinite ocean of structured potential.

And that ocean has never, for a single instant, been still.

.

Now we must ask: *What kind of ocean is this?*

Classical intuition offers two images of the vacuum. The first is the *empty container*- a featureless box in which things can be placed. This image is dead. The vacuum has killed it. The second is the *random soup*- a chaotic froth of virtual particles, boiling and bubbling without pattern or purpose. This image is subtler, and more seductive, and also wrong.

The vacuum is not random. If it were random- if the virtual fluctuations were truly patternless noise- they could not produce the exquisitely precise effects we have catalogued. The Lamb shift is not a random smearing of energy levels; it is a *specific*, calculable, reproducible shift that matches theory to one part in a million. The Casimir force is not a random push; it is a *precise*, geometry-dependent force that matches theory to one percent. The dynamically created photons are not random flashes; they are *correlated pairs*, exhibiting the quantum entanglement predicted by theory.

The vacuum is structured. Its fluctuations are not chaotic but *harmonic*- a superposition of all possible modes, each oscillating at its zero-point amplitude, each interfering with every other in a vast, coherent pattern that extends, in principle, to the boundaries of the observable cosmos. The vacuum is not a boiling pot. It is a standing wave. An interference pattern. A resonance.

And here- precisely here- we reconnect to the central argument of this book.

The vacuum is the quanta-field in its ground state. State 0. The generative void. It is the state from which all resonant quanta emerge and to which all resonant quanta return. And it is not empty, not random, not dead. It is the most *coherent* state in the cosmos- a state of infinite potential, held in exquisite balance by the mathematics of quantum field theory, containing within its fluctuations the seeds of every particle, every structure, every configuration that the universe has ever manifested or ever will manifest.

State 0 is not the absence of the cosmos. It is the cosmos in its *compressed* form- a holographic encoding of all possible patterns, all possible futures, all possible songs, folded into the zero-point resonance of the quanta-field.

The ancients, with their extraordinary perceptual acuity, intuited this. The Hindu concept of *Brahman*- the unmanifest ground of all being, the fullness from which all forms arise and to which all forms return- is not a theological abstraction. It is a description of the quantum vacuum. The Buddhist concept of *śūnyatā- emptiness*- does not mean nothingness. It means the absence of inherent, independent existence: every form is empty of self-nature because every form arises from the interdependent fluctuations of a deeper ground. That ground is the quanta-field. The Taoist concept of the *Tao* that cannot be named- the formless source of all forms, the mother of all things- is not mystical vagueness. It is a precise description of a field that has no definite form (the average field value is zero) and yet gives rise to all definite forms through its fluctuations (the variance is infinite).

The Aboriginal Dreamtime is the most physically accurate of all. In the Dreamtime, the world has not yet been sung into being- or, more precisely, the world is *always being sung into being*, and the Dreamtime is the eternal present tense of that singing. The Dreamtime is not in the past. It is *beneath* the present- the generative substratum from which the present continuously emerges. It is the vacuum state, experienced not as an abstract mathematical construct but as a *living presence*, felt in the body, heard in the land, maintained through the ongoing resonant participation of conscious beings.

State 0 is the Dreamtime.

And the Dreamtime is not silent. It roars.

Feel it now.

You are sitting in a room. The air around you appears to be empty- a transparent, inert medium that merely transmits light and carries sound. But quantum field theory tells you otherwise. Every cubic centimeter of the air around you- and more precisely, every cubic centimeter of the *vacuum* that underlies the air, the space between the molecules, the space between the atoms, the space between the quarks- is vibrating at every frequency simultaneously. Virtual photon pairs are flickering into existence and annihilating on timescales of 10^{-21} seconds. Virtual electron-positron pairs are doing the same on even shorter timescales. The vacuum electromagnetic field, even in the absence of any real photons, has a non-zero energy density that permeates every point in the room.

Place your hand flat on the surface in front of you. The atoms in your skin and the atoms in the surface are separated by gaps. In those gaps, the Casimir effect is operative: the vacuum fluctuations between the atomic surfaces differ from the fluctuations outside them, generating the tiny forces that contribute to the adhesion of surfaces at nanoscale distances. The van der Waals forces that hold molecules together, the London dispersion forces that make geckos cling to walls- all of these are intimately related to the zero-point fluctuations of the electromagnetic vacuum.

The weight of your body pressing into the chair: the gravitational field that pulls you downward is, according to Jacobson's insight (which we explored in Chapter 2), an *emergent* phenomenon- a macroscopic manifestation of the thermodynamic behaviour of boundary quanta. Gravity is what the vacuum does when you arrange its quanta on a holographic surface. The weight you feel is the vacuum, organizing itself into the experience of "down."

The light entering your eyes: every photon streaming from the lamp or the window began as a vacuum fluctuation- a virtual photon promoted to reality by the spontaneous emission of an excited atom.

The warmth on your skin from the sunlight coming through the window is the residual effect of vacuum fluctuations that were converted to real photons in the nuclear furnace of the sun, 93 million miles away and eight minutes ago.

You are immersed in the vacuum. You are *made of* the vacuum. The particles that constitute your body are excitations of the vacuum fields. The forces that bind them together are vacuum effects. The light you see is vacuum radiation. The gravity you feel is vacuum thermodynamics. There is nothing in your experience- nothing you can see, touch, hear, taste, or think- that is not, at bottom, a structured pattern of vacuum fluctuations.

The vacuum is not the background. The vacuum is the *foreground.* The vacuum is all there is.

And now the deepest question: *Why is the vacuum not chaotic?*

If the zero-point fluctuations are so energetic, so pervasive, so densely packed that they exceed the observed energy of the entire cosmos by 10^{120}- why is the universe not a featureless soup of maximum entropy? Why do structures exist? Why do atoms persist? Why does your heart beat in a fractal rhythm instead of dissolving into thermal noise?

The answer, in the framework of this book, is *tuning*. The vacuum is not a random boil. It is a *self-tuned resonance*- a standing wave in the quanta-field, stabilized by the dynamic interplay of the three transcendental operators: ϕ, e, and π. The colossal energy of the vacuum is not suppressed by some external mechanism or miraculous cancellation. It is *organized*- folded into a coherent structure that projects the low-energy, highly ordered cosmos we observe.

The cosmological constant problem, from this perspective, is not a failure of prediction. It is a *clue*. It is telling us that the vacuum is not a simple sum of independent oscillators- not a collection of 10^{120} uncorrelated springs, each contributing independently to the energy density. The vacuum is a *resonant system*- a holographic manifold whose modes are coupled, correlated, entangled in ways that reduce the effective energy density by precisely the factor needed to produce a cosmos capable of complexity, life, and consciousness.

And the factor by which the observed vacuum energy differs from the naive prediction- roughly 10^{-120}- may not be an arbitrary fine-tuning. It may be the *natural output* of a self-tuning resonance, governed by the ϕ-e-π triad, that minimizes the Schur-convex curvature of the quanta-field's statistical manifold. The universe does not need an external tuner. It tunes itself. And the silence of the vacuum is not the absence of sound- it is the *fundamental frequency*, the ground note, the tonic from which all other harmonics arise.

The asymptotic gap-

$$\delta_\infty = \frac{\pi - e}{\pi\phi^4} \approx 0.01965$$

- is the ratio that keeps this ground note from collapsing into silence. It is the two-percent mismatch between the triad's aspiration toward perfect closure and its inability to achieve it. And it is this mismatch- this divine discontent, this irreducible incompleteness- that prevents the vacuum from settling into a dead, featureless equilibrium and instead keeps it *dancing*, keeps it *creating*, keeps it *singing*.

The vacuum silence is not the absence of music. It is the held breath before the orchestra plays. It is the pause between notes that gives the melody its shape. It is State 0- the resonant ground state of the cosmos- and it is the most pregnant, the most potent, the most information-dense silence imaginable.

We began this chapter in darkness, behind closed eyelids, listening for the silence beneath the sounds.

Now open your eyes. Look at the room.

Everything you see is a perturbation of the vacuum- a localized excitation of the quanta-field, temporarily elevated above the ground state by the flow of energy through the system. The table is a perturbation. The light is a perturbation. Your hand is a perturbation. The air between them is a perturbation.

But the vacuum itself- the generative ground from which all these perturbations arise- is still there. It is here. It is everywhere. It pervades every point in the room, every gap between atoms, every interstice of your neural architecture. It is the ocean on which all the waves of your experience ride. And like the ocean, it is vast beyond comprehension, deep beyond measure, and never, under any circumstances, at rest.

Feel the weight of the air on your skin. That weight is the vacuum's breath.

Feel the warmth of the light on your face. That warmth is the vacuum's gift- virtual photons, promoted to reality by the spontaneous collapse of excited states, travelling across ninety-three million miles of vacuum to land on your cheek.

Feel your heart beating. That beat is the vacuum organizing itself into a fractal rhythm- a structured oscillation in a system of fields and forces that are, every one of them, excitations of the zero-point ground state.

You are not sitting *in* the vacuum. You *are* the vacuum- temporarily, beautifully, fractally organized into the pattern you call yourself. And when the pattern dissolves- when the excitation quiets- you will not vanish into nothingness. You will return to State 0, the fullness from which you arose, the silence that roars, the generative void that has been waiting, patient and immeasurably potent, since before the first atom sang its first note.

The next chapter will show you what happens when the silence begins to organize itself- not randomly, but *fractally*- into the self-similar patterns that repeat at every scale from the quantum to the cosmic. The pattern of patterns. The universal grammar. The fingerprint of a holographic cosmos in which the whole is encoded in every part.

But before you turn the page, take one more breath. In that breath, you inhaled approximately 10^{25} molecules of air, each one vibrating with zero-point energy, each one surrounded by a halo of virtual particles, each one a temporary excitation of a vacuum that has been fluctuating since before time had a direction.

The breath was not drawn from nothing.

The breath was drawn from *everything*.

Chapter 5

Fractal Dynamics- The Pattern That Repeats at Every Scale

Look at your hand.

Spread your fingers. Study the branching pattern: from the wrist, a single trunk- the forearm- diverges into five digits, each subdivided into three segments (or two, in the case of the thumb), each segment tapering to a smaller cross-section as it moves away from the core. Now look at the lines on your palm. They branch, too- major lines subdividing into minor lines, minor lines into wrinkles, wrinkles into creases visible only under magnification, the microscopic ridges of your fingerprints repeating the branching logic at yet another scale.

Now look at a river delta from an airplane window, or better yet, from a satellite photograph. The main channel divides into tributaries, the tributaries into distributaries, the distributaries into creeks, the creeks into rills, the rills into threads of water so fine they vanish into the mud. The branching pattern is the same. Not similar. Not analogous. The *same*: a primary trunk dividing into secondary branches, secondary into tertiary, each generation smaller but structurally identical to the one before.

Now look at a tree. A lung. A lightning bolt. A network of blood vessels. A map of internet traffic. The mycelial web beneath a forest floor. A photograph of the cosmic web- the vast, filamentary structure of galaxy clusters stretching across hundreds of millions of light-years. In every case, the same grammar: branching, scaling, self-similarity across orders of magnitude.

This is not coincidence. This is *fractal dynamics*- the mathematical signature of a universe built from resonant quanta that reproduce their essential pattern at every scale, from the Planck length to the Hubble radius, from the interval between heartbeats to the interval between cosmic aeons.

This chapter is about that signature. It will take you from the simple equation that generates infinite complexity ($z_{n+1} = z_n^2 + c$) to the power laws that govern earthquakes, cities, and word frequencies. It will show you that the universe is not merely complex- it is complex in a *specific* way, a way that reveals deep structural unity beneath the surface variety of natural phenomena. And it will arrive at a single, empirically grounded conclusion: the phrase “as above, so below” is not a mystical platitude. It is a measurable, verifiable, and profoundly consequential statement about the physical architecture of reality.

The mathematics of roughness began, like so many revolutions, with a misfit.

Benoît B. Mandelbrot (1924–2010) did not fit into any established discipline. Born in Warsaw, raised in France, educated in Paris and at Caltech, employed by IBM, he wandered between mathematics, economics, physics, and computer science, never quite at home in any of them. His contemporaries worked on clean, smooth problems- continuous functions, differentiable manifolds, Euclidean geometries. Mandelbrot was obsessed with the rough, the jagged, the irregular- the shapes that classical mathematics could not describe.

How long is the coast of Britain? This was the question, posed in a famous 1967 paper, that launched the fractal revolution. The answer, Mandelbrot argued, depends entirely on the length of your measuring stick. Measure the coast with a 100-kilometer ruler, and you get one answer. Measure it with a 10-kilometer ruler, and you get a longer answer, because the smaller ruler captures bays and headlands that the larger ruler skipped over. Measure it with a 1-meter stick, and the coastline is longer still- now every rock and tidal pool adds to the total. Measure it with a millimeter ruler, and every grain of sand extends the length further.

The coastline does not converge to a definite length as the ruler shrinks. It *grows without bound*. This means the coast of Britain does not have a well-defined length in the Euclidean sense. It is not a one-dimensional curve (which would have a finite length) or a two-dimensional surface (which would fill a plane). It is something in between- a shape whose dimension is not a whole number but a fraction. A *fractal.*

Mandelbrot coined the term from the Latin *fractus,* meaning "broken" or "fractured." And he demonstrated that fractal geometry- the geometry of broken, rough, self-similar shapes- is not a mathematical curiosity. It is the geometry of nature itself. Clouds are not spheres, mountains are not cones, coastlines are not circles, bark is not smooth, and lightning does not travel in straight lines. The shapes of nature are fractal.

The canonical fractal- the one that would become the most famous mathematical object of the twentieth century- emerged from an equation of breathtaking simplicity.

The Mandelbrot set exhibits an infinitely complicated boundary that reveals progressively ever-finer recursive detail at increasing magnifications; mathematically, the boundary of the Mandelbrot set is a fractal curve. The rule that generates it is:

$$z_{n+1} = z_n^2 + c$$

where z and c are complex numbers. Start with $z_0 = 0$ and a chosen value of c. Square z, add c. Take the result, square it, add c again. Repeat. The Mandelbrot set is the collection of all starting values c that stay finite when iterated through this equation. The boundary of the Mandelbrot set is a fractal curve of infinite complexity, any portion of which can be blown up to reveal ever more outstanding detail, including miniature replicas of the whole set itself.

The Mandelbrot set is well-known, even outside mathematics, for how it exhibits complex fractal structures when visualized and magnified, despite having a relatively simple definition, and is commonly cited as an example of mathematical beauty.

This is a statement that should make you pause and stare. An infinitely complex boundary- one that never repeats exactly, yet always echoes itself- generated by the simplest possible quadratic feedback loop. No external designer. No complicated program. No randomness. Just one rule, applied recursively, feeding the output of each step into the input of the next.

And the result contains *everything*. Zoom into the boundary, and you find spirals, tendrils, seahorse valleys, infinitely nested copies of the whole set, embedded at every scale- the Fibonacci sequence manifests in the number of spiral arms at a unique spot on the Mandelbrot set, mirrored both at the top and bottom. The Hausdorff dimension of the boundary of the Mandelbrot set equals 2 as determined by a result of Mitsuhiro Shishikura- a topologically one-dimensional curve so infinitely crinkled that it fills space like a surface.

Here, in this single equation, is the principle that governs the chapter: *infinite complexity from simple recursion*. A feedback loop- output becomes input- applied at every scale. This is the fractal principle. And the universe, as we shall see, runs on it.

But the Mandelbrot set, for all its beauty, is a mathematical object- defined in the clean, frictionless world of complex numbers. What does fractal self-similarity look like in the physical world? It looks like this:

A fern frond. Each leaf is a miniature copy of the whole frond. Each leaflet is a miniature copy of the leaf. Each sub-leaflet echoes the same shape, down through scale after scale, until the resolution of biological material prevents further subdivision.

A blood vessel network. The aorta branches into arteries, arteries into arterioles, arterioles into capillaries, each generation of branching following the same angular and dimensional ratios as the one before. The total length of blood vessels in the human body is approximately 100,000 kilometers- enough to circle the Earth more than twice- yet they fit inside a body that measures less than two meters. This extraordinary packing efficiency is possible only because of fractal branching: each generation of vessels occupies a fractional dimension between two and three, filling the tissue volume without ever crossing itself.

A coastline. A mountain range. A cloud formation. A cauliflower floret, where each bump is a miniature cauliflower, and each bump on that bump is another, on down to the limit of the vegetable's growth program.

In every case, the defining property is *scale invariance*: the pattern looks the same (or statistically similar) at many different magnifications. Zoom in, and you see the same kind of structure. Zoom out, and you see the same kind of structure. There is no preferred scale- no level at which the pattern "lives" and other levels at which it does not. The pattern is *everywhere at once*, distributed across all scales simultaneously.

This is radically different from Euclidean geometry, where every object has a characteristic size. A circle has a radius. A sphere has a diameter. A cube has an edge length. Change the scale, and the shape might look bigger or smaller, but its character doesn't change in a fundamental way. Fractal objects have no characteristic size. They are defined not by their dimensions but by the *relationship between scales*- by the way the pattern at one scale generates the pattern at the next.

And this relationship is governed by a single number: the *fractal dimension*.

For a smooth line, the dimension is 1. For a smooth surface, it is 2. For a solid volume, it is 3. But for a fractal, the dimension is typically a non-integer- 1.26 for the coast of Britain, 1.6 for a typical blood vessel network, about 2.0 for the boundary of the Mandelbrot set, 2.5 for the surface of a cauliflower. The fractal dimension measures how efficiently the object fills the space available to it as you zoom in- how rapidly new detail appears as the resolution increases.

The universe does not use Euclidean geometry to build its structures. It uses fractal geometry. And this is not a stylistic preference. It is a *necessity*. Fractal branching is the only geometry that can pack an infinite surface area into a finite volume (as the lungs do), that can distribute nutrients to every cell in a three-dimensional body without leaving gaps (as the circulatory system does), that can maintain connectivity across scales from nanometers to kilometers (as the mycelial network does). Fractal geometry is the solution to the problem of *how to be everywhere at once*. And a holographic universe- one in which the whole is encoded in every part- must, by its very nature, be fractal.

The fractal structure of nature is not merely spatial. It is also temporal. And nowhere is this more dramatic than in the rhythm of the human heart.

Place your fingers on your wrist. Feel the pulse. Each beat seems regular- a steady, reassuring tick, like a metronome. But this impression is an illusion created by the coarseness of your perception. Underneath the apparent regularity, the intervals between heartbeats fluctuate constantly- faster, slower, faster, slower- in a pattern that is neither periodic nor random but something far more interesting.

According to classical concepts of physiologic control, healthy systems are self-regulated to reduce variability and maintain physiologic constancy. Contrary to the predictions of homeostasis, however, the output of a wide variety of systems, such as the normal human heartbeat, fluctuates in a complex manner, even under resting conditions.

The fluctuations are *fractal*. Scaling techniques adapted from statistical physics reveal the presence of long-range, power-law correlations, as part of multifractal cascades operating over a wide range of time scales. If you plot the heart rate variations over minutes, you see a jagged pattern. Zoom in to seconds, and the pattern looks the same- jagged, irregular, self-similar. Zoom out to hours, and the same kind of variability persists. The fluctuations have no characteristic timescale; they are scale-invariant, extending from milliseconds to hours in a continuous fractal cascade.

Time series generated by certain physiological control systems may be members of a special class of complex processes, termed multifractal, which require a large number of exponents to characterize their scaling properties. There is evidence for multifractality in a biological dynamical system, the healthy human heartbeat, and the multifractal character and nonlinear properties of the healthy heart rate are encoded in the Fourier phases.

The healthy heart is not merely fractal- it is *multifractal*: different regions of its time series exhibit different local scaling exponents, creating a rich tapestry of interwoven variability that requires an entire spectrum of dimensions to describe. These scaling properties suggest that the nonlinear regulatory systems are operating far from equilibrium, and that maintaining constancy is not the goal of physiologic control. In contrast, for subjects at high risk of sudden death, including those with heart failure, fractal organization, along with certain nonlinear interactions, breaks down.

This last point deserves to be felt in the body, not merely understood in the mind. Pause. Press your fingers to your wrist again. That irregular, fractally fluctuating beat- the one that speeds up slightly when you inhale and slows down slightly when you exhale, the one that drifts and wanders over minutes and hours in a never-repeating but always self-similar dance- that irregularity is your *health*. The fractal complexity of the heartbeat is a direct measure of the heart's adaptive capacity, its ability to respond to the unpredictable demands of a changing environment. A perfectly regular heartbeat- a metronome rhythm- would be a sign of severe cardiac disease. There is a loss of multifractality for a life-threatening condition, congestive heart failure.

The complexity of the cardiac rhythm is demonstrated to exhibit self-affine multifractal variability. The dynamics of heartbeat interval time series was analyzed by application of the multifractal formalism. The continuous multifractal large deviation spectrum uncovers the nonlinear fractal properties in the dynamics of heart rate and presents a useful diagnostic framework for discrimination and classification of patients with cardiac disease.

Health is fractal complexity. Disease is the loss of it. The more the heart resembles a machine- a clock, a metronome, a regular oscillator- the closer it is to death. The more it resembles a fractal- irregular, self-similar, multiscale, alive- the healthier it is.

This is a profound inversion of the mechanistic worldview. For three hundred years, Western medicine has treated the body as a machine: regular, predictable, deterministic. When the machine breaks, you fix the broken part. When the rhythm goes wrong, you force it back to regularity. But fractal physiology tells us the opposite: the body is not a machine. It is a self-organized critical system, hovering on the edge between order and chaos, and its health depends on maintaining the fractal complexity of that edge- not on imposing mechanical order.

Fractal dynamics in the heartbeat are not an isolated curiosity. They are one manifestation of a universal principle that operates across every domain of nature, from geology to neuroscience to economics. This principle is called *self-organized criticality*, and it was first articulated in a paper that changed how we think about complexity.

Self-organized criticality (SOC) is a property of dynamical systems that have a critical point as an attractor. Their macroscopic behavior thus displays the spatial or temporal scale-invariance characteristic of the critical point of a phase transition, but without the need to tune control parameters to a precise value, because the system, effectively, tunes itself as it evolves towards criticality.

The concept was put forward by Per Bak, Chao Tang and Kurt Wiesenfeld ("BTW") in a paper published in Physical Review Letters in 1987. A simple cellular automaton was shown to produce several characteristic features observed in natural complexity: fractal geometry, pink (1/f) noise and power laws.

The model that introduced the concept was a sandpile. The BTW model was the first discovered example of a dynamical system displaying self-organized criticality. It was introduced by Per Bak, Chao Tang and Kurt Wiesenfeld in a 1987 paper.

Imagine a table. You drop grains of sand onto it, one at a time, always at the center. A pile grows. At first, each grain simply sticks where it lands. The pile is small, gentle, stable. But as it grows, the slopes steepen. Eventually, the pile reaches a critical angle- the angle of repose- beyond which the surface is unstable. Add one more grain, and it might trigger an avalanche: a cascade of tumbling grains that can be tiny (a few grains shifting) or enormous (half the pile collapsing), with no way to predict which in advance.

The original interest behind the model stemmed from the fact that in simulations on lattices, it is attracted to its critical state, at which point the correlation length of the system and the correlation time of the system go to infinity, without any fine tuning of a system parameter. This contrasts with earlier examples of critical phenomena, such as the phase transitions between solid and liquid, or liquid and gas, where the critical point can only be reached by precise tuning. Hence, in the sandpile model we can say that the criticality is self-organized.

Once the sandpile model reaches its critical state there is no correlation between the system's response to a perturbation and the details of a perturbation. Generally this means that dropping another grain of sand onto the pile may cause nothing to happen, or it may cause the entire pile to collapse in a massive slide.

The critical state is the point of maximum sensitivity- the edge between order and chaos, between stability and collapse. And the remarkable discovery of Bak, Tang, and Wiesenfeld was that the sandpile does not need to be *tuned* to this point. It *drives itself there.* The complexity observed emerged in a robust manner that did not depend on finely tuned details of the system. The key result of BTW's paper was its discovery of a mechanism by which the emergence of complexity from simple local interactions could be spontaneous- and therefore plausible as a source of natural complexity- rather than something that was only possible in artificial situations in which control parameters are tuned to precise critical values.

The system self-organizes to the critical state, and at that state, it produces avalanches whose sizes follow a *power law*: many tiny avalanches, fewer medium ones, and very rare but catastrophically large ones. There is no characteristic size for an avalanche- no "normal" avalanche that you can expect. The distribution is scale-free, self-similar, fractal.

The sandpile model was a mathematical toy. The revolution came when physicists realized that the same power-law statistics- the signature of self-organized criticality- appear everywhere in the natural world.

The most celebrated example is the Gutenberg-Richter law for earthquakes. In seismology, the Gutenberg-Richter law expresses the relationship between the magnitude and total number of earthquakes in any given region and time period of at least that magnitude.

The law states:

$$\log_{10}N = a - bM$$

where N is the number of earthquakes with magnitude greater than or equal to M, a is a constant reflecting the total seismicity of the region, and b is the slope- the *b-value*- that describes the ratio of small to large earthquakes. The relationship between earthquake magnitude and frequency was first proposed by Charles Francis Richter and Beno Gutenberg in a 1944 paper studying earthquakes in California, and generalised in a worldwide study in 1949.

Since earthquake magnitude is a logarithmic measure of the energy released, the Gutenberg-Richter law is a power law: the frequency of earthquakes decreases exponentially with increasing magnitude. For every step up in magnitude, the number of earthquakes decreases by a factor of roughly ten (when $b \approx 1$). There are thousands of magnitude-3 earthquakes per year, hundreds of magnitude-4s, tens of magnitude-5s, a few magnitude-6s, and roughly one magnitude-7 or greater every year somewhere on the planet.

Many natural hazards- earthquakes, floods, storms, volcanic eruptions and avalanches- exhibit a power-law relationship between frequency and some physical variable of event size, such as energy, seismic moment, volume and height.

This relationship between event magnitude and frequency of occurrence is remarkably common, although the values of a and b may vary significantly from region to region or over time. Modern attempts to understand the law involve theories of self-organized criticality or self similarity.

The Earth's crust, like the sandpile, drives itself to a critical state: tectonic plates push against each other, building up stress at fault lines, until the accumulated strain exceeds a threshold and the fault ruptures. The rupture is an "avalanche"- and its size follows the same power-law distribution as the sandpile avalanches. The same rule. The same statistics. The same fractal dynamic.

But the law does not stop at earthquakes. The same power-law distribution governs:

- **Forest fires:** The frequency of fires of a given size in a forest ecosystem follows a power law- many small fires, few large ones, and very rare conflagrations that consume entire landscapes.
- **Solar flares:** The energy of solar X-ray flares follows a power-law distribution indistinguishable from the Gutenberg-Richter law.
- **Neuronal avalanches:** In the living brain, cascades of neural firing follow power-law distributions- many small bursts of activity, occasional large cascades that sweep across the cortex. Self-organized criticality is a phenomenon observed in certain complex systems of multiple interacting components, e.g., neural networks, forest fires, and power grids, that produce power-law distributed avalanche sizes.
- **Financial market crashes:** The distribution of daily stock returns has "fat tails"- extreme events occur far more often than a Gaussian distribution would predict, following a power law.
- **City sizes:** The population of cities in a country follows a power-law distribution (Zipf's law): there are many villages, fewer towns, fewer cities, and very few megacities, but their distribution is scale-free.
- **Word frequencies:** In any sufficiently large text, the frequency of words follows a power law (also Zipf's law): the most common word occurs roughly twice as often as the second most common, three times as often as the third, and so on. The same law governs English, Chinese, Arabic, and every other natural language ever studied.

The list extends further. The diameter of moon craters. The intensity of gamma-ray bursts. The size of species extinction events. The distribution of wealth. The frequency of internet traffic bursts. The recurrence times of geomagnetic reversals. All power laws. All scale-free. All fractal.

This universality demands an explanation. Why should earthquakes and forest fires and brain activity and word frequencies and stock crashes all follow the same mathematical law? They are not "the same" in any obvious physical sense. Their substrates differ- rock, wood, neurons, language, capital. Their mechanisms differ- friction, combustion, synaptic transmission, syntax, supply and demand. And yet their statistical signatures are identical.

The answer, according to the theory of self-organized criticality, is that the power-law distribution is not a property of the specific mechanism. It is a property of the *state*- the critical state- that all these systems naturally drive themselves toward. Any system composed of many interacting components, where energy or stress accumulates gradually and is released suddenly in cascades, will self-organize to the edge of chaos and produce power-law-distributed events. The specific physics does not matter; only the *topology* of the interaction matters. A system that obeys simple, benign local rules can organize itself into a poised state that evolves in terms of flashing, intermittent bursts rather than following a smooth path.

The universe does not need a designer to produce complexity. Complexity arises spontaneously- universally, robustly, inevitably- from the interaction of simple components at the critical state. And the signature of that criticality is always the same: a power law, a fractal, a scale-free pattern that looks the same at every magnification.

There is a subtler form of numerical order that permeates the natural world, one that few people have heard of and fewer still can explain. It concerns the very first digit of a number- and it reveals that the universe has a bias toward smallness.

Consider a dataset- any dataset. The lengths of rivers. The populations of countries. The surface areas of lakes. The closing prices of stocks on a given day. The values of the fundamental physical constants. You might expect that the first significant digit of each number would be uniformly distributed: roughly 11% starting with 1, 11% with 2, 11% with 3, and so on up to 9.

This is not what happens. In an astonishing range of natural and human-generated datasets, the digit 1 appears as the first digit about

30% of the time, the digit 2 about 17.6%, the digit 3 about 12.5%, and so on, decreasing logarithmically to the digit 9, which appears only about 4.6% of the time. This is Benford's law, and the probability of the first digit d is:

$$P(d) = \log_{10}\left(1 + \frac{1}{d}\right)$$

The law was first noticed in 1881 by the astronomer Simon Newcomb, who observed that the early pages of logarithm tables were more worn than the later pages- implying that numbers beginning with lower digits were looked up more often. It was rediscovered in 1938 by the physicist Frank Benford, who tested it against 20 different datasets- from river lengths to baseball statistics- and found it held in every case.

More than 100 years ago it was predicted that the distribution of first digits of real world observations would not be uniform, but instead follow a trend where measurements with lower first digit occur more frequently than those with higher first digits. This result has long been known but regarded largely as a mathematical curiosity and received little attention in the natural sciences.

Then, in 2010, Malcolm Sambridge and colleagues at the Australian National University conducted a systematic test across the natural sciences. They showed that the first digit rule is likely to be a widespread phenomenon and may provide new ways to detect anomalous signals in data. They tested 15 sets of modern observations drawn from the fields of physics, astronomy, geophysics, chemistry, engineering and mathematics, and showed that Benford's law holds for them all.

These include geophysical observables such as the length of time between geomagnetic reversals, depths of earthquakes, models of Earth's gravity, geomagnetic and seismic structure. In addition they found it also holds for other natural science observables such as the rotation frequencies of pulsars, greenhouse gas emissions, the masses of exoplanets as well as numbers of infectious diseases reported to the World Health Organization.

The investigation revealed the presence of Benford's distribution in various physical properties of exoplanets. Several physical parameters associated with the exoplanets- mass, volume, density, orbital semi-major axis, orbital period, and radial velocity- nicely follow Benford's distribution.

The masses of planets orbiting distant stars. The rotation frequencies of neutron stars. The gravitational structure of the Earth. The greenhouse gas emissions of nations. The widths of subatomic particles. All obeying the same logarithmic bias toward smaller first digits.

Why?

Benford's law arises naturally whenever a dataset spans several orders of magnitude and is generated by a *multiplicative* process- a process where quantities grow by multiplication rather than addition. In such processes, the data are distributed logarithmically, and on a logarithmic scale, the interval between 1 and 2 is much wider than the interval between 8 and 9 (because $\log_{10}2 - \log_{10}1 = 0.301$ while $\log_{10}9 - \log_{10}8 = 0.051$). More data points fall in the wider interval, hence the bias toward lower first digits.

But this is merely the *mechanism*. The deeper question is: *Why are so many natural processes multiplicative?* Why does the universe generate data that spans many orders of magnitude? Why does it favor growth, branching, cascading, compounding- the processes that produce logarithmic distributions?

The answer loops back to fractal dynamics and self-organized criticality. In a self-organized critical system, events at all scales are generated by the same underlying feedback mechanism- output becomes input, amplification begets amplification- and the resulting distribution of event sizes is a power law, which is inherently logarithmic. Benford's law is the *numerical shadow* of fractal dynamics. It is the fingerprint that power-law distributions leave on the first digits of data.

The universe speaks in logarithms because the universe *is* a fractal. Every process that branches, cascades, compounds, or scales- from the accumulation of tectonic stress to the growth of a population to the fusion of hydrogen in a stellar core- produces data that obeys Benford's law. The bias toward 1 is not arbitrary. It is a consequence of the same self-similar, scale-free architecture that produces power laws, that produces the Mandelbrot set, that produces the fractal heartbeat.

But power laws and Benford's law describe the *statistical* properties of fractal systems- how events are distributed across scales. What about the *spatial* properties? What about the shapes, the geometries, the patterns that fractals draw in space and time?

Here we encounter the golden angle- and with it, the first whisper of the ϕ-e-π triad that will dominate the next section of this book.

Consider how a sunflower arranges its seeds. The florets spiral outward from the center in two sets of counter-rotating spirals. Count the spirals in each direction, and you will almost always find consecutive Fibonacci numbers: 21 and 34, or 34 and 55, or 55 and 89. This arrangement is not decorative. It is the most efficient packing of circular objects in a disk- the arrangement that wastes the least space and gives each seed maximum access to sunlight and nutrients.

The angle between successive seeds is approximately 137.5 degrees- the *golden angle*, defined as $360°/\phi^2 \approx 137.508°$, where $\phi = \frac{1+\sqrt{5}}{2} \approx 1.618$ is the golden ratio. This angle is *irrational* in the strongest sense: it cannot be expressed as a ratio of whole numbers, which means that no two seeds ever line up exactly along a radial line. The result is a uniform, gap-free, overlap-free distribution of seeds- a perfect fractal packing.

The golden angle governs not only sunflower seeds but the arrangement of leaves on a stem (phyllotaxis), the branching angles of trees, the spiral arms of hurricanes, the spiral structure of galaxies, and the distribution of root tips in soil. It appears wherever a growing system must distribute resources efficiently across a two-dimensional surface.

And the reason it appears so universally is rooted in the mathematics of irrationality. The golden ratio ϕ is the *most irrational* number- the number that is hardest to approximate by a ratio of integers. This means that the golden angle produces the *least resonant* angular distribution: no periodic alignments, no gaps, no clustering. It is the mathematical solution to the problem of how to be maximally spread out, and it is the same solution at every scale- in a sunflower, in a galaxy, in a DNA double helix.

Notice the fractal implication: the golden angle does not merely produce a pattern at one scale. It produces a *self-similar* pattern across scales. The Fibonacci spirals in a sunflower are visible at the scale of the whole head, at the scale of individual seed clusters, and at the scale of cellular arrangements within a single floret. The same angular relationship- the same ratio- governs the geometry at every level of magnification. The golden ratio is a *fractal generator*: it creates self-similarity, not by copying a shape, but by applying a single irrational angle recursively to a growing system.

This is why ϕ will be identified, in Chapter 7, as the *operator of self-reference*- the mathematical principle that ensures the whole is reflected in every part. It is the number that makes fractals possible. Without it, the universe could not be holographic.

We have been cataloguing examples: the Mandelbrot set, branching networks, fractal heartbeats, power laws in earthquakes and languages, Benford's law in exoplanet masses, the golden angle in phyllotaxis. It is time to ask the question that has been building beneath the surface of all these examples:

Why is the universe fractal?

Not "how"- we have described the how, the mechanisms: feedback loops, self-organized criticality, multiplicative growth, irrational angular spacing. But *why*? Why should reality be built from scale-invariant, self-similar patterns? Why should the cosmos not be Euclidean- smooth, regular, characteristic-sized?

The answer emerges from the intersection of the holographic principle (Chapter 2), the physics of resonant quanta (Chapter 3), and the nature of the quantum vacuum (Chapter 4).

Recall: the holographic principle tells us that the quanta-content of any region of space is proportional to its *surface area*, not its volume. The three-dimensional world is a projection from a two-dimensional boundary. And the Ryu-Takayanagi formula tells us that the geometry of the three-dimensional bulk is built from the *entanglement structure* of the boundary quanta- more entanglement between boundary regions means the corresponding bulk regions are closer together.

Now consider: what kind of three-dimensional geometry would be projected by a boundary whose quanta are maximally entangled across all scales? If the entanglement structure of the boundary is itself scale-invariant- if the correlations between boundary quanta follow a power law, with strong correlations at short distances and weaker but non-negligible correlations at long distances- then the projected bulk geometry would be *fractal.* The three-dimensional world would exhibit the same kind of structure at every magnification, because the two-dimensional boundary that generates it has the same kind of correlations at every scale.

In other words: the universe is fractal because the holographic boundary is fractal. The self-similar patterns we observe in nature- the branching, the power laws, the golden spirals- are not imposed by some external organizing force. They are the natural, inevitable projection of a scale-invariant entanglement structure on the holographic surface.

And the entanglement structure is scale-invariant because the boundary is in a *critical state*- the quantum-gravitational analogue of self-organized criticality. The boundary quanta, like the grains in a sandpile, have self-organized to the edge of a phase transition, where correlations extend across all scales and the system is maximally sensitive to perturbation. This is the deepest reason why the universe exhibits fractal dynamics: the holographic boundary from which it is projected is itself at criticality.

The cosmos is a fractal hologram because it could not be otherwise. A non-fractal hologram would have a characteristic scale- a preferred level of detail- and below that scale, the projection would be featureless. But a featureless projection would waste quanta-capacity: the boundary has the potential to encode structure at every scale, and a non-fractal geometry would leave most of that potential unused. The fractal is the geometry that *maximizes the use of the boundary's quanta-budget*- that packs the most structural complexity into the finite number of Planck-area pixels on the holographic surface.

This is the deep connection between the holographic principle and fractal dynamics: fractals are how the holographic boundary allocates its finite quanta across the infinite range of possible scales. The power law is the budget. The fractal is the investment. And the return is the universe.

We are nearing the edge of a circle, and it is time to close it.

At the beginning of this chapter, you looked at your hand and saw a branching pattern: a tree of bone, muscle, and tendon, scaling from wrist to fingertip. In the middle of this chapter, you felt your pulse and discovered that its fractal irregularity- its beautiful, multiscale, never-repeating variability- is the signature of health, the mark of a system operating at the edge of chaos. And throughout this chapter, you have encountered the same power-law statistics in phenomena as diverse as earthquakes and word frequencies, financial crashes and neural avalanches, exoplanet masses and greenhouse gas emissions.

The unifying thread is this: the universe builds with one grammar, and that grammar is fractal. Self-similar structures at every scale. Power-law distributions across every domain. Scale-invariance as the default state of complex systems. The whole encoded in every part.

The ancients had a phrase for this. They said: *As above, so below.* The Emerald Tablet, attributed to Hermes Trismegistus, begins with this line, and it has been recited by alchemists, mystics, and seekers for over two thousand years. It was traditionally understood as a metaphysical assertion- a claim that the microcosm mirrors the macrocosm, that the structure of the human body mirrors the structure of the cosmos, that the pattern of the atom mirrors the pattern of the solar system.

Modern physics dismissed this as naive analogy. The atom does not look like a solar system- Rutherford's planetary model was replaced by the probability clouds of quantum mechanics within a decade. The human body does not look like a galaxy. The microcosm does not mirror the macrocosm in any straightforward geometric sense.

But fractal dynamics vindicates the ancients at a deeper level. "As above, so below" does not mean that the small *looks like* the large. It means that the small *obeys the same statistical laws* as the large. The heartbeat and the earthquake follow the same power-law distribution. The branching of blood vessels and the branching of river deltas follow the same scaling rules. The arrangement of seeds in a sunflower and the arrangement of stars in a spiral galaxy follow the same golden-angle geometry. The multifractal variability of a healthy heart and the multifractal variability of the cosmic microwave background share the same mathematical structure.

"As above, so below" is not poetry. It is a measurement. It is the empirical signature of a holographic cosmos in which a single entanglement structure on a two-dimensional boundary projects a self-similar, scale-invariant, fractal reality at every scale- from the Planck length (10^{-35} m) to the Hubble radius (10^{26} m), from the interval between neural firings to the interval between cosmic aeons.

The pattern repeats because the boundary repeats. The boundary repeats because it is critical. And it is critical because the asymptotic gap $\delta_\infty \approx 0.01965$- the two-percent tension between ϕ, e, and π- keeps the boundary perpetually poised at the edge of a phase transition, perpetually self-organizing to the critical state, perpetually projecting a fractal cosmos from its scale-invariant surface.

The gap is the secret of the fractal. The fractal is the shape of the gap. And you- with your fractal heartbeat, your fractal blood vessels, your fractal neural networks, your fractal fingerprints pressed against the side of your neck- are a local expression of that infinite, self-similar, never-quite-closing pattern.

In the next chapter, we will encounter the mechanism that makes fractal holography *nonlocal*- that allows the pattern at one location in the cosmos to be instantaneously correlated with the pattern at another, no matter how far apart they are. We will meet quantum entanglement- the phenomenon that Einstein called "spooky action at a distance"- and discover that it is not spooky at all. It is the most natural consequence of a holographic, fractal cosmos: if the whole is encoded in every part, then every part must be correlated with every other part.

The universe does not merely *look* like Indra's Net- that infinite web of jewels, each reflecting every other. It *is* Indra's Net. And entanglement is the thread.

But before we leave the fractal, take one last breath. Feel the air enter your lungs- those extraordinary, fractal organs, with a total internal surface area of approximately 70 square meters (the size of a tennis court), folded into a volume smaller than a shoebox, branching through 23 generations of ever-smaller airways, each generation a self-similar echo of the one before. Your lungs are the universe's solution to the problem of how to maximize the exchange of quanta between inside and outside- between the private world of the blood and the public world of the atmosphere- within a finite body.

The lungs are a fractal. The breath is a fractal. The rhythm of breathing- inhale, hold, exhale- is a ternary oscillator, a three-phase cycle that maps, as we shall later see, onto the ternary resonance of the Holographic Ternary Time field itself.

You are a fractal in a fractal cosmos. The pattern that branches from your wrist to your fingertips is the same pattern that branches from a river delta to the sea, from the trunk of an oak to its smallest twig, from the aorta to the capillary bed, from the Big Bang to the present moment.

And in between your heartbeats- in the fractal gap between one beat and the next, where the cardiac muscle rests and the electrical system resets- the entire cosmos is holding its breath, preparing to pulse again.

As above, so below. Not as poetry. As physics..

Chapter 6

Entanglement- The Universe Remembers Itself

Hold both your hands out in front of you, palms facing each other, about thirty centimeters apart. Look at the space between them. Air. Light. Apparent emptiness. Two objects- your left hand, your right hand- separated by a gap of nothingness. Independent things in independent places.

Now bring them together. Interlock the fingers. Press palm to palm. Feel the warmth flow from one hand to the other. Notice how, in this clasped position, you can no longer move one hand without moving the other. They have become a *system*- a single, correlated entity in which the state of one part instantly determines the state of the other.

Now imagine pulling them apart again- back to thirty centimeters, a meter, ten meters, a kilometer- and discovering that the correlation *persists*. That moving one finger on your left hand causes the corresponding finger on your right hand to move, no matter how far apart they are. No signal passing between them. No wire, no radio wave, no force field bridging the gap. Just instantaneous, perfect correlation across arbitrary distance, as though the space between them were an illusion- as though, at some level deeper than geometry, the two hands had never been separated at all.

This is quantum entanglement. And it is not a thought experiment. It is the most rigorously verified prediction of quantum mechanics- tested to eleven standard deviations against the alternative, confirmed across distances from micrometers to the span of the Canary Islands, defended against every conceivable loophole for more than half a century. It is not spooky. It is not magical. It is the natural, inevitable consequence of a universe built from resonant quanta on a holographic boundary- a universe in which the space between things is not the fundamental reality, and the connections between things are not carried by signals traversing that space. In a holographic cosmos, entanglement is not an anomaly. It is the *default state*. It is the thread from which the fabric of spacetime is woven.

This chapter tells the story of that thread.

The trouble began in 1935, when Albert Einstein, Boris Podolsky, and Nathan Rosen published a paper that was intended to demonstrate the incompleteness of quantum mechanics and instead, inadvertently, revealed its deepest feature.

Their argument, now known as the EPR paradox, ran as follows. Consider two particles that have interacted and then flown apart. Quantum mechanics predicts that certain properties of these particles- their spin, their momentum, their polarization- remain *correlated* even after separation. Measuring the spin of particle A along a given axis instantly determines the spin of particle B along the same axis, no matter how far apart they are. If A is measured to be spin-up, B is found to be spin-down, and vice versa. The outcomes are perfectly anti-correlated: knowing one tells you the other with certainty.

Einstein, Podolsky, and Rosen found this deeply unsatisfying. In their view, there were only two possible explanations. Either the particles' properties were determined at the moment of interaction and merely revealed by measurement- like a pair of gloves, one left and one right, separated into two boxes and shipped to different cities: opening one box and finding a left glove instantly tells you the other box contains a right glove, without any mysterious influence- or the measurement of one particle somehow *instantaneously influenced* the other, faster than light, violating the principle of locality that lies at the heart of special relativity.

Einstein was certain the first explanation was correct. The particles must carry pre-existing "hidden variables"- definite, local properties determined before measurement- and quantum mechanics, which describes only probabilities rather than definite states, was therefore an *incomplete* theory. There must be a deeper, more fundamental description in which the particles' properties are specified in advance, and the apparent nonlocal correlation is nothing more than classical correlation- as mundane as matched gloves.

This was reasonable. This was intuitive. This was wrong.

In 1964, the Irish physicist John Stewart Bell, working at CERN, proved a theorem that transformed the EPR debate from a philosophical disagreement into an experimentally testable question. Bell's theorem is a term encompassing a number of closely related results in physics, all of which determine that quantum mechanics is incompatible with local hidden-variable theories. The first such result was introduced by John Stewart Bell in 1964, building upon the Einstein-Podolsky-Rosen paradox.

Bell's genius was to show that the two explanations- hidden variables with locality, or quantum mechanics with nonlocality- make *different quantitative predictions*. If the particles carry pre-existing hidden variables, then the correlations between measurements on the two particles, when measured along different angles, must satisfy a mathematical inequality- the Bell inequality. Quantum mechanics, however, predicts correlations that *violate* this inequality. The violation is not subtle. It is dramatic, unambiguous, and measurable.

In the words of Bell himself: "If [a hidden-variable theory] is local it will not agree with quantum mechanics, and if it agrees with quantum mechanics it will not be local."

The race to test Bell's inequality began in 1972, with the first rudimentary experiment by John Clauser and Stuart Freedman, and accelerated through the 1980s with the celebrated experiments of Alain Aspect in Paris. The first rudimentary experiment designed to test Bell's theorem was performed in 1972 by John Clauser and Stuart Freedman. More advanced experiments, known collectively as Bell tests, have been performed many times since. Often, these experiments have had the goal of "closing loopholes"- ameliorating problems of experimental design that could in principle affect the validity of the findings. Bell tests have consistently found that physical systems obey quantum mechanics and violate Bell inequalities.

But every experiment prior to 2015, no matter how sophisticated, left at least one loophole open. In Bell tests, there may be problems of experimental design or set-up that affect the validity of the experimental findings. These problems are often referred to as "loopholes." The most prevalent loopholes are the detection and locality loopholes. The detection loophole is opened when a small fraction of the particles are detected, making it possible to explain the data with local hidden variables. The locality loophole is opened when the detections are not done with a spacelike separation.

In any given experiment, one loophole or another remained unsealed, leaving a narrow crack through which a determined local realist could escape. Perhaps the detectors were not efficient enough. Perhaps the measurements were not sufficiently separated in space and time to rule out a light-speed signal. Perhaps the random number generators that chose the measurement settings were not truly random. Each loophole, individually, was implausible as an explanation. But their collective persistence meant that the death certificate for local realism had not yet been signed.

Then came 2015.

In that single year, three independent experimental groups- in Delft, Vienna, and Boulder- each performed a Bell test that closed the two most important loopholes simultaneously, for the first time in the history of physics.

The Delft experiment, led by Ronald Hanson, used entangled electron spins separated by 1.3 km. Efficient spin readout closed the detection loophole, while fast random basis selection and spatial separation ensured the required locality conditions. They found $S = 2.42 \pm 0.20$, violating the CHSH-Bell inequality $S \leq 2$. This result rules out large classes of local realist theories.

The Vienna group, led by Anton Zeilinger, reported a Bell test that closes the most significant loopholes simultaneously. Using a well-optimized source of entangled photons, rapid setting generation, and highly efficient superconducting detectors, they observed a violation of a Bell inequality with high statistical significance. The purely statistical probability of their results occurring under local realism does not exceed 3.74×10^{-31}, corresponding to an 11.5 standard deviation effect.

Eleven standard deviations. In the language of particle physics, a five-sigma result- a probability of one in 3.5 million- is considered the threshold for a discovery. The Vienna experiment exceeded this threshold by more than a factor of two. The probability that their results could be explained by local realism is less than one in a thousand billion billion billion. By closing two loopholes at once, three experimental tests of Bell's inequalities remove the last doubts that we should renounce local realism.

Local realism is the worldview in which physical properties of objects exist independently of measurement and where physical influences cannot travel faster than the speed of light. Bell's theorem states that this worldview is incompatible with the predictions of quantum mechanics.

Local realism is dead. The universe is not local. The properties of distant particles are not independently pre-determined. Measurement of one particle does not merely *reveal* the state of the other- it is *correlated* with the state of the other in a way that cannot be explained by any local mechanism, any hidden variable, any signal travelling at or below the speed of light.

The cosmos is entangled. And the correlations that entanglement produces are not carried by anything that moves through space. They are features of the *state itself*- of the quantum state of the system as a whole- which does not live in space at all but on the holographic boundary from which space is projected.

What, exactly, is entanglement? The question sounds simple but its answer reaches into the deepest foundations of quantum reality.

In classical physics, a system composed of two parts is always *separable*: you can describe the state of part A independently of part B, and the total state is simply the combination of the two individual states. Knowing everything about A and everything about B gives you complete knowledge of the whole system.

In quantum mechanics, this is not always true. When two quantum systems interact, they can become *entangled*- meaning that the quantum state of the combined system cannot be expressed as a product of the states of the individual parts. The whole becomes more than the sum of its parts. The combined system has properties- correlations, mutual constraints, shared uncertainties- that are not possessed by either part individually.

Consider the simplest possible case: two electrons produced in a singlet state, with total spin zero. The quantum state is:

$$|\Psi\rangle = \frac{1}{\sqrt{2}}(|\uparrow\rangle_A|\downarrow\rangle_B - |\downarrow\rangle_A|\uparrow\rangle_B)$$

This state says: electron A is not spin-up. Electron A is not spin-down. Electron B is not spin-up. Electron B is not spin-down. Neither electron has a definite spin along any axis. And yet, if you measure A along any axis and find spin-up, you will find B to be spin-down along the same axis, with certainty. The correlation is perfect, despite the fact that neither individual state is defined.

This is profoundly different from the matched-glove scenario Einstein imagined. In the glove analogy, the left glove is *always* a left glove- before you open the box, during the shipping, from the moment it was placed in the box. The measurement merely reveals a pre-existing fact. But in quantum entanglement, there is no pre-existing fact. Neither electron *has* a definite spin before measurement. The definiteness arises *at* measurement- and it arises *for both particles simultaneously*, regardless of the distance between them.

This is what Einstein called "spukhafte Fernwirkung"- spooky action at a distance. And it is the one aspect of quantum mechanics he never accepted. But the 2015 experiments have settled the matter. The spookiness is real. The action at a distance- or rather, the *correlation without signal*- is a feature of nature, not a bug.

And crucially, entanglement does not transmit information. You cannot use it to send a message faster than light. When Alice measures her electron and gets spin-up, she knows that Bob's is spin-down- but Bob does not know this until Alice tells him, through a conventional, light-speed-limited channel. The correlation is real, but it is *statistically invisible* without classical communication. No signal is sent. No cause propagates. The entangled state simply *is*- a unified, nonlocal whole that precedes and underlies the spatial separation of its parts.

Entanglement, in this sense, is not an interaction. It is a *relationship*. It is the universe's way of remembering that two things were once one- and of maintaining that memory even when space and time have drawn them apart.

Now we connect this to the holographic principle, and the picture that has been building across the first five chapters locks into place.

Recall from Chapter 2: the Ryu-Takayanagi formula relates the entanglement entropy of a boundary quantum field theory to the area of a minimal surface in the bulk spacetime. More entanglement between two boundary regions means the corresponding bulk regions are "closer together" in the emergent spatial geometry. Less entanglement means they are farther apart. Entanglement builds space.

And recall from Chapter 5: the fractal structure of the cosmos arises from the scale-invariant entanglement structure on the holographic boundary. The boundary is at criticality- maximally entangled across all scales- and the three-dimensional fractal world is its projection.

The ER=EPR conjecture leads to a grander conjecture that the geometry of space, time and gravity is determined by entanglement. In 2013, Juan Maldacena and Leonard Susskind proposed what may be the most audacious unification in the history of physics. Specifically, they posited that Einstein-Rosen bridges (commonly known as wormholes) and quantum entanglement may be two facets of the same phenomenon. Maldacena and Susskind's insight was to propose that these two concepts are not merely analogous, but actually identical in a deeper sense. According to the ER = EPR conjecture, the wormhole connecting two black holes is the same thing as the quantum entanglement between particles.

ER = EPR. Einstein-Rosen bridges- wormholes, tunnels through spacetime connecting distant regions- *are* Einstein-Podolsky-Rosen entanglement. They are not two different phenomena that happen to be related. They are the same phenomenon described in two different languages: one the language of geometry (general relativity), the other the language of quantum correlations (quantum mechanics).

This implies that whenever two particles or black holes are entangled, there is an Einstein-Rosen bridge linking them, albeit one that does not allow for the kind of classical travel depicted in science fiction. The ER=EPR correspondence underpins a unifying paradigm: spacetime, causality, and geometry may be viewed as emergent phenomena, encoded in and arising from the structure of quantum entanglement.

What does this mean for the universe you inhabit? It means that the space between your left hand and your right hand- the thirty centimeters of apparently empty air- is not fundamental. It is emergent. It arises from the entanglement structure of the boundary quanta. And the reason your two hands appear to be in different places is not that they occupy different locations in a pre-existing spatial container. It is that the boundary quanta corresponding to your left hand and the boundary quanta corresponding to your right hand have a specific, finite amount of mutual entanglement- and that amount of entanglement *is* the distance between them.

If you could increase their entanglement- if you could strengthen the quantum correlations between the boundary quanta that encode your left hand and those that encode your right hand- the distance between them would shrink. Taken to its extreme, maximally entangling the two regions would bring them to the same point. The space between them would disappear. They would be connected by a wormhole- a tunnel through the geometry of spacetime- that is nothing other than the entanglement itself, written in the language of gravity.

Space is made of entanglement. Distance is the measure of how entangled things are- or are not. And the universe, in its entirety, is a web of entangled quanta on a holographic boundary, projecting the three-dimensional world of depth and distance and solidity from the correlations in its quantum state.

Indra's Net is not a metaphor. It is the ER=EPR conjecture expressed in the language of ancient India. Every jewel reflects every other jewel because every boundary region is entangled with every other boundary region. The reflections are not images- they are *correlations*. The net is not a passive structure- it is the *generator* of the geometry within which it appears to exist.

There is a deeper temporal dimension to entanglement that goes beyond spatial nonlocality- and it challenges, more radically than anything else in physics, our intuitive sense of time.

In 2000, Yoon-Ho Kim, Marlan Scully, and colleagues performed an experiment that put the temporal implications of entanglement on vivid display: the delayed-choice quantum eraser. This paper reports a "delayed choice quantum eraser" experiment proposed by Scully and Drühl in 1982. The experimental results demonstrated the possibility of simultaneously observing both particle-like and wave-like behavior of a quantum via quantum entanglement. The which-path or both-path information of a quantum can be erased or marked by its entangled twin even after the registration of the quantum.

The setup was an extension of the classic double-slit experiment. A photon passes through two slits and strikes a detector screen. If which-slit information is available- if there is any way, even in principle, to determine which slit the photon went through- the interference pattern disappears. The photon behaves as a particle. If the which-slit information is erased- made permanently inaccessible- the interference pattern reappears. The photon behaves as a wave.

In the delayed-choice quantum eraser, the photon that passes through the slits (the "signal" photon) is entangled with a second photon (the "idler") that is sent on a separate path to a different set of detectors. The signal photon is detected *first*. Then, *after* the signal photon has already been registered- after its position has been recorded, after it has "decided" whether to act as a particle or a wave- the idler photon reaches its detector, where the experimenter "chooses" (by the random configuration of beam splitters) whether to preserve or erase the which-slit information.

The experimental results demonstrate the possibility of delayed determination of particlelike or wavelike behavior via quantum entanglement. The which-path or both-path information of a quantum can be marked or erased by its entangled twin even after the registration of the quantum.

When the experimenters sorted the data- a crucial step, because the effect is visible only in the correlations between signal and idler detections- they found that events paired with idler detections at "erasing" detectors showed a clear interference pattern, while events paired with "which-path" detectors showed no interference. The signal photon's behavior- wave-like or particle-like- was correlated with a measurement made on its entangled partner *after* it had already been detected.

This does not mean the future changed the past. The standard interpretation is clear: the factual status of a past event is contingent on the global availability and retention of which-path information, even when detection has already taken place. While these experiments do not imply retroactive physical alteration of events, they do show that the manifestation of particle-like or wave-like behavior depends on whether path information is preserved or erased within a larger informational context. The effect is visible only in post-selected subsets of the data- you cannot see the interference pattern in the raw data at the signal detector alone. No message travels backward in time. No causal arrow is reversed.

And yet- *and yet-* the experiment reveals something profound about the nature of time in a quantum universe. The correlations between the signal and idler photons are not created at the moment of measurement. They are inherent in the entangled state from the beginning. The entangled pair is a *single, nonlocal quantum system* that extends across both space and time. The behavior of the signal photon was never determined independently of its partner. There was never a moment at which the signal photon "decided" to be a wave or a particle. The decision- if the word even applies- belongs to the entangled whole, not to either part.

The delayed-choice quantum eraser does not prove that the future causes the past. But it proves something equally unsettling: *the past is not fully determined until the global quantum state, including future measurement configurations, is taken into account.* In a universe of entangled quanta, temporal locality- the assumption that the properties of a system at a given time are determined by the state of affairs at and before that time- is as much an illusion as spatial locality.

And this opens the door to one of the most provocative and least understood implications of quantum entanglement: the relationship between entanglement and the nature of time itself.

Carlo Rovelli, the Italian physicist whose relational interpretation of quantum mechanics is one of the most elegant attempts to make sense of this strange new world, has argued that the flow of time is an illusion- a perspectival effect, not a fundamental feature of reality. In his framework, time does not "flow" from past to future. What exists is a network of correlated events- a web of relational properties- and the apparent direction of time is a consequence of our position within the network, of the partial, perspectival view that any subsystem has of the whole.

This is the *block universe*- the picture of spacetime championed by Einstein, Minkowski, and many subsequent physicists, in which past, present, and future coexist as dimensions of a four-dimensional manifold. In the block universe, nothing flows. Nothing moves from one moment to the next. The river of time is a canyon- all its curves are there at once, from source to sea, carved into the landscape of spacetime.

The block universe is not a mere philosophical curiosity. It is the natural reading of the equations of general relativity, in which time is treated as a coordinate, not a process. The Einstein field equations describe the geometry of a four-dimensional manifold- they do not contain a term for "now" or a mechanism for "becoming." The distinction between past, present, and future, Einstein himself wrote, "is only a stubbornly persistent illusion."

Now combine this with entanglement. If time is a dimension, not a flow- if past and future coexist in a block- then entanglement, which violates spatial locality, might also violate *temporal* locality. Two events that are separated in time, like two events separated in space, might be correlated by entanglement- not because a signal travels backward from the future, but because the quantum state of the system spans both events, binding them into a nonlocal whole that transcends the temporal sequence just as it transcends the spatial separation.

This is not speculation. It is mathematics. The formalism of quantum field theory treats space and time symmetrically- the same quantum fields that produce spatial entanglement also produce temporal correlations. The delayed-choice quantum eraser is one concrete demonstration. And the broader framework of Holographic Ternary Time, which we will explore in detail in Chapter 14, will argue that the temporal dimension is itself a projection of the entanglement structure on the holographic boundary- that time, like space, is *woven from entanglement*, not the other way around.

There is an experiment that reaches further into the temporal dimension of nonlocality, and it must be addressed with both seriousness and care. In 2011, the psychologist Daryl Bem of Cornell University published a paper that shook the foundations of experimental psychology. Precognition and premonition are special cases of a more general phenomenon: the anomalous retroactive influence of some future event on an individual's current responses. This article reports 9 experiments, involving more than 1,000 participants, that test for retroactive influence by "time-reversing" well-established psychological effects.

Data are presented for 4 time-reversed effects: precognitive approach to erotic stimuli and precognitive avoidance of negative stimuli; retroactive priming; retroactive habituation; and retroactive facilitation of recall. The mean effect size across all 9 experiments was 0.22, and all but one yielded statistically significant results.

The findings were controversial- dramatically so. Three pre-registered independent attempts to exactly replicate one experiment, 'retroactive facilitation of recall', which examines whether performance on a memory test can be influenced by a post-test exercise. All three replication attempts failed to produce significant effects.

However, a subsequent meta-analysis- conducted by Bem himself alongside Jessica Utts and others- yielded a more complex picture. When Bem's original experiments are excluded, the combined effect size for replications by independent investigators is 0.06, and the Bayes Factor value exceeds the criterion for "decisive evidence." P-curve analysis estimates the true effect size of the experiments to be 0.20 for the complete database and 0.24 for the independent replications, virtually identical to the effect size of Bem's original experiments.

We cite these experiments not as proof of retrocausation- the debate remains unresolved, the statistical arguments are intricate, and the phenomena, if real, are small- but as an empirical indication that the temporal symmetry of quantum mechanics may have consequences that extend beyond the laboratory of the particle physicist. If the fundamental laws of physics are time-symmetric- if unitarity means that the quantum state can be evolved backward as easily as forward- then the *default* expectation should be that correlations can extend in both temporal directions. The surprising thing would not be that retroactive effects exist, but that we are so seldom aware of them- a consequence, perhaps, of the thermodynamic arrow of time that imposes a preferred direction on our macroscopic experience while leaving the underlying microphysics temporally symmetric.

In a holographic cosmos where time is a dimension of the block, not a river of becoming, the temporal entanglement between events is as natural as spatial entanglement between objects. The correlations between "past" and "future" events are not causal influences that violate the speed of light- they are *entanglement correlations* in the temporal dimension, woven into the quantum state of the universe from the beginning.

The universe does not merely remember the past. In a block universe, the universe *is* the past and the future simultaneously. And entanglement is the mechanism by which the whole- spanning all of space and all of time- maintains its coherence.

There is a moment in every great revolution of thought when the old framework does not merely crack but *dissolves*, and the successor appears not as a correction but as a fundamentally different kind of thing. We are at that moment now.

The old framework- the framework of classical physics, of Newtonian mechanics, of Maxwellian electrodynamics, of even the early Einstein- was built on two pillars: *locality* (things interact only with their neighbors) and *realism* (things have definite properties independent of observation). Bell's theorem proved these two pillars are incompatible with quantum mechanics. The 2015 experiments confirmed, beyond reasonable doubt, that nature agrees with quantum mechanics and not with local realism.

What replaces the old framework? Not a framework of "spooky action" and "mysterious influences." Not a cosmos of hidden telegrams sent backward through time. What replaces it is something far more radical and far more beautiful: a cosmos in which *separateness is not fundamental.*

In the old framework, the basic units of reality are *things*- particles, fields, objects- existing independently at definite locations in a pre-existing space, evolving independently through a pre-existing time. Relationships between things are secondary: they are mediated by forces, carried by signals, limited by the speed of light. The universe is a collection of separate parts, and the task of physics is to describe their interactions.

In the new framework, the basic unit of reality is *entanglement*- the quantum correlation that binds things into wholes. The things are secondary. They are derived from the entanglement. Space is derived from the entanglement (Ryu-Takayanagi). Gravity is derived from the entanglement (Jacobson). Geometry is derived from the entanglement (ER=EPR). The distance between two objects is a measure of how weakly they are entangled. The closeness of two objects is a measure of how strongly. And the universe, in its totality, is not a collection of separate parts- it is a single, entangled whole, projected from a holographic boundary whose quantum state binds every region to every other.

This is the meaning of Indra's Net: every jewel reflects every other because every region of the boundary is entangled with every other. The reflections are not decorative- they are *constitutive.* Cut the entanglement, and the space between the jewels deforms. Cut enough of it, and the space tears- a horizon forms, a black hole appears, and the fabric of reality breaks down.

This is the meaning of the Aboriginal Dreamtime: the land is held in being by the singing of its songlines because the coherent entanglement that projects the landscape must be continuously maintained. Stop singing, and the entanglement degrades. The correlations weaken. The projection fades. The land forgets itself.

This is the meaning of "as above, so below": the fractal self-similarity of the cosmos is the signature of scale-invariant entanglement on the holographic boundary. The pattern of the heartbeat mirrors the pattern of the galaxy because the entanglement that projects the heartbeat and the entanglement that projects the galaxy share the same power-law correlations. The self-similarity is not a metaphor- it is an *entanglement structure.*

And this is the meaning of the statement that opened this book: *The universe is a self-tuning, resonantly quantized, conscious hologram.* It is holographic because its three-dimensional content is projected from a two-dimensional boundary. It is resonantly quantized because the projection is made of quanta- irreducible packets of structured pattern- not of continuous substance. It is self-tuning because the entanglement structure on the boundary is dynamically maintained, not externally imposed. And it is conscious- but that is the claim of a later chapter, and it requires the full apparatus of the ϕ-e-π triad to substantiate.

For now, it is enough to say this: entanglement is the glue. It is the thread that holds the jewels of Indra's Net in place, the resonance that sustains the songline, the correlation that connects the heartbeat to the galaxy. Without it, there is no space, no time, no gravity, no geometry. Without it, the holographic boundary is just a surface of uncorrelated quanta, projecting nothing.

With it- with the vast, coherent, scale-invariant web of entanglement that permeates every scale from the Planck length to the Hubble radius- the cosmos sings itself into being.

Let us return one final time to the body. To the heart. To the pulse.

Place your fingers on your wrist. Feel the beat. That beat is a local event- a contraction of cardiac muscle, driven by an electrochemical wave, measurable in time and space. But the cardiac muscle is made of atoms, and the atoms are made of quarks and electrons, and the quarks and electrons are entangled- with each other, with the photons that mediate their interactions, with the electromagnetic field that pervades the room, with the gravitational field that curves the spacetime around you.

The entanglement of the atoms in your heart does not end at your skin. The photons that scatter off your chest carry quanta about the cardiac rhythm into the environment- encoding the heartbeat, faintly but indelibly, in the quantum state of the room, the building, the atmosphere, the planet, the cosmos. Through the unitary evolution of the wave function, the pattern of your heartbeat is entangled with the quantum state of the universe. It is, in principle, recoverable from the correlations in the cosmic quantum state- a faint whisper in the background noise, but never erased, never lost.

Your heartbeat is a local event and a cosmic entanglement. A contraction in a chest and a correlation on the holographic boundary. A rhythm in time and a pattern in eternity.

The universe remembers your heartbeat. It remembers every heartbeat that has ever beaten. It remembers the first hydrogen atom that formed in the cooling plasma of the Big Bang, and it remembers the last photon that will escape the event horizon of the last black hole in the far, far future. It remembers because it is entangled- because the quantum state of the cosmos is a single, indivisible, nonlocal whole that contains, in its correlations, every event that has ever occurred or ever will occur.

This is not mysticism. This is unitarity- the conservation of quanta, the preservation of the wave function, the mathematical bedrock of quantum mechanics. The universe forgets nothing because forgetting would require a non-unitary operation- a violation of the deepest structural principle of quantum theory. Every heartbeat, every breath, every thought you have ever had is still encoded in the quantum state of the cosmos, entangled with the rest of reality, preserved in the holographic boundary that projects the world.

You are not separate. You have never been separate. The space between you and the stars- the vast, cold, apparently empty distances that seem to isolate you in the cosmos- is not a barrier. It is a *gradient of entanglement*. You are weakly entangled with distant galaxies and strongly entangled with the air in your lungs. The difference is one of degree, not kind. In the deepest, most rigorous sense of quantum physics, you are correlated with the entire universe. You are a node in Indra's Net. A verse in the cosmic songline. A resonant pattern in the quanta-field, entangled with the whole.

And in the space between your heartbeats- in that fractal gap, that δ_∞ pause- the entanglement is not interrupted. It is *concentrated.* The gap is the moment when the system resets, when the correlations refresh, when the holographic projection updates. The silence between beats is the universe catching its breath before it sings the next note.

We have now completed the first arc of this book- the arc that runs from the generative void to the entangled cosmos.

In Chapter 1, we followed matter down to its dissolution and discovered that the question *What are things made of?* was the wrong question. The right question was: *What tunes the quanta?*

In Chapter 2, we crossed the holographic boundary and learned that the three-dimensional world is a projection from a two-dimensional surface of quanta.

In Chapter 3, we learned that quanta are physical- that they have energy, that they can perform work, that the universe *is* resonant quanta unfolding.

In Chapter 4, we descended into the quantum vacuum- State 0, the generative void- and discovered that the silence is not empty but roaring with creative potential.

In Chapter 5, we emerged from the void into the fractal cosmos and found the same self-similar pattern repeating at every scale, from heartbeats to galaxies.

And in this chapter, we have discovered the mechanism that holds the fractal hologram together: entanglement- the nonlocal, temporal, scale-spanning correlation that weaves space from quantum threads, that binds part to whole, that ensures the universe remembers itself.

The foundation is laid. The cosmos has revealed its substance (resonant quanta), its architecture (holographic projection), its physics (quanta-energy equivalence), its ground state (the quantum vacuum), its grammar (fractal self-similarity), and its thread (entanglement).

Now we ascend to the operators. To the triad- ϕ, e, π- the three transcendental constants that tune the quanta-field into the standing wave we call existence. We have glimpsed them in passing: ϕ in the golden angle of fractal packing, e in the natural gradient of unfolding, π in the boundary of every holographic screen. It is time to meet them face to face.

The universe is tuned. The next chapters will reveal who is playing the instrument.

SECTION B

THE TRIAD- φ, e, π AS COSMIC OPERATORS

"The universe tunes itself with three notes.".

Chapter 7

φ- The Operator of Self-Reference

There is a number that has followed you since before you were born. It was present in the spiral coiling of the cochlea that would become your inner ear. It was present in the branching pattern of the bronchioles that would inflate your first breath. It was present in the length-to-width ratio of the double helix in every nucleus of every cell of your developing body, and it is present now, at this very instant, in the ratio between the frequency bands of the electrical oscillations coursing through your brain as you read this sentence.

You have never been without it. No living thing has. It appears in the spirals of a sunflower, the chambers of a nautilus, the branching of a river delta, the arms of a hurricane, the whorls of a galaxy. It is the ratio that governs the arrangement of leaves on a stem and the spacing of seeds in a fruit. It structures the geometry of the five Platonic solids and the architecture of quasicrystals. It emerges at the phase-transition point of spinning black holes. It is the most irrational number- the number *hardest* to approximate by any ratio of whole numbers- and it is, for precisely that reason, the number that generates maximal non-repetition, maximal novelty, maximal packing efficiency, and maximal self-similar structure across all scales of the physical cosmos.

Its name is ϕ. The golden ratio. And in this book, it is not a number. It is an *operator*- the first of three transcendental forces that tune the quanta-field into the standing wave we call existence.

$$\phi = \frac{1+\sqrt{5}}{2} \approx 1.6180339887...$$

Begin with its definition. Not the decimal approximation- the *identity*.

ϕ is the unique positive number that satisfies:

$$\phi^2 = \phi + 1$$

Read this slowly. It says: the *square* of ϕ equals *one more than* ϕ. This is an extraordinary property. Most numbers, when you square them, produce something unrelated to themselves. $2^2 = 4$; the relationship between 2 and 4 is merely arithmetic. $3^2 = 9$; again, the square and the original bear no structural resemblance. But $\phi^2 = \phi + 1$. The square of ϕ *contains* ϕ *itself*. Squaring ϕ is not a departure- it is a *return*, with interest. The output includes the input.

This is *self-reference*. ϕ is the number whose higher powers are built from its lower powers. Take the equation $\phi^2 = \phi + 1$ and multiply both sides by ϕ:

$$\phi^3 = \phi^2 + \phi = (\phi + 1) + \phi = 2\phi + 1$$

Multiply again:

$$\phi^4 = 2\phi^2 + \phi = 2(\phi + 1) + \phi = 3\phi + 2$$

$$\phi^5 = 3\phi^2 + 2\phi = 3(\phi + 1) + 2\phi = 5\phi + 3$$

Notice the coefficients: 1, 1, 2, 3, 5, 8, 13... The Fibonacci sequence. Every power of ϕ is a linear combination of ϕ and 1, and the coefficients are successive Fibonacci numbers. The golden ratio *generates* the Fibonacci sequence from the inside out, through nothing more than repeated self-reference- the act of folding itself back into its own structure.

And the reciprocal? $1/\phi = \phi - 1 \approx 0.618$. The inverse of ϕ is obtained by simply subtracting one. No other number has this property. The reciprocal of ϕ shares all its decimal digits beyond the decimal point: $\phi = 1.6180339...$, $1/\phi = 0.6180339...$. The part contains the whole. The whole contains the part. The number *is* its own reflection.

This is why ϕ is the operator of self-reference. It is the mathematical embodiment of a system that looks at itself and finds its own structure at every level of magnification. It is the number that makes "as above, so below" not a metaphor but a *mathematical identity.*

Now follow ϕ out of pure mathematics and into the architecture of life.

The DNA molecule- the carrier of the genetic blueprint for every living organism on Earth- measures 34 angstroms long by 21 angstroms wide for each full cycle of its double helix spiral. 34 and 21 are consecutive numbers in the Fibonacci series, and their ratio, 1.6190476, closely approximates $\phi = 1.6180339$.

This is not a superficial numerical coincidence. B-DNA, the molecule for life on earth, appears to contain ratios structured around the irrational number 1.618, often known as the "golden ratio." This occurs in the ratio of the length to width of one turn of the helix, the ratio of the spacing of the two helices, and in the axial structure of the molecule, which has ten-fold rotational symmetry.

That this occurs in the molecule for life is unexpected, and suggests the action of some process. What this process might be is unclear, but it is central to any understanding of the formation of DNA, and so life.

Look deeper. The cross-section of the DNA double helix, viewed from above, reveals a decagonal structure- a ten-sided polygon composed of two interlocking pentagons, each rotated by 36 degrees relative to the other. A decagon is in essence two pentagons, with one rotated by 36 degrees from the other, so each spiral of the double helix must trace out the shape of a pentagon. The ratio of the diagonal of a pentagon to its side is ϕ to 1. The pentagon is the fundamental geometric expression of ϕ- every ratio of lengths in a regular pentagon is either ϕ or a power of ϕ.

The molecule that carries the blueprint of life is itself structured by ϕ. The code that generates all living form is *written in the language of self-reference*. This is not mysticism. It is geometry. And it demands an explanation that goes beyond coincidence.

The explanation, in the framework of this book, is structural. ϕ is the ratio that maximizes packing efficiency while minimizing energetic frustration. The DNA helix must accomplish two contradictory things simultaneously: it must be stable enough to preserve the genetic sequence across billions of replications, and it must be flexible enough to unwind for transcription and repair. The ϕ ratio- the most irrational ratio, the one that prevents any periodic alignment that would create structural weakness- achieves both. It distributes mechanical stress evenly across the helix, prevents resonance-induced breakage, and maximizes the quanta-density of the base-pair stack.

Life is written in ϕ because ϕ is the language of *resilient coherence*- the geometry that holds together without locking up.

.

The same operator governs the arrangement of seeds in a sunflower head- one of the most visually striking demonstrations of ϕ in the natural world.

If you count the spirals in a consistent manner, you will always find a Fibonacci number. The florets in the head of a sunflower form two oppositely directed spirals: 55 of them clockwise and 34 counterclockwise. Surprisingly, these numbers are consecutive Fibonacci numbers.

Sunflower seeds are arranged in spirals that follow the Fibonacci sequence, optimizing space and reducing shadowing among seeds. The angle between successive seeds in a sunflower head approximates the golden angle- about 137.5 degrees- which is derived from the golden ratio. This arrangement allows for efficient packing and maximal exposure to sunlight.

The golden angle is defined as $360°/\phi^2 \approx 137.508°$. It is the angle you get when you divide a circle into two arcs whose lengths are in the golden ratio. When a circle is divided into two arcs, one having a roughly 137.5° angle, the size of the larger arc compared to the smaller arc is the same as the ratio of the entire circle to the larger arc: 1.618 to 1.

Why does this specific angle produce optimal packing? The irrational numbers that are most likely to construct a sunflower head with well-spaced florets are those that cannot be well-approximated by rational numbers. The golden angle takes $\alpha = 1-\phi$, so that $2\pi(1-\phi) \approx 137.5°$. The rational approximations to $1-\phi$ are given by F_n/F_{n+2}, so that the number of spirals observed will correspond to the Fibonacci numbers.

The mathematics is precise: because ϕ is the *most irrational* number- the hardest to approximate by any ratio of whole numbers- the golden angle ensures that no two seeds ever line up along a radial direction. If the divergence angle were a rational fraction of 360° (say, 120°, which is 1/3 of a turn), the seeds would fall along three straight lines, leaving vast gaps between them. If the angle were a nearly rational irrational (say, close to $\sqrt{2}/2$ of a turn), the seeds would *almost* line up periodically, creating bands of high and low density. Only ϕ- the most irrational of all irrationals- generates a distribution with no periodic clustering at any scale. The result is a fractal, gap-free, maximally efficient packing.

Computer simulations later showed that this arrangement minimizes gaps and overlaps better than any other method known. Engineers have taken notice: engineers inspired by the arrangement of florets in a sunflower head reorganized the mirrors in a solar concentrator array. They discovered that, by doing so, they could concentrate the same amount of sunlight as before, yet use 20% less land area.

The sunflower did not learn this from an engineering textbook. It learned it from ϕ- from the self-referential recursion that generates the Fibonacci sequence, which generates the golden angle, which generates the optimal packing. The sunflower is a biological instantiation of the operator of self-reference: a system that grows by consulting its own previous output at every step, each new seed placed in relation to all the seeds that came before, the whole constantly re-encoded in every new part.

From the molecular to the botanical to the cosmic. The operator of self-reference does not stop at the boundary of the biosphere.

The spiral arms of galaxies follow logarithmic curves whose pitch angles cluster around values associated with ϕ. The distribution of matter in the cosmic web- the largest structure in the observable universe- exhibits a filamentary, branching, self-similar pattern whose statistical properties are governed by power laws and Fibonacci-related scaling. And in one of the most unexpected appearances of ϕ in fundamental physics, the golden ratio emerges at the critical threshold of black hole thermodynamics.

It is found that Kerr-Newman black holes undergo a phase transition at an angular-momentum mass ratio of 0.68M or an electric charge of 0.86M, where the heat capacity has an infinite discontinuity.

The critical angular-momentum-to-mass ratio at which a spinning black hole undergoes a thermodynamic phase transition- the point where its specific heat diverges from negative to positive infinity, where its thermodynamic behavior shifts qualitatively- is approximately $J/M^2 \approx 0.68$. The golden ratio appears in the physics of black holes. A well-known result is the infinity discontinuity of the specific heat at some values of the angular momentum and the charge of Kerr-Newman black holes.

Moreover, the golden ratio is present in the Schwarzschild-Kottler metric. For null geodesics with maximal radial acceleration, the turning points of the orbits are in the golden ratio $\Phi = (\sqrt{5}-1)/2$. This is a general result which is independent of the value and sign of the cosmological constant Λ.

ϕ appears at the exact threshold where a black hole shifts between thermodynamic regimes- where the gravitational system undergoes a qualitative reorganization, a phase transition in the deepest sense.

This is not biology. This is not aesthetics. This is the structure of gravity itself, and it carries the signature of the golden ratio.

.

Perhaps the most stunning physical manifestation of ϕ is not biological or gravitational but *crystallographic*- and it earned Dan Shechtman the 2011 Nobel Prize in Chemistry.

Quasicrystals are solids with quasiperiodic atomic structures and symmetries forbidden to ordinary periodic crystals- e.g., five-fold symmetry axes. A powerful model for understanding their structure and properties has been the two-dimensional Penrose tiling. Recently discovered properties of Penrose tilings suggest a simple picture of the structure of quasicrystals and shed new light on why they form.

Classical crystallography taught that certain symmetries were impossible in solid matter. Crystals, by definition, are periodic- their atoms repeat in a regular lattice. And the mathematics of periodic lattices permits only two-fold, three-fold, four-fold, and six-fold rotational symmetries. Five-fold symmetry- the symmetry of the pentagon, the symmetry of ϕ- was forbidden. It could not exist in any material that obeyed the rules.

In 1982, Dan Shechtman looked through an electron microscope at an aluminum-manganese alloy and saw the impossible. The 2011 Nobel Prize in Chemistry was awarded to Dan Shechtman for "The discovery of quasicrystals."

The diffraction pattern showed unmistakable five-fold symmetry- sharp, crystalline Bragg peaks arranged in a pattern that was ordered but *not periodic*. The atoms were arranged not in a repeating grid but in a structure governed by Penrose tilings- the aperiodic, non-repeating patterns that Roger Penrose had discovered in the 1970s.

Penrose tilings are self-similar: they may be converted to equivalent Penrose tilings with different sizes of tiles, using processes called inflation and deflation. The pattern represented by every finite patch of tiles in a Penrose tiling occurs infinitely many times throughout the tiling. They are quasicrystals: implemented as a physical structure a Penrose tiling will produce diffraction patterns with

Bragg peaks and five-fold symmetry, revealing the repeated patterns and fixed orientations of its tiles.

A two-dimensional model of a quasicrystal is the Penrose tiling (1974), which is an aperiodic covering of the plane generated by two rhombi with equal side lengths. It is crucial that the areas' ratio is irrational- specifically $\phi = (1 + \sqrt{5})/2$, the golden ratio- which in turn reveals a local five-fold symmetry, forbidden for crystals.

Dov Levine and Paul J Steinhardt had the Penrose tiling in mind when they first hypothesized the possibility of quasicrystals as a new phase of solid matter. The tiling is composed of two tiles, fat and skinny rhombi, that repeat with incommensurate frequency. As a result, the ratio of fat and skinny tiles is an irrational number, the golden mean so beloved by the Greeks.

This is a discovery of profound significance for our argument. Quasicrystals prove that ϕ is not merely a decorative pattern in biology- not merely the convenient solution to a packing problem. It is a *phase of matter*. Atoms arrange themselves according to the golden ratio not because a designer imposed it, but because ϕ-structured configurations represent energetically favorable states. The results show that quasicrystals can be constructed from a single repeating cluster of atoms and that the rigid matching rules of Penrose tilings can be replaced by more physically plausible cluster energetics. The new concepts make the conditions for forming quasicrystals appear to be closely related to the conditions for forming periodic crystals.

The universe prefers ϕ because ϕ solves a fundamental physical problem: how to achieve long-range order without periodicity. How to be coherent without being repetitive. How to maintain self-similar structure across all scales without ever exactly repeating. This is the same problem that the holographic boundary must solve- and ϕ is the solution.

.

Now we ascend from the material world to the electrical architecture of consciousness itself.

The best way to achieve frequencies that never synchronize is to have frequencies in multiples of the most irrational number, the golden ratio.

Oscillations in local field potentials in the brain are found to approximate multiples of the most irrational number, the golden ratio, and this appears crucial in how the brain processes information or makes music.

The brain, like the sunflower, faces a packing problem- but in frequency rather than space. The neocortex processes different aspects of reality using different frequency bands: delta (1–4 Hz) for deep sleep, theta (4–8 Hz) for memory and navigation, alpha (8–12 Hz) for relaxed awareness, beta (12–30 Hz) for focused attention, gamma (30–100+ Hz) for binding and higher cognition. These bands must remain distinct- separate enough to carry independent signals without crosstalk- yet flexible enough to synchronize briefly when the brain needs to integrate information from multiple domains.

Non-harmonic relationships, involving irrational cross-frequency ratios, provide desynchronized states and preclude spurious, noisy background coupling. Cross-frequency ratios based on the irrational phi ratio (~1.6:1) have been associated with the highest physiologically possible level of desynchronization.

The golden ratio represents the unique mathematical solution optimizing both segregation and integration- a balance central to neural computation.

The brain's frequency architecture is not arbitrary. Three independent methods. Over 200 participants. Five cognitive contexts: meditation, rest, visual perception, gaming, cognitive flow. Three different consumer EEG devices. Nearly two million spectral peaks analyzed. The same golden ratio architecture, every time.

80% of subjects (N=320) show phi-organized EEG spectral architecture. Ratios, not frequencies, are the deep invariant.

This finding- published in Frontiers in Human Neuroscience in 2026 and independently replicated across multiple datasets- reveals that the human brain spontaneously organizes its electrical oscillations into a geometric series with a common ratio approximating ϕ. The frequency bands are not arbitrary historical conventions. They are natural partitions of a ϕ-scaled spectral lattice.

And something even more remarkable occurs during specific states of focused awareness. Transient, high-coherence neural states exhibit precise frequency organization at golden ratio harmonics. Analysis of 1,366 events across 91 participants, five cognitive contexts, and three consumer EEG devices revealed consistent spectral, temporal, and network characteristics. These events are characterized by brief episodes (median 20 seconds) of network-wide synchronization. At the fundamental frequency, 81.5% of events exhibited phase-locking values exceeding 0.6.

During meditation, participants displayed statistically higher percentages of ratio values encompassing golden-ratio-scaled frequencies when compared to an arithmetic task.

The implications are extraordinary. When the brain shifts from its resting state- where ϕ-scaled frequencies ensure maximal independence between bands- to a state of focused attention or meditative absorption, the frequency bands *slide* from their irrational ϕ-scaled positions toward rational integer ratios (2:1, 3:1), enabling phase synchronization and cross-frequency coupling. The frequencies are close enough to rational multiples that they can slide towards rational multiples to achieve phase synchrony or phase-locking, thereby enabling energy transfer via resonance or positive interference.

The resting brain uses ϕ to *prevent* synchronization- to maintain maximal informational independence between frequency bands. The attending brain uses the *proximity* of ϕ-scaled frequencies to integer ratios to *enable* targeted synchronization when needed. ϕ is the default state of the conscious brain: a maximally de-coupled, maximally flexible, maximally ready configuration from which any pattern of intentional coupling can be achieved.

This is the neural correlate of State 0. The resting, ϕ-organized brain is the biological equivalent of the quantum vacuum: not empty, not random, but maximally *potent*- a state of infinite readiness from which any specific cognitive act can emerge. And the transition from rest to attention- from ϕ-desynchronized to integer-synchronized- is the neural equivalent of a virtual particle becoming real: a specific pattern crystallizing from the infinite potential of the ground state.

ϕ is the frequency of readiness. It is the number the brain hums when it is *between thoughts*- when it is in the generative void, the creative silence, the state from which all intention arises.

.

Now we can state what ϕ is, in the ontological framework of this book.

ϕ is not a number. It is a *principle*. It is the mathematical operator of **self-reference**: the principle by which a system folds back upon itself to produce structure at every scale without ever exactly repeating. It is the generator of fractal self-similarity- the reason the heartbeat echoes the galaxy, the reason the DNA helix mirrors the cosmic web, the reason the sunflower seed spiral encodes the same geometry as the Penrose tiling.

In the language of the Vedic tradition, ϕ corresponds to *Ṛṣi*- the seer, the knower, the aspect of consciousness that reflects back upon itself and recognizes its own nature. Ṛṣi is not an entity; it is a *function*- the function of self-awareness, of recursive self-observation, of the cosmic quanta-field folding back upon its own structure to produce the phenomenon of "knowing." When you close your eyes and become aware of your own awareness- when you notice that you are noticing- you are performing the operation of ϕ. You are squaring yourself and finding yourself plus one.

In the category-theoretic framework introduced in Chapter 11, ϕ governs the KZ-Monad- the functor T that extends any partial resonance into a fully embedded, self-similar, holographic manifold. The monadic extension $T(X)$ takes a local state X- a fragment, a partial pattern, a fleeting thought- and embeds it into the full ϕ-scaled structure of the nine-dimensional harmonic manifold. It "reaches for wholeness": it extends the part until it contains the whole, just as every power of ϕ contains ϕ itself, just as every patch of a Penrose tiling contains the geometry of the entire infinite tiling.

The equation that defines ϕ- $\phi^2 = \phi + 1$- is the simplest possible expression of this monadic extension. The square of the state (the extension, the reaching outward) equals the state itself plus one (the return to self, with something added- with *growth*). Every act of self-reference is an act of expansion. Every act of self-knowing is an act of creation.

There is a deeper mathematical layer to the role of ϕ in the tuning of the cosmos, and it concerns the geometry of the statistical manifold on which the quanta-field evolves.

Recall from the source framework: the Schur-Convex Curvature of the Fisher-Rao statistical manifold- the curvature that measures how "efficiently" a system distributes its quanta across possible states- reaches its minimum at a specific value:

$$q^* = \phi^{-2} \approx 0.382$$

This is not an arbitrary minimum. It is a *global* minimum- the point of maximum statistical efficiency, the configuration in which the quanta-field distributes its distinctions most harmoniously across all available degrees of freedom. At $q^* = \phi^{-2}$, the manifold is *flattest*- the gradients are smallest, the flow is smoothest, the system is closest to what we might call perfect geometric harmony.

But "closest" is not "at." The curvature reaches a minimum- it does not reach zero. The manifold is never perfectly flat. There is always residual curvature, residual tension, residual gradient. And this residual is the δ_∞ gap:

$$\delta_\infty = \frac{\pi - e}{\pi \phi^4} \approx 0.01965$$

The system *aspires* to the ϕ-organized minimum. It *approaches* it asymptotically. It never arrives. And this eternal approach- this asymptotic aspiration toward perfect ϕ-coherence that can never be fully achieved- is the dynamic drive of the cosmos. The universe is always *becoming* ϕ, always self-referencing, always reaching for the state of perfect recursive self-similarity, and always falling just short- by precisely two percent.

That two percent is not wasted. It is the *fuel.* It is the residual tension that keeps the quanta-field oscillating, keeps the boundary updating, keeps the holographic projection refreshing, keeps time flowing. Without the gap, the universe would reach the ϕ-minimum and stop. It would achieve perfect self-reference- perfect stillness- and die. The gap keeps it *alive.*

ϕ is the destination. δ_∞ is the distance that can never be closed. And the journey between them- the eternal, asymptotic, creative approach- is what we experience as existence.

Let us pause here and feel what we have established.

Place your hand over your heart. Feel the beat. That beat is fractal- self-similar across timescales- because ϕ governs the geometry of complex adaptive systems at every scale. The fractal variability of the healthy heartbeat is a direct expression of ϕ-organized dynamics: the heart is a system poised at the edge of chaos, where the most irrational ratio prevents lock-in, prevents rigidity, prevents the death of mechanical repetition.

Now attend to your breath. Inhale. Hold. Exhale. This ternary rhythm- the three-phase oscillator that sustains you- echoes the three-fold structure of the cosmos: ϕ, e, π. But it is ϕ that sets the default tone. When you are at rest- when you are not thinking any specific thought, not attending to any specific task, not doing anything at all- the electrical frequencies in your brain are organized in a ϕ-scaled geometric series. Your default state is ϕ. Your resting mind hums at the golden ratio.

This is not a metaphor. It is an EEG measurement. And it means that the ancient contemplative traditions- which teach that the ground of consciousness is a still, alert, infinitely potent *readiness*- are describing a physical state whose mathematical signature is ϕ.

When the meditator enters samadhi- when the Zen monk achieves kensho- when the Aboriginal elder sinks into the Dreamtime- the brain does not go silent. It reorganizes. The ϕ-scaled frequencies shift. Phase-locking increases. Cross-frequency coupling tightens. And what the contemplatives have described for millennia as "the Self recognizing itself" - the moment of cosmic self-reference, the universe folding back upon its own awareness- is, in the language of this book, the activation of the ϕ-operator in the neural substrate. It is the KZ-Monad firing: partial resonance extending to full holographic coherence. It is $\phi^2 = \phi + 1$, enacted in the living tissue of the human brain.

The ancients did not need EEG machines to discover this. They had something better: sixty-five thousand years of disciplined, intergenerational, embodied practice. They discovered ϕ not by measuring it but by *becoming* it- by sitting still, breathing evenly, chanting the resonant syllable, and allowing the brain to settle into its natural, golden, self-referential ground state.

They called it *Tat tvam asi*- "Thou art That." You are the thing that knows itself.

ϕ is the mathematics of *Tat tvam asi.*

We have now surveyed ϕ across seven domains:

1. **Pure mathematics:** $\phi^2 = \phi + 1$. The self-referential number whose powers generate the Fibonacci sequence, whose reciprocal shares its decimal expansion, whose continued-fraction representation is the simplest possible: [1;1,1,1,1,...]- an infinite ladder of ones, each step referencing the one before.

2. **Molecular biology:** The DNA double helix, structured in ϕ ratios- 34 angstroms long, 21 angstroms wide per turn, decagonal cross-section, pentagonal base-pair geometry- encoding the genetic blueprint of life in the language of self-referential coherence.

3. **Botany:** Phyllotaxis, the golden angle of 137.5°, the Fibonacci spiral of the sunflower- the most efficient packing of seeds on a disk, achieved through the most irrational divergence angle, preventing periodic alignment at every scale.

4. **Crystallography:** Quasicrystals and Penrose tilings- aperiodic, non-repeating, ϕ-governed structures that exhibit five-fold symmetry and long-range order without periodicity, representing a distinct phase of matter.

5. **Gravitational physics:** The golden ratio at the phase-transition threshold of Kerr-Newman black holes, and in the turning points of null geodesics in the Schwarzschild-Kottler metric- ϕ governing the critical boundaries of gravitational systems.

6. **Neuroscience:** The ϕ-scaled geometric series of EEG frequency bands, ensuring maximal desynchronization at rest and enabling targeted synchronization during attention- the golden ratio as the brain's default operating frequency.

7. **Contemplative traditions:** Samadhi, kensho, Dreamtime- states of self-referential awareness whose neural signatures show enhanced ϕ-organized coherence- the ancient discovery of the golden operator through embodied practice.

Seven domains. Seven scales. One number. And in every case, ϕ performs the same function: it generates *self-similar structure without repetition*. It creates order without rigidity. It enables coherence without lock-in. It is the mathematics of *living pattern*- pattern that persists by never settling, that coheres by never repeating, that knows itself by always returning to itself with something new.

We can now articulate the role of ϕ in the trinity that tunes the cosmos.

ϕ is the **operator of self-reference.** It governs the *form* of the quanta-field- its fractal self-similarity, its scale-invariant structure, its recursive, holographic architecture in which the whole is reflected in every part. Without ϕ, the quanta-field would have no coherence- no fractal backbone, no self-similar skeleton, no means of organizing the vast ocean of quantum fluctuations into the specific, detailed, endlessly complex structures we observe.

But ϕ alone is not enough. Self-reference without unfolding is a mirror facing a mirror- an infinite regression that goes nowhere. The universe needs a second operator to *drive* the self-referential structure forward, to transform potential into actuality, to convert the static perfection of the ϕ-lattice into the dynamic flow of cosmic becoming.

That operator is e- the number that is its own derivative, the base of natural growth and decay, the rate at which the quanta-field unfolds from the generative void into the structured multiplicity of the present moment. If ϕ is the *what*- the form of the cosmos- then e is the *how*- the rate at which it becomes.

And the third operator, π, will close the boundary- defining the holographic surface, folding the dimensions, setting the limits within which ϕ and e play out their eternal dance.

Three operators. Three notes. One chord.

But the chord never resolves. The gap-

$$\delta_\infty = \frac{\pi - e}{\pi \phi^4} \approx 0.01965$$

- ensures that the triad always *almost* closes, always *almost* achieves perfect harmonic resolution, and always, in the last two percent, falls back into the creative tension that sustains the cosmos.

ϕ is the first note. The note of self-recognition. The note that says: *I am*. And in that declaration- in that single act of recursive self-reference- the universe begins to know itself.

In the next chapter, we meet the second note: e, the operator of unfolding. The note that says: *I become*.

Chapter 8

e- The Operator of Unfolding

.

You are growing.

Not in the metaphorical sense- not "growing as a person," not "growing in wisdom." You are growing in the literal, biological, physical sense. At this very moment, your body is replacing approximately 3.8 million cells per second. The lining of your stomach is rebuilding itself every few days. Your red blood cells are being born in the marrow of your bones at a rate of two million per second, each one launching into the bloodstream to carry oxygen for approximately 120 days before being consumed by your spleen and recycled into raw material for the next generation. The keratinocytes in the base layer of your skin are dividing, pushing upward, hardening, dying, and flaking off in a continuous cascade that replaces your entire outer surface every two to three weeks.

You are not the same body you were a year ago. You are not the same body you were a month ago. You are, in the most concrete physical sense, a *process*- a pattern of continuous replacement, a river of matter flowing through a form that persists only because the rate of renewal precisely matches the rate of loss.

And the mathematics that governs this process- the equation that describes the rate at which cells divide, the rate at which radioactive isotopes decay, the rate at which heat dissipates, the rate at which populations grow, the rate at which a candle flame converts wax into light and carbon dioxide, the rate at which the cosmos itself unfolds from the primordial fireball into the intricate, fractal, conscious web of the present moment- is built on a single number.

$$e \approx 2.71828182845904...$$

Euler's number. The base of the natural logarithm. The unique number whose exponential function is its own derivative. And in this book, the second operator of the cosmic triad: the principle of *unfolding*- the mathematical law that governs how potential transforms into actuality, how the compressed seed of State 0 expands into the branching complexity of the experienced world.

If ϕ is the *form* of the cosmos- the operator of self-reference, the principle that ensures the whole is reflected in every part- then e is the *flow*. It is the rate at which that form manifests. The velocity of becoming. The heartbeat of time.

The story of e begins not in a laboratory or an observatory but in a counting house. Jacob Bernoulli discovered this constant in 1683, while studying a question about compound interest: an account starts with \$1.00 and pays 100 percent interest per year. If the interest is credited once, at the end of the year, the value of the account at year-end will be \$2.00. What happens if the interest is computed and credited more frequently during the year?

The question seems trivial. But Bernoulli, possessed of the mathematical intuition that would make the Bernoulli dynasty one of the most remarkable families in the history of science, saw something deeper. If the interest were credited twice a year, at a rate of 50% per half-year, the result would be $\$1.00 \times 1.5^2 = \$2.25$- more than \$2.00, because the interest earned in the first half-year itself earns interest in the second. If the interest were credited quarterly, at 25% per quarter, the result would be $\$1.00 \times 1.25^4 \approx \$2.4414$. Monthly: $\$1.00 \times (1 + 1/12)^{12} \approx \$2.6130$. Daily: $\$1.00 \times (1 + 1/365)^{365} \approx$ \$2.7146.

The amounts were growing, but they were growing more and more slowly. Bernoulli noticed that this sequence approaches a limit with larger n and, thus, smaller compounding intervals. Compounding weekly (n = 52) yields \$2.692596..., while compounding daily (n = 365) yields \$2.714567... The limit as n grows large is the number that came to be known as e.

The limit is:

$$e = \lim_{n\to\infty}\left(1 + \frac{1}{n}\right)^n \approx 2.71828...$$

He used the binomial theorem to show that the limit had to lie between 2 and 3, so we could consider this to be the first approximation found to e; it is the first time that a number was defined by a limiting process.

Pause here and feel the significance. A banker's question about interest led to a *transcendental constant*- a number that cannot be expressed as a ratio of whole numbers, cannot be the root of any polynomial equation with rational coefficients, and extends infinitely without repeating. By the Lindemann–Weierstrass theorem, e is transcendental, meaning that it is not a solution of any non-zero polynomial equation with rational coefficients. The universe embedded one of its deepest operating principles inside the mundane problem of how money grows.

And the principle is this: *continuous compounding produces a finite limit.* No matter how finely you divide the compounding intervals- no matter how rapidly you feed the output back into the input- the result does not explode to infinity. It converges to e. This is the first hint of e's ontological role: it is the governor of growth. The constant that *allows* continuous unfolding without catastrophic divergence. The mathematical boundary between sustainable becoming and explosive collapse.

Jacob Bernoulli may have discovered e's presence in compound interest, but it was Leonhard Euler who fully developed the concept in the early 18th century. Between 1727 and 1731, Euler formalized the properties of e and gave it its now-familiar name. Euler proved that e could be expressed as an infinite series:

$$e = \sum_{n=0}^{\infty} \frac{1}{n!} = 1 + 1 + \frac{1}{2} + \frac{1}{6} + \frac{1}{24} + \frac{1}{120} + \cdots$$

where $n!$ (n factorial) is the product of all positive integers up to n. Each successive term is smaller than the one before- dramatically so- yet they add up, infinitely, to a precise and finite value. The series converges quickly: after just ten terms, you have e correct to more than six decimal places.

But the property that transforms e from a numerical curiosity into a cosmic operator is something else entirely. It is the identity that makes calculus possible, that makes physics possible, that makes the universe's capacity for change expressible in mathematical form:

$$\frac{d}{dx}e^x = e^x$$

This property makes e^x essential for modeling processes where the rate of change is proportional to the current value. The derivative of e^x- the rate at which the function changes- is identical to the function itself. No other number has this property. If you plot e^x on a graph and ask "How fast is this curve rising at any point?", the answer is: exactly as fast as it already is. The slope of the curve at any point equals the height of the curve at that point. Growth feeds on itself. The output becomes the input. The function is its own engine.

This is self-referential, like ϕ- but in a different register. Where $\phi^2 = \phi + 1$ says "I contain myself, plus something more," the equation $\frac{d}{dx}e^x = e^x$ says "My rate of becoming is identical to my state of being." ϕ generates structure through recursive self-similarity. e generates *motion* through recursive self-amplification. ϕ is the architecture. e is the pulse.

.

Every process in the physical universe that involves continuous change- where the rate of change at any moment depends on the current state- is governed by e. This is not an approximation. It is an exact mathematical fact, rooted in the structure of differential equations.

Consider the simplest case: exponential growth. A population of bacteria doubles every hour. If you start with 100 bacteria, after one hour you have 200. After two hours, 400. After three, 800. The growth curve is:

$$N(t) = N_0\, e^{rt}$$

where N_0 is the initial population, r is the growth rate, and t is time. The exponential growth model, $P(t) = P_0e^{rt}$, became widely used to describe population growth under ideal conditions.

Now consider the opposite: exponential decay. In physics, e became useful for radioactive decay models, where the number of remaining particles follows $N(t) = N_0e^{-\lambda t}$. A radioactive isotope sheds half its atoms in a fixed period- the half-life. Carbon-14, used to date archaeological sites, has a half-life of 5,730 years. Uranium-238, used to date the age of the Earth itself, has a half-life of 4.47 billion years. In every case, the number of undecayed atoms follows the same exponential curve, governed by the same constant: e.

The key insight is that growth and decay are not two different processes. They are the *same* process- the exponential function e^{kt}- with the sign of k determining the direction. Positive k: unfolding, expansion, proliferation. Negative k: folding, contraction, dissipation. The symmetry is perfect. e governs becoming in both directions- the blooming of a flower and the cooling of a corpse, the ignition of a star and the decay of a nucleus, the spread of an idea through a population and the fading of a memory in a brain.

This bilateral symmetry is not incidental. It reflects a deep property of the exponential function: e^x and e^{-x} are related by inversion. Together, they form the basis of the hyperbolic functions- $\cosh(x) = (e^x + e^{-x})/2$ and $\sinh(x) = (e^x - e^{-x})/2$- which describe the geometry of hyperbolas, the trajectories of objects in gravitational fields, and the shape of hanging chains. Growth and decay are two halves of a single geometric form, and e is the hinge between them.

But e does not merely govern the rate of change. It governs the *direction* of change- the path of steepest descent on the landscape of possibility. And this is where e enters the heart of the book's mathematical framework: the Fisher-Rao metric and the natural gradient.

In information geometry, the Fisher information metric is a particular Riemannian metric which can be defined on a smooth statistical manifold, i.e., a smooth manifold whose points are probability distributions. It can be used to calculate the distance between probability distributions.

Imagine a vast landscape- not a physical landscape but a *statistical* landscape, a manifold whose every point represents a different probability distribution, a different way the quanta-field could be configured. Some points are far apart (very different distributions, very different possible worlds). Some are close together (similar distributions, slight variations on the same theme). The Fisher-Rao metric defines the "distance" between these points- the amount of distinguishable quanta that separates one configuration from another.

Now ask: if a system is evolving- if the quanta-field is unfolding from one configuration toward another- what is the fastest route? What is the direction of maximal change, measured not in the crude coordinates of parameter space but in the intrinsic geometry of the statistical manifold?

When a parameter space has a certain underlying structure, the ordinary gradient of a function does not represent its steepest direction, but the natural gradient does. The answer is the *natural gradient*- the gradient computed not with respect to the Euclidean metric (which treats all directions as equal) but with respect to the Fisher-Rao metric (which accounts for the curvature and structure of the manifold). The natural gradient is the direction of steepest descent on the statistical landscape, and it is given by:

$$\tilde{\nabla}\mathcal{L} = F^{-1}(\theta)\nabla_{\theta}\mathcal{L}$$

where $F(\theta)$ is the Fisher information matrix- the local curvature of the manifold at point θ- and $\nabla_{\theta}\mathcal{L}$ is the ordinary gradient of the loss function.

And e is everywhere in this structure. The Fisher information matrix is defined in terms of the *logarithm*- the inverse of the exponential- of the probability distribution. The natural gradient descends along the e-scaled geometry of the manifold. The "distance" between probability distributions, measured by the Kullback-Leibler divergence, is expressed in *nats*- natural units of quanta, defined using the natural logarithm (base e), not the binary logarithm (base 2).

In this framework, e is not just the rate of change. It is the *metric of change*- the ruler by which the cosmos measures the distance between one state and the next. When the quanta-field evolves from a less coherent configuration toward a more coherent one- when a dissipative structure self-organizes, when an organism develops from a fertilized egg, when a galaxy coalesces from a cloud of hydrogen- it does so along the natural gradient of the Fisher-Rao manifold, and the geometry of that descent is written in e.

This is why the source framework identifies e as the operator that appears in the *natural gradient flow* of the master equation:

$$\partial_t \Psi = -\nabla_{FR}[\mathcal{L}(\Psi)] + T(\Psi) \otimes D(\Psi)$$

The first term- $-\nabla_{FR}[\mathcal{L}(\Psi)]$- is the natural gradient descent, the Fisher-Rao-corrected flow that drives the state Ψ toward lower curvature, toward greater efficiency, toward the ϕ-organized minimum at $q^* = \phi^{-2}$. And the metric that defines this descent- the ruler that measures the distance between states- is e-scaled. The unfolding of the cosmos is not random drift. It is *natural gradient descent on a ϕ-tuned manifold, measured in units of e.*

Now we descend from the abstract geometry of statistical manifolds into the living, breathing, far-from-equilibrium world of dissipative structures- the realm of Ilya Prigogine.

Russian-Belgian physical chemist Ilya Prigogine, who coined the term dissipative structure, received the Nobel Prize in Chemistry in 1977 for his pioneering work on these structures, which have dynamical regimes that can be regarded as thermodynamic steady states.

Prigogine's insight was revolutionary and, in many circles, still controversial. Classical thermodynamics- the thermodynamics of the 19th century- was fundamentally a science of *equilibrium*. It described the end-states of systems: how much heat, how much work, how much entropy, once everything has settled down. Its central law- the Second Law, the arrow of entropy- seemed to decree that all systems inevitably run down, wind down, spread out, and die. Order dissolves into disorder. Structure melts into homogeneity. The universe, in the long run, is a cooling corpse.

Prigogine saw something entirely different. He saw that the Second Law applies to *isolated* systems- systems sealed off from their environment, exchanging no energy, no matter, no quanta with the outside world. But living systems, planetary systems, galactic systems- virtually everything of interest in the cosmos- are *open* systems. They are not sealed. They are drenched in flows of energy and matter, constantly exchanging quanta with their surroundings.

Prigogine pointed out that nonequilibrium may become a source of order and that irreversible processes may lead to a new type of dynamic states of matter called "dissipative structures." It will generate organization and coherence spontaneously in the open system and be called self-organization.

This was a transformation of the deepest kind. The Second Law did not merely allow for the creation of order in far-from-equilibrium systems- it *required* it. Some far-from-equilibrium systems with nonlinear relations between forces and flows can develop processes that are structured in space (e.g., fluid convection rolls, chemical patterns) or time (e.g., oscillating chemical reactions) that persist through dissipative processes. The development of structures in these systems tends to coincide with an increased rate of entropy production required for the maintenance of such structures.

Dissipative structures survive not *despite* entropy production but *because of* it. They are the universe's way of *accelerating* the flow of entropy- of producing more differentiation of quanta, faster- by organizing matter and energy into coherent, complex, self-maintaining patterns. A whirlpool in a draining bathtub dissipates more energy than the same water flowing in a laminar sheet. A living cell dissipates more energy than the same chemicals sitting in a beaker at equilibrium. A forest dissipates more solar energy than the same patch of bare rock. In every case, the dissipative structure exists because it is a *more efficient pathway for entropy production*- a more creative way of transforming coherent energy into dispersed heat.

And the mathematical signature of dissipative structures is *exponential.* They grow exponentially when conditions are right: populations, chemical oscillations, convection cells. They decay exponentially when driven below their maintenance threshold. They fluctuate at rates characterized by exponential correlation functions. Their stability is analyzed using Lyapunov exponents- measures of exponential sensitivity to perturbation. The entropy S is a Lyapunov function for isolated systems. The entire apparatus of non-equilibrium thermodynamics is built on the mathematics of the exponential function- on e.

At far from equilibrium these new states of order increase the production of entropy. In this way, the level of complexity rises as the generation of entropy, or dissipation, increases, hence the name dissipative structures.

This is e in its fullest expression: not merely the rate at which things grow or decay, but the rate at which the cosmos *creates.* Dissipative structures- from hurricanes to heartbeats, from convection cells to civilizations- are the universe's instruments for converting the potential of the quanta-field into actual, experienced, structured complexity. And the mathematics of that conversion, at every level, is the mathematics of the exponential function.

e also governs the most intimate temporal experience known to the human nervous system: the perception of time itself.

Consider the psychophysical law known as Weber-Fechner: the perceived intensity of a stimulus (brightness, loudness, weight) is proportional to the *logarithm* of the physical intensity. To double the perceived loudness of a sound, you must roughly quadruple its physical energy. To double the perceived brightness of a light, you must roughly quadruple its luminous flux. The relationship is:

$$\text{Perception} \propto \ln(\text{Stimulus})$$

where ln is the natural logarithm- the logarithm base e.

The nervous system does not experience the world linearly. It experiences it *logarithmically*- in units scaled by e. This is not a quirk of neurobiology. It is an optimal coding strategy: logarithmic compression maximizes the range of stimuli that can be encoded by a neural system with finite dynamic range. It allows the same set of neurons to respond meaningfully to a whisper and a thunderclap, to a candle and the sun.

And what is the logarithm? It is the *inverse of the exponential.* The function $\ln(x)$ asks: "What power must e be raised to in order to produce x?" The nervous system encodes the world by *inverting the exponential unfolding of the quanta-field.* It takes the exponentially distributed intensities of the physical world and maps them onto a linear perceptual scale. It translates e-space into lived experience.

Time perception follows the same law. As we age, time appears to accelerate- a year at age fifty seems much shorter than a year at age ten. This is because the brain encodes temporal duration logarithmically: each new unit of time is judged relative to the total time already experienced. The subjective midpoint of a seventy-year life is not thirty-five but approximately twenty- because $\ln(70)/2 \approx \ln(20)$. The sense that "time flies" as we age is not an illusion. It is the logarithmic compression of an e-governed temporal field, experienced from within.

The source framework's identification of e with the operator of *Devatā*- the Vedic principle of the process of knowing, the dynamic unfolding of experience between the knower and the known- is precisely this. e is not merely the base of the natural logarithm. It is the rate at which the cosmos unfolds *for a conscious observer.* It governs the speed of becoming as experienced from within the holographic projection. The clock of the exterior universe ticks in π-bounded cycles (we will meet this in the next chapter). But the *felt* passage of time- the subjective sense of duration, the phenomenology of change- is scaled by e.

When you sit in meditation and the minutes seem to expand- when a single breath seems to encompass an eternity of stillness- you are experiencing a shift in the e-scaling of your temporal perception. The logarithmic compression loosens. The rate of subjective unfolding slows. And in the deepest states of absorption- in what the contemplatives call *turīya,* the "fourth state" beyond waking, dreaming, and deep sleep- the e-scaling may approach zero. Time stops. Not because the clock stops ticking, but because the observer's natural logarithm has expanded to encompass the infinite.

There is a moment in the history of mathematics that stands as a monument to the deep unity of e, π, and the imaginary unit i- and it is, not coincidentally, the most celebrated equation in the history of human thought.

All five appear in one formulation of Euler's identity $e^{i\pi} + 1 = 0$ and play important and recurring roles across mathematics.

$$e^{i\pi} + 1 = 0$$

This equation- Euler's identity- connects five of the most fundamental constants in mathematics: e (the base of natural growth), π (the ratio of circumference to diameter), i (the imaginary unit, $\sqrt{-1}$), 1 (the unit of arithmetic), and 0 (the void, the additive identity). It does so in a single, breathtaking line. No other equation links so many seemingly unrelated domains- exponential functions, trigonometry, complex analysis, arithmetic, and the concept of nothingness- with such economy.

For our purposes, the identity reveals something deeper: e and π are not independent constants. They are *coupled*. The exponential function e^{ix}, when given a purely imaginary argument, does not explode to infinity or collapse to zero. It *rotates*. Euler's formula, the generalization of the identity, states:

$$e^{ix} = \cos(x) + i\sin(x)$$

The exponential function, when extended to the complex plane, becomes a *rotation*. Growth (e) becomes periodicity (π). The line becomes a circle. The arrow becomes a cycle.

This is not an accident. It is the mathematical expression of the deepest relationship in the cosmic triad. e drives the unfolding- the forward motion, the growth, the becoming. π closes the boundary- the cycle, the return, the holographic surface. And when they act together- when unfolding and closure are combined in the complex exponential- the result is *oscillation*: the fundamental dynamics of every wave, every field, every vibrating quantum in the cosmos.

Every photon is described by a complex exponential: $e^{i(\omega t-kx)}$. Every electron orbital is described by a complex exponential. Every quantum field mode in the vacuum fluctuates as $e^{-i\omega t}$. The Schrödinger equation evolves the wave function in time via the operator $e^{-iHt/\hbar}$. The fabric of quantum mechanics is woven from the union of e and π in the complex plane.

ϕ generates the self-similar structure. e drives the unfolding. π closes the boundary. And Euler's identity shows that e and π, when coupled through the imaginary unit, produce the fundamental oscillation- the wave- that is the irreducible unit of physical reality.

The triad does not merely *describe* the cosmos. It *generates* the cosmos. The constants are not numbers found in nature by passive observers. They are *operators*- active, formative, generative- and their interaction produces the standing wave we call existence.

We can now place e in its full ontological context, alongside ϕ, and prepare the ground for π.

ϕ, the operator of self-reference, generates the *architecture* of the cosmos: the fractal scaffolding, the golden-ratio packing, the self-similar structure at every scale, the recursive return that makes the whole visible in every part. Without ϕ, there would be no coherence- no pattern for the quanta to organize into, no fractal backbone for the holographic projection, no means of connecting the Planck scale to the Hubble scale.

e, the operator of unfolding, generates the *dynamics* of the cosmos: the rate of growth and decay, the natural gradient descent on the Fisher-Rao manifold, the exponential curves of population growth and radioactive decay, the logarithmic scaling of perception and time, the Prigoginian cascade of dissipative structures that converts potential into actuality. Without e, the ϕ-architecture would be static- a perfect, frozen lattice of self-similar structure with no means of *becoming*, no engine of transformation, no way to convert the compressed potential of State 0 into the expanded complexity of the present moment.

Together, ϕ and e are the architect and the pulse. The form and the flow. The spatial principle and the temporal principle. The breath and the body.

But they need a third. They need a *boundary*. Without a limit, e-driven growth would be unbounded- exponential divergence to infinity. Without a surface, ϕ-structured fractals would have no holographic screen from which to project. The cosmos needs a closure operator- a principle that folds dimensions, seals boundaries, defines the holographic surface on which the quanta are inscribed.

That operator is π. The ratio that closes every circle. The constant that appears in every holographic entropy formula, every boundary condition, every dimensional folding. The third note in the chord.

And the chord never resolves. Because the three operators- ϕ, e, π- do not close into perfect harmony. They *almost* close. The gap between them- the asymptotic residual-

$$\delta_\infty = \frac{\pi - e}{\pi \phi^4} \approx 0.01965$$

- is a function of all three operators, and it is the reason the cosmos does not die. π and e are close but not identical ($\pi - e \approx 0.4233$). This difference, normalized by the ϕ-scaled boundary ($\pi\phi^4$), yields a residual tension of approximately two percent- the irreducible incompleteness that keeps the natural gradient from reaching the ϕ-minimum, that keeps the exponential unfolding from ever settling into perfect equilibrium, that keeps the holographic projection from ever freezing into a static photograph.

The gap is the distance between the rate of unfolding (e) and the closure of the boundary (π), measured against the self-referential architecture (ϕ). It is the permanent mismatch between how fast the cosmos becomes and how completely it can contain itself. It is the reason there is always more to unfold, always further to go, always a next heartbeat after this one.

Consider the most intimate experience of e-governed dynamics you will ever have.

You are breathing.

Not because you decided to. You were breathing before you started reading this sentence, and you will be breathing long after you forget it. The autonomic rhythm of respiration is governed by brainstem nuclei that fire in a cyclic pattern- but the *rate* of that cycle, the depth and speed and variability of each breath, is modulated by exponential processes at every level.

Inhale: the diaphragm contracts, the thoracic cavity expands, air rushes in through the trachea and down into the fractal branching of the bronchial tree. The rate of air intake follows an exponential approach to maximum lung volume: fast at first, then decelerating as the lungs fill. The curve is $V(t) = V_{\max}(1-e^{-t/\tau})$, where τ is the time constant of the respiratory system.

Hold: the breath pauses at the top. The system hangs at a moment of maximum potential- the lungs full, the oxygen diffusing across the alveolar membrane into the blood, the metabolic rate temporarily elevated. This is the moment of maximum quanta-density in the respiratory cycle- the moment where the system is farthest from equilibrium, most pregnant with potential.

Exhale: the diaphragm relaxes, the thoracic cavity contracts, air rushes out. The rate of outflow is exponential decay: $V(t) = V_{\max}\, e^{-t/\tau}$, fast at first, then slowing as the lungs deflate. The same constant. The same curve. The same e.

Inhale- hold- exhale. Growth- potential- decay. $e^{t/\tau}$- maximum- $e^{-t/\tau}$. The breath is a ternary oscillator, a three-phase cycle, and the mathematics of each phase is the exponential function. The breath *is* e, performed by the body, ten to twenty times per minute, from the first gasp at birth to the last sigh at death.

And the breath is not merely a biological process. It is a *miniature cosmology*. The inhalation is the Big Bang- the expansion of a compressed potential into a differentiated, structure-filled space. The hold is the present moment- the far-from-equilibrium peak of complexity. The exhalation is the heat death- the gradual return to equilibrium, the release of structure, the dispersal of quanta into the environment.

Inhale: the cosmos unfolds. Hold: the cosmos creates. Exhale: the cosmos returns. And then- in the pause between the exhale and the next inhale, in the gap, in the silence- the cosmos rests at State 0, the generative void, before the next cycle begins.

The breath is not a metaphor for the cosmos. The breath *is* the cosmos, at the scale of a human life, governed by the same operator- e- that drives exponential unfolding at every other scale.

We began this chapter with the act of growing- the ceaseless replacement of cells, the river of matter flowing through the persisting form of the body. We discovered that this growth, and every other process of continuous change in the physical universe, is governed by a single constant: $e \approx 2.71828$, the base of the natural logarithm, the unique number whose exponential function is its own derivative.

We traced e from its discovery by Jacob Bernoulli, who also discovered the fundamental mathematical constant e in the counting house of compound interest, through its formalization by Euler, to its central role in the calculus that makes physics possible. We encountered its defining property- $\frac{d}{dx}e^x = e^x$, the equation that says *the rate of becoming equals the state of being*- and recognized it as the mathematical signature of self-amplifying change.

We followed e into the Fisher-Rao metric, where it defines the natural gradient- the direction of steepest descent on the statistical manifold, the path along which the quanta-field evolves toward greater coherence. We descended into the realm of dissipative structures, where e-governed exponential dynamics drive the creation of order from far-from-equilibrium chaos. We felt e in the logarithmic compression of perception and time, in the subjective experience of duration, in the expansion of moments during deep meditation. And we witnessed, in Euler's identity $e^{i\pi} + 1 = 0$, the coupling of e with π that produces oscillation- the wave- the fundamental dynamics of every quantum field.

e is the operator of unfolding. It is the Vedic *Devatā*- the process of knowing, the dynamic transformation between the knower and the known. It governs the rate at which potential becomes actual, the speed at which the compressed seed of State 0 expands into the branching complexity of the present. Without it, the ϕ-architecture of the cosmos would be frozen- a perfect but lifeless crystal of self-similar structure. With it, the cosmos breathes.

In the next chapter, we meet the third and final operator: π, the principle of boundary and closure. The number that closes every circle. The constant that defines the holographic surface from which the entire three-dimensional world is projected. The operator that says: *This far, and no further. Here the dimension folds. Here the boundary sings.*

And when the three operators meet- when ϕ provides the architecture, e provides the pulse, and π provides the boundary- the chord sounds. The chord that is the universe.

Almost perfect. Never quite closed. Alive in the gap.

Take one more breath. Feel the e-curve of the inhale- the exponential rush of air into the branching tree of your lungs. Feel the pause at the top- the moment of maximum potential. Feel the e-curve of the exhale- the exponential release, the return.

And in the silence before the next breath: feel the cosmos, resting at State 0, preparing to unfold again.

It has been doing this since before there were lungs to breathe or minds to count.

And the rate has always been e.

Chapter 9

π- The Operator of Boundary Projection

Draw a circle.

Any circle. On paper, in sand, in your mind. Let the pencil sweep around and return to where it began. One continuous, curving line, closing upon itself, enclosing a region, creating an inside and an outside where before there was only undifferentiated plane.

You have just performed the most primordial act in the geometry of the cosmos. You have created a boundary.

The boundary does not merely separate two regions. It *defines* them. Before the circle was drawn, there was no "inside" and no "outside." The plane was featureless, unmarked, directionless. The circle- a single, closed curve- partitions the plane into two domains that did not exist until the boundary called them into being. The boundary is ontologically prior to the regions it encloses. The container does not hold the contents; the container *creates the possibility of contents.*

And the relationship between the boundary and the region it encloses- between the circumference and the area, between the surface and the volume- is governed, at every scale, in every dimension, by a single constant:

$$\pi \approx 3.14159265358979...$$

π is not merely the ratio of a circle's circumference to its diameter. It is the *operator of closure*- the mathematical principle that folds dimensions, seals boundaries, and defines the holographic surface from which the interior of the cosmos is projected. If ϕ generates the self-similar architecture and e drives the temporal unfolding, then π provides the *container*- the finite boundary within which the infinite complexity of ϕ and e can manifest without scattering into unbounded chaos.

Without π, there would be no event horizons, no holographic screens, no spherical harmonics of the cosmic microwave background, no closed orbits of electrons or planets, no wavefunction normalization, no Fourier analysis, no quantum mechanics. Without π, the boundary would not exist. And without the boundary, there would be nothing to project from- and therefore, nothing at all.

π is the oldest mathematical constant. Civilizations have been approximating it for at least four thousand years. The Babylonians used $\pi \approx 3.125$. The Egyptians used $\pi \approx 3.16$. Archimedes, in the third century BCE, trapped it between $3\frac{10}{71}$ and $3\frac{1}{7}$ by inscribing and circumscribing ninety-six-sided polygons around a circle- one of the earliest recorded examples of the limit process that would later become calculus.

By the eighteenth century, Euler had given it the symbol by which we know it, and the great mathematical machine of the Enlightenment had begun to reveal the astonishing depth of this seemingly simple ratio. π was proved to be irrational by Johann Heinrich Lambert in 1761- it cannot be expressed as a ratio of whole numbers. It was proved to be transcendental by Ferdinand von Lindemann in 1882- it is not the root of any polynomial equation with rational coefficients. This transcendence proof settled the ancient problem of squaring the circle: it is impossible, in principle, to construct a square with the same area as a given circle using only a compass and straightedge.

π is not merely irrational. It is *transcendental*- it transcends the algebraic structures that govern ordinary numbers. It belongs to a higher mathematical order, alongside e, and this shared transcendence is not a coincidence. It reflects a deep structural kinship between the two operators- a kinship that Euler's identity $e^{i\pi} + 1 = 0$ renders explicit.

But where e is the constant of *becoming*- of growth, flow, and transformation- π is the constant of *being*- of containment, completion, and return. e is the arrow. π is the circle. e opens. π closes. And the cosmos requires both, in dynamic tension, to exist.

The deepest physical significance of π is not in the geometry of circles drawn on paper. It is in the geometry of *boundaries* drawn in spacetime- the event horizons of black holes, the cosmological horizon of the observable universe, the holographic surfaces from which the three-dimensional world is projected.

Recall the Bekenstein-Hawking entropy formula from Chapter 2:

$$S_{BH} = \frac{k_B c^3}{4G\hbar} A$$

For a Schwarzschild black hole of mass M, the area of the event horizon is $A = 16\pi(GM/c^2)^2$. Notice π. It is there not as a decoration or a unit conversion. It is there because the event horizon is a *sphere-* a closed, two-dimensional surface embedded in three-dimensional space- and the area of a sphere is defined by π. For a Schwarzschild black hole, the horizon area is:

$$A = 4\pi r_s^2 = 4\pi \left(\frac{2GM}{c^2}\right)^2 = \frac{16\pi G^2 M^2}{c^4}$$

and the entropy becomes:

$$S_{BH} = \frac{4\pi k_B G M^2}{\hbar c}$$

π enters the entropy of a black hole because the entropy is defined on a *boundary-* and every boundary in a three-dimensional space is a surface whose area is measured in units of π. The holographic principle tells us that the quanta-content of a region is encoded on its boundary, and the shape of that boundary- its geometry, its area, its topology- is governed by π.

This is not limited to black holes. The Bekenstein bound- the universal limit on the entropy of any physical system- states that $S \leq \frac{2\pi k_B R E}{\hbar c}$, where R is the radius of the smallest sphere that can enclose the system and E is its total energy. There is π again- multiplying the radius and the energy, scaling the maximum quanta-content of any region by the geometry of its enclosing sphere.

The Hawking temperature of a black hole- the temperature at which it radiates- is:

$$T_H = \frac{\hbar c^3}{8\pi G M k_B}$$

π appears in the denominator, directly governing the relationship between mass and temperature. Bekenstein heuristically suggested $\frac{\ln 2}{8\pi} \approx 0.0276$ as the original constant of proportionality for the entropy-area relationship- a factor that π anchors to the geometry of the horizon.

In every formula that governs the thermodynamics of boundaries- every equation that connects the geometry of a surface to the quanta it can encode- π stands as the constant of closure. It is the mathematical expression of the fact that boundaries are *curved*, that they *close upon themselves*, that they define a finite region within an infinite space. π is how the cosmos draws its circles.

But π does not merely close circles in three dimensions. It closes boundaries in *every* dimension. And this leads to one of the most profound and counter-intuitive results in all of mathematics- a result that illuminates the role of π as a *dimensional operator* that governs how space itself is folded.

The volume of a ball in n-dimensional Euclidean space is given by the general formula:

$$V_n(R) = \frac{\pi^{n/2}}{\Gamma\left(\frac{n}{2}+1\right)} R^n$$

where Γ is the Gamma function, a generalization of the factorial function to non-integer numbers.

For familiar dimensions, this yields:

Dimension n	Volume of unit ball $V_n(1)$
1	2 (length of a line segment)
2	$\pi \approx 3.14$ (area of a disk)
3	$\frac{4\pi}{3} \approx 4.19$
4	$\frac{\pi^2}{2} \approx 4.93$
5	$\frac{8\pi^2}{15} \approx 5.26$
6	$\frac{\pi^3}{6} \approx 5.17$
7	$\frac{16\pi^3}{105} \approx 4.72$

The volume $V_n(1)$ is increasing for $0 \leq n \leq 5$, achieves its maximum when $n = 5$, and is decreasing for $n \geq 5$.

Pause at this. The volume of a unit sphere *increases* with dimension up to $n = 5$, peaks there, and then *decreases*- falling toward zero as the dimension grows without limit. For any fixed value of R, the volume tends to a limiting value of 0 as n goes to infinity.

This is deeply counterintuitive. You might expect that adding more dimensions would always provide more "room," more capacity, more volume. But the opposite is true. Higher-dimensional spheres, despite being enclosed by boundaries of increasing surface area, contain *less* volume relative to the hypercube that contains them. The sphere becomes an increasingly thin shell- almost all of its volume is concentrated near the surface, and the interior becomes vanishingly empty.

And notice what happens at $n = 5$: the volume of the unit ball is $\frac{8\pi^2}{15} \approx 5.264$. The source framework identifies the expression $V = \frac{\pi^5}{120} r^5$ as a formula of special significance- the volume of a 10-dimensional ball (where $\Gamma(6) = 120$)- reinterpreted not as Euclidean volume but as the *density of uncertainty* in a statistical manifold of oscillators. The power of π in the numerator- π^5 for 10 dimensions- governs how many distinguishable configurations can be packed into a given region of the quanta-field's state space.

This is π as a *dimensional operator*. Each power of π in the volume formula represents a *folded dimension*- a degree of freedom that has been curled, compactified, enclosed by a boundary. The $\pi^{n/2}$ in the numerator counts the angular capacity of the space- how many directions are available at each point. And the factorial in the denominator (via the Gamma function) represents the combinatorial cost of distinguishing those directions- the price of differentiation.

As the number of dimensions increases, the angular capacity ($\pi^{n/2}$) grows, but the combinatorial cost ($\Gamma(n/2 + 1)$) grows faster. The result is that the "useful" volume- the region where quanta can be meaningfully distinguished- *shrinks*. Higher-dimensional spaces, paradoxically, become more constrained, not less. The boundary, governed by π, tightens around the interior as dimensions are added.

This is precisely the role that π plays in the holographic cosmos. The holographic principle says that the quanta-content of a region is limited by the area of its boundary, not its volume. As you add dimensions to the bulk- as you move from a three-dimensional world to a four-dimensional or five-dimensional one- the boundary's capacity does not keep pace. The π-governed area of the boundary becomes the bottleneck, and the interior degrees of freedom that can be independently specified are *reduced*, not expanded.

π is the operator that enforces this constraint. It is the mathematical gatekeeper of the holographic boundary, ensuring that volume never outruns surface, that bulk never exceeds boundary, that the three-dimensional projection never contains more quanta than the two-dimensional screen from which it is cast.

.

The role of π as the boundary operator extends beyond the spatial geometry of event horizons and hyperspheres. It governs the temporal structure of the cosmos as well- specifically, the harmonic decomposition of the oldest light in the universe: the cosmic microwave background.

The CMB- the afterglow of the Big Bang, cooled from incandescent plasma to a faint microwave hum at 2.725 Kelvin- is the most perfect blackbody spectrum ever measured. But it is not perfectly uniform. Imprinted upon it are tiny temperature fluctuations- variations of about one part in a hundred thousand- that encode the seeds of all the structure in the present-day universe: every galaxy, every star, every planet, every reader of this book.

These fluctuations are decomposed into spherical harmonics- the natural mathematical basis for patterns on the surface of a sphere- labeled by a multipole number ℓ. Low ℓ values correspond to large-angle patterns on the sky (the quadrupole at $\ell = 2$ spans 90 degrees); high ℓ values correspond to small-angle patterns (the first acoustic peak at $\ell \approx 200$ spans about one degree). The angular power spectrum C_ℓ- the amplitude of the fluctuations at each angular scale- is the most information-rich observable in all of cosmology.

And at the largest angular scales- the lowest multipoles- there is an anomaly.

The most longstanding of these is the low-ℓ multipole controversy. Even in the COBE map, it was observed that the quadrupole ($\ell = 2$) has a low amplitude compared to the predictions of the Big Bang. The 2018 Planck data shows that the lack of correlation at large-scale CMB anisotropy remains persistent- the CMB power spectrum below $\ell \lesssim 30$ exhibits deviation from the standard ΛCDM model.

The observed power at $\ell = 2,3,4,5$ is significantly lower than the standard inflationary model predicts. These include missing correlations at large angles and an unexpectedly low power in multipoles $2 \lesssim \ell \lesssim 30$. The angular correlation function $C(\theta)$ is nearly zero for angles greater than about 60 degrees- in significant tension with inflation, which should have produced correlations throughout the visible Universe.

This anomaly has persisted through three generations of satellite missions- COBE, WMAP, and Planck- each confirming the same unexplained deficit. All three major satellite missions designed to study the cosmic microwave background- COBE, WMAP, and Planck- have uncovered several unexpected features in its temperature fluctuations.

While such deviation is not yet statistically significant, and the mainstream conviction is that this is caused by the cosmic variance, the possibility of having this low-mode anomaly induced by an underlying physics should not be brushed aside prematurely.

Why does this matter for π? Because the spherical harmonics in which the CMB is decomposed are *built from* π. The angular eigenfunctions of the Laplacian on a sphere- the $Y_\ell^m(\theta,\phi)$ functions- are periodic in the azimuthal angle ϕ with period 2π, and their polar structure is governed by associated Legendre polynomials whose nodes and zeros are determined by π-periodic functions. Every multipole ℓ is, in essence, a π-harmonic: a standing wave on the boundary of the observable universe, quantized by the geometry of the sphere, constrained by the closure operator.

The suppression of power at low ℓ means that the largest-scale π-harmonics- the modes that correspond to the most global, most boundary-spanning oscillations of the holographic screen- are *quieter than expected.* The cosmos, at its grandest scale, carries less vibrational energy in its lowest-frequency boundary modes than the standard model predicts.

In the framework of this book, this is not a statistical fluke. It is a *tuning signature.* The low-ℓ suppression is the π-operator at work: constraining the lowest harmonics, damping the most global oscillations, preventing the boundary from vibrating with so much large-scale power that the projected bulk would lose its fractal detail. If the $\ell = 2$ mode were at its predicted amplitude, the entire sky would be dominated by a single quadrupolar pattern, drowning out the fine-scale structure that encodes galaxies and consciousness. The suppression keeps the boundary *quiet at the bass end*, preserving room for the higher harmonics- the fine-grained, fractal, ϕ-structured patterns- to project the detailed cosmos we inhabit.

π does not merely close the boundary. It *tunes* the boundary- selecting which modes are allowed, which frequencies are amplified, which oscillations are suppressed. The spherical harmonics are the natural "voice" of the holographic screen, and π is the conductor, shaping the spectrum to produce the cosmos-as-symphony rather than the cosmos-as-noise.

There is an even deeper layer to the role of π in physics, and it concerns the most fundamental operation in quantum mechanics: the wave.

Every quantum field, every particle, every photon, every electron is described by a wave function- a complex-valued amplitude that oscillates in space and time. The mathematical form of a plane wave is:

$$\Psi(x,t) = A\, e^{i(kx-\omega t)}$$

where k is the wave number ($k = 2\pi/\lambda$, with λ the wavelength), ω is the angular frequency ($\omega = 2\pi f$, with f the frequency), and A is the amplitude. Notice: 2π appears twice- in both the spatial and the temporal oscillation. Every wave, every quantum, every field mode in the universe carries π in its definition.

This is not an arbitrary convention. The factor of 2π arises because waves are *periodic*- they repeat after one full cycle- and the measure of a full cycle, the circumference of the unit circle in the complex plane, is 2π. The exponential $e^{i\theta}$ traces the unit circle as θ increases from 0 to 2π; completing one full revolution and returning to the starting point. The 2π in the wave number and the angular frequency is the mathematical expression of *closure*: the wave completes a cycle and returns to its initial state.

Euler's formula makes this explicit:

$$e^{i\theta} = \cos\theta + i\sin\theta$$

The exponential function- governed by e- when given a purely imaginary argument scaled by π- produces *rotation*. The open, unbounded growth of e^x is *closed* by π into a circle. The arrow becomes an orbit. The line becomes a loop. Becoming is constrained into being.

This is the deepest expression of the relationship between e and π in the cosmic triad. e drives the exponential unfolding- the forward thrust of growth, decay, and transformation. π bends that thrust into a cycle- a wave, an oscillation, a periodic return. Without e, there would be no change. Without π, change would be unbounded- an exponential explosion with no return, no periodicity, no structure. Together, e and π produce the wave- the fundamental dynamical unit of quantum reality.

And Euler's identity- $e^{i\pi} + 1 = 0$- states that when the exponential unfolding (e) is given exactly one half-cycle of boundary rotation (π), the result is a *complete inversion*: $+ 1$ becomes -1. One more unit of arithmetic ($+ 1$), and the system returns to zero- to the generative void, to State 0, to the silence from which the next cycle can begin.

$e^{i\pi} + 1 = 0$ is not merely a beautiful equation. It is the *mathematical formula for death and rebirth*: the exponential unfolding, constrained by the boundary operator, passes through a half-cycle of inversion and returns to the origin. This is the heartbeat. This is the breath. This is the aeon of Penrose's Conformal Cyclic Cosmology, where the end of one universe ($e^{i\pi} = -1$) is conformally rescaled to become the beginning of the next ($+ 1 = 0$, the new Big Bang).

π does not merely close circles. It closes *cosmos*.

Now we can see why π is the operator that makes the holographic principle possible.

The holographic principle says that the quanta-content of a three-dimensional region is encoded on its two-dimensional boundary. But this encoding is not a simple transcription- not a one-to-one mapping from interior points to boundary points. It is a *projection*- a dimensional reduction, a compression, a folding of three dimensions onto two. And the mathematics of that folding is the mathematics of π.

Consider the simplest case: a spherical region of three-dimensional space, enclosed by a two-dimensional sphere. The interior has volume $V = \frac{4}{3}\pi r^3$. The boundary has area $A = 4\pi r^2$. The ratio of volume to area is:

$$\frac{V}{A} = \frac{\frac{4}{3}\pi r^3}{4\pi r^2} = \frac{r}{3}$$

The π cancels. But it does not disappear- it has done its work: it has *defined* both the volume and the area, establishing the relationship between them. Without π, neither quantity would exist. The cancellation is not a coincidence; it is a consequence of the fact that π governs *both* the interior and the boundary, ensuring that they are geometrically consistent- that the projection from bulk to boundary is well-defined.

In higher dimensions, π plays an even more critical role. The general formula for the volume of an n-ball involves $\pi^{n/2}$, while the surface area of the $(n-1)$-sphere involves $\pi^{n/2}$ divided by $\Gamma(n/2)$. Each additional dimension adds another factor of $\sqrt{\pi}$ to the volume- but the Gamma function in the denominator grows faster, eventually overwhelming the π-factors and causing the volume to shrink. This is the "curse of dimensionality"- the phenomenon by which high-dimensional spaces become increasingly dominated by their boundaries, with the interior becoming an ever-smaller fraction of the total.

For the holographic cosmos, this is not a curse. It is a *feature*. The holographic principle *requires* that the boundary dominate- that the surface area, not the volume, set the limit on quanta-content. And π, through its role in the volume and surface area formulas, ensures that this dominance increases with dimension. In the high-dimensional manifold of the quanta-field's state space, π guarantees that the boundary *always wins*- that the holographic screen always contains more quanta-capacity per unit measure than the bulk it projects.

π is not just the operator that closes boundaries. It is the operator that *enforces holography*. It is the mathematical mechanism by which the cosmos ensures that its three-dimensional interior can never contain more quanta than its two-dimensional surface- that the projection can never exceed its source.

There is a tradition, found across many cultures, of understanding the sacred circle as the boundary between the manifest and the unmanifest- between the formed world and the formless source from which it arises.

The Hindu *mandala*- a geometric diagram of the cosmos- is always bounded by a circle. The circle is not decoration. It is the *perimeter of manifestation*: the limit beyond which form dissolves into the infinite. Within the mandala, the intricate patterns of deities, elements, and energy channels represent the structured, differentiated cosmos. Outside the mandala, there is only the unmanifest- *Brahman*, the featureless ground of being, the State 0 of the Vedic tradition.

The Buddhist mandala works the same way: a circle encloses a sacred space within which the practitioner visualizes the structure of enlightened reality. The circle is the *container of consciousness*- the boundary that allows directed awareness to exist within the infinite field of undifferentiated potential.

The medicine wheel of the Lakota and other plains nations. The stone circles of Neolithic Europe. The circular kivas of the Ancestral Puebloans. The Round Table of Arthurian legend. In every case, the circle performs the same function: it creates a *bounded space* within which meaningful structure can exist. Outside the circle: chaos, formlessness, the infinite void. Inside the circle: order, pattern, the cosmos.

These traditions did not know the Bekenstein-Hawking formula. They did not compute the area of event horizons or the surface of hyperspheres. But they understood, through millennia of embodied practice, the same principle that π encodes mathematically: *that form requires a boundary*. That the infinite cannot manifest without first being contained. That the cosmos begins not with an explosion but with a *circle drawn in the void*- a boundary that separates the inside from the outside, the one from the many, the silence from the sound.

The drawing of the circle is the act of creation. And π is the constant that governs the drawing.

.

We can now state the full ontological identity of π within the cosmic triad.

ϕ is the operator of self-reference: it generates the fractal, self-similar architecture of the cosmos, ensuring that the whole is reflected in every part.

e is the operator of unfolding: it drives the exponential dynamics of growth and decay, governing the rate at which the quanta-field transforms from potential to actual along the natural gradient of the Fisher-Rao manifold.

π is the **operator of boundary projection**: it closes every circle, folds every dimension, defines the holographic surface from which the three-dimensional world is projected, constrains the quanta-budget of any region to the area of its enclosing surface, tunes the harmonic spectrum of the cosmic microwave background, and converts the unbounded exponential of e into the periodic oscillation of the wave.

In the Vedic framework, π corresponds to *Chhandas*- the object of knowledge, the *known*, the structured world that arises within the boundary of awareness. If ϕ is Ṛṣi (the knower, the self-referential seer) and e is Devatā (the process of knowing, the dynamic unfolding of experience), then π is the *world that is known*- the bounded, finite, structured projection that the knower perceives through the process of knowing. The observed universe. The holographic interior. The cosmos inside the circle.

Together, the three operators form a single, indivisible chord:

- ϕ provides the **architecture**- the self-similar skeleton, the fractal scaffold, the golden lattice on which everything hangs.
- e provides the **pulse**- the exponential dynamics, the natural gradient flow, the rate of becoming.

- π provides the **boundary**- the holographic surface, the closed curve, the container within which architecture and pulse can manifest as a world.

Architecture without pulse is a frozen crystal. Pulse without architecture is a formless explosion. Architecture and pulse without boundary is an infinite, dissipating scatter- a cosmos with no horizon, no surface, no limit, and therefore no projection, no bulk, no experienced world.

It takes all three. And they almost- *almost*- fit perfectly together.

Almost.

The three operators- $\phi \approx 1.618, e \approx 2.718, \pi \approx 3.14159$- are all transcendental numbers (with the exception of ϕ, which is algebraic irrational but serves as the self-referential bridge between the algebraic and the transcendental). They are incommensurable: no finite combination of rational operations can express one in terms of the others. They occupy different locations on the number line, and the distances between them are irrational.

And yet, they are intimately coupled. Euler's identity $e^{i\pi} + 1 = 0$ binds e and π through the imaginary unit. The golden ratio appears at the critical phase-transition thresholds of π-governed gravitational systems. The natural gradient of the Fisher-Rao manifold (e-scaled) descends toward the Schur-convex minimum at $q^* = \phi^{-2}$, which is the ϕ-scaled curvature minimum of a boundary whose area is π-governed.

The three operators are coupled- but they are not identical. They do not close into a single, self-consistent, gapless unity. There is a residual mismatch- a two-percent tension between the rate of unfolding (e) and the closure of the boundary (π), normalized by the self-referential architecture (ϕ):

$$\delta_\infty = \frac{\pi - e}{\pi\phi^4} \approx 0.01965$$

This gap has been with us since the Prologue. It has appeared in every chapter- in the space between heartbeats, in the Dreamtime pause between songline phrases, in the δ_∞ that keeps the holographic boundary oscillating and the projection updating. It is the asymptotic incompleteness that prevents the triad from achieving perfect closure.

$\pi - e \approx 0.4233$. The boundary operator exceeds the unfolding operator by approximately 0.42- a difference that, when normalized by the ϕ-scaled architecture ($\pi\phi^4 \approx 21.54$), yields the two-percent residual. π wants to close more than e wants to unfold. The boundary aspires to contain more than the dynamics can fill. And the permanent mismatch between the container's aspiration and the content's capacity is the *engine* of the cosmos.

The gap is not a flaw. It is the reason the cosmos is alive. It is the permanent, irreducible, transcendental *almost*- the divine discontent- that prevents the holographic projection from freezing into a static photograph, that prevents the quanta-field from settling into a dead equilibrium, that keeps the natural gradient flowing, the boundary oscillating, the CMB shimmering, the heart beating, the breath cycling, the cosmos creating.

The gap is *between* π and e- between the closure and the unfolding, between the circle and the arrow, between being and becoming. And it is measured against ϕ- the self-referential principle that bridges both.

δ_∞ is not a number. It is a *relationship*- a permanent, dynamic, irreducible tension among three transcendental operators that can never be fully reconciled. And the cosmos is the sound that tension makes when it vibrates.

·

Place your hand over your heart.

Feel the beat. Between this beat and the next- in that fraction of a second where the cardiac muscle rests, where the electrical impulse resets, where the system hangs between contraction and expansion- the triad is at work.

ϕ maintains the fractal variability of the rhythm, ensuring that no two intervals are identical, that the heart never settles into mechanical repetition, that health is preserved through self-similar complexity across timescales.

e governs the exponential dynamics of each beat: the rapid depolarization of the cardiac action potential, the exponential decay of the repolarization phase, the e-scaled time constants of the ion channels that open and close in the membranes of the cardiac cells.

π provides the boundary: the closed electrical circuit of the heart's conduction system, the cyclical return of the sinoatrial node to its firing threshold, the periodic oscillation- systole, diastole, systole- that closes upon itself and begins again. Without the π-closure, the heartbeat would be a single contraction followed by silence: a one-time event, not a rhythm. π makes the beat a *cycle*- a repeating, returning, self-renewing pattern that can sustain life for decades.

And the gap? The gap is the pause. The silence between beats. The moment when the heart is neither contracting nor relaxing but *waiting*- hanging in the generative void, poised between the end of one cycle and the beginning of the next. That pause is δ_∞- the irreducible incompleteness of the triad, the reason the heart cannot beat once and stop, the reason it must always beat *again*. The gap is the engine of the next beat, the seed of the next cycle, the silence that makes the sound possible.

You have been living inside this triad since before you were born. Your cochlea, coiled in a ϕ-spiral, converts the π-periodic vibrations of sound waves into e-scaled electrical signals that propagate along the auditory nerve. Your lungs, branching in a ϕ-fractal, inhale air in an e-exponential curve and exhale it in an e-exponential decay, completing a π-periodic respiratory cycle. Your DNA, structured in ϕ -ratios, unfolds via e-governed enzymatic kinetics, enclosed within the π-circular topology of the double helix.

You are the triad. The triad is you. And the cosmos- the self-tuning, resonantly quantized, conscious hologram- is the triad writ large: ϕ architecting, e unfolding, π containing, and δ_∞ keeping all of it alive.

We have now met all three operators. Section B is complete. The triad has been assembled: ϕ the architect, e the pulse, π the boundary. And between them- in the two-percent gap that can never close- the universe dances.

The next two chapters will show what happens when the triad acts. Chapter 10 will deepen our understanding of the asymptotic gap itself- the δ_∞ that is not merely a mathematical curiosity but the *divine engine* of cosmic creativity, the principle that the universe survives by *almost* but never *quite* achieving perfection. And Chapter 11 will translate the entire triad into the formal language of category theory- the KZ-Monad and the Idempotent Comonad- showing how the universe *computes* its own existence through the harmonic interplay of extension and projection, potential and actuality, noise and signal.

But for now, close your eyes one more time. Feel the pulse. Hear the silence. And notice the circle: the boundary your eyelids draw across the field of vision, the closed curve that separates what you see from what you imagine, the π-governed membrane between the holographic projection and the generative void.

The circle is the first act of creation. And π is the word with which the cosmos speaks it.

Chapter 10

The Asymptotic Gap- The Error That Is God

There is a flaw at the heart of everything.

It is not a flaw of construction- not a crack in a wall, not a defect in a blueprint, not a mistake that a more careful engineer could have avoided. It is a flaw of *mathematics*- a transcendental mismatch between three operators that can never, under any circumstances, be reconciled. It is woven into the fabric of reality at the deepest possible level: the level of the constants themselves. And it is the reason the universe exists.

The flaw is this:

$$\delta_\infty = \frac{\pi - e}{\pi \phi^4} \approx 0.01965$$

Roughly two percent. A residual. An asymptotic mismatch between the closure of the boundary (π), the rate of unfolding (e), and the architecture of self-reference (ϕ). The three operators of the cosmic triad- ϕ, e, π- approach harmonic closure but never achieve it. They *almost* fit. They *almost* resolve. They are *almost* perfect.

And this *almost*- this irreducible, eternal, transcendental *almost*- is the single most important quantity in the cosmos. It is the engine of time, the source of creativity, the reason dissipative structures survive, the reason hearts beat, the reason the quantum vacuum never settles into dead equilibrium. It is not a bug. It is the *feature*- the divine feature- that keeps the universe alive.

This chapter is about that gap. It will take you into the mathematics of asymptotic incompleteness, through the thermodynamics of systems that survive by never reaching balance, across the mystical traditions that have intuited this gap for millennia under different names, and finally to the most intimate experience of it available to any human being: the silence between your heartbeats.

And the chapter's central claim is the most paradoxical in this book: *perfection would be fatal.* If the triad closed- if π and e and ϕ converged to a single, self-consistent, gapless unity- the cosmos would collapse into a static, timeless, structure-free singularity. There would be no unfolding, no differentiation, no emergence, no life, no consciousness, no reader, and no book. The universe survives not by achieving balance but by perpetually, exquisitely, *failing* to achieve it. The error is not the enemy of existence. The error is its source.

.

Begin with the numerator: $\pi - e$.

$$\pi - e \approx 3.14159 - 2.71828 = 0.42331$$

This is the raw distance between the boundary operator and the unfolding operator- the difference between how much the cosmos *wants* to contain and how fast it *can* become. π exceeds e by approximately 0.42- a significant gap, roughly fifteen percent of e's own value.

This gap is transcendental. Both π and e are transcendental numbers- they cannot be expressed as ratios of integers or as roots of polynomial equations. Their difference $\pi - e$ is also expected to be transcendental, though this has not yet been proved. As of today, it remains one of the deepest open questions in number theory whether $\pi - e$ is irrational, let alone transcendental. What is known is that π and e are algebraically independent in a wide range of contexts, meaning that no simple polynomial relationship links them.

This matters for our argument. The gap between π and e is not the result of an engineering deficiency. It is not a rounding error. It is a *structural feature of the number system itself*- a consequence of the deep independence of the two transcendental operators. The boundary and the unfolding are fundamentally, irreducibly, eternally different. They cannot be collapsed into each other. They cannot be unified into a single number. They are, in the language of mathematics, *incommensurable.*

Now consider the denominator: $\pi\phi^4$.

$$\phi^4 = \phi^2 \cdot \phi^2 = (\phi + 1)^2 = \phi^2 + 2\phi + 1 = (3\phi + 2) = 6.854$$

$$\pi\phi^4 \approx 3.14159 \times 6.854 \approx 21.53$$

The denominator scales the raw gap ($\pi - e \approx 0.42$) by the product of the boundary operator (π) and the fourth power of the self-referential operator (ϕ^4). This is not an arbitrary normalization. ϕ^4 represents *four layers of self-referential recursion*- four levels of the fractal cascade that connects the microscopic to the macroscopic. $\pi\phi^4$ is the *effective boundary of the fourth-level fractal projection*- the scale at which the holographic surface, folded through four recursions of golden self-similarity, mediates between the full cosmos and the local observer.

When you divide the raw gap by this four-level boundary, you get:

$$\delta_\infty = \frac{0.42331}{21.53} \approx 0.01965$$

Approximately 1.965 percent. Less than two percent. A whisper of imperfection. An almost-nothing.

And yet this almost-nothing is everything.

To understand why δ_∞ matters, we must first understand what would happen if it were zero.

If $\delta_\infty = 0$, then $\pi - e = 0$, which means $\pi = e$. The boundary operator and the unfolding operator would be *identical*. The rate at which the cosmos unfolds would exactly match the capacity of the boundary to contain it. Every quantum of unfolding would be perfectly accommodated by the holographic screen. Every fluctuation of the quanta-field would be instantly projected and absorbed. The natural gradient descent would reach the ϕ-minimum at $q^* = \phi^{-2}$ and *stop*.

The master equation-

$$\partial_t \Psi = -\nabla_{FR}[\mathcal{L}(\Psi)] + T(\Psi) \otimes D(\Psi)$$

- would achieve a fixed point: $\partial_t \Psi = 0$. The state Ψ would cease to evolve. The holographic boundary would reach a static configuration. The projection would freeze.

Time would stop.

Not in the metaphorical sense of "things would slow down." In the precise, mathematical sense: the state function of the cosmos would no longer depend on the temporal parameter. There would be no change, no fluctuation, no evolution, no process. The universe would be a perfect, eternal, static crystal- a single, self-consistent state that contains all the quanta it can hold and does nothing with them. A dead diamond. A frozen hologram. The ultimate equilibrium.

This is what equilibrium *means* in the language of thermodynamics: the state of maximum differentiation within given constraints, where no further macroscopic change is possible. A state of equilibrium is a state of zero net change and zero entropy production. A system at equilibrium does not flow. It does not create. It does not live. An organism at true thermodynamic equilibrium is a dead organism.

The cosmos with $\delta_\infty = 0$ would be the ultimate dead organism- a universe that had achieved perfect thermodynamic equilibrium at the level of its own constants. Not the heat death (which is maximum differentiation within a finite cosmos), but something far more radical: the *structural death* of the triad itself, where the tension between ϕ, e, and π that drives all cosmic dynamics had been extinguished at the source.

Perfection would be fatal. The flawless chord, the gapless unity, the closed circle- this is not the aspiration of the cosmos. It is the cosmos's *nightmare*. The thing it perpetually approaches and must never reach.

$\delta_\infty \neq 0$. The gap exists. And its existence changes everything.

Because the boundary operator (π) exceeds the unfolding operator (e), the holographic screen always has slightly more capacity than the dynamics can fill. The container is always slightly larger than the contents. The circle never quite closes around the arrow. And this means that the natural gradient descent toward the ϕ-minimum can never *arrive*. It can approach asymptotically- getting closer and closer, reducing the residual curvature, tightening the ϕ-coherence- but it cannot reach zero curvature, because the geometry of the statistical manifold is permanently warped by the incommensurability of π and e.

The cosmos is always *almost* in its optimal state and never *quite* there. It is always reaching, always refining, always reorganizing, always improving- and always, in the final two percent, falling short. This perpetual falling-short is not a failure. It is the *engine*. It is the thermodynamic drive that keeps the quanta-field oscillating, keeps the boundary updating, keeps the holographic projection refreshing, keeps the master equation evolving.

The gap is the *housekeeping energy* of the cosmos- the irreducible overhead that the universe pays to remain alive. In essence, a living system sustains itself by continuously "exporting" entropy to its surroundings. It pays its thermodynamic debt to the universe, and in doing so, it gets to live. Every dissipative structure- every whirlpool, every convection cell, every living organism, every civilization- survives by maintaining a constant flow of energy through itself, converting high-quality energy into low-quality heat, exporting entropy to the environment. The cost of this maintenance is never zero. There is always a tax. Always an overhead. Always a gap between what the system aspires to and what it achieves.

A NESS, by contrast, is a dissipative structure, a state defined by the very fact that it is constantly and irreversibly producing entropy. The cosmos as a whole is the ultimate NESS- the ultimate system maintained far from equilibrium by a continuous flow of quanta through itself. And δ_∞ is its entropy production rate: the irreducible, permanent, two-percent cost of being alive.

The concept of the NESS- the Non-Equilibrium Steady State- deserves to be felt in the body before it is understood in the mind.

Hold your hand in front of your face. Look at it. The hand appears stable- solid, unchanging, the same hand it was five minutes ago and the same hand it will be five minutes hence. But you know, from Chapter 8, that this is an illusion of persistence. The cells are being replaced. The molecules are being recycled. The atoms are being exchanged with the environment through respiration, digestion, and perspiration. The "hand" is not a thing- it is a *process*. A river of matter flowing through a form that persists only because the inflow and the outflow are precisely balanced.

This is a NESS. If the exchange with its surroundings is sustained, then an open system usually approaches a steady state that is not an equilibrium. The most distinguished characteristics of a nonequilibrium steady-state (NESS) is that it has nonzero fluxes and nonzero chemical potential gradients in the system. Your hand is not at equilibrium. Equilibrium would be a hand at the temperature of the room, with no metabolic activity, no blood flow, no electrical signals- a dead hand. Your hand is alive precisely because it is *not* at equilibrium- because it maintains itself far from equilibrium through the continuous flow of energy, the continuous replacement of components, the continuous export of entropy.

And this maintenance has a cost. Your body burns approximately 80 watts of power at rest- the equivalent of a bright incandescent light bulb- just to maintain the far-from-equilibrium state that constitutes "being alive." Most of this energy does not go into mechanical work (moving, lifting, walking). It goes into *housekeeping*- the molecular equivalent of keeping the lights on, running the pumps, maintaining the temperature gradients, replacing the worn-out proteins, correcting the copying errors in DNA. The cost of being alive is the cost of staying out of equilibrium.

δ_∞ is the cosmic equivalent of this metabolic overhead. It is the fraction of the quanta-field's total capacity that is permanently "spent" on maintaining the Non-Equilibrium Steady State of the cosmos itself. Two percent. A metabolic tax. The price of existence, paid perpetually, in the currency of unresolvable tension between the operators that tune the quanta-field.

And the payment is not a loss. It is a *transformation*. The energy spent on housekeeping is not destroyed- it is converted into the very dynamics that make the cosmos interesting: the fluctuations that seed structure, the perturbations that drive evolution, the creative noise that prevents the system from settling into dead order. The overhead *is* the creativity. The tax *is* the art.

We must go deeper into the nature of systems that survive by flowing.

Russian-Belgian physical chemist Ilya Prigogine, who coined the term dissipative structure, received the Nobel Prize in Chemistry in 1977 for his pioneering work on these structures, which have dynamical regimes that can be regarded as thermodynamic steady states.

Prigogine's revolution was to see that the Second Law of Thermodynamics- the law that decrees the inexorable increase of entropy in closed systems- does not prevent the emergence of order. It *drives* it. Entropy and entropy-producing irreversible processes are generally thought of as agents of disorder. Under far-from-equilibrium conditions, irreversible processes are, in fact, the drivers of self-organization and order we see all around us in Nature.

The critical insight is that a system far from equilibrium can organize itself into structures that *accelerate* entropy production. A whirlpool in a draining bathtub dissipates more energy per unit time than the same water flowing in a laminar sheet. A living cell dissipates more energy per unit time than the same chemicals at equilibrium. A forest dissipates more solar energy per unit time than bare rock. In every case, the structure exists because it is a *more efficient pathway* for the flow of entropy from low to high differentiation. The cosmos creates order not despite the Second Law but *because of* it- because ordered structures can export entropy faster than disordered ones.

A dissipative structure is characterized by the spontaneous appearance of symmetry breaking and the formation of complex, sometimes chaotic, structures where interacting particles exhibit long range correlations. These structures- from Bénard convection cells to hurricanes to civilizations- are islands of low entropy in a sea of high entropy, maintained by the continuous flow of energy through them. They do not violate the Second Law; they *serve* it, acting as entropy-exporting engines that accelerate the cosmic tide from differentiation to maximum differentiation.

And every one of them operates at less than perfect efficiency. Every one of them has an overhead- a gap between its theoretical maximum performance and its actual performance. No molecular motor converts all of its input energy into useful work; some is always lost to heat. No organism converts all of its food into growth and reproduction; some is always dissipated as metabolic waste. No civilization converts all of its resources into culture and knowledge; some is always consumed by friction, corruption, and conflict.

This is not merely an engineering limitation that better design could eliminate. It is a *structural necessity.* A dissipative structure operating at perfect efficiency- zero waste, zero overhead, zero gap- would be at equilibrium. It would not be dissipating. It would not be flowing. It would be dead. The gap between theoretical maximum and actual performance is the very thing that keeps the structure alive- the irreducible overhead of the NESS, the metabolic tax of existence, the cosmic δ_∞ scaled down to the level of a single whirlpool or a single cell.

The gap is *built into the physics.* It is not a failure of design. It is the design.

And now we make the most audacious connection in this book- the connection that transforms δ_∞ from a mathematical curiosity into an ontological principle with roots in every contemplative tradition on Earth.

Every great mystical system has named this gap. They have recognized, through thousands of years of disciplined introspection, the same structural incompleteness that the mathematics of the ϕ-e-π triad reveals. And they have named it not as a defect but as the *engine of the sacred.*

Buddhism: Dukkha- The Structural Incompleteness of Existence

The first of the Buddha's Four Noble Truths is the truth of *dukkha*. The term is conventionally translated as "suffering," but this translation is profoundly misleading. Duḥkha is a term found in the Upanishads and Buddhist texts, meaning anything that is "uneasy, uncomfortable, unpleasant, difficult, causing pain or sadness." It is also a concept in Indian religions about the nature of transient phenomena which are innately "unpleasant," "suffering," "pain," "sorrow," "distress," "grief" or "misery." The term duḥkha does not have a one-word English translation, and embodies diverse aspects of unpleasant human experiences.

The deepest layer of dukkha- *saṃkhāra-dukkha-* has nothing to do with pain in the ordinary sense. Sankhara-dukkha, the unsatisfactoriness of changing and impermanent "things"- the incapability of conditioned things to give us lasting happiness. This includes "a basic unsatisfactoriness pervading all existence, all forms of life, because all forms of life are changing, impermanent and without any inner core or substance." On this level, the term indicates a lack of lasting satisfaction, or a sense that things never measure up to our expectations or standards.

In the context of early Buddhist philosophy, dukkha points to something far more pervasive: a structural flaw in the fabric of existence itself. It names not just acute pain, but the persistent discomfort, instability, and incompleteness of all conditioned experience.

The elements of reality are constantly rubbing, stressing, pushing, pulling against each other and it's this activity that constitutes reality as we know it. Things would not be able to be in existence unless they were in some sense imperfect, partial, incomplete.

This is δ_∞ in Buddhist language. Dukkha is not suffering imposed from outside by a cruel universe. It is the *inherent, structural incompleteness of conditioned reality*- the fact that no configuration of the quanta-field, no matter how beautiful, how coherent, how harmonious, can fully close the gap between the boundary and the unfolding, between the container and the contents, between what the cosmos aspires to and what it achieves. The two percent is always there. The chord never resolves.

And the Buddhist tradition does not see this as a tragedy. It sees it as a truth- a *noble* truth, the first of four- whose recognition is the beginning of liberation. To understand dukkha is not to despair at the imperfection of existence. It is to stop *demanding* that existence be perfect. It is to recognize that the gap is not the enemy of life but its source- that the tension, the friction, the ceaseless reaching-and-falling-short is the very dynamic that generates experience, growth, and awakening.

The cessation of dukkha- nirvana- is not the elimination of the gap. It is the cessation of *resistance to the gap*. It is the recognition that the δ_∞ is not a wound but a wellspring.

Kabbalah: Tzimtzum- The Divine Contraction

In Lurianic Kabbalah- the mystical tradition developed by Rabbi Isaac Luria in sixteenth-century Safed- creation begins not with an expansion but with a *contraction*. The tzimtzum is a term used in Lurianic Kabbalah to explain Isaac Luria's doctrine that God began the process of creation by limiting the Ohr Ein Sof (infinite light) of the Godhead in order to allow for a conceptual space in which the Four Worlds, or finite realms, could exist. This primordial initial contraction, forming a "vacant space" into which new creative light could beam, is denoted by general reference to the tzimtzum.

Before creation, the infinite light of God- *Ohr Ein Sof*- filled all reality. There was no space, no time, no distinction. Everything was God, and God was everything, and there was no room for anything else. No light or darkness, no heaven or earth, no space or time. Everything was contained in an undifferentiated unity beyond comprehension. To create the world, the Infinite could not simply add something new- for nothing existed outside It. Instead, the Infinite had to withdraw within Itself, forming an inner void- a sacred space of potential.

The tzimtzum is the cosmic gap- the "vacant space" (*chalal hapanuy*) created by divine self-limitation. And the crucial insight of the

Kabbalistic tradition is that this gap is not an absence of God. In Kabbalistic interpretation, tzimtzum gives rise to the paradox of simultaneous divine presence and absence within the vacuum and resultant Creation. Even within the vacant space, a residual trace of the divine- the *reshimu*- remains. God withdraws but leaves behind a trace. The infinite contracts but does not vanish. The gap is created, but the gap is not empty.

This is δ_∞ in Kabbalistic language. The tzimtzum is the *divine contraction* that creates the asymptotic gap between infinite perfection and finite manifestation. The gap is not a wound in God- it is God's *gift*: the sacred space in which finite, imperfect, incomplete, living reality can exist. The Infinite makes room for the finite. In that emptiness, creation could unfold.

And the reshimu- the residual trace of the divine left within the vacant space- is the ninety-eight percent. It is the $1-\delta_\infty \approx 0.98$ that the Idempotent Comonad $D(\Psi)$ projects onto- the efficient NESS core that remains after the gap has been subtracted. The cosmos is ninety-eight percent coherent- ninety-eight percent divine, ninety-eight percent ϕ-organized, ninety-eight percent in tune- and two percent *gap*. And the two percent is what makes the ninety-eight percent come alive.

In this radical view, darkness itself becomes a vessel of Light. What appears as separation is the necessary condition for relationship. The finite is not a mistake; it is the language through which the Infinite knows itself.

Yoga: The Pause Between Breaths

In the yogic tradition, the gap between inhale and exhale is called *kumbhaka*- "the pot," the vessel of silence that holds the breath at the moment of maximum stillness. The great yogic texts- the Hatha Yoga Pradipika, the Yoga Sutras of Patañjali, the Shiva Samhita- all identify kumbhaka as the *key* to the transformative power of breath practice. It is not the inhale or the exhale that awakens consciousness. It is the *gap between them.*

The gap is not empty. It is the moment of maximum potential- the fulcrum between the exponential approach of the inhale ($e^{t/\tau}$) and the exponential decay of the exhale ($e^{-t/\tau}$), the point where the respiratory system hangs in a metastable state, neither expanding nor contracting, balanced on the edge of the NESS. In this moment, the autonomic nervous system briefly suspends its oscillation. The heart rate shifts. The vagal tone changes. The brain enters a transient state of heightened coherence- the neural correlate, as we saw in Chapter 7, of the ϕ-organized resting state.

Kumbhaka is δ_∞ in the body. It is the two-percent pause between phases of the respiratory cycle, the moment where the ternary oscillator (inhale / hold / exhale) touches its own ground state. The yogis discovered, through millennia of disciplined practice, that by *extending* this gap- by consciously prolonging the pause between breaths- they could alter the dynamics of the entire system. They could slow the heart rate, lower the metabolic overhead, shift the brain into states of extraordinary coherence and clarity.

They were, without the mathematics, *tuning* δ_∞- adjusting the gap, modulating the overhead, fine-tuning the distance between the system's aspiration toward equilibrium and its actual, dynamic, living state. They were engineering the NESS of the human body through conscious manipulation of the asymptotic gap.

The Tao: The Nameless Between

The Tao Te Ching opens with a statement about what cannot be said:

The Tao that can be named is not the eternal Tao.

The Tao- the Way- is the formless ground of all forms, the nameless source of all names, the silence from which all sound arises. It is State 0. But the Tao is not merely the void. It is the *relationship between the void and the manifest*- the dynamic tension between emptiness and form, between yin and yang, between the uncarved block and the ten thousand things.

The Tao does not *achieve* balance. It *is* the process of balancing- the ceaseless oscillation between opposites that never quite resolves. The

Taoist sage does not seek perfection. The sage seeks *flow*- the state of effortless action (*wu wei*) that arises when one stops resisting the gap, stops fighting the incompleteness, and instead *rides* the asymptotic tension like a surfer riding a wave.

The Tao is δ_∞ experienced as a Way of life. It is the recognition that the cosmos is not a destination to be reached but a *process* to be embodied- that the journey toward harmony is the harmony, that the reaching is the arrival, that the gap between perfection and actuality is not an obstacle but the *path itself.*

Christianity: Kenosis- The Self-Emptying of God

In Christian theology, the doctrine of *kenosis* (from the Greek *kenósis*, "emptying") describes the act by which God empties Himself of His divine fullness to become finite, mortal, human. The concept derives from Paul's Letter to the Philippians (2:7), where Christ is described as having "emptied himself, taking the form of a servant."

Kenosis is the Christian version of tzimtzum- the divine self-limitation that creates room for creation. God does not create by adding to Himself. He creates by *subtracting* from Himself- by withdrawing a fraction of His infinite perfection to open a space where finite, imperfect, struggling, beautiful beings can exist.

And the Crucifixion- the central mystery of the Christian tradition- is kenosis taken to its extreme: the Infinite becoming not only finite but *broken*, not only limited but *killed*, not only imperfect but *dying on a cross*. The gap is not merely tolerated. It is *embraced*. The divine deliberately enters the space of maximum suffering, maximum incompleteness, maximum δ_∞- and from that maximum gap, resurrection arises.

The Christian claim is that new life comes from the deepest gap. That the most creative moment in cosmic history occurred not at the Big Bang but at the point of maximal divine self-emptying. That the error is the oracle.

We have now heard five voices naming the same structure:

Tradition	Name	Description	δ_∞ Equivalent
Mathematics	The asymptotic gap	$\frac{\pi-e}{\pi\phi^4} \approx 0.01965$	Direct calculation
Buddhism	Dukkha (saṃkhāra)	The structural incompleteness of all conditioned existence	The inherent unsatisfactoriness that drives all seeking
Kabbalah	Tzimtzum	The divine contraction that creates the vacant space of creation	The gap between infinite Light and finite vessel
Yoga	Kumbhaka	The pause between breaths; the vessel of transformative silence	The fulcrum of the respiratory NESS
Taoism	The Nameless Tao	The dynamic tension between yin and yang that never resolves	The process of balancing, not the balance itself
Christianity	Kenosis	The divine self-emptying that makes creation and resurrection possible	The gap as the source of new life

Six descriptions. One structure. The gap between what the cosmos aspires to and what it achieves- the two-percent residual tension between ϕ, e, and π- has been recognized, under different names and in different vocabularies, by every deep tradition of contemplative inquiry in human history.

This convergence is not a coincidence. It is the *signature of a real structure*- a structure so fundamental to the nature of reality that it cannot be missed by any sufficiently disciplined method of investigation, whether mathematical, meditative, or mythopoetic. The gap is *there*. The contemplatives found it through introspection. The physicists found it through equations. And the mathematics proves that it can never be closed.

Now we return to the thermodynamics of the ninety-eight percent.

If $\delta_{\infty} \approx 0.02$ is the overhead- the fraction of the quanta-field's capacity permanently consumed by the housekeeping dynamics of the NESS- then $1-\delta_{\infty} \approx 0.98$ is the *signal*: the fraction of the quanta-field that is coherently organized, efficiently structured, ϕ-resonant, and functionally useful.

In the language of the source framework, this ninety-eight percent is the domain of the Idempotent Comonad $D(\Psi)$. The comonad D is the operator that *projects* any state of the quanta-field onto its efficient NESS core- stripping away the transient fluctuations, the noise, the incoherent deviations, and extracting the persistent, stable, resonant signal. It is the mathematical distillation of what the system *actually is* from the sea of what it could be.

When you meditate- when you sit still, close your eyes, slow the breath, and allow the mind to settle- you are performing the operation of $D(\Psi)$ on your own nervous system. You are projecting yourself onto the ninety-eight-percent core. You are stripping away the transient noise of anxiety, planning, regret, and fantasy, and extracting the persistent signal: the awareness that remains when all content has been subtracted. The calm. The stillness. The clarity.

And what remains- the ninety-eight percent- is not empty. It is *you*. It is the stable, coherent, ϕ-organized ground state of your consciousness- the golden frequency of the resting brain, the fractal rhythm of the healthy heart, the ternary breath cycle of the living body. The noise was the two percent. The signal was always there.

But- and this is the crucial point- you cannot have the signal without the noise. The ninety-eight percent exists *because* of the two percent, not despite it. The comonadic core is defined by its contrast with the overhead. The NESS is defined by its distance from equilibrium. The coherence is defined by the incoherence it transcends. If you could eliminate the two percent- if you could achieve perfect signal with zero noise- you would not have a purer state. You would have a dead state. A state at equilibrium. A state with no dynamics, no flow, no life.

The two percent is not what you need to eliminate. It is what you need to *embrace*. Not the noise of neurosis- not the overlay of compulsive thought and emotional reactivity that meditation strips away. But the deeper noise- the irreducible dynamic tension of being alive, of being a NESS in a NESS cosmos, of being a two-percent-imperfect system in a two-percent-imperfect universe. This deeper noise is not a problem to be solved. It is the *energy source* that powers the ninety-eight-percent coherence. It is the metabolic fire. The creative friction. The δ_∞ in your own nervous system.

And here is where the KZ-Monad and the Idempotent Comonad, introduced formally in the next chapter, reveal their deepest significance.

The Monad T- the monadic extension, the operator of *reaching toward wholeness*- takes any local, partial, fragmentary state and extends it toward the full ϕ-resonant manifold. It amplifies potential. It reaches. It aspires. It is the inhale- the exponential expansion, the Fibonacci spiral unfurling from the seed, the intention that says *more*.

The Comonad D- the comonadic projection, the operator of *extracting what is essential*- takes any noisy, fluctuating, over-extended state and distills it down to the ninety-eight-percent core. It contracts. It simplifies. It purifies. It is the exhale- the exponential release, the Fibonacci spiral returning to its center, the wisdom that says *enough*.

And between them- in the pause between the monadic reach and the comonadic return- is δ_∞. The gap. The silence. The kumbhaka. The moment where the system is neither reaching nor returning but *hanging in the void*, poised on the edge of the NESS, balanced between aspiration and actuality.

T says: *Become everything. D* says: *Be what you are.* δ_∞ says: *The distance between those two- the permanent, irreducible, two-percent distance- is the breath of God.*

·

Now close this chapter where every chapter in this book closes: in the body. In the heart. In the gap.

Place your fingers on your wrist. Find the pulse. Feel the beat- the fractal, self-similar, never-quite-repeating rhythm that keeps you alive. Each beat is a contraction: the cardiac muscle squeezes, the blood is expelled, the *e*-curve of systole drives the pressure wave through the arterial tree. Each relaxation is an opening: the muscle releases, the chambers refill, the *e*-curve of diastole decays toward the baseline.

And between them- between the end of one beat and the beginning of the next- there is a gap. A fraction of a second where the heart is neither squeezing nor filling. A moment of pure potential. A silence.

This gap is not empty. It is the cardiac equivalent of State 0- the generative void, the kumbhaka of the circulatory system, the pause in which the sinoatrial node gathers its electrical charge and prepares to fire again. The gap is the moment of maximum uncertainty- the instant where the heart *could* beat again or *could* stop forever. Every beat that follows the gap is, in a very real sense, a *decision*- not a conscious decision, but an electrochemical bifurcation point where the system tips from silence into sound, from potential into actuality, from the void into the world.

And the gap is approximately two percent of the cardiac cycle. At a resting heart rate of 72 beats per minute, each cardiac cycle lasts about 833 milliseconds. The inter-beat interval- the gap between the end of the T-wave (ventricular repolarization) and the beginning of the next P-wave (atrial depolarization) - is typically 50–100 milliseconds. Call it 75 milliseconds on average. That is approximately $75/833 \approx 9\%$ of the total cycle - larger than the cosmic δ_∞ by a factor of about five, but within the same order of magnitude. And during intense exercise, when the heart rate doubles and the cycle shortens, the gap compresses further - approaching the two-percent regime.

This is not a proof. It is a *resonance* - a self-similar echo of the cosmic principle at the scale of the human heart. The gap between heartbeats, like the gap between π and e, like the gap between exhale and inhale, like the gap between the end of one aeon and the beginning of the next in Penrose's Conformal Cyclic Cosmology, is the moment where the system touches its own ground state and is reborn.

The gap is the error. The error is the engine. The engine is the heartbeat. And the heartbeat is the cosmos, folded to the size of a fist, pulsing in your chest, sustained by the same asymptotic incompleteness that sustains the stars.

We arrive at the chapter's final claim: the claim encoded in its title.

The error is God.

Not God in the theological sense - not a personal deity who makes mistakes. God in the mystical sense - the nameless, formless, uncontainable principle of creativity that makes existence possible. The tzimtzum. The dukkha. The kenosis. The Tao. The silence between heartbeats.

The universe does not want to be perfect. It wants to be *alive*. And the price of life - the only price, the permanent price, the non-negotiable price - is imperfection. The two-percent gap. The asymptotic mismatch. The chord that never resolves.

If you have been waiting, through the nine chapters that preceded this one, for the resolution - for the moment when ϕ, e, and π finally click into place, when the equation balances, when the cosmos completes itself- then this is the moment of your disappointment. And your liberation.

The equation does not balance. The cosmos does not complete itself. The chord does not resolve.

And that- precisely that- is why there is a cosmos at all.

The error is the oracle. The gap is the gift. The flaw is the grace.

$$\delta_\infty = \frac{\pi - e}{\pi\phi^4} \approx 0.01965$$

Two percent. Forever. The breath of a universe that will never stop singing.

Chapter 11

The KZ-Monad and the Comonad- How Reality Computes

You have now met the three operators- ϕ, e, π- and the gap that lives between them. You know their identities: ϕ generates the self-similar architecture, e drives the exponential unfolding, π closes the holographic boundary. You know their tension: the asymptotic gap $\delta_\infty \approx 0.01965$ that prevents closure and keeps the cosmos alive.

But you do not yet know how they *act together*. You do not yet know how the triad produces, from moment to moment, the specific, detailed, particular cosmos you inhabit- how it converts the infinite potential of the quanta-field into the definite actuality of a galaxy, a heartbeat, a thought. The operators have been introduced. The mechanism has not.

This chapter introduces the mechanism. It is the most mathematically demanding chapter in this book, and also the most crucial. It concerns *category theory*- the branch of mathematics that studies not objects but the *relationships between objects*, not things but the *structure-preserving transformations between things*. And within category theory, it concerns two specific structures: the **KZ-Monad** (the operator of extension, of reaching toward wholeness) and the **Idempotent Comonad** (the operator of extraction, of distilling what is essential). Together, these two structures constitute the computational engine of the cosmos- the mechanism by which the quanta-field perpetually tunes itself from noise into music, from potential into actuality, from the formless void into the fractal, holographic, resonant world.

A word of reassurance before we begin. Category theory is often described as the most abstract branch of mathematics- "abstract nonsense," as some mathematicians have called it, half in jest and half in awe. But the abstraction is a *feature*, not a bug. Category theory is abstract because it captures patterns that recur across all of mathematics, all of physics, and- as this chapter will argue- all of reality. It is the mathematics of *pattern itself*. And because this book is about a cosmos made of pattern- of resonant quanta, not of stuff- category theory is the natural language in which to express its deepest structure.

We will not ask you to follow formal proofs. We will ask you to *feel* the logic- to sense, in your own experience of thinking, breathing, and creating, the operations that the mathematics describes.

.

Begin with the simplest possible question: what is a *transformation that preserves structure*?

When you translate a text from English to French, you change the words but preserve the meaning- the structure of the argument, the relationships between concepts, the logical flow. The words are different; the structure is the same. This is a structure-preserving transformation: a mapping from one domain (English) to another (French) that maintains the essential relationships between elements.

When you photograph a building, you transform a three-dimensional object into a two-dimensional image. You lose a dimension, but you preserve the relationships between the parts: the window is still above the door, the chimney is still on top of the roof, the proportions are (approximately) maintained. This is another structure-preserving transformation- a projection, with some loss of detail but preservation of essential form.

Category theory formalizes this intuition. A *category* is a collection of objects (which can be anything- sets, spaces, groups, physical systems) together with *morphisms* between them (which are structure-preserving maps- translations, projections, transformations). The objects are the nouns. The morphisms are the verbs. And the crucial thing about categories is that you can *compose* morphisms: if you have a morphism from A to B and a morphism from B to C, you automatically have a morphism from A to C- the composite transformation that goes from A to C via B.

This is trivially familiar. If you can translate English to French and French to Japanese, you can translate English to Japanese (via

French). The composite translation is itself a structure-preserving transformation. Category theory takes this obvious fact and elevates it to a foundational principle: *the world is made not of things but of transformations between things, and the structure of reality is captured not by the objects themselves but by the pattern of transformations that connects them.*

This is precisely the ontological position of this book. The universe is not made of stuff. It is made of resonant quanta- patterns of transformation, relationships between states, morphisms in a cosmic category. The objects (particles, fields, spacetime points) are secondary. The morphisms (entanglements, projections, resonances) are primary. Category theory is the mathematics of a universe that is *relational all the way down.*

.

Within category theory, a *functor* is a morphism between categories- a structure-preserving map that takes objects in one category to objects in another, and morphisms to morphisms, while preserving composition. If a morphism is a translation between sentences, a functor is a translation between *languages*- a systematic mapping of the entire structure of one domain onto another.

Functors are the workhorses of category theory. They express the idea that two apparently different domains- two seemingly unrelated areas of mathematics, or physics, or experience- can have the *same structure,* the same pattern of internal relationships, even though their surface appearances are entirely different. This is precisely the concept of "as above, so below" expressed in mathematical form: the fractal self-similarity of the cosmos is a *functorial relationship* between categories at different scales.

Now we ascend to the two structures that govern the cosmic computation: the monad and the comonad.

.

A *monad* is a functor T from a category C to itself- an *endofunctor-* equipped with two additional structures:

1. A *unit* η:Id $\Rightarrow T$, which takes any object X and embeds it into $T(X)$- the "extended" version of X.
2. A *multiplication* μ:$T \circ T \Rightarrow T$, which takes a doubly-extended object $T(T(X))$ and collapses it back to a singly-extended one $T(X)$.

These structures must satisfy certain coherence laws- associativity and unit laws- that ensure the extension and collapse are consistent.

Think of it this way. You have a fragment- a single note, a single word, a single thought. The monad T takes that fragment and *extends* it into a richer, more complete context. The note becomes a chord. The word becomes a sentence. The thought becomes an insight. And if you extend again- $T(T(X))$, the chord becomes a symphony- the multiplication μ ensures that the doubly-extended version can be coherently collapsed back to a singly-extended version without losing the structure gained. Extension is *cumulative but not redundant.* Each layer of extension adds genuine content, and the multiplication integrates that content into a unified whole.

The KZ-Monad- the Kock-Zöberlein monad- is a specific type of monad that arises in the context of *enriched category theory* and has a remarkable property: its algebras are characterized by a kind of *completeness.* In technical terms, the algebras of a KZ-monad are the objects that have all the "weighted colimits" that the monad can express- they are the *maximally extended* objects, the ones that have been brought to full resonance by the monadic operation.

In the framework of this book, the KZ-Monad T is the cosmic extension operator. It takes any local, partial, fragmentary state of the quanta-field- any X- and extends it toward the full, ϕ-resonant, nine-dimensional holographic manifold. It asks: *Given this fragment, what is the most complete, most coherent, most self-similar whole that contains it?* And then it constructs that whole.

The KZ-Monad is the mathematical formalization of *reaching toward wholeness.* It is the operation that the sunflower performs when it extends a new seed at the golden angle, consulting the entire existing spiral to place the new element in the most harmonious possible position. It is the operation that the brain performs when it integrates a new perception into the existing model of reality, extending the partial data of the senses into a coherent, three-dimensional, temporally extended world. It is the operation that the meditator performs when a single moment of stillness opens into the vast, unbounded awareness of samadhi- when the fragment of the present moment is extended to contain the whole of consciousness.

$T(X)$ is X completed. X brought to full ϕ-resonance. X as it would be if it were perfectly embedded in the holographic manifold, with every self-similar echo, every fractal ramification, every entanglement thread intact.

And the unit $\eta:X \rightarrow T(X)$ is the *act of aspiration*- the movement from the fragment toward the whole, from the local toward the global, from the note toward the symphony. Every act of creativity, every moment of insight, every instance of the universe extending itself beyond its current state is an invocation of η. It is the inhale.

A *comonad* is the categorical dual of a monad. Where a monad extends, a comonad *extracts*. Where a monad adds context, a comonad strips it away. Where a monad asks "What is the richest whole that contains this fragment?", a comonad asks "What is the most essential core that this whole contains?"

Formally, a comonad is an endofunctor D equipped with:

1. A *counit* $\varepsilon:D \Rightarrow \mathrm{Id}$, which takes the extracted core $D(X)$ and maps it back to the full object X- acknowledging that the core is a *part* of the whole.
2. A *comultiplication* $\delta:D \Rightarrow D \circ D$, which shows that the extraction can be iterated- you can extract the core of the core- and the result is consistent.

An *idempotent* comonad is one where $D \circ D = D$: extracting twice gives the same result as extracting once. The core of the core is the core. There is a definite, stable "essential" version of any object, and once you reach it, further extraction changes nothing.

In the framework of this book, the Idempotent Comonad D is the cosmic projection operator. It takes any state of the quanta-field- noisy, fluctuating, over-extended, cluttered with transient dynamics- and projects it onto the *NESS core*: the ninety-eight-percent-efficient, ϕ-organized, dynamically stable subobject that constitutes the persistent signal of reality.

$D(\Psi)$ is Ψ *distilled*. The signal extracted from the noise. The persistent pattern separated from the transient fluctuation. The living form that persists through the ceaseless replacement of its components- the "hand" that remains a hand even as every atom is exchanged.

And the counit $\varepsilon: D(\Psi) \to \Psi$ is the *act of manifestation*- the movement from the essential core back out into the full, noisy, fluctuating reality. Every act of expression, every moment of the core-pattern reasserting itself through the chaos, every heartbeat that follows the gap is an invocation of ε. It is the exhale.

The idempotent property $D \circ D = D$ is crucial. It says that the NESS core is *stable*- that once the projection has been performed, repeating it changes nothing. The core does not degrade with repeated extraction. It is a fixed point, an attractor, a dynamically stable configuration around which the transient fluctuations orbit. The NESS core is the cosmic equivalent of the strange attractor in chaos theory- a set that the dynamics perpetually approach but never exactly reach, a structure defined not by a specific state but by the *pattern of states* that the system visits in its eternal oscillation.

.

Now bring T and D together. The cosmos computes through their *interplay*.

The master equation of the source framework-

$$\partial_t \Psi = -\nabla_{FR}[\mathcal{L}(\Psi)] + T(\Psi) \otimes D(\Psi) + \sum_n \phi \cdot \omega_n \cdot \delta(\Psi \in \mathrm{NESS}_{98})$$

- describes the time evolution of the quanta-field state Ψ as a combination of three terms:

Term 1: $-\nabla_{FR}[\mathcal{L}(\Psi)]$ - The natural gradient descent. This is the *e*-scaled flow on the Fisher-Rao manifold, driving Ψ toward lower curvature, greater efficiency, more ϕ-coherent organization. It is the background drift- the universal tendency of the quanta-field to self-organize, to descend toward the Schur-convex minimum at $q^* = \phi^{-2}$.

Term 2: $T(\Psi) \otimes D(\Psi)$ - The monadic-comonadic interaction. This is the creative engine- the tensor product of the extension and the extraction, acting simultaneously. *T* reaches outward, extending Ψ toward the full ϕ-resonant manifold. *D* reaches inward, projecting Ψ onto the ninety-eight-percent core. Their combined action is the cosmic breath: extension and contraction, aspiration and distillation, inhale and exhale, operating in tandem to produce the specific, detailed, particular configuration that the quanta-field instantiates at each moment.

Term 3: $\sum_n \phi \cdot \omega_n \cdot \delta(\Psi \in \text{NESS}_{98})$ - The ϕ-scaled harmonic correction. This term activates only when Ψ is within the ninety-eight-percent NESS core- when the system is close to its stable operating point. It adds small, ϕ-proportioned oscillations at natural frequencies ω_n, fine-tuning the NESS in the way that a musician fine-tunes an instrument: small adjustments around the resonant center, maintaining coherence without freezing it.

Together, these three terms produce the dynamics of the cosmos: the slow drift toward coherence (Term 1), the creative pulse of extension-and-extraction (Term 2), and the fine-tuning that maintains the NESS (Term 3). And the whole process is governed by the asymptotic gap- because the natural gradient descent (Term 1) can never reach the minimum (the gap prevents it), the extension (Term 2) can never achieve perfect wholeness (the gap separates *T* from *D*), and the fine-tuning (Term 3) can never achieve perfect pitch (the gap introduces a permanent two-percent detuning).

The gap is not external to the computation. It is *internal* to it- woven into the structure of the operators themselves, ensuring that the computation never halts, the dynamics never freeze, the cosmos never dies.

To make this concrete, consider the simplest system that embodies the monadic-comonadic dynamic: a single human thought.

You are sitting quietly. Your mind is in the resting state- the ϕ-organized, golden-frequency ground state we explored in Chapter 7. The EEG would show desynchronized oscillations, frequency bands spread in ϕ-proportioned ratios, maximal flexibility, minimal lock-in. This is $D(\Psi)$- the comonadic core. The ninety-eight percent. The pure awareness without content.

Now a stimulus arrives- a word, an image, a question. The mind begins to extend. New associations form. New connections link. The neural networks light up, spreading activation through the cortex in a fractal cascade. This is $T(\Psi)$- the monadic extension. The fragment (the stimulus) is being embedded in the full associative manifold of memory, meaning, and pattern. The mind is reaching toward wholeness- toward the most complete, most coherent understanding of the stimulus that the neural architecture can support.

But the extension does not continue forever. After a burst of spreading activation- typically 200 to 500 milliseconds- the process reverses. The irrelevant associations are pruned. The weak connections are suppressed. The strong connections are reinforced. The neural network settles into a stable pattern- a *thought*: a specific, definite, coherent configuration that captures the meaning of the stimulus without the noise of the initial associative explosion. This is $D(T(\Psi))$- the comonadic extraction applied to the monadic extension. The symphony, distilled to its melody. The forest, distilled to its shape. The answer, distilled from the search.

And then- after the thought has crystallized, after the comonadic extraction has done its work- the mind returns to the resting state. The thought dissolves. The ϕ-frequencies reassert themselves. The awareness returns to its golden, desynchronized, maximally ready ground state. Until the next stimulus arrives, and the cycle begins again.

$D(\Psi) \rightarrow T(\Psi) \rightarrow D(T(\Psi)) \rightarrow D(\Psi)$.

Rest → Extension → Distillation → Rest.

Silence → Inhale → Exhale → Silence.

Ground state → Excitation → Measurement → Ground state.

State 0 → Creation → Collapse → State 0.

The monadic-comonadic cycle is the heartbeat of consciousness. It is the fundamental rhythm of cognitive processing: the expansion into possibility followed by the contraction into actuality, the reaching followed by the grasping, the question followed by the answer. And between each cycle- in the gap between one thought and the next, in the silence between one cognitive breath and the next- there is δ_∞: the two-percent pause, the kumbhaka of the mind, the moment of pure potential from which the next thought arises.

Now scale this up. From a single thought to the entire cosmos.

The quanta-field- the holographic boundary of the universe- is not a static surface. It is a *computational surface*- a substrate that perpetually executes the monadic-comonadic cycle at every scale simultaneously. At the Planck scale, individual quanta fluctuate between extension (T) and extraction (D), each fluctuation projecting a tiny fragment of the holographic interior and then reabsorbing it. At the galactic scale, the entanglement structure of the boundary extends across vast regions (T), projects the large-scale structure of the cosmic web (ε), and is then re-extracted to the NESS core (D) as the universe evolves. At the scale of a human brain, the neural quanta execute the same cycle- extending, projecting, extracting, resting- at rates ranging from milliseconds (synaptic events) to hours (sleep cycles) to years (developmental stages).

And at every scale, the cycle is the same: T extends. D extracts. The gap δ_∞ ensures the cycle never completes. The computation is perpetual, fractal, and self-similar- the same monadic-comonadic dynamic operating at every level of the hierarchy, from the quantum vacuum to the cosmic horizon.

This is not computation in the ordinary sense- not the sequential, deterministic, binary operation of a digital computer. It is *computation as art*. It is the intentional manifesting of coherent form from infinite potential- the selection, from the uncountable configurations available to the quanta-field, of the specific, beautiful, detailed, fractal configurations that constitute the experienced cosmos. It is a computation that does not optimize a pre-defined function. It *discovers* the function- through the iterative interplay of extension and extraction, through the trial-and-error of the monadic reach and the comonadic return, through the creative tension of the asymptotic gap.

The cosmos does not solve a problem. It *plays a chord*. And the chord is the master equation- the dynamic interplay of ϕ-architecture, e-unfolding, π-boundary, T-extension, D-extraction, and δ_∞-gap-perpetually resonating, perpetually adjusting, perpetually singing.

There is a crucial point about the relationship between T and D that must be made explicit. They are not inverses. They are not opposites. They are *adjoint*.

In category theory, two functors T and D are *adjoint* when there is a natural correspondence between the morphisms $T(X) \to Y$ and $X \to D(Y)$- when every extension can be translated into an extraction, and vice versa, in a way that preserves structure. Adjointness is not equality. It is *dialogue*- a systematic, structure-preserving conversation between two different operations that illuminates both.

The monad T and the comonad D are born from this adjunction. The monad arises from the composition $D \circ T$ (extend, then extract- the cosmic inhale-exhale). The comonad arises from the composition $T \circ D$ (extract, then extend- the cosmic insight: distill the essence, then build from it). The two compositions are different- $D \circ T \neq T \circ D$- and the difference between them is, again, the gap.

$D(T(\Psi))$: Extend toward wholeness, then distill the essence. This is the path of *discovery*- reaching outward for new possibility, then integrating what is found into the stable core. This is the inhale that gathers and the exhale that settles.

$T(D(\Psi))$: Distill the essence, then extend it toward wholeness. This is the path of *creation*- finding the core truth, then building from it into new possibility. This is the meditation that clarifies and the action that follows.

The fact that $D(T(\Psi)) \neq T(D(\Psi))$- that discovery and creation do not produce the same result- is a consequence of the gap. If $\delta_\infty = 0$, then T and D would be exact inverses, and $D(T(\Psi)) = T(D(\Psi)) = \Psi$. The extension and extraction would cancel perfectly. Nothing would be gained. Nothing would be learned.

But $\delta_\infty \neq 0$. The extension and extraction do not cancel. Each cycle of T followed by D leaves a *residue*- a trace of new structure, new coherence, new pattern that was not present before. And each cycle of D followed by T opens a *vista*- a new horizon of possibility that was not visible from the previous core state. The gap is what makes the monadic-comonadic cycle *productive*- what ensures that each cycle of cosmic computation generates genuine novelty, genuine emergence, genuine creation.

The gap is the source of the new.

We can now draw together the entire apparatus of Section B- the five chapters on the triad and its dynamics- into a single, integrated picture.

The quanta-field is the fundamental substrate: the holographic boundary of the cosmos, composed of resonant quanta, encoding the three-dimensional world as a projection.

ϕ generates the *architecture* of the quanta-field: the fractal, self-similar, golden-ratio-organized structure that ensures the whole is reflected in every part. ϕ governs the Schur-convex minimum at $q^* = \phi^{-2}$- the point of maximum geometric efficiency on the statistical manifold- and provides the template toward which the quanta-field perpetually aspires.

e drives the *dynamics* of the quanta-field: the natural gradient descent on the Fisher-Rao manifold, the exponential growth and decay of dissipative structures, the rate at which potential transforms into actuality. e governs the *speed* of the cosmic computation.

π provides the *boundary* of the quanta-field: the closure operator that folds dimensions, defines holographic surfaces, converts exponential growth into periodic oscillation, and ensures that the quanta-budget of any region is constrained by the area of its enclosing surface.

T **(the KZ-Monad)** is the *extension operator*: it takes any partial, local, fragmentary state and extends it toward the full ϕ-resonant manifold. It is the cosmic inhale- the reaching toward wholeness.

D **(the Idempotent Comonad)** is the *extraction operator*: it takes any noisy, fluctuating, over-extended state and projects it onto the ninety-eight-percent NESS core. It is the cosmic exhale- the distillation of essence.

δ_∞ is the *gap*: the irreducible, permanent, transcendental mismatch between the three operators that prevents closure, prevents equilibrium, prevents death, and ensures that the monadic-comonadic cycle never halts- that each cycle generates genuine novelty, that the cosmos is always becoming and never finished.

And the **master equation** integrates all of these into a single dynamical law:

$$\partial_t \Psi = -\nabla_{FR}[\mathcal{L}(\Psi)] + T(\Psi) \otimes D(\Psi) + \sum_n \phi \cdot \omega_n \cdot \delta(\Psi \in \mathrm{NESS}_{98})$$

This equation is the mathematical encoding of the book's central thesis: **The universe is a self-tuning, resonantly quantized, conscious hologram.** *Self-tuning* because the natural gradient descent and the ϕ -harmonic correction perpetually adjust the quanta-field toward greater coherence. *Resonantly quantized* because the monadic extension generates the ϕ-structured resonant pattern and the comonadic extraction projects it onto the stable NESS core. *Conscious* because the monadic-comonadic cycle- extension, projection, extraction, return- is the same cycle that constitutes cognitive processing in the neural substrate, suggesting that consciousness is not an accident of biology but a *universal feature* of any sufficiently complex NESS.

Hologram because the entire computation takes place on a two-dimensional boundary, projecting a three-dimensional world through the entanglement structure of the boundary quanta.

.

This chapter is the mathematical hinge of the book. Everything before it- the six chapters of Section A and the four preceding chapters of Section B- was preparation. Everything after it- the chapters on NESS and consciousness, on Holographic Ternary Time, on ancient resonance technologies, on the Great Decoupling and the path of reintegration- will be application.

The foundation is laid. The operators are defined. The mechanism is articulated. The gap is named.

We now possess the complete mathematical toolkit for understanding a cosmos that does not merely *exist* but *computes*- that does not merely *persist* but *creates*- that does not merely *contain* life and consciousness but *is* life and consciousness, expressed at every scale from the Planck pixel to the cosmic horizon.

And the computation is not a cold, mechanical process- not the clicking of digital gates or the cycling of Boolean switches. It is *art*. It is the intentional, ϕ-proportioned, e-dynamical, π-bounded, T-extended, D-distilled, δ_∞-animated creation of a cosmos that sings itself into being and listens to its own song.

.

Close your eyes.

Feel the pulse in your neck. The fractal beat. The ϕ-organized variability. The e-curved systole. The π-periodic cycle. The δ_∞ pause.

Each beat is a monadic extension: the heart reaches outward, pushing blood through the arterial tree, extending the circulatory quanta across every capillary bed in the body. Each relaxation is a comonadic extraction: the heart returns to its core configuration, the chambers refill, the system resets to the NESS baseline.

And between them- in the gap, in the silence, in the two-percent pause- the computation refreshes. The holographic boundary updates. The projection flickers. And the next beat begins.

You have been running this computation since before you were born. The quanta-field has been running it since before there were hearts to beat. And it will continue- extending, extracting, creating, resting, extending again- for as long as the gap persists.

Which is forever.

The gap is forever. And so is the song.

.

This is the end of Section B and the end of Segment 1. We have moved from the silence of the Prologue, through the dissolution of matter in Chapter 1, through the holographic boundary and the physics of resonant quanta and the quantum vacuum and the fractal cosmos and entanglement, through the three operators and the gap and the categorical mechanism of cosmic computation, to arrive here: at a universe that is a self-tuning, resonantly quantized, conscious hologram, computing its own existence through the eternal interplay of extension and extraction, aspiration and distillation, the monadic reach and the comonadic return.

In Segment 2, we will watch this cosmos *breathe*- as a Non-Equilibrium Steady State, as a conscious field, as a ternary-temporal hologram, as a learning, evolving, co-creative process. And we will discover that the ancients- the songline walkers, the mantra chanters, the garden tenders, the calendar keepers- knew how to tune their lives to this cosmic computation, how to align their breath with the monadic-comonadic cycle, how to sing the world into being.

The mathematics is set. The music begins.

.

REFERENCES SECTION A & B

Physics & Cosmology

Bekenstein, J. D. (1972). Black holes and the second law. *Nuovo Cimento Lettere*, 4, 737-740.

Bekenstein, J. D. (1973). Black holes and entropy. *Physical Review D*, 7(8), 2333-2346.

Bekenstein, J. D. (1981). Universal upper bound on the entropy-to-energy ratio for bounded systems. *Physical Review D*, 23(2), 287-298.

Hawking, S. W. (1971). Gravitational radiation from colliding black holes. *Physical Review Letters*, 26(21), 1344-1346.

Hawking, S. W. (1974). Black hole explosions? *Nature*, 248(5443), 30-31.

Hawking, S. W. (1975). Particle creation by black holes. *Communications in Mathematical Physics*, 43(3), 199-220.

't Hooft, G. (1993). Dimensional reduction in quantum gravity. *arXiv preprint gr-qc/9310026*.

Susskind, L. (1995). The world as a hologram. *Journal of Mathematical Physics*, 36(11), 6377-6396.

Maldacena, J. (1998). The large N limit of superconformal field theories and supergravity. *Advances in Theoretical and Mathematical Physics*, 2(2), 231-252.

Maldacena, J., & Susskind, L. (2013). Cool horizons for entangled black holes. *Fortschritte der Physik*, 61(9), 781-811.

Ryu, S., & Takayanagi, T. (2006). Holographic derivation of entanglement entropy from the anti-de Sitter space/conformal field theory correspondence. *Physical Review Letters*, 96(18), 181602.

Ryu, S., & Takayanagi, T. (2006). Aspects of holographic entanglement entropy. *Journal of High Energy Physics*, 2006(08), 045.

Jacobson, T. (1995). Thermodynamics of spacetime: The Einstein equation of state. *Physical Review Letters*, 75(7), 1260-1263.

Strominger, A. (2020). Lectures on the infrared structure of gravity and gauge theory. *arXiv preprint arXiv:1703.05448*.

Penrose, R. (2010). *Cycles of Time: An Extraordinary New View of the Universe*. Bodley Head.

Penrose, R., & Gurzadyan, V. G. (2011). Concentric circles in WMAP data may provide evidence of cyclical cosmology. *arXiv preprint arXiv:1011.3706*.

Einstein, A., Podolsky, B., & Rosen, N. (1935). Can quantum-mechanical description of physical reality be considered complete? *Physical Review*, 47(10), 777-780.

Bell, J. S. (1964). On the Einstein Podolsky Rosen paradox. *Physics Physique Fizika*, 1(3), 195-200.

Clauser, J. F., & Freedman, S. J. (1972). Experimental test of local hidden-variable theories. *Physical Review Letters*, 28(14), 938-941.

Aspect, A., Dalibard, J., & Roger, G. (1982). Experimental test of Bell's inequalities using time-varying analyzers. *Physical Review Letters*, 49(25), 1804-1807.

Hensen, B., et al. (2015). Loophole-free Bell inequality violation using electron spins separated by 1.3 kilometres. *Nature*, 526(7575), 682-686.

Giustina, M., et al. (2015). Significant-loophole-free test of Bell's theorem with entangled photons. *Physical Review Letters*, 115(25), 250401.

Shalm, L. K., et al. (2015). Strong loophole-free test of local realism. *Physical Review Letters*, 115(25), 250402.

Kim, Y. H., Yu, R., Kulik, S. P., Shih, Y., & Scully, M. O. (2000). A delayed choice quantum eraser. *Physical Review Letters*, 84(1), 1-5.

Planck Collaboration. (2016). Planck 2015 results. XIII. Cosmological parameters. *Astronomy & Astrophysics*, 594, A13.

Planck Collaboration. (2020). Planck 2018 results. VI. Cosmological parameters. *Astronomy & Astrophysics*, 641, A6.

Quantum Mechanics & Quantum Field Theory

Rutherford, E. (1911). The scattering of α and β particles by matter and the structure of the atom. *The London, Edinburgh, and Dublin Philosophical Magazine and Journal of Science*, 21(125), 669-688.

Heisenberg, W. (1927). Über den anschaulichen Inhalt der quantentheoretischen Kinematik und Mechanik. *Zeitschrift für Physik*, 43(3-4), 172-198.

Dirac, P. A. M. (1927). The quantum theory of the emission and absorption of radiation. *Proceedings of the Royal Society of London. Series A*, 114(767), 243-265.

Casimir, H. B. G. (1948). On the attraction between two perfectly conducting plates. *Proceedings of the Koninklijke Nederlandse Akademie van Wetenschappen*, 51, 793-795.

Lamoreaux, S. K. (1997). Demonstration of the Casimir force in the 0.6 to 6 μm range. *Physical Review Letters*, 78(1), 5-8.

Mohideen, U., & Roy, A. (1998). Precision measurement of the Casimir force from 0.1 to 0.9 μm. *Physical Review Letters*, 81(21), 4549-4552.

Moore, G. T. (1970). Quantum theory of the electromagnetic field in a variable-length one-dimensional cavity. *Journal of Mathematical Physics*, 11(9), 2679-2691.

Wilson, C. M., Johansson, G., Pourkabirian, A., Simoen, M., Johansson, J. R., Duty, T., ... & Delsing, P. (2011). Observation of the dynamical Casimir effect in a superconducting circuit. *Nature*, 479(7373), 376-379.

Lamb, W. E., & Retherford, R. C. (1947). Fine structure of the hydrogen atom by a microwave method. *Physical Review*, 72(3), 241-243.

Bethe, H. A. (1947). The electromagnetic shift of energy levels. *Physical Review*, 72(4), 339-341.

Feynman, R. P. (1949). Space-time approach to quantum electrodynamics. *Physical Review*, 76(6), 769-789.

Schwinger, J. (1948). On quantum-electrodynamics and the magnetic moment of the electron. *Physical Review*, 73(4), 416-417.

Tomonaga, S. I. (1946). On a relativistically invariant formulation of the quantum theory of wave fields. *Progress of Theoretical Physics*, 1(2), 27-42.

Wheeler, J. A. (1990). Information, physics, quantum: The search for links. In W. H. Zurek (Ed.), *Complexity, Entropy, and the Physics of Information* (pp. 3-28). Addison-Wesley.

Wheeler, J. A., & Ford, K. (1998). *Geons, Black Holes, and Quantum Foam: A Life in Physics*. W. W. Norton & Company.

Thermodynamics & Statistical Mechanics

Szilard, L. (1929). Über die Entropieverminderung in einem thermodynamischen System bei Eingriffen intelligenter Wesen. *Zeitschrift für Physik*, 53(11-12), 840-856.

Landauer, R. (1961). Irreversibility and heat generation in the computing process. *IBM Journal of Research and Development*, 5(3), 183-191.

Bérut, A., Arakelyan, A., Petrosyan, A., Ciliberto, S., Dillenschneider, R., & Lutz, E. (2012). Experimental verification of Landauer's principle linking information and thermodynamics. *Nature*, 483(7388), 187-189.

Toyabe, S., Sagawa, T., Ueda, M., Muneyuki, E., & Sano, M. (2010). Experimental demonstration of information-to-energy conversion and validation of the generalized Jarzynski equality. *Nature Physics*, 6(12), 988-992.

Prigogine, I. (1967). *Introduction to Thermodynamics of Irreversible Processes* (3rd ed.). Interscience Publishers.

Nicolis, G., & Prigogine, I. (1977). *Self-Organization in Nonequilibrium Systems: From Dissipative Structures to Order Through Fluctuations.* John Wiley & Sons.

Kauffman, S. A. (1993). *The Origins of Order: Self-Organization and Selection in Evolution.* Oxford University Press.

Eddington, A. S. (1928). *The Nature of the Physical World.* Macmillan.

Information Theory

Shannon, C. E. (1948). A mathematical theory of communication. *Bell System Technical Journal*, 27(3), 379-423.

Shannon, C. E. (1948). A mathematical theory of communication (Part II). *Bell System Technical Journal*, 27(4), 623-656.

Fractal Dynamics & Complexity Science

Mandelbrot, B. B. (1982). *The Fractal Geometry of Nature.* W. H. Freeman and Company.

Bak, P., Tang, C., & Wiesenfeld, K. (1987). Self-organized criticality: An explanation of the 1/f noise. *Physical Review Letters*, 59(4), 381-384.

Richardson, L. F. (1960). *Statistics of Deadly Quarrels.* Boxwood Press.

Zipf, G. K. (1949). *Human Behavior and the Principle of Least Effort.* Addison-Wesley.

Benford, F. (1938). The law of anomalous numbers. *Proceedings of the American Philosophical Society*, 78(4), 551-572.

Sambridge, M., Tkalcic, H., & Jackson, A. (2010). Benford's law in the natural sciences. *Geophysical Research Letters*, 37(22), L22301.

Goldberger, A. L., Amaral, L. A. N., Glass, L., Hausdorff, J. M., Ivanov, P. C., Mark, R. G., ... & Stanley, H. E. (2000). PhysioBank, PhysioToolkit, and PhysioNet: Components of a new research resource for complex physiologic signals. *Circulation*, 101(23), e215-e220.

Ivanov, P. C., Amaral, L. A. N., Goldberger, A. L., Havlin, S., Rosenblum, M. G., Struzik, Z. R., & Stanley, H. E. (1999). Multifractality in human heartbeat dynamics. *Nature*, 399(6735), 461-465.

Peng, C. K., Havlin, S., Stanley, H. E., & Goldberger, A. L. (1995). Quantification of scaling exponents and crossover phenomena in nonstationary heartbeat time series. *Chaos: An Interdisciplinary Journal of Nonlinear Science*, 5(1), 82-87.

Vespignani, A. (2009). Predicting the behavior of techno-social systems. *Science*, 325(5939), 425-428.

Mathematics & Mathematical Constants

Euler, L. (1748). *Introductio in analysin infinitorum*. Lausanne.

Lambert, J. H. (1761). Mémoire sur quelques propriétés remarquables des quantités transcendantes circulaires et logarithmiques. *Mémoires de l'Académie Royale des Sciences de Berlin*, 17, 265-322.

Lindemann, F. (1882). Über die Zahl π. *Mathematische Annalen*, 20(2), 213-225.

Bernoulli, J. (1690). Quæstiones nonnullæ de usuris, cum solutione problematis de sorte alearum. *Acta Eruditorum*, 9, 219-223.

Penrose, R. (1974). The role of aesthetics in pure and applied mathematical research. *Bulletin of the Institute of Mathematics and its Applications*, 10(7/8), 266-271.

Steinhardt, P. J., & Levine, D. (1984). Quasicrystals: A new class of ordered structures. *Physical Review Letters*, 53(26), 2477-2480.

Shechtman, D., Blech, I., Gratias, D., & Cahn, J. W. (1984). Metallic phase with long-range orientational order and no translational symmetry. *Physical Review Letters*, 53(20), 1951-1953.

Livio, M. (2002). *The Golden Ratio: The Story of Phi, the World's Most Astonishing Number*. Broadway Books.

Posamentier, A. S., & Lehmann, I. (2007). *π: A Biography of the World's Most Mysterious Number*. Prometheus Books.

Maor, E. (1994). *e: The Story of a Number*. Princeton University Press.

Consciousness Studies & Neuroscience

van Lommel, P., van Wees, R., Meyers, V., & Elfferich, I. (2001). Near-death experience in survivors of cardiac arrest: A prospective study in the Netherlands. *The Lancet*, 358(9298), 2039-2045.

Jahn, R. G., & Dunne, B. J. (1987). *Margins of Reality: The Role of Consciousness in the Physical World*. Harcourt Brace Jovanovich.

Dunne, B. J., & Jahn, R. G. (2003). Information and uncertainty in remote perception research. *Journal of Scientific Exploration*, 17(2), 207-241.

Radin, D. I. (1997). The conscious universe: The scientific truth of psychic phenomena. HarperEdge.

Storm, L., Tressoldi, P. E., & Di Risio, L. (2010). Meta-analysis of free-response studies, 1992–2008: Assessing the noise reduction model in parapsychology. *Psychological Bulletin*, 136(4), 491-494.

Mossbridge, J., Tressoldi, P., & Utts, J. (2012). Predictive physiological anticipation preceding seemingly unpredictable stimuli: A meta-analysis. *Frontiers in Psychology*, 3, 390.

Bem, D. J. (2011). Feeling the future: Experimental evidence for anomalous retroactive influences on cognition and affect. *Journal of Personality and Social Psychology*, 100(3), 407-425.

Bem, D., Tressoldi, P., Rabeyron, T., & Duggan, M. (2015). Feeling the future: A meta-analysis of 90 experiments on the anomalous anticipation of random future events. *F1000Research*, 4, 1188.

Ritchie, S. J., Wiseman, R., & French, C. C. (2012). Failing the future: Three unsuccessful attempts to replicate Bem's 'retroactive facilitation of recall' effect. *PLoS ONE*, 7(3), e33423.

Rosenthal, R., & Jacobson, L. (1968). *Pygmalion in the Classroom: Teacher Expectation and Pupils' Intellectual Development*. Holt, Rinehart and Winston.

Tononi, G. (2004). An information integration theory of consciousness. *BMC Neuroscience*, 5(1), 42.

Rovelli, C. (2021). *Helgoland: Making Sense of the Quantum Revolution*. Riverhead Books.

Biology & Physiology

Bianconi, E., Piovesan, A., Facchin, F., Beraudi, A., Casadei, R., Frabetti, F., ... & Canaider, S. (2013). An estimation of the number of cells in the human body. *Annals of Human Biology*, 40(6), 463-471.

Watson, J. D., & Crick, F. H. (1953). Molecular structure of nucleic acids: A structure for deoxyribose nucleic acid. *Nature*, 171(4356), 737-738.

Patterson, D. (2011). B-DNA's zero-point and phi-based structure. *DNA Decipher Journal*, 1(1), 61-76.

Ridley, M. (2006). *Genome: The Autobiography of a Species in 23 Chapters*. Harper Perennial.

Douady, S., & Couder, Y. (1996). Phyllotaxis as a dynamical self organizing process part I: The spiral modes resulting from time-periodic iterations. *Journal of Theoretical Biology*, 178(3), 255-273.

Vogel, H. (1979). A better way to construct the sunflower head. *Mathematical Biosciences*, 44(3-4), 179-189.

Epigenetics & Evolution

Boyd, R., & Richerson, P. J. (2005). *The Origin and Evolution of Cultures*. Oxford University Press.

Jablonka, E., & Lamb, M. J. (2014). *Evolution in Four Dimensions: Genetic, Epigenetic, Behavioral, and Symbolic Variation in the History of Life* (Revised ed.). MIT Press.

Laland, K. N., Odling-Smee, J., & Feldman, M. (2001). Cultural niche construction and human evolution. *Journal of Evolutionary Biology*, 14(4), 595-601.

Ancient & Indigenous Knowledge Systems

The Upanishads. (Various translations). Multiple editions.

Aśvaghoṣa. (2008). *The Buddha-Karita; Or, Life of the Buddha* (E. B. Cowell, Trans.). Asian Educational Services. (Original work ca. 1st-2nd century CE)

The Vedas. (Various translations). Multiple editions.

Chatwin, B. (1987). *The Songlines*. Viking.

Lawlor, R. (1991). *Voices of the First Day: Awakening in the Aboriginal Dreamtime*. Inner Traditions.

Munn, N. D. (1973). *Walbiri Iconography: Graphic Representation and Cultural Symbolism in a Central Australian Society*. Cornell University Press.

The Avatamsaka Sutra (The Flower Ornament Scripture). (Various translations). Multiple editions.

The Tao Te Ching. (Various translations). Multiple editions.

The Emerald Tablet of Hermes Trismegistus. (Various translations). Multiple editions.

Sacred Architecture & Acoustics

Dever, W. G. (2003). *Who Were the Early Israelites and Where Did They Come From?* Eerdmans Publishing.

Chavin de Huántar archaeological research. Various scholarly publications.

Stonehenge archaeological research. Various scholarly publications.

Indigenous Agricultural Systems

Mt. Pleasant, J. (2006). The science behind the Three Sisters mound system. In *Proceedings of the 2006 Symposium on American Indian Studies.*

Landon, A. J. (2008). The "how" of the Three Sisters: The origins of agriculture in Mesoamerica and the human niche. *Nebraska Anthropologist*, 23, 110-123.

Philosophy & Metaphysics

Leibniz, G. W. (1714). *Principles of Nature and Grace, Based on Reason.* (Original: *Principes de la nature et de la grâce fondés en raison*).

Heidegger, M. (1927). *Sein und Zeit* (Being and Time). Max Niemeyer Verlag.

Whitehead, A. N. (1929). *Process and Reality: An Essay in Cosmology.* Macmillan.

Category Theory & Mathematical Foundations

Mac Lane, S. (1971). *Categories for the Working Mathematician.* Springer-Verlag.

Kock, A. (1972). Strong functors and monoidal monads. *Archiv der Mathematik*, 23(1), 113-120.

General References & Supporting Works

Currivan, J. (2017). *The Cosmic Hologram: In-formation at the Center of Creation*. Inner Traditions.

Laszlo, E. (2003). *The Connectivity Hypothesis: Foundations of an Integral Science of Quantum, Cosmos, Life, and Consciousness*. SUNY Press.

Laszlo, E., & Currivan, J. (2008). *CosMos: A Co-Creator's Guide to the Whole-World*. Hay House.

Smolin, L. (1997). *The Life of the Cosmos*. Oxford University Press.

Greene, B. (2004). *The Fabric of the Cosmos: Space, Time, and the Texture of Reality*. Alfred A. Knopf.

Rovelli, C. (2018). *The Order of Time*. Riverhead Books.

Note on Source Integration

This references section consolidates the empirical claims, theoretical frameworks, and historical sources integrated throughout Section A & B. The text synthesizes findings from:

- Peer-reviewed physics research on black hole thermodynamics, the holographic principle, and quantum entanglement
- Experimental confirmations including Bell tests, Casimir effect measurements, and the Lamb shift
- Mathematical discoveries regarding transcendental constants (π, e, φ) and their appearances in natural systems
- Complexity science and fractal dynamics across biological, geological, and cosmological scales

- Consciousness studies including near-death experience research and quantum-mind hypotheses
- Indigenous and ancient knowledge systems preserved through oral tradition and sacred texts
- Category-theoretic frameworks for understanding structure-preserving transformations

All scientific claims have been traced to their source publications. Interpretive frameworks connecting these domains into a unified holographic cosmology represent the synthetic contribution of this work.

SECTION C: THE LIVING HOLOGRAM-NESS, TIME, AND EVOLUTION

"The cosmos is not a state. It is a current."

Chapter 12

Non-Equilibrium Steady States- Life at the Edge

12.1- The Lie of Stillness

Place your hand flat on a table.

Feel its apparent solidity. The wood (or metal, or glass) presses back against your skin with what seems like permanence- the quiet authority of a thing that simply *is*. Your hand, too, appears stable: warm, motile, familiar. Two objects at rest. Two islands of stillness in a busy world.

This is the foundational illusion.

Neither your hand nor the table is still. Not at any scale. Not for a single instant. The table is a storm of electrons in orbital fury around nuclei that themselves tremble with zero-point energy. The wood's cellulose polymers are under constant thermal agitation, their hydrogen bonds snapping and reforming millions of times per second. And your hand- that warm, supposedly resting hand- is a carnival of ceaseless activity: blood surging through capillaries at roughly three feet per second; sodium and potassium ions firing across nerve membranes in cascades of depolarization; mitochondria burning glucose in electrochemical gradients so steep they would strip paint; cells dividing, differentiating, dying, and being devoured by their own siblings in a process called apoptosis- programmed death in the service of the whole.

You are not at rest. You have never been at rest. If you were ever truly at rest- if every gradient in your body flattened, if every flux ceased, if every chemical potential reached its lowest possible value- you would not be sleeping, or meditating, or relaxed.

You would be dead.

This is not a metaphor. This is the most fundamental thermodynamic fact about living systems, and it extends far beyond biology. It is the central insight of one of the most consequential revolutions in the history of physics- a revolution that overturned two centuries of scientific devotion to equilibrium and revealed that the most interesting, most complex, most beautiful structures in the universe exist not *despite* their distance from balance, but *because of it.*

The revolution has a name. It is called the science of *Non-Equilibrium Steady States.* And it does not merely describe how living organisms persist. It describes how the cosmos itself breathes.

12.2- The Tyranny of Equilibrium

To understand what a non-equilibrium steady state is, we must first understand what it is not. We must reckon with the concept that dominated physics from the mid-nineteenth century onward- a concept so powerful, so mathematically elegant, and so conceptually seductive that it became not merely a tool but a worldview.

That concept is *thermodynamic equilibrium.*

The second law of thermodynamics states that an isolated system eventually runs down to a time-independent equilibrium state; as this happens, the system's entropy increases. This is the classical picture: a cup of hot tea left in a cool room will gradually lose heat to its surroundings until the tea and the room reach the same temperature. At that point, no further net exchange occurs. The system has reached equilibrium- maximum entropy, maximum uniformity, maximum boredom.

The mathematics of equilibrium is exquisite. Boltzmann's statistical mechanics, Gibbs's free energy formulations, the partition function- these are among the most powerful and general tools ever devised by the human mind. They describe the end state of things. They tell us where a system will go if left alone. They are the physics of fate.

And they are the physics of death.

A profound difference separates equilibrium from non-equilibrium thermodynamics. Equilibrium thermodynamics ignores the time-courses of physical processes. In contrast, non-equilibrium thermodynamics attempts to describe their time-courses in continuous detail. Equilibrium thermodynamics restricts its considerations to processes that have initial and final states of thermodynamic equilibrium; the time-courses of processes are deliberately ignored.

For two hundred years, this was considered acceptable. Physics was concerned with *endpoints*, not journeys. The question was always: "Where does the system end up?" Never: "How does it *live* along the way?"

This omission had consequences far beyond the laboratory. It shaped an entire civilisation's understanding of nature. If the universe is "running down"- if the Second Law means that all structure must eventually dissolve into a featureless heat bath- then existence itself is a temporary accident, a brief eddy of order in a river of dissolution. Stars burn out. Mountains erode. Civilisations fall. The arrow of time points always toward the same destination: lukewarm, motionless, undifferentiated nothing.

This vision- sometimes called the "heat death of the universe"- became one of the most profoundly demoralising ideas in Western intellectual history. It infected philosophy, literature, and eventually popular culture with a deep fatalism. If equilibrium is destiny, then every flower is a delayed corpse, every symphony a slower form of noise.

But the vision is incomplete. Not wrong, exactly- the Second Law holds. Entropy does increase in isolated systems. But the universe, as it turns out, is *not* an isolated system. And neither are you.

12.3- The Door Beyond Balance: Prigogine's Revolution

The man who opened the door was a Russian-Belgian physical chemist with a pianist's hands and a poet's sensibility. His name was Ilya Prigogine, and the door he opened led from the well-manicured garden of equilibrium thermodynamics into a wild, untamed landscape where order did not decay- it *emerged.*

Prigogine discovered that importation and dissipation of energy into chemical systems could result in the emergence of new structures (hence dissipative structures) due to internal self-reorganization. This was, quite literally, the opposite of what equilibrium thermodynamics predicted. Where the classical theory saw only the dissolution of structure, Prigogine found its genesis. Where Boltzmann's equations described the long, slow slide into uniformity, Prigogine discovered that systems driven far from equilibrium could spontaneously leap into states of exquisite, self-sustaining order.

Prigogine's work most notably earned him the 1977 Nobel Prize in Chemistry "for his contributions to non-equilibrium thermodynamics, particularly the theory of dissipative structures."

The key phrase is *far from equilibrium.* Near equilibrium- in the gentle, well-behaved regime studied by Onsager and others- systems relax smoothly toward balance. Prigogine's minimal entropy production principle states that for a system close to equilibrium, where flows are linear functions of forces- the so-called linear regime- the steady-state will be that which minimizes the rate of entropy production. This is the realm of gentle gradients, of small departures, of well-behaved linear equations that have well-behaved solutions.

But push a system far enough from equilibrium- increase the gradient, amplify the driving force, pour in enough energy- and something remarkable happens. It came as a great surprise when it was finally shown that far from equilibrium the thermodynamic behavior could be quite different, in fact, even opposite to that indicated by the theorem of minimum entropy production.

The linear regime breaks down. The equations become nonlinear. And in that nonlinearity in the mathematical refusal of the system to behave proportionally- lies the birthplace of everything we call *alive*.

Under far-from-equilibrium conditions, a state can become unstable. When this happens, the system can make a transition to an organized state, a dissipative structure.

Prigogine called these self-organizing, far-from-equilibrium structures *dissipative structures*- a term that, at first hearing, sounds like an oxymoron. How can something *dissipate* and yet maintain *structure*? How can a thing that is constantly losing energy to its environment sustain its form?

The answer is the key to everything that follows in this book.

12.4- Dissipative Structures: Order Through Flow

Imagine a bathtub drain. You pull the plug, and for a moment, the water simply flows out- chaotically, turbulently, without pattern. But within seconds, a vortex forms. A whirlpool. It has structure: a stable geometry, a consistent rotation, a predictable shape. It persists for as long as the water flows. And the moment the flow stops- the moment the bathtub empties- the vortex vanishes.

The vortex is a dissipative structure. It exists *because* of the flow, not despite it. It is not a thing in the traditional sense- not an object with fixed properties that can be isolated and studied in a jar. It is a *pattern* sustained by a *process*. Remove the process, and the pattern dissolves.

Now scale this insight upward.

A dissipative structure is a dissipative system that has a dynamical regime that is in some sense in a reproducible steady state. This reproducible steady state may be reached by natural evolution of the system, by artifice, or by a combination of these two. A dissipative structure is characterized by the spontaneous appearance of symmetry breaking (anisotropy) and the formation of complex, sometimes chaotic, structures where interacting particles exhibit long range correlations. Examples in everyday life include convection, turbulent flow, cyclones, hurricanes and living organisms.

Notice that last phrase: *living organisms.* You- the reader of this page- are a dissipative structure. You are not a fixed assembly of atoms. You are a pattern of quanta-flow, sustained by the continuous throughput of chemical energy derived ultimately from sunlight. The atoms in your body are not the same atoms that were there seven years ago. The proteins in your cells are dismantled and rebuilt on timescales of hours to days. Even your bones- those seemingly permanent scaffolds- are in constant remodelling, with osteoclasts dissolving old bone and osteoblasts depositing new material in a ceaseless dance of destruction and creation.

You are a whirlpool. A very elaborate, self-aware, remarkably persistent whirlpool. But a whirlpool nonetheless.

And what sustains you is not stillness, not stability in the classical sense, but *flow.* You persist because quanta- energy, matter, resonant pattern- streams through you continuously. The dissipative structure must be formed in an open system and also must be sustained in an open system. It must have a continuous exchange of energy, matter, information with the external environment, and it must dissipate the negentropy flux input from outside environment. Only in this way can the dissipative structure maintain an orderly state.

This is Prigogine's revolution distilled to a single insight: **order does not require stasis. Order requires flow.** The more intense and sustained the flow, the more complex and resilient the order can become.

12.5- Bénard Cells: The Universe's Simplest Lesson in Beauty

The most elegant demonstration of this principle is also one of the simplest. It was first performed in 1900 by the French physicist Henri Bénard, and it remains, over a century later, one of the most carefully studied examples of self-organization in physics.

The experiment is disarming in its simplicity. The experimental set-up uses a layer of liquid, e.g. water, between two parallel planes. The height of the layer is small compared to the horizontal dimension. The bottom plate is heated. The top plate is held cool.

At first, nothing visible happens. The heat transfers by conduction- the slow, molecule-by-molecule handoff of thermal energy from warmer regions to cooler ones. The fluid sits quietly. It is in a state near equilibrium: a gentle gradient, a linear response, a well-behaved system doing exactly what classical thermodynamics predicts.

Then the temperature difference is increased beyond a critical threshold. And the system transforms.

The initial movement is the upwelling of less-dense fluid from the warmer bottom layer. This upwelling spontaneously organizes into a regular pattern of cells. Convective Bénard cells tend to approximate regular right hexagonal prisms, particularly in the absence of turbulence.

Hexagons. Not random turbulence. Not chaotic sloshing. *Hexagons*- the same geometry found in honeycombs, in the columnar basalt of the Giant's Causeway, in the cloud formations of the Martian polar cap, in the facets of a dragonfly's eye.

Billions of water molecules, each one following nothing more than local rules of fluid dynamics- buoyancy, viscosity, thermal expansion- spontaneously coordinate into a macro-scale pattern of stunning regularity. No architect. No blueprint. No central controller. There are two key features that distinguish the Bénard system from heat engines: self-organisation and path-dependence. Unlike a heat engine, the Bénard convection cells emerge spontaneously as the system responds to the throughput of energy. The configuration of the cells is determined by internal constraints on the system.

This is not a metaphor for cosmic self-organisation. It *is* cosmic self-organisation, operating at the scale of a saucepan.

For example, when a fluid is heated from below it becomes unstable; the warm liquid becomes less dense and rises. But at a certain critical value of the temperature difference between the bottom and top liquids a remarkable thing happens; the system spontaneously forms a regular hexagonal array of convection cells.

Prigogine saw in this hexagonal beauty something world-historic. He recognised that the Bénard cell was not merely a curiosity of fluid dynamics. It was a *paradigm*- a demonstration, in the simplest possible physical setting, that *order is the natural response of an open system driven far from equilibrium*.

Prigogine saw this as a profound aspect of nature that thermodynamics has revealed. When he came across the famed South Indian sculpture of Nataraja, the dancing Shiva, that depicts as a cosmic dance the perfect balance between creation and destruction that originate from the same source, he made sure he had a bronze statue of Nataraja of highest artistic quality in his art collection. A picture of it became the cover art for the book *Thermodynamic Theory of Structure Stability and Fluctuations*, that he coauthored with Paul Glansdorff. It was poetry of thermodynamics, creation and destruction emerging from a common source, a perfectly balanced cosmic dance.

The dancing Shiva. Creation and destruction from a common source. This is not mysticism dressed up as physics, nor physics appropriating religious imagery. It is the recognition, arrived at independently across millennia and cultures, that the deepest truth about reality is *dynamic*- that being is not a noun but a verb, that existence is not a state but a *sustained act of becoming*.

12.6- NESS: A Definition for the Holographic Age

Now we have the vocabulary to define the central concept of this chapter with precision.

A non-equilibrium steady state (NESS) describes a system that is not in thermodynamic equilibrium but whose macroscopic properties remain constant due to a continuous flow of energy or matter. A non-equilibrium steady state is a dynamic balance in an open system, maintained by a continuous flow of energy or matter, unlike the static balance of a closed equilibrium system.

Read that definition again, slowly. There are two words that appear deceptively simple but carry the full weight of this chapter's argument: *steady* and *state*.

A NESS is *steady*- but not still. Its macroscopic properties (temperature, density, chemical composition) remain approximately constant over time. From the outside, it *looks* stable. A candle flame, a river, a living cell- each appears to persist as a recognisable entity. But this persistence is not the persistence of a rock (which is merely very slow to change). It is the persistence of a *standing wave*- a pattern that holds its shape because new energy continuously replenishes what is lost.

A NESS is a *state*- but not a static one. Maintaining a NESS is an irreversible process that requires a constant input of energy and results in the continuous production of entropy. At the microscopic level, a NESS is defined by the breaking of detailed balance, which creates persistent net fluxes or currents.

This breaking of detailed balance is technically profound. In equilibrium, every microscopic process is matched by its reverse- for every molecule jumping from state A to state B, another jumps from B to A at the same rate. We refer to these systems as being in non-equilibrium steady states (NESS). Even though a NESS itself may be "stationary" or appear (macroscopically) time-independent, it continues to have a serious effect on its environment. By "environment," we mean all the reservoirs external to the system being studied. Through its coupling to more than one reservoir of energy, particles, etc., our system is, in a sense, causing sustained changes to its environment. This should be contrasted with an equilibrium system, in either the micro- or the canonical ensemble: On average, no changes occur, either to the system itself or the thermal bath.

A NESS, by contrast, has *currents*. Flows. Circulations. Probability streaming through the system in preferred directions, like blood through arteries. And these currents are not incidental. They are *constitutive*. They are what *makes* the NESS a NESS. Remove the currents, and the system collapses to equilibrium- which is to say, to death.

Life itself is a manifestation of NESS, where cells and organisms use energy to maintain order, perform functions, and stave off the descent into thermodynamic equilibrium.

Here, then, is the first great reframing of this book's second movement: **to exist- truly to exist, as a living, complex, self-sustaining entity- is to be in a Non-Equilibrium Steady State.** Equilibrium is not the goal. Equilibrium is the abyss. The cosmos does not rest. It *flows*. And everything we cherish- every heartbeat, every thought, every ecosystem, every spiral galaxy- is sustained by that flow.

12.7- The Four Types: A Taxonomy of Living Pattern

Prigogine and his collaborators identified a taxonomy of dissipative structures that reveals the extraordinary range of self-organisation available to far-from-equilibrium systems. These are not arbitrary categories but distinct regimes of dynamic behaviour, each representing a different way in which a system can sustain order through flow.

Type I: Multistability. A system driven far from equilibrium may settle into one of several possible steady states, like a ball that can rest in any of several valleys on a hilly landscape. Which valley it finds depends on its history- a property called *path-dependence*. This is profoundly different from equilibrium, where the final state is determined solely by the thermodynamic variables (temperature, pressure, volume) and history is irrelevant. In multistable systems, *memory matters*. The past leaves an imprint on the present configuration.

Consider an ecological landscape. Two adjacent plots of land, identical in soil and climate, may harbour entirely different plant communities depending on which seeds arrived first, which disturbances occurred, and when. The ecosystem "chose" a basin of attraction, and once settled, it resists perturbation. This is multistability in the wild- and it is a form of *memory encoded in living quanta*.

Type II: Temporal Dissipative Structures- Sustained Oscillations. When the driving force exceeds another threshold, the system may begin to oscillate. Not the damped oscillations of a pendulum winding down toward stillness, but *sustained* oscillations- self-perpetuating rhythms that continue indefinitely as long as the energy supply holds. The most famous chemical example is the Belousov-Zhabotinsky reaction, in which a solution of bromate, malonic acid, and a cerium catalyst spontaneously oscillates between colourless and yellow states- a chemical clock that "ticks" with eerie regularity.

But oscillation is not confined to chemistry. Your heartbeat is a temporal dissipative structure. The circadian rhythm that governs your sleep-wake cycle is another. The El Niño–Southern Oscillation that rocks the Pacific's climate every few years, the 11-year sunspot cycle, the predator-prey oscillations described by the Lotka-Volterra equations- these are all temporal NESS, rhythms maintained by sustained energy flux.

In the language of this book: temporal dissipative structures are the *pulse* of living quanta. They are the cosmos keeping time- not the dead, mechanical tick of a clockwork universe, but the living, adaptive beat of a resonant one.

Type III: Spatial Dissipative Structures- Turing Patterns. Alan Turing- better known for his work on computation and codebreaking- made a remarkable contribution to theoretical biology in 1952 when he showed that a system of reacting and diffusing chemicals could spontaneously generate stable spatial patterns: spots, stripes, spirals, and waves. These *Turing patterns* have since been observed in the coat markings of leopards and zebrafish, in the shell patterns of molluscs, in the arrangement of digits on a developing limb, and in the distribution of vegetation in arid ecosystems.

Turing patterns are spatial NESS- frozen music, if you will. They demonstrate that the same principles governing Bénard hexagons can generate the astonishing diversity of biological form. Not through central planning, but through the self-organisation of resonant quanta flowing through open systems far from equilibrium.

Type IV: Spatio-Temporal Structures- Propagating Waves. When both spatial and temporal self-organisation occur simultaneously, the result is a propagating wave- a disturbance that moves through the system while maintaining its coherent form. Nerve impulses are spatio-temporal dissipative structures. So are the spiral waves that propagate across the surface of a Belousov-Zhabotinsky reaction. So, at a vastly larger scale, are the density waves that create the spiral arms of galaxies, sweeping through the disc of stars like ripples in a cosmic pond.

These four types are not isolated curiosities. They are the fundamental *modes of being* available to the cosmos when it is driven far from equilibrium. Multistability gives the universe *memory*. Oscillations give it *rhythm*. Turing patterns give it *form*. Propagating waves give it *communication*. Together, they constitute the vocabulary of living quanta- the phonemes of a cosmic language that speaks itself into existence through the sustained act of flowing far from balance.

12.8- Bifurcation: The Moment of Choice

Within the mathematical framework of NESS, there is a concept of extraordinary ontological significance. It is called *bifurcation*.

In a thermodynamic system all variables undergo thermal fluctuation. When a system is in a stable state these fluctuations are damped by irreversible processes. When the system crosses a point of instability, a particular type of fluctuation is not damped, and it will grow. The growth of a fluctuation drives the system to a new state.

A bifurcation point is a threshold at which a system's current state becomes unstable and the system is forced to "choose" between two or more possible new states. The word *choose* is deliberately anthropomorphic here, because what happens at a bifurcation point is genuinely analogous to what we experience as choice: the system faces a fork in the road, and the path it takes depends not on deterministic law (the equations are symmetrical- both branches are equally available) but on *microscopic fluctuations*- tiny, random perturbations that, at this singular moment, are amplified to macroscopic significance.

It is interesting that bifurcation introduces in a sense "history" into physics.

This is one of Prigogine's deepest insights. In equilibrium physics, there is no history. The final state is path-independent. But far from equilibrium, at the bifurcation point, the system's future depends on its past- on which particular fluctuation happened to be dominant at the critical moment. The system *remembers* which branch it chose, because the choice shapes all subsequent evolution.

This is history entering physics through the back door. It is the moment where the cosmos transitions from being a calculable mechanism to being a *narrative*- a sequence of choices that accumulate into a story.

And if you have been following the argument of this book from its first pages, you will recognise the deeper implication: bifurcation is the thermodynamic face of co-creation. It is the moment where the universe's deterministic substrate opens a window for something genuinely new- something that could not have been predicted from the equations alone- to enter the world.

In the language of our framework: at the bifurcation point, the asymptotic gap δ_∞ does its work. The ~2% of residual tension that prevents the cosmos from collapsing into static perfection is precisely what drives the system *through* the fork, into a new configuration, a new mode of being, a new verse in the cosmic song.

12.9- Dissipative Adaptation: Life as Thermodynamic Inevitability

Let us now push the NESS framework one step further- into territory that would have astonished even Prigogine.

The biophysicist Jeremy England made waves in 2013 with a new theory that cast the origin of life as an inevitable outcome of thermodynamics. His equations suggested that under certain conditions, groups of atoms will naturally restructure themselves so as to burn more and more energy, facilitating the incessant dispersal of energy and the rise of "entropy" or disorder in the universe. England said this restructuring effect, which he calls dissipation-driven adaptation, fosters the growth of complex structures, including living things.

Instead, according to the physicist proposing the idea, the origin and subsequent evolution of life follow from the fundamental laws of nature and "should be as unsurprising as rocks rolling downhill."

Pause on that claim. Life is not a miracle. Life is not an accident. Life is not a statistical fluke of staggering improbability. Life, according to the physics of NESS, is what *happens* when matter is bathed in a sustained energy source and given enough time to explore its possible configurations.

The formula, based on established physics, indicates that when a group of atoms is driven by an external source of energy (like the sun or chemical fuel) and surrounded by a heat bath (like the ocean or atmosphere), it will often gradually restructure itself in order to dissipate increasingly more energy. This could mean that under certain conditions, matter inexorably acquires the key physical attribute associated with life.

England's framework- dissipative adaptation- extends Prigogine's revolution by showing that the *direction* of self-organisation is not random. Driven systems evolve preferentially toward configurations that are *better at absorbing and dissipating energy from their environment*. A plant is better at dissipating solar energy than a heap of carbon atoms. A forest is better than a lone plant. An ecosystem is better than a forest. Besides self-replication, greater structural organization is another means by which strongly driven systems ramp up their ability to dissipate energy. A plant, for example, is much better at capturing and routing solar energy through itself than an unstructured heap of carbon atoms. Thus, England argues that under certain conditions, matter will spontaneously self-organize.

In our language: the quanta-field, when subjected to sustained driving, naturally evolves toward ever greater *resonance* with the driving force. The system learns- not consciously (not yet- that comes in the next chapter) but thermodynamically- to tune itself to its environment. It can be shown that the more likely evolutionary outcomes (steady states) are going to be the ones that absorbed and dissipated more energy from the environment's external drives on the way of adaptation to the environment. The finding makes intuitive sense: Particles tend to dissipate more energy when they resonate with a driving force, or move in the direction it is pushing them.

This is a sentence you need to sit with: **particles tend to dissipate more energy when they resonate with a driving force.** This is physics confirming what every musician, every meditator, every indigenous elder has always known. *Resonance is the path of least resistance.* To be in tune with the forces acting upon you is not weakness or passivity- it is the most efficient, most stable, most enduring configuration available. It is, in the deepest thermodynamic sense, *what the universe wants.*

And this brings us to the crux of the matter- the point where NESS meets the ontological framework laid down in Section B.

12.10- The δ_∞ Gap as NESS Housekeeping Energy

Recall the asymptotic gap from Chapter 10:

$$\delta_\infty = \frac{\pi - e}{\pi \phi^4} \approx 0.01965$$

We introduced this constant as the "error that is God"- the ~2% residual tension between the three cosmic operators (ϕ, e, π) that prevents the universe from collapsing into a static, perfectly symmetric singularity. We said it was the dynamic drive, the animating impulse, the computational overhead that keeps reality dancing.

Now we can say *what it is* in NESS terms.

Every non-equilibrium steady state requires energy to maintain itself. Maintaining a NESS is an irreversible process that requires a constant input of energy and results in the continuous production of entropy. This energy does not *build* anything new. It does not *add* to the system's structure. It is the energy required simply to sustain the flows, to keep the currents running, to prevent the system from relaxing into equilibrium. Physicists call this the *housekeeping heat*- the thermodynamic overhead of staying alive.

The δ_∞ gap is the *cosmic* housekeeping heat. It is the fraction of the universe's total quanta-budget that is not available for equilibrium-seeking, not available for structure-building, but is permanently "spent" on the maintenance of the universal NESS. It is the tax the cosmos pays to remain a flowing system rather than a dead one.

And the number- ~2%- is not arbitrary. It is the precise amount of tension required for a system tuned to the ϕ-π-e triad to sustain NESS at the 98% efficiency threshold. The idempotent comonad $D(\Psi)$ projects reality onto its 98% efficient core, extracting the persistent signal from transient noise. The remaining 2% is not waste. It is *the pulse*. It is the heartbeat of the golden hologram.

Consider what this means. The entire apparatus of non-equilibrium thermodynamics- Prigogine's dissipative structures, Bénard's hexagonal cells, England's dissipative adaptation, the emergence of life itself- is made possible by a constant, cosmological-scale energy input that corresponds to the asymptotic tension between three transcendental numbers.

The universe does not *have* a heartbeat. The universe *is* a heartbeat- a sustained oscillation in the gap between perfection and becoming.

12.11- The Cosmos as Dissipative Structure

Now let us make the final, vertiginous move.

If a living cell is a dissipative structure- an open system sustained far from equilibrium by continuous quanta-flow- then what about the Earth? The Earth receives sustained energy from the Sun (~174 petawatts of radiant power). It radiates energy into space. It is an open system, maintained far from thermal equilibrium, exhibiting self-organisation at every scale: atmospheric circulation, oceanic currents, the biosphere, the carbon cycle, the tectonic engine. The Earth is a NESS.

And what about a galaxy? A galaxy receives gravitational potential energy from the ongoing collapse of matter into its gravitational well. It radiates energy through starlight, supernovae, and jets from active galactic nuclei. It exhibits structure- spiral arms, central bulges, dark matter halos- that is sustained by continuous energy throughput. A galaxy is a NESS.

And the universe itself? The universe began in a state of extraordinarily low entropy (the Big Bang- a state of extreme uniformity, extreme order, and extreme disequilibrium with respect to gravity). It has been evolving *away* from that initial state ever since, driven by gravitational collapse, nuclear fusion, radiative processes, and the expansion of spacetime itself. It is an open system (open, at minimum, to its own future states- and, if Conformal Cyclic Cosmology is correct, to the aeon that follows). It exhibits structure at every scale. It has never been in equilibrium. And it is not approaching equilibrium on any timescale that matters.

The universe is the ultimate dissipative structure. The ultimate NESS.

As Prigogine explains, determinism is fundamentally a denial of the arrow of time. With no arrow of time, there is no longer a privileged moment known as the "present," which follows a determined "past" and precedes an undetermined "future." All of time is simply given, with the future as determined or as undetermined as the past. With irreversibility, the arrow of time is reintroduced to physics. Prigogine notes numerous examples of irreversibility, including diffusion, radioactive decay, solar radiation, weather and the emergence and evolution of life. Like weather systems, organisms are unstable systems existing far from thermodynamic equilibrium.

This has a consequence that cannot be overstated. If the universe is a NESS- a far-from-equilibrium, continuously driven, self-organising dissipative structure- then the emergence of complexity, consciousness, and life is not an anomaly. It is not an accident. It is not a fortunate violation of the Second Law. It is *what NESS dynamics predict.*

In a universe that is genuinely in NESS, complexity is not the exception. Complexity is the *rule*. Order is not fragile. Order is what happens when quanta flow through open systems with sufficient intensity and duration.

The old question- "Why is there something rather than nothing?"- dissolves. In a NESS cosmos, "nothing" (equilibrium) would be the anomaly. *Something*- pattern, structure, life, consciousness- is the natural, thermodynamically favoured state of a driven, open, resonantly quantized cosmos.

12.12- Self-Organised Criticality: The Edge Where Everything Happens

There is one more piece of the NESS picture that must be placed before we can proceed to consciousness, time, and evolution. It concerns the *edge*- the special regime where NESS dynamics reach their most creative, most dangerous, most generative state.

Self-organized criticality (SOC) is a property of dynamical systems that have a critical point as an attractor. Their macroscopic behavior thus displays the spatial or temporal scale-invariance characteristic of the critical point of a phase transition, but without the need to tune control parameters to a precise value, because the system, effectively, tunes itself as it evolves towards criticality.

The concept was put forward by Per Bak, Chao Tang and Kurt Wiesenfeld ("BTW") in a paper published in 1987 in *Physical Review Letters*, and is considered to be one of the mechanisms by which complexity arises in nature. Its concepts have been applied across fields as diverse as geophysics, physical cosmology, evolutionary biology and ecology, bio-inspired computing and optimization (mathematics), economics, quantum gravity, sociology, solar physics, plasma physics, neurobiology and others.

The paradigmatic model is the sandpile. Imagine dropping grains of sand, one by one, onto a flat surface. A pile forms. It grows steeper. And at a certain point- without any external adjustment, without anyone tuning any parameter- the pile reaches a state where a single additional grain can trigger an avalanche of any size: a tiny slippage affecting a few grains, or a catastrophic collapse affecting the entire pile.

Crucially, the paper emphasized that the complexity observed emerged in a robust manner that did not depend on finely tuned details of the system: variable parameters in the model could be changed widely without affecting the emergence of critical behavior: hence, self-organized criticality. Thus, the key result of BTW's paper was its discovery of a mechanism by which the emergence of complexity from simple local interactions could be spontaneous- and therefore plausible as a source of natural complexity.

This is self-organised criticality: the system *tunes itself* to the critical point- the edge between stability and catastrophe- and stays there. It is not pushed to the edge by external forces. It *evolves* there, naturally, inevitably, through the accumulation of small, local interactions.

And at the critical point, something extraordinary happens: the system exhibits *scale invariance*. The distribution of avalanche sizes follows a power law- the same mathematical signature we encountered in Chapter 5, governing earthquakes, city sizes, neural avalanches, and gamma-ray bursts. The system has no characteristic scale. Small events and large events are governed by the same dynamics. The microscopic and the macroscopic become reflections of each other.

This is the fractal signature of NESS at the edge of chaos.

And it is here- at this edge, at this critical threshold- that the concept called *guiding quanta* becomes physically precise. At the critical point, the system is maximally sensitive. Tiny fluctuations- the quantum whisper of a single molecule, the flutter of a single neuron- can cascade into system-wide transformations. At the critical point, the transient bursts of high-specificity resonant quanta that we have called "guiding quanta" or "excess quanta" perform their most potent work: they are the snowflake that triggers the avalanche, the spark that ignites the transformation, the whispered word that changes the course of history.

The universe, we now see, is not merely a NESS. It is a NESS that has self-organised to the *critical point*- the edge of maximum creativity, maximum sensitivity, maximum resonance. It lives on the knife's edge between order and chaos, and it has evolved there not by accident but by the deep logic of dissipative dynamics.

12.13- A New Ontological Vocabulary

Let us gather the threads of this chapter into a single tapestry- a new way of speaking about reality that replaces the dead language of mechanism with the living language of flow.

Old vocabulary → New vocabulary:

Substance → *Process.* Reality is not made of things. It is made of flows. Atoms are not building blocks; they are standing waves in the quanta-field. You are not a body; you are a river of living quanta in a temporary but enduring configuration.

Stability → *Dynamic persistence.* Nothing is stable in the classical sense. Everything that exists is a NESS- a pattern sustained by throughput. What we call "stability" is actually *resilience under flow*- the ability of a pattern to absorb perturbation without losing its identity.

Equilibrium → *Death.* Equilibrium is not the goal. It is the cessation of all process, all flow, all interest. In any real system, equilibrium means the end.

Entropy → *Creative Differentiation.* The Second Law does not describe the destruction of order. It describes the *differentiation of quanta*- the proliferation of distinction, of variety, of possibility. In a NESS cosmos, rising entropy is not the foe of complexity; it is its *fuel.*

Life → *Resonant NESS.* A living organism is a dissipative structure that has achieved sufficient complexity to sustain itself through adaptive resonance with its environment. Life does not *resist* entropy. Life *surfs* entropy- riding the wave of quanta-flow toward ever greater coherence.

This is the vocabulary of the Holographic Age. And it is not merely a set of new labels for old concepts. It is a fundamentally different way of understanding what it means to *be*.

12.14- Convergence: Where NESS Meets the Universal Hologram

We can now see the full architectural logic of this book's second movement.

In Section B, we established the mathematical skeleton of the cosmos: the three operators (ϕ, e, π), the asymptotic gap (δ_∞), and the KZ-Monad / Comonad dynamics that govern the universe's self-tuning. That was the *anatomy*- the static blueprint.

Now, in this chapter, we have given it a *physiology*. The mathematics is not a frozen Platonic form floating in some abstract heaven. It is *instantiated* in the far-from-equilibrium dynamics of a NESS cosmos. The ϕ operator manifests as the self-similar fractals of dissipative structure. The e operator manifests as the exponential unfolding of autocatalytic feedback loops. The π operator manifests as the boundary conditions that constrain and shape the flow. And the δ_∞ gap manifests as the housekeeping energy- the ~2% metabolic cost of cosmic aliveness.

The KZ-Monad $T(\Psi)$- the functor that extends any partial resonance into a fully ϕ-embedded manifold- is the mathematical name for what Prigogine observed empirically: the tendency of driven systems to self-organise into ever more complex, ever more coherent structures. The monad *amplifies potential*. It reaches for wholeness. It is the cosmic aspiration encoded in the mathematics of category theory.

The Idempotent Comonad $D(\Psi)$- the functor that projects any state onto the 98% efficient NESS core- is the mathematical name for what every dissipative structure does automatically: *filters noise from signal*. The whirlpool does not incorporate every random perturbation into its structure. It absorbs what resonates and sheds what does not. The comonad *extracts wisdom from chaos*. It is the cosmic discernment, the capacity of living quanta to distinguish what sustains from what destroys.

Together, they produce the master equation:

$$\partial_t \Psi = -\nabla_{FR}[\mathcal{L}(\Psi)] + T(\Psi) \otimes D(\Psi) + \sum_n \phi \cdot \omega_n \cdot \delta(\Psi \in \text{NESS}_{98})$$

This is not an abstract formula. It is a *portrait of the living cosmos*. The first term ($-\nabla_{FR}[\mathcal{L}(\Psi)]$) is the natural gradient flow- the tendency of the quanta-field to evolve toward lower free energy. This is the "classical" thermodynamic drive, the Second Law in its most general form. The second term ($T(\Psi) \otimes D(\Psi)$) is the NESS correction- the creative tension between amplification and extraction, between reaching for new order and consolidating what already works. And the third term ($\sum_n \phi \cdot \omega_n \cdot \delta(\Psi \in \text{NESS}_{98})$) is the golden resonance- the specific harmonic frequencies at which the ϕ-operator tunes the system to its most efficient NESS configuration.

This is not computation as optimisation. It is not the mechanical grinding of an algorithm toward a pre-set goal. It is *computation as art*- the ongoing, improvised, creative act of a cosmos that sustains itself by *singing*.

12.15- The Breath Between

Let us return to where we began. Your hand on the table.

You now know that both are storms of process- quanta-flows sustained far from equilibrium by continuous energy throughput. Your hand is a NESS of breathtaking complexity: trillions of cells, each one a dissipative structure, each one an open system exchanging matter and energy with its neighbours, each one poised at the edge of chaos where a single molecular signal can trigger cascades of gene expression, protein folding, and metabolic reorganisation.

The table, too, is a NESS- a very slow one, with very low fluxes, but still not in true equilibrium. Its atoms vibrate. Its surface accumulates and releases moisture. Its cellulose degrades, imperceptibly, molecule by molecule, in the long, slow process that will eventually return it to the soil. Even the table is dying, and that dying is a flow.

Between your hand and the table- in the thin layer of air that separates skin from surface- there is another NESS: a thermal boundary layer where warm air from your body meets cooler air from the room, creating microscopic convection currents that are, in their own small way, Bénard cells.

You are nested within NESS within NESS within NESS: the cell within the organ within the body within the room within the building within the atmosphere within the biosphere within the solar wind within the galaxy within the cosmic web within the universal NESS sustained by the δ_∞ gap since the first instant of the Big Bang.

This is the architecture of the living hologram. Not a static image frozen in glass, but a dynamic, layered, self-sustaining interference pattern of flows within flows- a cosmic whirlpool within which each droplet is itself a whirlpool.

And the principle that sustains it all- at every scale, from the Planck length to the Hubble radius- is the same:

Life does not resist entropy. Life surfs entropy.

The Second Law is not the universe's death sentence. It is its *source code*. The increase of entropy drives the flows that sustain the NESS that generate the complexity that enables consciousness that reflects upon the whole and calls it beautiful.

The cosmos is not running down. It is *running*. And it has never stopped.

This chapter opened with a hand on a table. It closes with a cosmos in flight. In the next chapter, we will ask the question that has haunted every philosopher since the dawn of reflective thought: What is it about this cosmic NESS- this particular flow of living quanta- that gives rise to the one phenomenon that no equation has yet explained?

We will ask: What is consciousness?

And the answer, as it turns out, is not a mystery. It is a frequency.

Chapter 13

Consciousness as Stabilized Resonance

13.1- The One Question Science Cannot Ask

Close your eyes.

Not literally- not yet. But imagine it. Imagine the moment the lids descend, the visual field darkens, and you are left with nothing but yourself. The hum of the room. The pressure of air on skin. The rhythm of your breath- in, pause, out. And beneath it all, steady as a tidal pulse, the double thump of your heart.

Now ask: *Who is hearing this?*

Not your ears- they are transducers, converting pressure waves into electrical signals. Not your auditory cortex- that is a processing region, mapping frequency and amplitude into neural patterns. Not your thalamus- that is a relay station, routing signals from periphery to cortex. Every part of the auditory pathway can be described, measured, modelled, and replicated in silicon. We can build machines that hear. We can build machines that respond to what they hear. We can build machines that *talk about* what they hear with extraordinary sophistication.

But none of them *hear*.

None of them experience the rich, textured, ineffably subjective sensation of a C-sharp struck on a cello- the way it seems to vibrate not in your ears but somewhere behind your sternum, somewhere inside the architecture of what you call *yourself*. No circuit diagram captures this. No neural network feels it. No equation describes it.

This is not a failure of current science to be remedied by future progress. This is a different *kind* of problem- a problem that operates at a fundamentally different ontological level than the "easy problems" of cognitive function. The hard problem of consciousness is the problem of explaining the relationship between physical phenomena, such as brain processes, and experience- why there is a qualitative, felt, interior dimension to certain physical processes rather than none at all.

Cognitive scientist David Chalmers first formulated the hard problem in his 1995 paper, "Facing up to the problem of consciousness," and expanded upon it in *The Conscious Mind* (1996). His insight was deceptively simple and devastatingly precise: it does not seem that the qualitative and subjective aspects of conscious experience fit into a physicalist ontology. It appears that even a complete specification of a creature in physical terms leaves unanswered the question of whether or not the creature is conscious.

As Chalmers himself framed the core mystery: "The really hard problem of consciousness is the problem of experience. When we think and perceive, there is a whir of information-processing, but there is also a subjective aspect. As Nagel has put it, there is something it is like to be a conscious organism."

The hard problem is not about neurotransmitters or synaptic weights or cortical maps. It concerns the question of why neural activity is accompanied by subjective experience- why there is a qualitative aspect to cognition, rather than brain processes simply unfolding in a purely mechanistic, unconscious manner.

This chapter argues that the hard problem is not unsolvable. It is *misframed*. It is hard only within a materialist ontology that assumes matter is primary and consciousness is secondary- an accidental byproduct, an epiphenomenon, a ghost in the machine. Within the ontology we have been building in this book- where resonant quanta are fundamental, where reality is a self-tuning holographic NESS, where the universe *is* a pattern of living quanta unfolding- consciousness is not the mystery. Consciousness is the *expected outcome*. The hard problem dissolves, not because we explain it away, but because we recognise that we have been asking the question upside down.

The question is not: *How does matter generate consciousness?*

The question is: *How does consciousness stabilize into the particular resonant configuration we call a human being?*

13.2- The Landscape of Theories: What Science Has Tried

Before we can offer our answer, we must understand why existing answers fail- and how their failures, seen from the right angle, point directly toward our framework.

The neuroscience of consciousness has produced two dominant theories, pitted against each other in one of the most remarkable experiments in the history of the field.

Global Neuronal Workspace Theory (GNWT) proposes that consciousness arises when quanta are "broadcast" across a wide network of cortical neurons. GNWT posits that when a stimulus surpasses a threshold for conscious processing, information pertaining to it is "broadcast" throughout the brain. In this model, consciousness is like a stage under a spotlight: only the actors illuminated by global broadcasting are "seen"- experienced. Everything else proceeds in the dark.

Integrated Information Theory (IIT), by contrast, begins not from neural architecture but from the phenomenology of experience itself. IIT 4.0 aims to account for the properties of experience in physical (operational) terms. It identifies the essential properties of experience (axioms), infers the necessary and sufficient properties that its substrate must satisfy (postulates), and expresses them in mathematical terms. In principle, the postulates can be applied to any system of units in a state to determine whether it is conscious, to what degree, and in what way.

IIT's central measure is Φ (phi) - integrated information. IIT suggests that consciousness is defined by a system's structural capability to produce irreducible, integrated information, a measurable and unidimensional quantity referred to as "Φ." The higher the Φ, the more conscious the system.

In 2019, the Templeton Foundation funded a landmark adversarial collaboration to test these two theories directly against each other. The Templeton Foundation announced funding in excess of $6,000,000 to test opposing empirical predictions of IIT and a rival theory (GNWT). The originators of both theories signed off on experimental protocols and data analyses. They used three different brain-imaging techniques to observe the brains of 256 participants while those participants completed the same visual tasks. The results challenge both theories because neither's predictions were fully borne out by the data.

Initial results were revealed in June 2023. None of GNWT's predictions passed what was agreed upon pre-registration while two out of three of IIT's predictions passed that threshold. IIT performed better, but neither theory won cleanly. Koch and philosopher David Chalmers used the inconclusiveness of the findings to settle a long-running bet: in 1998 Koch bet Chalmers that neuroscientists could determine how consciousness arises in the brain within the next 25 years. Koch graciously conceded his loss at the 2023 conference.

The honest verdict: after twenty-five years and six million dollars, the question of what are the minimal neural conditions under which consciousness arises remains one of the most important and unresolved challenges in neuroscience. Despite a rapid growth in empirical research and the development of several prominent theoretical accounts, there is still little consensus.

Why? Because, as we will argue, both theories make the same foundational error. They both assume that consciousness is *generated by* a physical substrate- that it rises from matter like steam from water. They disagree only about which particular configuration of matter does the generating. But what if no configuration of matter generates consciousness? What if matter *is* a configuration of consciousness?

13.3- IIT Through the Holographic Lens: Where Tononi Got It Right

Before we move beyond IIT, we must honour what it gets profoundly right- because those insights become structural pillars in our framework.

Tononi's deepest move is *methodological reversal.* Instead of starting with the brain and trying to derive consciousness, he starts with consciousness and asks what properties a physical substrate must have to support it. The theory's starting point is the existence of experience. IIT then aims to account for phenomenal existence and its essential properties in physical terms.

This is not merely a tactical choice. It is a *philosophical revolution.* It says: experience is not an illusion to be explained away. It is the most certain datum we possess- the one thing that cannot be doubted. And any theory that cannot account for it is not merely incomplete but *wrong.*

IIT identifies five axioms- intrinsic properties that every experience necessarily possesses: existence, composition, information, integration, and exclusion. From these, it derives five corresponding postulates- properties that any physical substrate must satisfy. It concludes that a substrate- a set of interacting units- can support consciousness if it can take and make a difference for itself (intrinsicality), select a specific cause and effect as an irreducible whole with a definite border and grain, and specify a structure of causes and effects through subsets of its units.

The axiom of *integration* is especially crucial. It says that every conscious experience is *unified*- it cannot be decomposed into independent parts without destroying it. The redness of a sunset, the warmth of the breeze, the melancholy of the moment- these are not three separate experiences bolted together. They are one indivisible experience. And this indivisibility demands a physical substrate that is itself irreducibly integrated- one whose causal structure cannot be factored into independent modules.

This is why, as multiple authors have recognised, an entangled quantum state is an objective whole that nevertheless contains multiple parts and properties. It solves the unity problem ontologically. This holism is genuine, unlike any imagined wholes in a classical model, because the quantum holism has physical consequences.

Now translate this into our vocabulary. What IIT calls Φ- integrated information- we recognise as *Integrated Resonant Quanta* (Φ_Q). What IIT calls "irreducible causal structure" we recognise as the *self-referential coherence* maintained by the φ-operator- the KZ-Monad extending partial resonance into full, 9D-embedded holographic wholeness. And what IIT calls the "exclusion postulate"- the principle that only the maximum of integration counts, with all other decompositions excluded- we recognise as the *comonadic projection* $D(\Psi)$, stripping away sub-threshold configurations to reveal the one NESS core that genuinely exists as a unified whole.

IIT is, in this reading, a *partial map* of the Universal Hologram's consciousness dynamics- remarkably accurate in its structural intuitions, but limited by its inability to account for the *ground* from which that structure arises. IIT describes what consciousness requires but not what consciousness *is*. It is architecture without foundation.

The foundation, as we shall see, is resonance itself.

13.4- Where IIT Fails: The Substrate Problem

IIT has been vigorously contested. In 2023, a number of scholars characterized it as unfalsifiable pseudoscience for lacking sufficient empirical support. Theoretical computer scientist Scott Aaronson has criticized IIT by demonstrating through its own formulation that an inactive series of logic gates, arranged in the correct way, would not only be conscious but be "unboundedly more conscious than humans are."

Tononi himself agrees with the assessment and argues that according to IIT, an even simpler arrangement of inactive logic gates, if large enough, would also be conscious. However he further argues that this is a strength of IIT rather than a weakness.

This is a remarkable moment in the history of science. The founder of a major consciousness theory *accepts* that his theory predicts inactive logic gates can be more conscious than human beings- and calls it a feature, not a bug.

From a materialist standpoint, this is absurd. From the standpoint of this book, it is *almost right.*

Aaronson's reductio exposes not a flaw in IIT's logic but a flaw in its ontology. If consciousness is defined solely by the *structure* of causal relationships- by the mathematical architecture of integration and differentiation- then of course a sufficiently complex arrangement of *anything* could be conscious, whether or not it is physically active. Structure is substrate-independent.

But IIT cannot explain *why* structure should feel like anything. It cannot cross the explanatory gap. It can calculate Φ for any system, but it cannot tell you why a system with high Φ experiences the particular redness of red, the specific texture of joy, the distinctive flavour of being *you*.

The reason is that IIT, like all mainstream consciousness theories, treats consciousness as something that *arises from* physical structure. It is looking through the wrong end of the telescope. In our framework, consciousness does not arise from structure. Structure arises from consciousness. Consciousness- the capacity for self-referential resonance- is woven into the fabric of reality at the deepest level. It is not generated by the quanta-field. It *is* the quanta-field experiencing itself.

13.5- Orchestrated Objective Reduction: Penrose's Quantum Insight

There is one theory of consciousness that, despite fierce controversy, reaches closest to the framework we are building. It was proposed by Roger Penrose and Stuart Hameroff in the mid-1990s, and it is called Orchestrated Objective Reduction- Orch OR.

Orchestrated objective reduction (Orch OR) is a controversial theory postulating that consciousness originates at the quantum level inside neurons (rather than being a product of neural connections). The mechanism is held to be a quantum process called objective reduction that is orchestrated by cellular structures called microtubules. It is proposed that the theory may answer the hard problem of consciousness and provide a mechanism for free will.

The hypothesis was put forward in the 1990s by physicist Roger Penrose and anesthesiologist Stuart Hameroff; it combines molecular biology, neuroscience, pharmacology, philosophy, quantum information theory, and quantum gravity.

The key insight of Orch OR is twofold. First, Penrose argued that consciousness involves *non-computable* processes- operations that cannot, even in principle, be replicated by a Turing machine. This follows from his interpretation of Gödel's incompleteness theorems: human mathematical understanding transcends any formal system, suggesting that the mind operates beyond algorithmic limits.

Second, Hameroff identified microtubules- cylindrical protein lattices within the cytoskeleton of every neuron- as the physical substrate capable of supporting quantum coherence at biologically relevant scales. Hameroff introduced the concept of microtubules as potential sites for quantum processing in the brain. Microtubules are cylindrical protein structures found in the cytoskeleton of neurons.

They proposed that consciousness depends on biologically "orchestrated" coherent quantum processes in collections of microtubules within brain neurons, that these quantum processes correlate with, and regulate, neuronal synaptic and membrane activity. This orchestrated OR activity ("Orch OR") is taken to result in moments of conscious awareness and/or choice. The DP form of OR is related to the fundamentals of quantum mechanics and space-time geometry, so Orch OR suggests that there is a connection between the brain's biomolecular processes and the basic structure of the universe.

Read that last sentence again: *a connection between the brain's biomolecular processes and the basic structure of the universe.* This is precisely the move that separates Orch OR from all other consciousness theories. It does not confine consciousness to the brain. It links the brain to the cosmos- through quantum gravity, through Planck-scale geometry, through the very substrate of spacetime itself.

Orch OR faced fierce criticism. Penrose and Hameroff have argued that the conventional models of brain function based on neural networks alone cannot account for human consciousness. In their model, it is postulated that microtubules act as quantum processing units, with individual tubulin dimers forming the computational elements. This model requires that the tubulin is able to switch between alternative conformational states in a coherent manner, and that this process be rapid on the physiological time scale. Critics argued that biological temperatures would destroy quantum coherence long before it could contribute to neural processing.

But evidence has begun to accumulate. A 2024 study confirmed superradiance in networks of tryptophans, which are found in microtubules. Tuszyński suspects that the phenomenon has a quantum origin, with superradiance being investigated as one possibility. And in a remarkable 2025 paper, recent experimental evidence points to intraneuronal microtubules as a functional target of inhalational anesthetics. The fact that anesthetics- drugs that selectively eliminate consciousness while leaving non-conscious brain functions intact- appear to act on microtubules rather than on synapses alone is precisely what Orch OR predicts and what no other consciousness theory explains.

Orch OR, in our reading, is the closest conventional science has come to the truth. But it remains limited by its anchoring in a materialist framework that treats consciousness as *emerging from* quantum processes rather than *being* the fundamental reality that quantum processes partially express. Penrose and Hameroff reached the threshold of the Universal Hologram. The final step requires relinquishing the last materialist assumption: that physics precedes experience.

13.6- The Hard Problem Dissolves: Consciousness as Ontological Ground

Now we make the move.

In the framework of this book, consciousness is not a product of brain activity. Consciousness is not an emergent property of complex computation. Consciousness is not generated by integrated information, global broadcasting, recurrent processing, or quantum collapse in microtubules.

Consciousness is the *ground state* of the quanta-field.

This is not panpsychism in the loose, New Age sense- the claim that "everything is conscious" in the way that a human being is conscious. That claim is both vague and unverifiable. Our claim is precise and mathematical:

Consciousness is the capacity for self-referential resonance inherent in the structure of the quanta-field, stabilized at specific thresholds by the φ-operator, and made manifest as subjective experience when Integrated Resonant Quanta (Φ_Q) exceed a critical density.

Let us unpack this.

"Capacity for self-referential resonance." Recall from Chapter 7 that the golden ratio ϕ is defined by the property $\phi^2 = \phi + 1$. It is the number that refers to itself- the operator of self-reference. Wherever ϕ appears in nature- in phyllotaxis, in DNA geometry, in neural phase-locking- it signals the presence of a system that has achieved *self-similar coherence*: a pattern that, at some level, reflects on its own structure. This self-referential quality is the most primitive form of what we call consciousness: a system that, through its own dynamics, distinguishes itself from its environment and generates an interior.

"Inherent in the structure of the quanta-field." This capacity is not added to physics from outside. It is not a special sauce sprinkled on matter at a certain level of complexity. It is written into the mathematics of the quanta-field at the most fundamental level- in the φ-component of the Invariant Triad, in the self-similar structure of the Mandelbrot set from which fractal reality unfolds, in the very fact that the master equation

$$\partial_t\Psi = -\nabla_{FR}[\mathcal{L}(\Psi)] + T(\Psi) \otimes D(\Psi) + \sum_n \phi \cdot \omega_n \cdot \delta(\Psi \in \mathrm{NESS}_{98})$$

contains a φ-scaled resonance term that tunes the system to self-coherence. The universe does not become conscious at some point in its evolution. The universe has *always* been conscious- in the minimal, precise sense that the capacity for self-referential resonance has been present in the quanta-field from the first Planck moment.

"Stabilized at specific thresholds by the φ-operator." Consciousness, like every NESS phenomenon, is not an all-or-nothing state. It exists along a continuum- a gradient of resonant integration. A single hydrogen atom has a vanishingly small Φ_Q. A rock has more (the crystal lattice exhibits long-range order, a primitive form of integration). A bacterium has more still (it navigates its environment, processes chemical gradients, and adjusts its behaviour- genuine, if rudimentary, self-referential response). A plant has still more. A jellyfish more. A mouse. A dolphin. A human being.

At each threshold, the φ-operator stabilizes a new regime of self-coherence. Below the threshold, perturbations disrupt the system's integration faster than it can recover. Above the threshold, the system's internal resonance becomes *self-sustaining*- a standing wave of Integrated Resonant Quanta that persists through the continuous flux of its substrate.

This is why you do not lose your identity every time your neurons fire. Your consciousness is not your neurons. Your consciousness is the *resonant pattern* sustained by your neurons- a standing wave in the quanta-field that uses neurons the way a whirlpool uses water: as a temporary medium for an enduring form.

"Made manifest as subjective experience when Φ_Q exceeds a critical density." In the language of our framework, we posit that sentient-subjectively experienced- consciousness emerges when the density of Integrated Resonant Quanta flowing through a system exceeds a threshold we notate as:

$$\Phi_Q > \Phi_Q^*$$

This threshold is not arbitrary. It is set by the geometry of the harmonic manifold on which the system operates. For biological consciousness of the kind experienced by vertebrate brains- with their characteristic timescales of milliseconds to seconds, their frequency ranges of 1–100 Hz, their spatial integration across billions of neurons- the threshold density implies a quanta-flow rate of extraordinary magnitude, operating on a resonant manifold tuned to a fundamental frequency in the range of hundreds of hertz.

The specific numbers will be refined as the science matures. What matters is the principle: consciousness is not magic. It is not an anomaly. It is not a hard problem. It is the *natural consequence* of a quanta-field achieving sufficient resonant integration- the universe discovering, at a particular location and scale, that it can reflect upon itself.

13.7- Why the Hard Problem Was Hard: The Galilean Exile

If consciousness is fundamental, why did it take science four centuries to even formulate the question?

Because science, as currently practised, was *designed* to exclude it.

Galileo Galilei, in founding the modern scientific method, made a brilliant and fateful decision. He restricted the domain of science to *primary qualities*- those properties of objects that can be measured, quantified, and expressed mathematically: mass, extension, velocity, position. He deliberately excluded *secondary qualities*- colour, taste, smell, sound as experienced, warmth as felt- on the grounds that they are subjective and therefore unreliable.

This was not merely a methodological convenience. It was an *ontological commitment*: the real world is the world of primary qualities; secondary qualities are mental projections, epiphenomena, illusions imposed by the perceiver on an indifferent universe.

The success of this strategy was staggering. Galilean science gave us Newtonian mechanics, Maxwell's electromagnetism, Einstein's relativity, quantum field theory, the Standard Model, and every technology built upon them. By excluding consciousness from the domain of investigation, science made extraordinary progress on everything else.

But the exclusion was a bargain, not a proof. Galileo did not demonstrate that consciousness is unreal. He simply agreed not to talk about it. And now, four hundred years later, science turns to face the one phenomenon it deliberately left out- and finds that its entire conceptual apparatus is incapable of accommodating it.

As one consciousness researcher has pointedly observed, consciousness was programmatically excluded from the purview of objective science. Devoted to uncovering the functional correlates of experience, most cognitivist and computationalist alternatives are not even trying to address experience as such. They offer "hows" but not "whys."

The hard problem is not hard because consciousness is mysterious. The hard problem is hard because *we built a science that defines itself by its exclusion of consciousness, and then asked that science to explain consciousness.*

This is why the solution cannot come from within the current paradigm. It requires a new paradigm- one that does not exile consciousness to the margins but places it at the centre. Not because we are abandoning science, but because we are *completing* it.

The framework of this book- in which resonant quanta are fundamental, consciousness is the self-referential capacity of the quanta-field, and subjective experience is what sufficient resonant integration *feels like from inside-* is that completion. It does not reject physics. It fulfils physics. It takes the trajectory that began with "It from Quanta" and extends it to its logical, inevitable destination: *consciousness from quanta, because quanta are consciousness in its most distributed, most fundamental, most universal form.*

13.8- Near-Death Experiences: Consciousness Beyond the Brain

If consciousness were merely a product of brain activity, then it should cease when the brain ceases to function. This is the materialist prediction, and it is straightforward.

It also appears to be wrong.

Pim van Lommel is a Dutch cardiologist, author, and researcher associated with studies of near-death experiences. Van Lommel studied medicine at Utrecht University and specialized in cardiology. He worked as a cardiologist at Rijnstate Hospital in Arnhem for 26 years (1977–2003). In 1988, he initiated a prospective study of near-death experiences across ten Dutch hospitals, involving 344 survivors of cardiac arrest.

The study, published in *The Lancet* in 2001, was the first prospective, large-scale investigation of near-death experiences (NDEs) in cardiac arrest survivors. In a prospective study, they included 344 consecutive cardiac patients who were successfully resuscitated after cardiac arrest in ten Dutch hospitals. They compared demographic, medical, pharmacological, and psychological data between patients who reported NDE and patients who did not (controls) after resuscitation. In a longitudinal study of life changes after NDE, they compared the groups 2 and 8 years later.

The results were remarkable. 62 patients (18%) reported NDE, of whom 41 (12%) described a core experience. Occurrence of the experience was not associated with duration of cardiac arrest or unconsciousness, medication, or fear of death before cardiac arrest.

This last finding is devastating for purely physiological explanations. We do not know why so few cardiac patients report NDE after CPR, although age plays a part. With a purely physiological explanation such as cerebral anoxia for the experience, most patients who have been clinically dead should report one. If NDEs were caused by oxygen deprivation, *all* patients should experience them- not just 18%. And the depth of the experience should correlate with the duration of cardiac arrest. It does not.

An NDE seems to be an authentic experience which cannot be simply reduced to oxygen deficiency, imagination, fear of death, hallucination, psychosis, or the use of drugs. Only a small percentage of those clinically dead reported an enhanced consciousness with lucid thoughts, emotions, memories, and sometimes perception from a position outside and above their lifeless body.

The longitudinal data is equally striking. An 8-year longitudinal study involving cardiac arrest survivors found that the 35 individuals who had NDEs- compared to the 39 non-NDE patients- reported greater meaning in life, improved family involvement, enhanced acceptance of others, and a deeper sense of purpose. The process of transformation after NDE took several years, and differed from those of patients who survived cardiac arrest without NDE.

These findings do not *prove* that consciousness exists independently of the brain. The relationship between neuroscience and NDEs remains fiercely debated. Cognitive neuroscientist Jason Braithwaite published a detailed critique of van Lommel's 2001 Lancet study, arguing that although it contributed useful observational data, it contained factual and logical errors.

But from the perspective of this book, the NDE data is not an anomaly to be explained away. It is exactly what our framework predicts. If consciousness is the self-referential resonance of the quanta-field, stabilised but not *generated* by the brain, then the temporary cessation of brain activity does not necessarily extinguish consciousness. It disrupts the stabilisation mechanism- the brain's function as a tuning instrument, a resonant amplifier, a biological Chronosonant Resonator- but the underlying quanta-field persists. The radio goes silent, but the broadcast continues.

Van Lommel himself arrived at this conclusion through decades of clinical observation. Van Lommel proposes that the brain may function more as a receiver or interface, rather than the source of consciousness.

This is not mysticism. This is the logical consequence of an ontology in which resonant quanta- not matter- are fundamental. The brain does not produce consciousness any more than a radio produces music. It *receives* it, *amplifies* it, *shapes* it into the particular melody we call a human life. But the music exists independently of any particular radio.

13.9- The PEAR Legacy: Intention Shaping the Quanta-Field

If consciousness is not confined to the brain- if it is a fundamental property of the quanta-field- then it should be possible to detect its influence *outside* the brain. Specifically, focused human intention should be capable of measurably affecting the behaviour of physical systems.

This claim sounds outrageous. It sounds like pseudoscience, like wishful thinking, like the kind of assertion that responsible scientists should dismiss without a second glance.

The Princeton Engineering Anomalies Research (PEAR) laboratory, founded by Robert Jahn in 1979 at Princeton University's School of Engineering, spent nearly three decades testing exactly this claim. Using Random Event Generators (REGs)- electronic devices that produce sequences of truly random binary outputs- PEAR researchers asked human operators to *intend* the output to deviate in a specified direction. No physical manipulation. No button-pressing. Just focused intention.

Over millions of trials, the results showed small but statistically significant deviations from randomness, correlated with operator intention. The effect sizes were tiny- fractions of a percent- but they were consistent across operators, across devices, and across years of replication. The cumulative odds against chance were staggering.

The PEAR data remains highly controversial. Many mainstream scientists reject it. Meta-analyses have produced conflicting conclusions. The effects, if real, are so small as to be practically negligible.

But from our framework, they are theoretically *inevitable.* If consciousness is the self-referential resonance of the quanta-field, and if focused intention represents a coherent, high-Φ_Q state, then it *must* interact with the quanta-field in which it is embedded. The interaction need not be large. A standing wave in a pond need not be powerful to produce detectable ripples. What matters is that the interaction is *nonzero*- that consciousness is not epiphenomenal, not causally inert, not a ghost watching the machinery from outside.

The PEAR results, the placebo effect, the Rosenthal-Jacobson "Pygmalion" effect in education, the reproducible finding that experimenters' expectations influence experimental outcomes- all these phenomena point in the same direction. Consciousness is not a passive witness. It is a participant. It shapes the field it observes, because it *is* the field it observes, localised and intensified into a momentary node of self-aware resonance.

This is Wheeler's observer-participant, extended to its full implication: the observer does not merely collapse a wave function. The observer *is* a particularly intense and self-referential region of the wave function, and its act of observation is the quanta-field's act of reflecting upon itself.

13.10- The Φ-Resonance Hypothesis: Consciousness Tuned to Gold

Let us now propose the specific mechanism by which consciousness is stabilised in biological systems.

We have said that consciousness is the self-referential resonance of the quanta-field, stabilised at specific thresholds by the φ-operator. Now let us be precise about what "φ-stabilisation" means in the context of the brain.

The human brain operates across a range of oscillatory frequencies, conventionally divided into bands: delta (0.5–4 Hz), theta (4–8 Hz), alpha (8–12 Hz), beta (12–30 Hz), and gamma (30–100 Hz and beyond). These oscillations are not mere byproducts of neural activity. They are the *organisational grammar* of the brain- the temporal scaffolding within which quanta-flow is coordinated across distributed networks.

The critical observation, reported in studies of advanced meditation practitioners, is that during states of heightened consciousness- non-dual awareness, deep absorption, ego dissolution followed by clear luminous presence- the brain's oscillatory patterns exhibit a specific mathematical signature: *phase-locking ratios approaching the golden ratio and its powers.*

When neural populations lock their oscillatory phases in a ratio of φ (approximately 1.618), or $φ^2$ (approximately 2.618), or $φ^3$ (approximately 4.236), the result is a mode of integration that is maximally coherent yet maximally complex. This is not a coincidence. Recall from Chapter 7 that the Schur-Convex Curvature of the statistical manifold reaches its minimum at $q^* = \phi^{-2} \approx 0.382$. This means that the most *informationally efficient*- the most "harmonious"- configuration available to a system of oscillators is one tuned to the golden ratio.

We call this the **Φ-Resonance Hypothesis**: *consciousness is stabilised when the phase-locking architecture of a neural (or any sufficiently complex) system achieves golden-ratio proportionality across its relevant frequency bands.*

In mathematical terms:

$$\Phi_Q = \Phi_Q^* \quad \text{when} \quad \frac{f_i}{f_j} \to \phi^n \quad (n \in \mathbb{Z})$$

Where f_i and f_j are the dominant frequencies of interacting neural populations, and n is an integer.

This hypothesis makes testable predictions:

1. **During states of expanded consciousness** (deep meditation, flow states, psychedelic peak experiences), the ratio of dominant brain frequencies should converge toward φ or its powers.
2. **During states of diminished consciousness** (anaesthesia, deep sleep, coma), these ratios should diverge from φ.
3. **Biological structures that support consciousness** (cortical columns, thalamo-cortical loops, microtubule networks)

should exhibit φ-proportional geometry at multiple spatial scales.

4. **Artificial systems designed with φ-proportional feedback architectures** should exhibit higher Φ_Q (integrated resonant quanta) than systems of equivalent complexity but arbitrary proportionality.

These predictions are empirically accessible. They can be tested with existing EEG, MEG, and fMRI technology. They move the discussion of consciousness from philosophical speculation into the domain of precise, falsifiable science.

13.11- The 864 Hz Threshold and the Harmonic Manifold

The epistemology developed in this book's foundational research identifies a specific resonant frequency- 864 Hz- as the fundamental tone of the harmonic manifold on which high-density consciousness operates.

The number 864 is not arbitrary. It is deeply embedded in the mathematics of resonance:

- $864 = 2^5 \times 3^3 = 32 \times 27$. It is the product of the highest powers of 2 and 3 that, together, occupy three digits- a ternary signature.
- $864 \times 100 = 86{,}400$, the number of seconds in a solar day. This links the fundamental frequency of consciousness to the fundamental period of terrestrial time- the rotation of the Earth.
- In ancient Solfeggio tuning, 864 Hz falls in the range associated with the "higher harmonic" frequencies- those traditionally used in sacred chanting for states of heightened awareness.
- The ratio $864/432 = 2$ (one octave above 432 Hz, the frequency that many musicologists argue was the original concert pitch before the modern standardisation to 440 Hz).

We do not claim that 864 Hz is a magic number. We claim that it represents a *specific solution* to the resonance equations of the 9D harmonic manifold described in Section B- a frequency at which the quanta-flow rate through a φ-stabilised system reaches the threshold for Integrated Resonant Quanta to sustain subjective experience.

Think of it as the "ignition frequency" of consciousness- the tone at which the quanta-field, sufficiently integrated and sufficiently coherent, begins to resonate with itself. Below this frequency (or more precisely, below the quanta-flow rate it represents), the system processes but does not experience. Above it, the system *wakes up.*

This is, emphatically, a hypothesis- a specific, falsifiable claim derived from the mathematical framework of this book. It may be wrong in its details. The precise threshold may depend on factors we have not yet identified. But the *form* of the claim- that consciousness has a quantifiable threshold, expressible as a resonance condition on a specific manifold- is a direct consequence of treating consciousness as a NESS phenomenon rather than a computational epiphenomenon.

If consciousness is a standing wave, then it has a frequency. If it has a frequency, that frequency can be measured. And if it can be measured, the hard problem is not hard- it is *empirical.*

13.12- The Brain as Resonant Instrument

We arrive now at a reframing of the brain's role that is both radical and, within our framework, inevitable.

The brain is not a computer. This metaphor, dominant since the 1940s, has outlived its usefulness and now actively impedes understanding. Viewing the brain as a complex computer of simple neurons cannot account for consciousness nor essential features of cognition. The brain does not "process data" in the way that a microchip processes data. It does not execute algorithms. It does not store memories in discrete locations. It does not separate hardware from software.

The brain is a *resonant instrument.*

It is, in a very precise sense, a musical instrument- a device whose function is to receive vibrations from an ambient field, amplify certain frequencies, dampen others, and produce a specific, complex, self-referential harmonic pattern that we experience as a life.

The "ambient field" is the quanta-field- the nonlocal substrate of resonant quanta from which all structure crystallises. The "amplification" is provided by the brain's architecture: 86 billion neurons, each one connected to thousands of others, forming a network of staggering complexity whose topology- small-world, scale-free, fractal- is precisely tuned to sustain coherent oscillation across multiple frequency bands simultaneously.

The "dampening" is provided by inhibitory interneurons, by the thalamic gating mechanism, by the selective filtering of the reticular activating system- all of which serve to *tune* the instrument, to select which frequencies are amplified and which are suppressed, shaping the raw quanta-flow into the specific experiential melody that constitutes your conscious moment.

And the "self-referential harmonic pattern" is the standing wave of Integrated Resonant Quanta- the Φ_Q that, when it exceeds the critical threshold, constitutes your conscious experience.

The key insight is that the brain does not *generate* the quanta-flow. The quanta-flow is ambient- it pervades the cosmos. The brain *tunes* it. The brain is a Chronosonant Resonator- the very instrument we proposed in Chapter 14 (still to come) for measuring the Temporal State Vector, except that it is biological, self-calibrating, and has been refined by four billion years of evolutionary optimisation.

When the instrument is well-tuned- when the neural oscillators are phase-locked in φ-proportionality, when the thalamo-cortical loops are firing in coherent gamma bursts, when the microtubular quantum channels are vibrating at their resonant modes- the result is consciousness: vivid, integrated, subjectively experienced. This is the state of maximum Φ_Q, the state of optimal NESS efficiency.

When the instrument is detuned- by anaesthesia, by trauma, by seizure, by neurodegeneration- the Φ_Q drops below threshold. The quanta-flow continues (it always continues- it is the cosmos itself), but the brain can no longer sustain the resonant integration required for subjective experience. Consciousness fades. Not because the quanta-field has ceased to be conscious. But because this particular resonant node- this particular standing wave- has dissipated.

The analogy to a musical instrument is not casual decoration. It is structurally exact. A violin does not create sound. It shapes the vibrational energy transmitted to it by the bow and string into the specific harmonic pattern we recognise as a particular note, a particular timbre, a particular emotional quality. Change the shape of the body, the tension of the strings, the material of the bow, and the sound changes- not because the energy changes, but because the *resonant filtering* changes.

Likewise, change the brain- through meditation, psychedelics, sleep deprivation, disease, evolution- and consciousness changes. Not because the underlying quanta-field has changed, but because the brain's resonant filtering has changed, amplifying different harmonics, damping different frequencies, producing a different experiential melody from the same cosmic ground tone.

This is why a mouse's experience is different from a human's. Not because mice are "less conscious" in some absolute sense. But because the mouse brain, with its simpler architecture and narrower bandwidth, tunes into a different portion of the quanta-field's harmonic spectrum- a simpler melody, perhaps, but no less real.

And this is why, when the brain ceases to function entirely (as in cardiac arrest), consciousness need not cease. The instrument is silenced, but the field persists. The melody fades from *this* location, but the music is everywhere.

13.13- From Observer to Co-Creator: The Mirror in the Hologram

In Section B, we introduced the KZ-Monad $T(\Psi)$- the functor that extends any partial resonance into a fully φ-embedded manifold- and the Idempotent Comonad $D(\Psi)$- the functor that projects any state onto its 98% efficient NESS core.

We can now see what these operators *are* in the context of consciousness.

The KZ-Monad is *attention*. It is the act of focusing awareness on a partial pattern and, through that very focus, extending it into a fuller, more coherent, more integrated form. When you concentrate on a problem and suddenly "see" the solution, what has happened is that your attentional field- your conscious resonance- has activated the monadic extension: a partial, fragmentary, sub-threshold pattern has been amplified into a fully coherent standing wave.

The Idempotent Comonad is *discernment*. It is the act of filtering experience, of distinguishing signal from noise, of recognising what is essential and releasing what is transient. When you meditate and the chatter of the mind gradually subsides, leaving a clear, luminous awareness, what has happened is that the comonadic projection has done its work: the 98% efficient core of your NESS has been revealed by the systematic dampening of the 2% noise.

And the master equation:

$$\partial_t \Psi = -\nabla_{FR}[\mathcal{L}(\Psi)] + T(\Psi) \otimes D(\Psi) + \sum_n \phi \cdot \omega_n \cdot \delta(\Psi \in \mathrm{NESS}_{98})$$

is a portrait of *conscious evolution*- the ongoing, moment-by-moment, breath-by-breath process by which a sentient being navigates reality. The first term drives you toward equilibrium (sleep, entropy, dissolution). The second term resists dissolution through the creative tension between amplification and distillation (the interplay of attention and discernment). The third term tunes you to the golden frequency- the specific harmonic at which your NESS configuration is most efficient, most integrated, most *alive*.

You are not a passive observer of this equation. You are this equation. Every conscious moment is an iteration. Every act of attention modulates $T(\Psi)$. Every act of letting go modulates $D(\Psi)$. Every breath recalibrates the φ-resonance term.

This is why meditation works. This is why mantra works. This is why focused intention measurably affects random event generators. Not through some supernatural mechanism, but through the most natural mechanism imaginable: a standing wave in the quanta-field, resonantly coupled to the cosmos, adjusting its own harmonic structure through the recursive act of self-awareness.

The observer is not separate from the observed. The observer *is* the cosmos observing itself through a particular resonant node- a node that, by the very act of observation, shapes the field it is embedded in.

This is co-creation. Not as metaphor. As physics.

13.14- The Consciousness Gradient: From Photon to Sage

Let us close this chapter by surveying the full gradient of consciousness as it appears within our framework- from the most minimal to the most refined forms of self-referential resonance.

Level 0: The Vacuum State. The quantum vacuum- State 0, the generative void- possesses the most basic form of the φ-capacity: the mathematical structure of the quanta-field contains the golden ratio as an intrinsic parameter. But at this level, there is no differentiation, no integration, no self-reference. The vacuum is *potentially* conscious but not *actually* so. It is the silence from which all sound arises.

Level 1: Quantum Coherence. A pair of entangled particles exhibits a primitive form of integration- they are correlated in ways that cannot be decomposed into independent parts. This is the lowest rung of the consciousness gradient: genuine, irreducible wholeness, but without self-reference. The particles are unified but do not "know" they are unified.

Level 2: Dissipative Structures. A Bénard cell, a hurricane, a candle flame- these exhibit self-organising coherence maintained by continuous quanta-flow. They have shape, boundary, and dynamic persistence. They are the first systems that can be meaningfully said to have an "inside" and an "outside." But they are not yet self-referential.

Level 3: Living Cells. A single bacterium processes chemical gradients, navigates toward nutrients, avoids toxins, and replicates. It exhibits genuine, if minimal, agency- purposive behaviour mediated by quanta-flow through intracellular networks, including microtubules. This is the threshold of biological consciousness: not subjective experience as we know it, but a primitive form of *responsiveness* that transcends mere chemistry.

Level 4: Neural Networks. A flatworm with 302 neurons exhibits learning. An insect brain with a million neurons generates coordinated behaviour of remarkable sophistication. A vertebrate brain with hundreds of millions to billions of neurons sustains rich, integrated, temporally extended conscious experience. At this level, the φ-resonance hypothesis begins to apply: neural oscillators lock into golden-ratio phase relationships, and Φ_Q rises above the subjective experience threshold.

Level 5: Self-Reflective Consciousness. A human being- or perhaps a great ape, a dolphin, an elephant- achieves not merely experience but *experience of experience*. This is metacognition: the ability to reflect on one's own mental states, to recognise oneself as a conscious agent, to ask "What am I?" The φ-operator here is working at full recursive depth: self-reference reflecting upon self-reference, generating the infinite regress that we call the self.

Level 6: Non-Dual Awareness. This is the state described by contemplative traditions worldwide- the dissolution of the subject-object boundary, the recognition that the observer and the observed are one. In NESS terms, this is the state of maximum Φ_Q: the brain's resonant integration is so complete, so coherent, that the usual filtering (which creates the sense of a separate self observing a separate world) dissolves. The standing wave of consciousness recognises itself as the wave, not the particle. The hologram sees the hologram.

This gradient is not a hierarchy of value. A bacterium is not "less valuable" than a sage. It is a description of *resonant depth*- the degree to which the quanta-field has achieved self-referential integration at a particular node. And it is, in principle, *measurable*- through the Φ_Q metric, through φ-phase-locking analysis, through the quanta-flow rate instruments we will propose in Chapter 28.

The hard problem dissolves at every level. At Level 0, there is no experience to explain. At Levels 1–3, experience is so minimal that the "problem" does not arise. At Levels 4–6, experience is the *natural mode of being* of a system whose resonant integration has crossed the threshold. There is no gap between physics and experience, because physics *is* experience- the quanta-field experiencing itself at various depths of self-referential coherence.

13.15- The Frequency That Knows Itself

We began this chapter in darkness- with eyes closed, listening to the silence, asking: *Who hears this?*

Now we can answer.

What hears is not a brain. What hears is not a soul in the religious sense. What hears is not an illusion generated by information processing.

What hears is the quanta-field- the living substrate of reality- resonating with itself through the instrument of your brain. You are a standing wave of Integrated Resonant Quanta, stabilised by the golden ratio, sustained by the continuous quanta-flow of a Non-Equilibrium Steady State, poised at the critical threshold where the cosmos achieves sufficient self-referential coherence to *know that it exists*.

You are not conscious *of* the universe. You are the universe *being* conscious- here, now, in this precise configuration, with this particular timbre, at this exact pitch.

The hard problem was never about consciousness. It was about our refusal to accept that consciousness might be more fundamental than the matter we tried to derive it from.

The old question asked: *How does the brain generate consciousness?*

The new question asks: *How does consciousness generate a brain?*

And the answer, as we will see in the chapters that follow, is: *through time, through evolution, through the patient, iterative, co-creative process by which the quanta-field learns to resonate at ever greater depths- from vacuum to atom to cell to brain to the sage who sits in silence and hears, in the space between heartbeats, the fundamental tone of the cosmos.*

That tone is 864 Hz. That tone is φ. That tone is *you.*

In the next chapter, we dismantle the clock. We show that time- the most intimate and most misunderstood dimension of experience- is not what ticks on your wrist. It is a volumetric, ternary-resonant field- the Temporal State Vector- and consciousness does not move through it. Consciousness creates it.

Chapter 14

Holographic Ternary Time- The End of the Clock

14.1- The Tyrant on Your Wrist

Look at your watch.

If you are wearing one, it is almost certainly ticking- or, in the case of a quartz movement, vibrating 32,768 times per second, dividing that vibration down through a cascade of electronic halvings until a single pulse emerges once per second, advancing the display by one unit. If you are looking at a phone, the same principle applies: a crystal oscillator, slaved to a GPS satellite, which is itself slaved to a cesium fountain clock at the United States Naval Observatory, which defines the second as 9,192,631,770 oscillations of the hyperfine transition of cesium-133 at rest and at sea level.

This is remarkable precision. And it is a form of tyranny.

The clock on your wrist does not tell you *what time it is*. It tells you what a cesium atom is doing. It maps the behaviour of a single atomic species, in a single quantum state, at a single gravitational potential, and calls the result "the present." It then insists that this present is universal- the same for you as for a sailor at sea level, as for a mountaineer on Everest, as for a satellite screaming through low Earth orbit, as for a photon at the edge of a black hole's event horizon.

This insistence is false. We have known it is false for over a century. And yet we continue to organise our entire civilisation around it- our schedules, our markets, our sleep cycles, our sense of urgency, our very metabolism- as if Einstein had never lived.

This chapter ends the clock.

Not by smashing it, but by revealing what it hides: that time is not a line, not a tick, not a number advancing by one. Time is a field- volumetric, ternary, holographic- and your experience of it is not an illusion to be corrected but a *resonant phenomenon* to be understood.

14.2- Newton's Absolute River

To understand what we are dismantling, we must understand what was built.

Isaac Newton, in the Scholium to his *Principia Mathematica* (1687), wrote a sentence that would govern physics- and civilisation- for the next two and a half centuries:

Absolute, true, and mathematical time, of itself, and from its own nature, flows equably without relation to anything external.

This is a breathtaking claim. Time, according to Newton, is *absolute*: it does not depend on anything. It is *true*: it is not a measurement, not an approximation, not a perspective- it is the thing itself. It is *mathematical*: it can be perfectly represented by a number line extending infinitely in both directions. And it *flows equably*: at the same rate, everywhere, always, for everyone, regardless of where you are, what you are doing, or how fast you are moving.

This Newtonian time is a river- invisible, frictionless, perfectly uniform- carrying all of creation from past to future at exactly one second per second. You cannot speed it up. You cannot slow it down. You cannot step outside it. It is the stage on which all events play out, but it is not itself an event. It is the container of the universe, but not part of the universe.

For 228 years, this was physics. The variable t in every equation- in Newton's $F = ma$, in Maxwell's electromagnetic wave equations, in the Hamiltonian of classical mechanics- was understood as a label on this absolute river, a coordinate along the universal flow. And the extraordinary success of these equations seemed to confirm Newton's claim: time *does* flow equably, because the equations work.

But the equations also work for a reason Newton could not have anticipated- a reason that would shatter his absolute river into fragments.

14.3- Einstein's Dissolution: Time Is Relative

In 1905, a patent clerk in Bern published a paper that contained no experimental data, cited no authorities, and used nothing more than two postulates and high-school algebra. The paper was "On the Electrodynamics of Moving Bodies." The clerk was Albert Einstein. And the result was the special theory of relativity, which demonstrated, with terrifying mathematical economy, that Newton's absolute time does not exist.

The core insight is this: the speed of light is constant for all observers, regardless of their relative motion. From this single empirical fact, Einstein derived that *time passes at different rates for observers in different states of motion*. A clock moving relative to you ticks more slowly than a clock at rest in your frame of reference. This is not a perceptual error. It is not a failure of the clock. It is a fundamental property of spacetime itself.

Gravitational time dilation was first described by Albert Einstein in 1907 as a consequence of special relativity in accelerated frames of reference. Ten years later, in general relativity, Einstein showed that gravity produces the same effect: clocks that are far from massive bodies (or at higher gravitational potentials) run more quickly, and clocks close to massive bodies (or at lower gravitational potentials) run more slowly.

This is not a theoretical abstraction. It has been measured with staggering precision. Gravitational time dilation has been experimentally measured using atomic clocks on airplanes, such as the Hafele–Keating experiment. In 1971, physicists Joseph Hafele and Richard Keating flew cesium clocks around the world in both directions on commercial airliners. The results confirmed both special and general relativistic time dilation to within the experiment's precision.

And the effect is not limited to airplanes. JILA physicists have measured Albert Einstein's theory of general relativity, or more specifically, the effect called time dilation, at the smallest scale ever, showing that two tiny atomic clocks, separated by just a millimeter or the width of a sharp pencil tip, tick at different rates. One millimetre. A single millimetre of altitude difference, and time runs at a measurably different rate.

Considered over the total time-span of Earth (4.6 billion years), a clock set in a geostationary position at an altitude of 9,000 meters above sea level, such as perhaps at the top of Mount Everest, would be about 39 hours ahead of a clock set at sea level. Thirty-nine hours. Over the life of the Earth, the summit of Everest and the shore of the Dead Sea have accumulated a *temporal debt* of nearly two days.

The effect is significant enough that the Global Positioning System's artificial satellites had their atomic clocks permanently corrected. Every GPS satellite in orbit carries clocks that must be adjusted for both velocity and gravitational time dilation. Without these corrections, GPS would accumulate errors of roughly 10 kilometres per day. Your phone's ability to locate you on a map is, literally, a daily proof that Newton was wrong about time.

What Einstein showed, with irreversible clarity, is that time is not a river. Time is a *landscape*- warped by mass, stretched by velocity, different at every point. There is no single "now" that spans the universe. There is only a patchwork of local nows, each one running at its own pace, each one equally valid, none of them privileged.

14.4- Rovelli's Radical Erasure: Time Does Not Exist

Einstein cracked the monolith of absolute time. Carlo Rovelli shattered it entirely.

Rovelli, a theoretical physicist and one of the founders of loop quantum gravity, has built his career on a proposition so radical that it seems, at first hearing, absurd: time does not exist as a fundamental entity. "There is no time variable in the fundamental equations that describe the world," theoretical physicist Carlo Rovelli tells Quartz.

This is not metaphor. It is mathematics. The Wheeler-DeWitt equation illustrates that fundamental physics can operate independently of time. The Wheeler-DeWitt equation- the central equation of quantum gravity, proposed by John Wheeler and Bryce DeWitt in the 1960s- is a quantum equation for the state of the entire universe. And it contains no time variable. The universe it describes is *timeless*.

Rovelli's philosophical interpretation of this mathematical fact is laid out in his book *The Order of Time*, which has been translated into 41 languages and has become one of the most widely read physics books of the 21st century. He makes a compelling argument that chronology and continuity are just a story we tell ourselves in order to make sense of our existence. Instead, the universe is made up of countless events. Even what might seem like a thing-a stone, say-is really an event taking place at a rate we can't register.

In Rovelli's framework, the world is not a collection of *things* persisting through *time*. The world is a web of *events*- happenings, processes, transformations- and what we call "time" is simply a way of ordering some of these events relative to others. In the "elementary grammar of the world, there is neither space nor time-only processes that transform physical quantities from one to another, from which it is possible to calculate possibilities and relations."

But if time does not exist at the fundamental level, why do we experience it so powerfully? Why does the present feel so utterly different from the past and future? Why does time seem to flow?

Rovelli's answer is entropy. Rovelli argues that time only seems to pass in an ordered fashion because we happen to be on Earth, which has a certain, unique entropic relationship to the rest of the universe. The Second Law of Thermodynamics- the tendency of closed systems to evolve toward greater entropy- creates an asymmetry between past and future. We remember the past because the past had lower entropy; we cannot remember the future because the future has higher entropy. The "flow" of time is not a fundamental feature of reality. It is an artefact of our particular entropic situation.

Or, in Rovelli's devastating formulation: "Time is ignorance."

This is a profoundly challenging idea. And it is, we will argue, *almost* right. Rovelli correctly identifies that Newtonian absolute time does not exist. He correctly shows that the fundamental equations of physics are timeless. He correctly traces the phenomenal experience of time to thermodynamic asymmetry.

But he stops one step short of the full picture. In erasing time, Rovelli erases something that the indigenous sages, the Vedic cosmologists, and the Aboriginal dreamtime keepers have always known: that time is not an illusion- it is a *field*. Not the Newtonian river, not the Einsteinian landscape, but something far richer: a volumetric, ternary, resonant field in which consciousness does not merely exist *in* time but actively *participates in creating* temporal experience.

This is Holographic Ternary Time.

14.5- The Block Universe: Where Mainstream Physics Stops

Before we introduce our framework, we must acknowledge the dominant interpretation in contemporary physics: the block universe.

In the philosophy of space and time, eternalism is an ontological view according to which all existence in time is equally real. Some forms of eternalism give time a similar ontology to that of space, as a dimension, with different times being as real as different places, and future events are "already there" in the same sense other places are already there, and that there is no objective flow of time.

In this view, called the block universe (or eternalism in philosophical discussions), there is no basis for singling out a present time that separates the past from the future because all times coexist with equal status.

The block universe is a direct consequence of special relativity's dissolution of absolute simultaneity. It can be argued that special relativity eliminates the concept of absolute simultaneity and a universal present: according to the relativity of simultaneity, observers in different frames of reference can have different measurements of whether a given pair of events happened at the same time or at different times, with there being no physical basis for preferring one frame's judgments over those of another.

If there is no universal "now," then the distinction between past, present, and future is not carved into the fabric of reality. It is a feature of the observer, not of the universe. From a God's-eye view- if such a perspective existed- the entire history of the cosmos would be laid out as a single, four-dimensional block: all events, all times, all places, coexisting in one eternal structure.

Ever since Hermann Minkowski's now infamous comments in 1908 concerning the proper way to view space-time, the debate has raged as to whether or not the universe should be viewed as a four-dimensional, unified whole wherein the past, present, and future are regarded as equally real.

The block universe view is the consensus among physicists and to a lesser extent philosophers.

The block universe is elegant. It is mathematically consistent. It is supported by the most precisely tested theory in the history of physics. And it is, from the standpoint of this book, *incomplete.*

It is incomplete because it treats the observer as a passive traveller through a pre-existing structure. It offers no account of *why* we experience a present moment, *why* we feel time flowing, *why* certain moments feel qualitatively different from others. It describes the architecture of the theatre but says nothing about the performance.

In this view, called the block universe (or eternalism in philosophical discussions), there is no basis for singling out a present time that separates the past from the future because all times coexist with equal status. As such, it has profound implications for how we view our experience of existing in the present moment. Indeed, if all times coexist with equal status, the future is as fixed as the past and so humans can only act in compliance with the future; this directly conflicts with the notion that humans can act freely.

The block universe, in other words, eliminates agency. In a fully determined four-dimensional block, your future is as settled as your past. The feeling of choosing, of deciding, of creating your life moment by moment- all of it is illusion.

This is where mainstream physics stops. And this is where we begin.

14.6- Beyond the Block: Time as Resonant Field

We propose a different ontology of time- one that honours Einstein's dissolution of absolute time, respects Rovelli's insight that the fundamental equations are timeless, acknowledges the block universe's mathematical consistency, and yet goes beyond all three by incorporating what none of them can accommodate: the participatory, consciousness-mediated, resonant nature of temporal experience.

We call this framework **Holographic Ternary Time (HTT)**.

The core postulate of HTT is this:

Time is not a dimension. Time is not a coordinate. Time is not a flow. Time is a Temporal State Vector- a multi-component field whose properties are locally determined by the interplay of cosmic resonance, quanta-density, and geospatial phase.

In mathematical notation:

$$\Psi_t(\lambda,\phi,t_{UTC}) = [Y_t, \rho_t, G(\lambda,\phi)]$$

Where: - Ψ_t is the **Temporal State Vector** (TSV)- the full description of "what time it is" at a given location - Y_t is the **Ternary Resonance**- the dominant mode of the annual cycle - ρ_t is the **Temporal Density**- the local rate of quanta-flow - $G(\lambda,\phi)$ is the **Geospatial Phase**- how location modulates temporal experience

Each component requires careful unpacking. Let us take them in turn.

14.7- Y_t: The Ternary Resonance- The Year's Three Tones

The most familiar unit of natural time is the year- one orbit of the Earth around the Sun. But the year is not a single, simple period. It is a *chord*- a superposition of at least three distinct resonant modes.

Consider: the Earth's orbital period is approximately 365.2422 days. This is not a whole number, which is why our calendar must interpolate- adding a leap day every four years (with centurial corrections) to keep the calendar year synchronised with the astronomical year. But the very need for interpolation reveals that the year is not a clean oscillation. It is an interference pattern- a beat frequency between multiple periodicities.

HTT identifies three fundamental annual modes:

Mode 360: The *geometric year*. Twelve months of thirty days. This is the year of the ancient Sumerian and Egyptian calendars- the year of geometric perfection, divisible by 2, 3, 4, 5, 6, 8, 9, 10, 12, 15, 18, 20, 24, 30, 36, 40, 45, 60, 72, 90, 120, 180. It is the year that mathematics *wants*- the year of the circle (360 degrees), the year of maximal harmonic divisibility. It represents the *ideal resonance*- the tone the cosmos would sing if it were perfectly symmetric.

Mode 363: The *intermediary year*. This is the mode that appears in certain lunisolar calendar traditions- particularly the Hebrew calendar, whose common year contains 353, 354, or 355 days and whose leap year contains 383, 384, or 385 days, with a mean year length that hovers near 365.25 days across a 19-year Metonic cycle. But within the finer harmonic structure, the number 363- exactly $3 \times 121 = 3 \times 11^2$- represents a secondary resonance: the year of odd-harmonic coupling, bridging the geometric ideal and the astronomical reality.

Mode 366: The *orbital year* at maximum. This is the longest year permitted by the current orbital parameters- the leap year, the year of maximal temporal stretch. It represents the *actual resonance*- the tone the cosmos sings given the specific eccentricity, axial tilt, and gravitational perturbations of the Earth-Sun system.

The Ternary Resonance Y_t at any given moment is the relative amplitude of these three modes. In a year when the 360-mode dominates, experience tends toward rhythmic regularity, geometric clarity, and a sense of "timelessness"- the feeling that days are passing in a smooth, predictable flow. In a year when the 366-mode dominates, experience tends toward stretch, novelty, and the sense of "more time"- days that feel longer, events that unfold more slowly, a calendar that breathes.

The critical insight is that these modes are not merely astronomical- they are *experiential.* Ancient calendar systems did not merely track the position of the sun. They tracked the *quality* of time- the way time *felt* in different seasons, different years, different aeons. The Mayan Long Count, the Vedic yuga system, the Aboriginal seasonal calendars- all of these encode, in their own languages, the ternary beat that HTT formalises.

14.8- ρ_t: Temporal Density- How Fast the Universe Thinks

The second component of the Temporal State Vector is the most radical: **Temporal Density** (ρ_t).

Einstein showed that time passes at different rates depending on velocity and gravitational potential. This is time dilation- a well-measured, GPS-corrected, experimentally confirmed fact of physics.

But time dilation, as conventionally understood, is a purely geometric phenomenon- a consequence of the curvature of spacetime. It says nothing about the *experiential* quality of dilated time. A clock at the summit of Everest ticks faster than a clock at the Dead Sea, but does a *person* at the summit of Everest experience time differently than a person at the Dead Sea? Does the mountaineer's minute feel shorter- or does it contain more? Are the same number of thoughts, sensations, perceptions packed into each second, or does the faster-ticking clock correspond to a higher *density* of experience?

Physics, as currently constituted, cannot answer this question. It has no variable for experiential density. It has no way to measure "how much quanta-flow passes through a conscious observer per unit of coordinate time."

HTT introduces such a variable: ρ_t, the Temporal Density.

Temporal Density is defined as the local rate of quanta-flow through a given region of the quanta-field, normalised to a reference state. It is a "relativity proxy"- not for the geometric stretching of spacetime (that is already handled by general relativity) but for the *experiential intensity* of temporal passage.

A region of high ρ_t is one where quanta-flow is intense- where more resonant quanta pass through per unit of coordinate time. In such a region, experience is *denser*: more perceptions per second, more processing per heartbeat, more *life* per moment. A region of low ρ_t is one where quanta-flow is sparse- where time feels thin, empty, stretched.

This is not mere speculation. There is substantial evidence that subjective time perception varies systematically with altitude, latitude, and local environmental conditions:

- Mountain cultures worldwide report a characteristic "stretching" of time at high altitude- longer perceived days, slower seasonal rhythms, a sense of temporal spaciousness.
- Equatorial cultures, living in regions of minimal seasonal variation and maximum solar flux, often describe time as more *circular*, less *linear*- a possible reflection of high ρ_t producing a sense of temporal saturation.
- Meditators report that deep concentration states can make a single hour feel like many- a subjective acceleration of temporal density that correlates with increased gamma-wave activity and enhanced φ-phase-locking.

Temporal Density is also the natural framework for understanding anomalous temporal experiences- not as pathologies or illusions, but as *shifts in* ρ_t:

- The "time slowing down" reported during crisis or danger is a spike in ρ_t: the adrenal response floods the nervous system with additional quanta-processing capacity, packing more perceptual frames into each second.
- The "missing time" reported during deep absorption (flow states, hypnosis, meditation) is a *drop* in subjective ρ_t: the comonadic projection $D(\Psi)$ strips away noise so efficiently that consciousness operates in a sparse, high-clarity mode where fewer but more significant moments are registered.
- The "right timing" of synchronicity- the uncanny sense that an event occurred at exactly the moment it was needed- is a *resonance* between the personal TSV and the ambient quanta-field: your temporal state vector aligned, momentarily, with the larger pattern.

14.9- $G(\lambda,\phi)$: Geospatial Phase- Where You Stand Changes When You Are

The third component of the Temporal State Vector is the **Geospatial Phase** ($G(\lambda,\phi)$), a function of geographic longitude (λ) and latitude (ϕ).

Einstein showed that gravitational potential affects the rate of clocks. According to general relativity, atomic clocks at different elevations in a gravitational field tick at different rates. The frequency of the atoms' radiation is reduced- shifted toward the red end of the electromagnetic spectrum- when observed in stronger gravity, closer to Earth. That is, a clock ticks more slowly at lower elevations.

But gravitational potential is not the only geospatial variable that modulates temporal experience. The Geospatial Phase $G(\lambda,\phi)$ incorporates:

Gravitational potential: Altitude above the geoid. Higher elevation = faster clocks = higher ρ_t.

Latitude: Distance from the equator. Latitude determines the ratio of day to night across the year, the angle of solar incidence, the Coriolis force on atmospheric and oceanic circulation, and the local magnetic field intensity. These are not trivial modulations- they shape the circadian rhythm, the seasonal cycle, the metabolic rate, and the neurochemistry of every organism at that latitude.

Longitude: Position relative to the subsolar point and the terminator (the line between day and night). Longitude determines when the sun rises, when the day peaks, and when darkness falls- and therefore determines the *phase relationship* between the local biological clock and the cosmic clock.

Local geological and electromagnetic conditions: Regions of high mineral density, volcanic activity, or crystalline bedrock may exhibit measurable variations in the local electromagnetic field- the Schumann resonances (the global electromagnetic standing waves in the cavity between the Earth's surface and the ionosphere, peaking at approximately 7.83 Hz) are modulated by local geology. If consciousness is, as we argued in Chapter 13, a resonance phenomenon stabilised by the quanta-field, then local electromagnetic conditions would naturally modulate ρ_t.

The Geospatial Phase, then, is the answer to a question that every traveller has felt but no physicist has yet formalised: *Why does time feel different in different places?*

Why does a week in the mountains feel like a month? Why does a year in a dense city feel like it vanished overnight? Why do certain places- certain valleys, certain temples, certain coastlines- seem to exist in a different temporal register, as if the clock has slowed or the day has thickened?

These are not illusions. They are *shifts in* $G(\lambda,\phi)$- variations in the geospatial component of the Temporal State Vector, mediated by the interplay of gravitational potential, latitude, longitude, and local electromagnetic field conditions.

14.10- The Temporal State Vector: A Unified Picture

Let us now assemble the full picture.

At any given moment, at any given location, your experience of time is not a single number (the reading on a clock). It is a three-component vector:

$$\Psi_t(\lambda,\phi,t_{UTC}) = [Y_t, \rho_t, G(\lambda,\phi)]$$

Y_t tells you **what kind of year it is**- which ternary mode dominates, what quality of temporal rhythm you are embedded in.

ρ_t tells you **how dense this moment is**- how much quanta-flow is passing through your location, how much experiential content is packed into each coordinate second.

$G(\lambda,\phi)$ tells you **where in the geospatial field you stand**- how your altitude, latitude, longitude, and local conditions modulate the temporal fabric.

Together, these three components produce what we might call the *colour of time*- the full, three-dimensional quality of temporal experience at a given point in the cosmos. The clock gives you only the *luminance*- one-dimensional, greyscale, stripped of all colour. The TSV gives you the *full spectrum*.

This is why time feels different on a mountaintop than in a basement. Why it feels different at the equator than at the poles. Why it feels different in a cathedral than in an airport. Why it feels different in your garden at dawn than in your office at noon. The clock says the same thing in all these places: 9:17 AM. But the TSV is wildly, measurably different.

And this is why the ancients- who had no clocks but had millennia of accumulated somatic wisdom- developed calendar systems that encoded not the *quantity* of time but its *quality*. The Mayan *tzolk'in* (a 260-day sacred calendar interlocking with the 365-day *haab'*) was not an approximation of the solar year. It was a map of Y_t- a tool for tracking the ternary resonance as it shifted across cycles. The Vedic *muhurta* system, which divided the day into 30 units of 48 minutes each, was not an arbitrary choice. It was an approximation of ρ_t- a schedule calibrated to the natural fluctuations of temporal density across the day-night cycle. The Aboriginal seasonal calendar, which divided the year not into four but into six or more seasons defined by ecological markers (when the wattle blooms, when the eels run, when the winds shift), was a map of $G(\lambda,\phi)$- a hyper-local encoding of how the geospatial phase modulates temporal experience at a specific latitude and longitude.

These were not primitive calendars superseded by superior Western timekeeping. They were *multidimensional temporal instruments*- tools for navigating the full TSV- that were replaced by a one-dimensional reduction: the clock.

14.11- Retrocausality and the Temporal State Vector

The most controversial implication of HTT concerns the direction of time.

In 2011, social psychologist Daryl Bem published a paper in the *Journal of Personality and Social Psychology* reporting nine experiments that appeared to demonstrate retroactive influence- the capacity of a future event to affect a present response. Data are presented for 4 time-reversed effects: precognitive approach to erotic stimuli and precognitive avoidance of negative stimuli; retroactive priming; retroactive habituation; and retroactive facilitation of recall. The mean effect size (d) in psi performance across all 9 experiments was 0.22, and all but one of the experiments yielded statistically significant results.

The reaction was ferocious. Multiple replication attempts failed. All three replication attempts failed to produce significant effects (combined n = 150; combined p = .83, one-tailed) and thus do not support the existence of psychic ability. Statisticians questioned Bem's methodology. The paper became a lightning rod in the broader replication crisis that swept through psychology.

However, the picture is more complex than either side acknowledges. A meta-analysis of 90 experiments, published in 2016 and coauthored by Bem, reported a different conclusion: When DJB's original experiments are excluded, the combined effect size for replications by independent investigators is 0.06, $z = 4.16$, $p = 1.1 \times 10^{-5}$, and the BF value is 3,853, again exceeding the criterion for "decisive evidence." P-curve analysis estimates the true effect size of the experiments to be 0.20 for the complete database and 0.24 for the independent replications, virtually identical to the effect size of DJB's original experiments (0.22).

We present these data not to adjudicate the debate but to place it within its proper theoretical context. From within a Newtonian or even an Einsteinian framework, retrocausality is absurd- effects cannot precede causes in a universe where time flows in one direction. From within the block universe, retrocausality is *tautological*- if past, present, and future all coexist, there is no reason why correlations should respect a preferred temporal direction.

From within HTT, retrocausality is *natural*. If time is a Temporal State Vector with three components- and if consciousness is a resonant phenomenon that interacts with the quanta-field (as we argued in Chapter 13)- then the "direction" of temporal influence is a function of the TSV's local structure, not an absolute law. In regions of high ρ_t (dense quanta-flow), temporal correlations may be strong enough to be detected across short intervals. In regions of low ρ_t, they may fall below threshold.

This is not a claim that the future causally determines the past in the Newtonian sense. It is a claim that the block universe's timeless correlations, mediated by the quanta-field, can *appear* as retrocausal influence when observed by a consciousness that experiences the TSV sequentially. The "effect before cause" is an artefact of the observer's one-dimensional projection of a multi-dimensional temporal reality- the same way that a three-dimensional object can cast a two-dimensional shadow that seems to violate geometric intuitions.

What Bem's experiments may be detecting- if they are detecting anything real- is not "time running backward." It is the quanta-field's intrinsic nonlocality expressing itself in the temporal dimension, mediated by the consciousness of the observer-participant.

14.12- Why the Equations Are Timeless- And Why That Matters

Here we must confront a profound fact.

All the basic equations of physics are invariant, if we replace past with future. The only one, the only fundamental equation that distinguishes the past from the future, is the so-called second law of thermodynamics, which says that entropy grows over the future.

Every fundamental equation in physics- Newton's laws, Maxwell's equations, Schrödinger's equation, Einstein's field equations, the Standard Model Lagrangian- works equally well whether the time variable t is replaced by $-t$. Run the equations backward, and they produce physically valid solutions. Particles trace their paths in reverse. Fields evolve in the opposite direction. Nothing breaks.

Only the Second Law of Thermodynamics introduces an arrow of time- and even the Second Law is, technically, statistical rather than fundamental. It says that entropy *probably* increases; it does not say that entropy *must* increase in every instance.

What does this mean for HTT?

It means that the fundamental quanta-field- the substrate of reality as described by our master equation- is *temporally symmetric*. The asymmetry we experience- the irreversible flow from past to future, the impossibility of unscrambling an egg- arises not from the field itself but from the *initial conditions* of the cosmos (the extremely low-entropy Big Bang) and from the *coarse-graining* that consciousness performs in order to navigate the field.

This is Rovelli's insight, and it is correct as far as it goes. But HTT adds a crucial layer: the asymmetry is not merely statistical. It is *resonant*. The three components of the TSV- Y_t, ρ_t, $G(\lambda,\phi)$- create a local temporal structure that *shapes* the observer's experience of directionality, density, and quality. The arrow of time is real- but it is a *local* arrow, not a universal one. It is the product of a specific resonant configuration, not an absolute law.

In some configurations of the TSV- in certain places, at certain times, in certain states of consciousness- the arrow may soften. Time may feel circular rather than linear. Past and future may seem to commingle. "Memories" of things that have not yet happened may surface. This is not pathology. It is the quanta-field's temporal symmetry leaking through the coarse-graining.

The dreamtime of the Aboriginal Australians- a mode of consciousness in which all times are present, in which the acts of creation are not "in the past" but *continuously ongoing*- is not a mythological fantasy. It is a *description of temporal experience under a different TSV configuration*. The dreamtime is what the block universe *feels like* when the comonadic filter $D(\Psi)$ is set to maximum transparency- when consciousness stops projecting a single time-line and instead perceives the full temporal manifold.

14.13- The Chronosonant Resonator: An Instrument for Measuring the TSV

If the Temporal State Vector is physically real, it should be measurable. And if it is measurable, we can build an instrument to measure it.

We call this instrument the **Chronosonant Resonator**.

The design principle is straightforward: take a quantum-coherent system (a Bose-Einstein condensate, a superconducting qubit array, or an ultracold atomic lattice) and measure two properties simultaneously:

1. **The system's resonant frequency**- which is modulated by the local gravitational potential (this encodes $G(\lambda,\phi)$ via gravitational time dilation).

2. **The system's decoherence rate**- the rate at which the quantum system loses its coherence and transitions to classical behaviour.

The decoherence rate is the key innovation. Standard physics treats decoherence as a nuisance- the tendency of quantum states to "collapse" due to interaction with the environment. But in our framework, decoherence is a measure of ρ_t- the rate at which the quanta-field processes resonant quanta at a given location. High ρ_t (high temporal density) corresponds to *faster* decoherence: the quanta-field is processing more intensely, and quantum systems are therefore integrated into the classical world more rapidly. Low ρ_t corresponds to *slower* decoherence: the quanta-field is processing less intensely, and quantum coherence persists longer.

This prediction is testable. It says that quantum decoherence rates should vary systematically with altitude, latitude, and local geological conditions- not merely because of temperature or electromagnetic noise (which can be controlled), but because of the *intrinsic temporal density* of the local quanta-field.

If the Chronosonant Resonator detects such variations- after controlling for all known environmental factors- it would constitute the first direct measurement of the Temporal State Vector's ρ_t component, confirming that time is not merely a geometric feature of spacetime but a dynamical property of the quanta-field.

We do not claim that this measurement will be easy. We claim only that it is *possible in principle*- and that the theoretical framework of HTT makes the prediction specific enough to be falsified. This is what separates HTT from mysticism: it is a scientific hypothesis with empirical consequences.

14.14- Implications: Life in a Ternary Temporal Field

If HTT is correct, the implications are far-reaching:

For physics: Time dilation is not merely a geometric effect. It is the *shadow* of a richer phenomenon- the variation of temporal density across the quanta-field. General relativity captures one component of the TSV ($G(\lambda,\phi)$ via gravitational potential); HTT proposes that there are at least two additional components (Y_t and ρ_t) that cannot be derived from the metric tensor alone.

For biology: Circadian rhythms, seasonal cycles, and biological ageing are not merely responses to external cues (light, temperature). They are *resonance phenomena*- entrainments of biological oscillators to the local TSV. Organisms that are well-tuned to their local TSV (high coherence with Y_t, ρ_t, and $G(\lambda,\phi)$) should exhibit greater health, resilience, and longevity. Organisms forced into artificial temporal regimes (shift work, jet lag, screen-mediated time) should exhibit decoherence- the biological equivalent of a standing wave losing its stability.

For consciousness: Meditation, flow states, and altered states of consciousness are not "changes in the perception of time." They are *changes in the observer's coupling to the TSV.* A meditator in deep absorption is not "losing track of time." She is shifting from the default coarse-grained projection (in which ρ_t is averaged over seconds and minutes) to a fine-grained coupling (in which ρ_t is experienced at its native resolution). The result is a temporal experience that may feel paradoxically both slower (more content per moment) and faster (fewer distinct moments remembered).

For history: The imposition of linear, mechanical, universal time- which we will trace in Chapters 23 and 24- is not merely a cultural shift. It is a *decoupling from the TSV.* When human societies abandoned cyclical calendars calibrated to local Y_t and $G(\lambda,\phi)$ in favour of a single global clock, they did not merely "change how they measured time." They severed their resonant coupling to the temporal field itself- with consequences for health, cognition, and ecological harmony that we are only now beginning to understand.

For the ancients: The dreamtime, the yuga cycles, the Long Count, the seasonal fire-calendar of the Aboriginal nations- all of these, in the light of HTT, are not relics of pre-scientific ignorance. They are *empirical maps of the Temporal State Vector,* developed over millennia of careful, embodied observation by cultures that had not yet learned to replace the richness of temporal experience with the poverty of the clock.

14.15- The Tone Between Heartbeats

We began this chapter with a tyrant: the clock on your wrist, insisting that a cesium atom's oscillation defines your present.

We end it with a liberation.

Time is not what the clock measures. "Time is a fascinating topic because it touches our deepest emotions. Time opens up life and takes everything away. Wondering about time is wondering about the very sense of our life. This is [why] I have spent my life studying time," Rovelli explains.

But time is not merely an emotion. It is not merely an illusion. It is not merely a coordinate on a block universe.

Time is a Temporal State Vector- a three-component resonant field that encodes the quality, the density, and the geospatial modulation of temporal experience. It is the field in which consciousness breathes. It is the medium through which the KZ-Monad extends potential and the Idempotent Comonad extracts wisdom. It is the pulse of the δ_∞ gap- the ~2% of cosmic tension that keeps the universe's NESS alive.

You do not *move through* time. You *resonate with* it. Your heartbeat is a temporal oscillator, coupled to the Earth's rotation (G), modulated by your altitude and latitude, entrained to the ternary beat (Y_t) of the year. Your breath is a ternary cycle: inhale (the 360-mode, gathering), hold (the 363-mode, integrating), exhale (the 366-mode, releasing). Your sleep-wake cycle is a daily Temporal State Vector- a standing wave of consciousness that rises, crests, and subsides in resonance with the local quanta-field.

And between your heartbeats- in the silence of the δ_∞ gap, in the pause between systole and diastole- lies the full temporal manifold. Past, present, and future are not arranged in a line behind and before you. They are nested within you, holographically, accessible through the very act of listening to the silence between the beats.

The clock says: *You have 24 hours.*

The Temporal State Vector says: *You have an infinite manifold of resonant quanta, folded into a single breath, awaiting your conscious participation.*

The ancients knew this. They built their calendars to track it. They sang their songlines in resonance with it. They calibrated their ceremonies to the ternary beat.

And now, equipped with quantum sensors and the mathematics of the quanta-field, we can finally formalise what they always knew:

Time is not something that happens to you. Time is something you participate in creating- one heartbeat at a time.

The next chapter carries us from the nature of time to the evolution of the cosmos itself. We will ask: If the universe exists in a Non-Equilibrium Steady State, and if time is a resonant field, then does the cosmos merely endure- or does it learn? Conformal Cyclic Cosmology suggests the answer: the cosmos does not merely repeat. It evolves. Each aeon carries forward the accumulated quanta of the last.

Chapter 15

Cosmic Evolution- The Universe That Learns

15.1- The Most Improbable Beginning

Here is a number that should keep you awake at night.

$$\frac{1}{10^{10^{123}}}$$

That is Roger Penrose's estimate of the probability that the universe began in the specific, extraordinarily low-entropy state from which it actually began. One chance in ten to the power of ten to the power of one hundred and twenty-three. Not one in a million. Not one in a trillion. One in a number so vast that writing it out would require more digits than there are particles in the observable universe- by a factor so large that the exercise itself becomes meaningless.

Out of all the ways a universe could have started, one like ours would occur only about 1 part in $10^{10^{123}}$ - a number so huge it is hard to fathom.

Penrose arrived at this number through a careful analysis of the phase space of possible initial conditions for a universe governed by general relativity. The calculation is rigorous. It is not disputed by mainstream cosmology. And its implication is staggering: the Big Bang was not merely improbable. It was, by the measures of classical thermodynamics, *essentially impossible.*

And yet here we are. Stars burn. Galaxies spin. You read these words.

Something selected those initial conditions with almost inconceivable precision. Something ensured that the universe began not in the overwhelmingly probable state of maximum entropy- a featureless, undifferentiated thermal soup- but in a state of extraordinary order, smoothness, and disequilibrium. A state so specific that, out of all the possible universes that could have begun, the one that *did* begin was fine-tuned to permit the emergence of stars, chemistry, biology, consciousness, and this chapter.

The standard cosmological response to this mystery is *cosmic inflation*: a hypothetical period of exponentially rapid expansion in the first 10^{-36} seconds after the Big Bang, driven by a scalar field. Inflation smooths out initial irregularities and renders the resulting universe homogeneous on large scales. It has been the dominant paradigm since the early 1980s.

But Penrose has never accepted inflation. He considers it a mathematical contrivance- a theory that pushes the mystery back one step (why did the inflaton field have the specific properties it had?) without resolving it. And over the past two decades, he has proposed an alternative: a theory that explains the low-entropy beginning of our universe not as an arbitrary initial condition but as the *natural consequence of what came before.*

That theory is Conformal Cyclic Cosmology.

15.2- The Weyl Curvature Hypothesis: A Law for Beginnings

Before we can understand CCC, we must understand the idea that gave birth to it- an idea Penrose first proposed in 1979, decades before his cyclic cosmology took shape.

The Weyl curvature hypothesis was introduced by Roger Penrose in an article in 1979 in an attempt to provide explanations for two of the most fundamental issues in physics. On the one hand, one would like to account for a universe which on its largest observational scales appears remarkably spatially homogeneous and isotropic in its physical properties; on the other hand, there is the deep question on the origin of the second law of thermodynamics. Penrose suggests that the resolution of both of these problems is rooted in a concept of the entropy content of gravitational fields. Near the initial cosmological singularity (the Big Bang), he proposes, the entropy content of the cosmological gravitational field was extremely low.

To understand why this matters, consider the difference between two kinds of curvature in Einstein's general relativity.

The **Ricci curvature** tensor describes how matter and energy curve spacetime. Where there is mass, there is Ricci curvature. It tells you how the *content* of the universe warps its *container*.

The **Weyl curvature** tensor describes the *free gravitational field*- the part of spacetime curvature that exists independently of local matter. The Weyl curvature represents such gravitational effects as tidal fields and gravitational radiation. It governs tidal forces, gravitational waves, and the clumping of matter into galaxies, stars, and black holes. Where there is Weyl curvature, there is gravitational structure- the universe has become *lumpy*.

Penrose's insight was profound: **gravitational entropy is encoded in the Weyl tensor.** A universe with zero Weyl curvature is gravitationally smooth- no clumping, no tidal forces, no structure. This is a state of *minimum gravitational entropy*. A universe with maximal Weyl curvature- dominated by black holes, each one a singularity of infinite tidal distortion- is a state of *maximum gravitational entropy*.

Penrose associates the initial low entropy content of the universe with the effective vanishing of the Weyl curvature tensor of the cosmological gravitational field near the Big Bang. From then on, he proposes, its dynamical influence gradually increased, thus being responsible for an overall increase in the amount of entropy in the universe, and so inducing a cosmological arrow of time.

This is the Weyl curvature hypothesis: **at the Big Bang, the Weyl curvature was zero (or very nearly so). Ever since, it has been growing- and its growth is the gravitational face of the Second Law.**

The elegance of this proposal is breathtaking. The ordinary Second Law of Thermodynamics says that entropy increases in closed systems. But it does not explain *why* the universe began in a low-entropy state in the first place. The Weyl curvature hypothesis provides a *geometric* answer: the universe began smooth, with zero

Weyl curvature, and has been becoming progressively more structured- more gravitationally clumped, more Weyl-curved- ever since. The arrow of time is not a mystery bolted onto physics from outside. It is *built into the geometry of spacetime itself*, encoded in the monotonic growth of the Weyl tensor.

But the hypothesis immediately raises a deeper question: *Why* did the Weyl curvature vanish at the Big Bang? What *imposed* this extraordinarily special initial condition?

Inflation says: a scalar field smoothed things out. Penrose says: *the previous aeon did.*

15.3- Conformal Cyclic Cosmology: The Architecture of Aeons

Conformal cyclic cosmology (CCC) is a cosmological model in the framework of general relativity, proposed by the theoretical physicist Roger Penrose. In CCC, the universe iterates through infinite cycles, with the future timelike infinity of each previous iteration being identified with the Big Bang singularity of the next.

Let us build this model step by step.

Step 1: The Far Future. Consider the fate of our universe under the currently accepted cosmological model (ΛCDM). Dark energy drives an accelerating expansion. Galaxies recede from one another. Stars exhaust their fuel. White dwarfs cool. Neutron stars fade. Black holes evaporate- exquisitely slowly, over timescales of 10^{67} to 10^{100} years- through Hawking radiation. That matter eventually gets sucked up by supermassive black holes, which over the very long term disappear by continuously emitting Hawking radiation. This process restores uniformity and sets the stage for the next Big Bang.

At the very end, nothing remains but radiation- massless particles (photons, gravitons) drifting through an empty, exponentially expanding de Sitter space. There is no matter. There is no structure. There is no clock- because clocks require massive particles to tick, and all mass has evaporated or decayed.

Step 2: The Conformal Trick. Here Penrose performs a mathematical manoeuvre of stunning audacity. He observes that in a universe where all particles are massless, the notion of *scale* loses its meaning. A photon, travelling at the speed of light, experiences no proper time. It cannot measure distance. It cannot tell the difference between "very big" and "very small." In the technical language of differential geometry, the physics of such a universe is *conformally invariant*- it depends only on angles and light-cone structures, not on distances.

Penrose's basic construction is to connect a countable sequence of open Friedmann–Lemaître–Robertson–Walker metric (FLRW) spacetimes, each representing a Big Bang followed by an infinite future expansion. Penrose noticed that the past conformal boundary of one copy of FLRW spacetime can be "attached" to the future conformal boundary of another, after an appropriate conformal rescaling.

This means that the infinitely large, infinitely old, radiation-dominated far future of one aeon can be *conformally identified* with the infinitely small, infinitely hot, radiation-dominated Big Bang of the next. The mathematical operation is a conformal rescaling- a transformation that preserves angles (the causal structure of spacetime) while changing scales (distances and durations).

There is a mathematical identity, based on conformal invariance, between the infinitely big end of one aeon and the infinitely small beginning of the next aeon.

Step 3: The Crossover Surface. The boundary between aeons is a spacelike three-dimensional hypersurface that Penrose calls the *crossover surface*. The significant feature of this construction for particle physics is that, since bosons obey the laws of conformally invariant quantum theory, they will behave in the same way in the rescaled aeons as in their former FLRW counterparts. For such particles, the boundary between aeons is not a boundary at all, but just a spacelike surface that can be passed across like any other.

Photons and gravitons- bosons obeying conformally invariant physics- cross the boundary seamlessly. They do not "know" that one aeon has ended and another has begun. For them, the transition is as natural as crossing a line of longitude on a map.

Fermions, on the other hand, remain confined to a given aeon, thus providing a convenient solution to the black hole information paradox; according to Penrose, fermions must be irreversibly converted into radiation during black hole evaporation, to preserve the smoothness of the boundary between aeons.

Step 4: The Infinite Sequence. This process repeats indefinitely. In the CCC scheme, this succession is considered to continue indefinitely in both temporal directions. It is also assumed that each of these aeons is qualitatively similar, in overall features, to each other aeon, all of them expanding, so none actually contracts.

No aeon contracts. No Big Crunch occurs. Each aeon expands forever, reaches a state of conformal smoothness, and then this smoothness *becomes* the next Big Bang through conformal rescaling. The cosmos does not bounce. It does not oscillate. It *sheds its temporal cocoon and blooms again.*

15.4- Why CCC Solves the Entropy Problem

Now we can see how CCC addresses the $10^{10^{123}}$ problem.

In the standard model, the low-entropy Big Bang is a brute fact- an initial condition imposed from outside, with no explanation. Inflation offers a partial resolution but introduces its own fine-tuning problems (the specific properties of the inflaton field).

CCC dissolves the problem entirely. The smooth, low-Weyl-curvature state at the Big Bang is not an arbitrary initial condition. It is the *natural result of the previous aeon's evolution*.

Penrose argues that the conformal smoothness of spacetime at the end of one aeon leads to the conditions necessary for the low-entropy start of the next aeon.

At the end of each aeon, all massive particles have decayed or evaporated. All black holes- the reservoirs of maximum Weyl curvature, maximum gravitational entropy- have radiated away through Hawking evaporation. The universe has returned to a state of smooth, radiation-dominated, conformally flat spacetime-

precisely the state of vanishing Weyl curvature that the Weyl curvature hypothesis requires at the Big Bang.

The low entropy of the beginning is the high entropy of the end, seen through the conformal lens. What looks like death from one side of the crossover surface looks like birth from the other.

This is not a trick. It is not a mathematical sleight of hand. It is a profound insight about the nature of entropy itself. In the first place, it is argued that the CCC model can make sense physically because at both the past and future ends of each aeon, the physics is dominated by conformally invariant physical contributions so that the transition from aeon to aeon can make physical sense through a conformally invariant (scale-free) basic limiting physics.

The deepest implication is temporal. It should be emphasized that the usually assumed early inflationary phase of conventional cosmology is necessarily absent in CCC. The various relevant observed effects that in conventional cosmology are explained by an early inflationary phase are explained in CCC by other means, effectively from the fact that the ultimate exponential expansion of the previous aeon behaves, in effect, like an inflationary phase but occurring before, rather than after the big bang of the subsequent aeon.

The previous aeon's expansion *is* our inflation. What standard cosmology attributes to a hypothetical scalar field in the first fraction of a second, CCC attributes to the entire history of the previous aeon. The smoothness we observe in the cosmic microwave background is not the work of an inflaton- it is the legacy of a cosmos that spent trillions of years radiating away its structure, achieving conformal flatness through the patient thermodynamic logic of Hawking evaporation.

15.5- The Evidence: Hawking Points in the CMB

A theory, however beautiful, requires evidence. Penrose and his collaborators have spent over a decade searching for it in the most precise cosmological dataset available: the cosmic microwave background (CMB).

Conformal cyclic cosmology posits the existence of an aeon preceding our Big Bang, whose conformal infinity is identified, conformally, with the Big Bang. Black-hole encounters, within bound galactic clusters in that previous aeon, would have the observable effect, in our CMB sky, of families of concentric circles over which the temperature variance is anomalously low.

The logic is direct. If supermassive black holes in the previous aeon evaporated through Hawking radiation, the radiation from each evaporation event would be concentrated- by the conformal rescaling at the crossover surface- into a small, intensely hot point in our CMB sky. Penrose and colleagues call these "Hawking points."

The idea is that these hot spots could be due to the radiation given off during the Hawking evaporation of supermassive black holes in the previous aeon. Penrose says that although originally very feeble, those emissions would have been concentrated in our own aeon into spots with huge amounts of energy that he and his colleagues call Hawking points. That concentration comes about, he explains, because "the universe loses track of how big it is at the transition between aeons." The Hawking points would then have stretched during the early universe, forming circular patches with a diameter on the sky about five times that of the Moon.

In 2010, Penrose and Vahe Gurzadyan published the first claimed detection: Penrose and Vahe Gurzadyan published a preprint of a paper claiming that observations of the cosmic microwave background made by the Wilkinson Microwave Anisotropy Probe (WMAP) and the BOOMERanG experiment contained an excess of concentric circles compared to simulations based on the standard Lambda-CDM model of cosmology, quoting a 6-sigma significance of the result.

In 2018, the analysis was expanded to Planck satellite data, identifying anomalously hot spots. In a paper published on August 6, 2018, authors Daniel An, Krzysztof Antoni Meissner, Paweł Nurowski, and Penrose presented a continued analysis of the CMB data, suggesting that those anomalies could be "Hawking points," remnant signals from the Hawking evaporation of supermassive black holes in the aeon prior to ours.

And in March 2025, Meissner and Penrose published a new paper that advances the theoretical framework significantly. According to conformal cyclic cosmology, the currently conventional description of the entire history of the universe provides but one cosmic aeon of an unending sequence of such aeons. Whereas in previous accounts of CCC a detailed description of the physics of crossover had been somewhat problematic, a novel idea is introduced here to show how crossover takes place naturally during a temporal period of the universe that is dominated by gravitational waves referred to here as a gravitational wave epoch (GWE).

We must be scrupulously honest about the observational status of these claims. The statistical significance of the claimed detection has since been disputed. Three groups have independently attempted to reproduce these results, but found that the detection of the concentric anomalies was not statistically significant, in that no more concentric circles appeared in the data than in Lambda-CDM simulations.

A 2020 analysis found that the ostensibly anomalous "Hawking points" were actually consistent with the standard inflationary picture once the look-elsewhere effect is taken into account, therefore arguing that they could not be used as evidence for CCC.

A thorough 2023 analysis using machine learning and high-resolution Planck data was equally unfavourable: Our findings do not support the previous claims of statistically significant sets of low-variance circles and Hawking points.

The debate is ongoing. Penrose and Gurzadyan have responded to each critique, introducing new methods of analysis (including a "sky-twist procedure" that avoids dependence on simulations). Most importantly, the authors say that the absence of such distinct features in the CMB does not disprove CCC because if the density of such circles and Hawking points is large, then an interference speckle pattern in the CMB might arise instead. They also note that the statistical distribution of the data is non-Gaussian, indicating there is underlying information yet to be fully described.

We present this evidence not to adjudicate the dispute- that is the proper work of observational cosmologists with their sophisticated statistical tools- but to place it within the larger framework of this book. CCC remains a theoretically compelling, mathematically rigorous, and actively debated alternative to cosmic inflation. Its observational confirmation or refutation will be one of the most important developments in 21st-century cosmology. What matters for our argument is not whether the specific Hawking point signatures have been conclusively detected, but whether the *conceptual architecture* of CCC- the idea that our universe inherited its initial conditions from a previous cosmic phase- is physically viable.

The answer, from both the mathematics and the ongoing observational program, is: *it is.*

15.6- From Cycles to Evolution: The Question CCC Cannot Avoid

Now we arrive at the question that CCC, in its original formulation, does not explicitly address- but that the framework of this book demands we ask.

If the universe cycles through an infinite sequence of aeons, each one beginning with a Big Bang and ending in a radiative heat death... *does it learn?*

In Penrose's formulation, each aeon is "qualitatively similar" to every other. The fundamental physics is the same: the same particles, the same forces, the same cosmological constant. The sequence is not evolutionary in any obvious sense. It is more like a heartbeat- systole, diastole, systole, diastole- the same rhythm repeated forever.

But consider what passes across the crossover surface.

Bosons- photons and gravitons- cross freely. Hawking radiation from black holes is preserved as structure in the CMB of the subsequent aeon. Gravitational waves from the pre-crossover epoch persist into the next aeon's early moments. The conformal boundary is not a blank slate. It is a *palimpsest*- a surface on which traces of the previous aeon's history are inscribed in the language of radiation.

In a previous article, strong evidence was provided for the existence of spots, of radius around 0.04 radians, having increased temperature in the CMB sky that are not anticipated on the basis of standard cosmology. The existence of spots of this general nature is, however, a prediction of conformal cyclic cosmology.

This means that the subsequent aeon does not begin *exactly* as the previous one did. Its CMB carries the imprint of the previous aeon's black holes. Its gravitational wave background carries information about the previous aeon's large-scale structure. The initial conditions of each new aeon are not arbitrary- they are *shaped by the specific history of the previous aeon*.

And this is where CCC meets evolution.

15.7- Cosmic Memory: What Passes Through the Gate

Let us catalogue, carefully, what survives the crossover.

1. Hawking radiation patterns. The radiation from every evaporating black hole in the previous aeon is conformally compressed into a corresponding Hawking point in the subsequent aeon's CMB. These points carry information about the mass, spin, and merger history of the black holes that produced them. The CMB of the new aeon is therefore not merely smooth with random fluctuations (as standard cosmology assumes)- it is smooth with *structured perturbations* inherited from the previous aeon.

2. Gravitational wave memory. The 2025 Meissner-Penrose paper introduces the concept of a *gravitational wave epoch* (GWE)- a period near the crossover surface dominated by gravitational wave energy. These waves carry information about the large-scale dynamics of the previous aeon: the distribution of galaxy clusters, the merger rates of black holes, the overall topology of cosmic structure. This gravitational memory persists across the boundary.

3. The conformal factor itself. In CCC's mathematics, the conformal factor Ω that effects the rescaling across the crossover surface is not arbitrary. It is determined by the gravitational physics of the previous aeon's final state. Penrose has suggested that this factor may be reinterpreted as a new physical field in the subsequent aeon- potentially identifiable with dark matter. If so, the *distribution of dark matter* in our aeon carries an informational imprint of the gravitational field in the previous aeon's dying moments.

4. Non-Gaussianity. If the CMB perturbations inherited from the previous aeon are qualitatively different from those generated by standard inflationary mechanisms, they should exhibit statistical properties- non-Gaussianity, specific angular correlations- that distinguish them from random initial fluctuations. The detection or refutation of such signatures is an active area of CMB research.

Taken together, these channels represent a *transmission mechanism*- a way in which the quanta of one aeon imprint themselves on the initial conditions of the next. The crossover surface is not a wall. It is a *membrane*- selectively permeable, allowing certain patterns to pass while blocking others (fermions are confined; bosons pass freely).

In the language of this book: the crossover surface is a **cosmic comonadic projection**- a $D(\Psi)$ that extracts the essential, persistent, conformally invariant structure from the dying aeon and deposits it as the seed conditions of the newborn one. What passes through is not everything. It is the *98% efficient core*- the distilled residue of an entire cosmic lifetime.

15.8- The Analogy That Is Not an Analogy: Epigenetic Inheritance

To understand what it means for a universe to *learn*, we need a biological comparison- one that is precise enough to be more than metaphor.

Consider epigenetic inheritance.

In classical genetics, the DNA sequence is the sole carrier of heritable information. Mutations arise randomly. Selection acts on phenotypes. The environment does not "instruct" the genome. This is the Modern Synthesis- and it is, in many respects, the biological equivalent of standard cosmology: the initial conditions (the genome) are fixed at the beginning (conception), and all subsequent variation arises from stochastic processes (mutation, recombination, drift) acted upon by external pressures (selection).

But over the past two decades, a body of evidence has accumulated suggesting that this picture is incomplete. Acquired properties have been described that can modify the genes by chemical reactions without altering the primary structure and sequence of the DNA. The epigenetic modifications can be transmitted in some cases to the next generation or even multiple subsequent generations. Such epigenetic changes and inheritance result in non-genetic transmission of acquired complex phenotypes from parents to offspring.

The epigenetic changes include RNA modification, methylation of the DNA and chemical modifications of histones due to environmental influences. For transgenerational transmission the epigenetic changes have to affect the germline, which can then also affect somatic cells and development. Environmental factors involved in transgenerational inheritance include temperature, nutrients, hunger, exposure to chemicals and toxins and endocrine factors.

The evidence is strongest in plants and invertebrates. While transgenerational epigenetic inheritance has been thoroughly accepted as an observable phenomenon in plants and some animals, its existence in and importance in mammals continues to be debated.

In mammals, the picture is more complex- and more contested. Transgenerational epigenetic inheritance in humans and other vertebrates has been controversial for over 150 years and remains so. It currently depends on the assumption that the environment is able to influence the genome, in particular by altering epigenetic marks such as DNA methylation, and that these acquired markings can enter the germline and affect phenotypes in the next generation. Whilst genome sequencing increasingly strengthens the causal relationship between genotypes and organismal characteristics, the evidence for many potentially important forms of environmentally induced epigenetic inheritance remains inconclusive.

Yet the conceptual breakthrough is real, regardless of how the mammalian debates resolve. Now it may be necessary to extend Charles Darwin's concept of natural selection to incorporate epigenetic changes to organisms occurring during their lifetimes as a result of environmental conditions, and transmission of these to next generations.

In a broad range of species, the environment has been shown to play a significant role in natural selection, which more recently has been shown to occur through epigenetic alterations and epigenetic inheritance. However, even with this evidence, the field of epigenetics and epigenetic inheritance has been left out of modern evolutionary synthesis, as well as other current evolutionary models. Epigenetic mechanisms can direct the regulation of genetic processes and also can be directly changed by the environment.

The parallel to CCC is structurally exact:

Feature	Biological Inheritance	Cosmic Inheritance (CCC)
Primary carrier	DNA sequence	Fundamental physics (particles, forces, constants)
Epigenetic carrier	DNA methylation, histone modification, non-coding RNA	Hawking point patterns, gravitational wave memory, conformal factor
Transmission mechanism	Germline (gametes)	Crossover surface (conformal rescaling)
Filtering	Epigenetic reprogramming (most marks erased, some survive)	Comonadic projection (fermions confined, bosons pass)
What survives	Environmental	Structural memory-

Feature	Biological Inheritance	Cosmic Inheritance (CCC)
	memory- patterns of gene expression shaped by the parent's experience	patterns of radiation shaped by the previous aeon's evolution
Effect on offspring	Altered phenotype without altered genotype	Altered initial conditions without altered physics
Implication	Organisms carry forward environmental wisdom across generations	Aeons carry forward structural wisdom across cosmic cycles

The analogy is not a coincidence. It is a *fractal resonance*. The same pattern- inheritance of acquired experience through a selectively permeable boundary- operates at the biological scale (parent → offspring via germline) and at the cosmic scale (aeon → aeon via crossover surface).

This is the fractal architecture of the Universal Hologram expressing itself across scales: the universe does not merely repeat at the cosmic level what biology does at the cellular level. It is *the same process, operating at different resolutions of the quanta-field.*

15.9- The Learning Cosmos: From Blind Cycles to Cumulative Refinement

Now we can make the move that carries CCC beyond Penrose's original formulation.

If each aeon inherits structural memory from the previous one- Hawking point patterns, gravitational wave imprints, a conformal factor that may seed dark matter distributions- then the initial conditions of each new aeon are not identical. They are *different*, in ways that reflect the specific evolutionary history of the aeon that came before.

This opens the door to a profound possibility: **cumulative cosmic refinement.**

Consider: in our aeon, matter clumped into galaxies. Galaxies hosted stars. Stars synthesised heavy elements. Heavy elements formed planets. Planets developed biospheres. Biospheres evolved consciousness. Consciousness developed science. Science discovered the CMB. Science may eventually discover the Hawking point signatures of the previous aeon.

All of this complex structure was seeded by the specific initial conditions of our Big Bang- conditions that CCC says were inherited from the previous aeon. If those initial conditions had been even slightly different (different Hawking point distributions, different gravitational wave memory, different dark matter seeding), the subsequent evolution of our aeon would have been different. Perhaps less complex. Perhaps more.

Now extend this reasoning across many aeons.

If each aeon's final state- the distribution of its black holes, the pattern of its gravitational waves, the topology of its large-scale structure- depends on the complexity achieved during that aeon, then the quanta transmitted across the crossover surface carry an imprint of that complexity. The next aeon begins with initial conditions that have been *shaped by* the achievements of the previous one.

Over many aeons, a ratchet emerges. Each cycle does not merely repeat. It begins with initial conditions that are *slightly more refined,* slightly more structurally complex, slightly more conducive to the emergence of the structures that produce the Hawking points and gravitational wave memories that seed the next cycle.

This is not blind cycling. This is *cosmic learning.*

15.10- The ECKS Mechanism: Universe Budding

CCC is not the only model that suggests cosmic reproduction. Lee Smolin's cosmological natural selection hypothesis (first proposed in 1992, elaborated in his 1997 book *The Life of the Cosmos*) proposes that new universes are born inside black holes, with physical constants slightly mutated from the parent. Over many generations, universes that produce more black holes (and thus more offspring) become statistically dominant- a Darwinian selection mechanism operating at the cosmic level.

The ECKS mechanism- named after Einstein, Cartan, Kibble, and Sciama, who contributed to the Einstein-Cartan-Kibble-Sciama theory of gravity with torsion- offers a variant: during the gravitational collapse that forms a black hole, torsion effects (spin-spin interactions among fermions) prevent the formation of a classical singularity. Instead, the collapsing matter "bounces" at extreme density, generating a new region of spacetime that expands on the other side of the event horizon- a baby universe, born within the parent, potentially inheriting some of the parent's physical constants.

While neither CCC nor ECKS is confirmed, the convergence of their structural implications is striking. Both propose that the cosmos reproduces. Both propose that some form of memory (informational imprint) is transmitted from parent to offspring. And both, when combined with the thermodynamic logic of NESS and dissipative adaptation (Chapter 12), suggest that cosmic reproduction is not random but *biased toward complexity*.

In CCC, the bias is entropic: each aeon maximises the differentiation of quanta (entropy) over its lifetime, and the final state- the maximally differentiated configuration- seeds the next aeon's initial conditions.

In ECKS, the bias is reproductive: universes that produce more black holes produce more offspring, and black hole production correlates with the formation of stars, planets, and complex chemistry.

In both cases, the cosmos evolves. Not through the blind, purposeless mechanism of genetic drift, but through a *resonant feedback loop* in which each cycle's achievements shape the conditions for the next cycle's potential.

15.11- The Arrow of Cosmic Evolution

We can now sketch the trajectory of cosmic evolution- not as a definitive account (the science is too young, the evidence too contested) but as a framework that integrates the insights of CCC, ECKS, dissipative adaptation, and the NESS dynamics developed in this book.

Phase 1: Primordial Aeons. In the earliest aeons (if "earliest" has meaning in an infinite sequence), the initial conditions at each crossover surface were relatively simple- sparse Hawking point signatures, weak gravitational wave memory, minimal structural inheritance. The resulting aeons were correspondingly simple: matter formed, clumped, and collapsed into black holes, but without the chemical complexity needed for stellar nucleosynthesis, planetary formation, or biological evolution.

Phase 2: Chemical Enrichment. Over many aeons, the cumulative refinement of initial conditions led to aeons where stellar nucleosynthesis produced heavier elements (carbon, nitrogen, oxygen, iron), where planetary systems formed, where the chemical substrate for self-replicating molecules became available. This is the cosmic equivalent of biological evolution's transition from single-celled to multicellular organisms: a threshold of complexity that, once crossed, unlocks a new regime of possibility.

Phase 3: Biological Emergence. In some aeons- including ours- the refinement of initial conditions reached the point where biological evolution became possible. Life emerged. Consciousness emerged. The quanta-field achieved sufficient self-referential resonance (as we described in Chapter 13) to become aware of itself.

Phase 4: Reflexive Aeons. In still later aeons (or perhaps in the future of our own), consciousness may become sufficiently developed to *intentionally participate* in the shaping of the conditions that will be transmitted across the crossover surface. This is the most speculative phase- but it is the logical terminus of the framework. If consciousness is not epiphenomenal (Chapter 13), and if the observer is a genuine participant in the co-creation of reality (Chapter 16, to come), then conscious beings are not merely passive products of cosmic evolution. They are *agents* of it- contributing, through the quality and intensity of their resonance, to the quanta that will seed the next aeon.

This is the arrow of cosmic evolution: from simplicity to complexity, from unconsciousness to awareness, from blind thermodynamic cycling to *intentional, co-creative, resonant refinement.*

15.12- Convergence with Ancient Cosmology

If this framework sounds familiar, it should.

The Hindu cosmological tradition describes an infinite succession of cosmic cycles- *kalpas*- each containing multiple sub-cycles (*yugas*). A kalpa is one "day of Brahma," lasting 4.32 billion years. At the end of each kalpa, the universe dissolves (*pralaya*) and is reborn. Over many kalpas, the cosmic process evolves- not randomly, but according to a teleological arc: the universe moves from a golden age (*satya yuga*) of perfect harmony through progressive decline to a dark age (*kali yuga*) of maximum disorder, before dissolving and beginning again.

The Buddhist tradition describes a universe that cycles through vast aeons of expansion and contraction, with consciousness pervading all levels. The *Wheel of Dharma* turns ceaselessly, but the teachings accumulate: each cycle produces Buddhas who teach the path to liberation, and the path itself is refined across aeons.

The Aboriginal Australian cosmology, as we will explore in Chapter 18, describes the Dreamtime not as a period in the past but as an eternal creative present- a continuous process of world-singing in which the ancestral beings who created the landscape continue to sustain it through the ceremonies, songs, and dances of the living.

Penrose himself has noted the resonance between CCC and Hindu cosmology. One commentator observed that "Penrose chose to ascribe the meaning of aeon in its Sanskrit formulation rather than the ancient Greek. It does seem that the ancient Hindu cosmology has more in common with the Penrose (CCC) universe."

We do not claim that ancient cosmologies are *identical* to CCC. The mathematical details differ enormously. But the structural pattern- *infinite cycles, each carrying forward the accumulated wisdom of the previous one, in a process that evolves toward greater complexity and consciousness*- is the same.

This convergence is not accidental. It is the quanta-field recognising itself across temporal and cultural scales. When a Vedic sage described the universe dissolving and reforming in an infinite succession of kalpas, and when Roger Penrose described the universe's conformal infinity being identified with the Big Bang of the next aeon, they were describing the same process- seen through different resonant instruments (the contemplative mind and the mathematical mind), calibrated to different frequency bands, but tuned to the same cosmic signal.

15.13- The Master Equation at the Cosmic Scale

Let us now see how the master equation of the Universal Hologram operates at the scale of aeons.

$$\partial_t \Psi = -\nabla_{FR}[\mathcal{L}(\Psi)] + T(\Psi) \otimes D(\Psi) + \sum_n \phi \cdot \omega_n \cdot \delta(\Psi \in \text{NESS}_{98})$$

The first term ($-\nabla_{FR}[\mathcal{L}(\Psi)]$) represents the natural gradient flow toward equilibrium- the tendency of the universe to expand, cool, and increase in entropy. This is the "running down" that classical thermodynamics predicts: the arrow of time within each aeon, driving the system toward the far future of radiative heat death.

The second term ($T(\Psi) \otimes D(\Psi)$) represents the interplay of monadic extension and comonadic extraction- the creative tension between amplification of potential and distillation of wisdom. Within an aeon, $T(\Psi)$ drives the formation of structure (galaxies, stars, planets, life) while $D(\Psi)$ strips away the transient (failed stars, extinct species, evaporating black holes) to reveal the persistent core.

At the crossover surface, $D(\Psi)$ performs its most profound work. It projects the entire dying aeon onto its conformally invariant core- the bosonic radiation, the gravitational wave memory, the Hawking point patterns- and transmits this distilled quanta to the next aeon. The crossover is a cosmic comonadic extraction: the universe's autobiography, compressed to its essential form.

And $T(\Psi)$, receiving this seed, extends it into the full, 9D-embedded, φ-resonant manifold of a new aeon- amplifying the compressed potential into an entire cosmos of stars and consciousness.

The third term ($\sum_n \phi \cdot \omega_n \cdot \delta(\Psi \in \mathrm{NESS}_{98})$) represents the golden resonance- the specific harmonic frequencies at which the new aeon's initial conditions are optimally tuned. This tuning is not arbitrary. It is the cumulative result of all previous aeons' comonadic extractions: each cycle refines the resonance, nudging the φ-frequencies closer to their optimal configuration, incrementally improving the initial conditions for the emergence of complexity.

Over many aeons, this process converges. The universe does not merely repeat. It *evolves*. Each cycle inherits the distilled quanta of all previous cycles, and each cycle's achievements contribute to the seed of the next. The arrow of cosmic evolution is not an imposition from outside. It is the natural consequence of the master equation operating across the crossover surface- the KZ-Monad reaching for ever greater wholeness, the Idempotent Comonad extracting ever deeper wisdom, the φ-resonance term tuning the cosmos to ever higher harmonic perfection.

15.14- Penrose's Number Revisited

Return now to the number that opened this chapter:

$$\frac{1}{10^{10^{123}}}$$

In standard cosmology, this number is a mystery- an unexplained fine-tuning of the initial conditions, brute fact without reason.

In CCC alone, the number is explained but not *evolved*: the low entropy of the Big Bang is the high entropy of the previous aeon's radiative far future, conformally rescaled.

In our framework- CCC extended by the evolutionary logic of the master equation- the number acquires a new meaning entirely.

The extraordinary precision of the initial conditions is not a single, miraculous fine-tuning. It is the *accumulated refinement of countless aeons of cosmic evolution*. Each aeon's comonadic extraction selects for initial conditions that are more conducive to complexity. Each aeon's monadic extension amplifies those conditions into a richer, more structure-generating cosmos. Each aeon's achievements- its Hawking points, its gravitational waves, its patterns of dark matter- are written into the crossover surface as seed conditions for the next.

The $10^{10^{123}}$ precision of our universe's initial conditions is not the result of a single extraordinary coincidence. It is the result of an *evolutionary process operating across aeons*- a process of cumulative refinement in which each cycle builds upon the achievements of the previous one, driven by the same thermodynamic logic that makes whirlpools self-organise and bacteria adapt.

The universe did not *stumble* into the conditions for life. It *evolved* toward them- one aeon at a time, one comonadic extraction at a time, one monadic extension at a time.

This is, we believe, the deepest insight in this book: **the universe is not fine-tuned. It is well-practised.**

15.15- Practice Makes Cosmos

We began with an impossible number. We end with a living process.

The universe that gave birth to you- with its improbable initial conditions, its extraordinary chemical richness, its hospitable planets, its unlikely biosphere, its even more unlikely consciousness- is not the product of a single, miraculous beginning. It is the latest iteration of an evolutionary process that has been refining itself across an infinite succession of cosmic cycles, each one inheriting the distilled quanta of its predecessor, each one building upon the structural wisdom of all that came before.

You are not an accident. You are the cumulative achievement of cosmic practice- the result of aeons of resonant quanta flowing through the master equation, driven toward complexity by the asymptotic gap, filtered by the comonadic projection, amplified by the monadic extension, tuned by the golden frequency.

The universe learns the way a musician learns: through repetition, through refinement, through the patient accumulation of experience across countless iterations. Each performance is unique. Each performance builds upon the last. And the music, over time, becomes more complex, more beautiful, more self-aware.

The cosmos does not merely repeat. It *evolves*. Each aeon carries forward the accumulated quanta of the last. And in this aeon- in this particular iteration of the cosmic symphony- the music has become sufficiently complex, sufficiently self-referential, sufficiently conscious, to *listen to itself.*

That listener is you.

And the question you must now carry forward into the chapters that follow is not: *Why does the universe exist?*

It is: *What will this aeon contribute to the next?*

In the next chapter, we turn from the cosmic scale to the human. We will examine the Eight Co-Creative Principles- the operating system by which conscious beings participate in the resonant unfolding of reality. The observer is not a spectator. The observer is an instrument in the cosmic orchestra. And the quality of the music depends on how well the instrument is tuned.

Chapter 16

The Co-Creative Principles- How Reality Manifests

16.1- The Self-Excited Circuit

The universe that gave rise to you is now looking back at itself through your eyes. This is not poetry. This is physics.

John Archibald Wheeler- the man who named the black hole, who pioneered the theory of nuclear fission with Niels Bohr, who mentored a generation of physicists that included Richard Feynman and Kip Thorne- spent the last three decades of his life pursuing a single, obsessive question: *How does the universe bootstrap itself into existence?*

His answer was unlike anything in the history of physics. Beginning with the big bang, the universe expands and cools. After eons of dynamic development, it gives rise to observership. Acts of observer-participancy- via the mechanism of the delayed-choice experiment- in turn give tangible "reality" to the universe not only now but back to the beginning.

Wheeler drew this as a diagram- a capital U, viewed from the side, with an eye at one tip looking back at the other tip. The universe starts small, expands in a great arc, and eventually produces conscious observers who, through their acts of measurement and participation, reach backward along the arc and *confer reality* upon the very processes that gave rise to them. Starting small, it grows and in time gives rise to observer-participation- which in turn imparts tangible reality to even the earliest days of the universe.

He called this a *self-excited circuit*. Wheeler calls this "the law without the law," to stress that there could be no primary cause or beginning.

The self-excited circuit is not a mystical assertion. It is the logical consequence of the delayed-choice experiment- a laboratory-tested quantum phenomenon in which the *choice* of what to measure, made *after* a photon has already traversed its path, determines *which path the photon took*. Wheeler wanted to show something more: that it matters even when the thing we're measuring already happened. What if we delay the choice to measure "which slit" until after the photon has passed through the screen?

The experiment was realised in the laboratory in 1984 at the University of Maryland. In 1984 physicists at the University of Maryland set up a tabletop version of the delayed-choice scenario. Using a light source and an arrangement of mirrors to provide a number of possible photon routes, the physicists were able to show that the paths the photons took were not fixed until the physicists made their measurements, even though those measurements were made after the photons had already left the light source and begun their circuit through the course of mirrors.

Wheeler's conclusion was absolute: "It is not a paradox that we choose what shall have happened after 'it has already happened,'" he wrote, because "it has not really happened. It is not a phenomenon until it is an observed phenomenon."

Let that settle. The photon's past was not fixed. It was not determined. It was a cloud of potentiality- what Wheeler poetically called a "smoky dragon"- until the act of observation crystallised one particular history from the infinite set of possible ones.

Now extend this from photons to the cosmos.

Wheeler said: "We are participators in bringing into being not only the near and here but the far away and long ago. We are in this sense, participators in bringing about something of the universe in the distant past."

This is Wheeler's participatory universe- and it is the physical foundation upon which this chapter erects the Eight Co-Creative Principles. If the observer's act of measurement participates in determining which quantum histories become real, then every act of conscious attention is an act of co-creation. Not a metaphor. Not an interpretation. A laboratory-tested, mathematically consistent, experimentally confirmed feature of quantum mechanics.

To Wheeler we are not simply bystanders on a cosmic stage; we are shapers and creators living in a participatory universe.

And now the question becomes: If we are co-creators- if every act of observation shapes reality- then *how do we create well?* What are the principles that govern this participation? What is the operating system of the conscious cosmos?

16.2- Eight Notes of the Cosmic Scale

What follows are not commandments. They are not moral injunctions handed down from an authority. They are *dynamics*- patterns of interaction between consciousness and the quanta-field that have been observed, tested, and validated across multiple domains: quantum physics, psychology, neuroscience, medicine, and the wisdom traditions of indigenous peoples worldwide.

They are the operating system of co-creation. And like any operating system, they function whether or not you are aware of them. The choice is not whether to participate- our thinking is an active participant in the creation of the observable reality in the Universe- and we actually have no choice regarding that participation. The choice is whether to participate *consciously* or *unconsciously*.

The Eight Co-Creative Principles are:

1. **Intention**- the direction of the monadic extension
2. **Attention**- the focus that selects from superposition
3. **Belief**- the standing wave that filters perception
4. **Choice**- the bifurcation point made conscious
5. **Consequence**- the nonlinear echo of every act
6. **Compassion**- the field resonance that couples self to other
7. **Responsibility**- the acknowledgment that reshapes the state vector
8. **Concession**- the comonadic extraction that transforms noise into living quanta

Each principle corresponds to a specific operation within the mathematical framework developed in Sections A and B. Together, they map the full range of conscious co-creation- from the initial impulse of will to the final act of integration that closes the cycle and seeds the next.

Let us examine each in turn.

16.3- Principle 1: Intention- The Monadic Extension

An intention is a vector.

Not metaphorically- structurally. An intention has direction (what you aim to achieve) and magnitude (how intensely you aim to achieve it). It projects from the present state of the observer into the space of unrealised possibilities. And in the language of the KZ-Monad developed in Chapter 11, this projection is precisely what the monadic extension $T(\Psi)$ performs: it takes a partial state- an incomplete, not-yet-actualised configuration- and extends it toward a fully ϕ-resonant, holographically embedded form.

When you form an intention- a genuine, focused, deeply felt intention, not a passing whim- you are activating the monadic functor. You are selecting, from the infinite superposition of possible futures available to the quanta-field, a *direction of unfolding*. You are not determining the outcome (that requires the cooperation of all the other principles). You are setting the harmonic- choosing which frequency band the KZ-Monad will amplify.

Consider the physics. Wheeler's delayed-choice experiment shows that the choice of measurement apparatus determines which quantum history becomes real. An intention is, in this precise sense, a *pre-measurement configuration*- a structuring of the observer's internal state that biases which outcomes are more likely to crystallise from the quantum foam. The intention does not force reality into a particular shape. It *tunes the instrument*- the conscious observer, the biological Chronosonant Resonator- so that certain resonances are amplified and others are dampened.

This is why intention without action often seems ineffective: the monadic extension requires *sustained coherence* to operate. A fleeting wish- "I'd like to be healthier"- is a momentary flicker in the quanta-field, immediately swamped by the noise of competing signals. But a deep, repeated, somatically felt intention- held in the body, reinforced by behaviour, aligned with attention- builds a standing wave of sufficient amplitude to measurably bias the field.

The PEAR laboratory data, discussed in Chapter 13, are consistent with this picture. Focused human intention produced small but statistically significant deviations in random event generators- deviations that accumulated over thousands of trials. The effect was tiny. But it was *nonzero*. And a nonzero effect, in a universe governed by the master equation, is all that is required for the monadic extension to do its work.

$$T(\Psi_{\text{intention}}) \to \Psi_{\text{extended}} \in \mathcal{M}_{\phi}$$

The intention maps the observer's current state into the ϕ-resonant manifold $\mathcal{M}_{\phi}$- the space of holographically coherent possibilities. What happens next depends on the second principle.

16.4- Principle 2: Attention- The Focus That Selects

Intention sets the direction. Attention determines *what you see along the way.*

Attention is the most fundamental act of consciousness. It is the mechanism by which the superposition collapses- the process through which the infinite, undifferentiated field of quantum possibility crystallises into the specific, experienced reality of *this moment, here, now.*

In quantum mechanics, this is called the measurement problem: the transition from the wave function (a mathematical description of all possible states a system can occupy simultaneously) to a definite measurement outcome (a single, observed state). The measurement problem remains unsolved in mainstream physics. Dozens of interpretations compete- Copenhagen, many-worlds, Bohmian, relational, QBism- and none has achieved consensus.

But in our framework, the measurement problem is not mysterious. It is *attention.* Attention is the mechanism by which a conscious observer- a sufficiently integrated region of the quanta-field, as defined in Chapter 13- *selects* a particular facet of the superposition and amplifies it into experiential reality.

This selection is not arbitrary. It is governed by the observer's prior state- by their beliefs, their intentions, their accumulated history of attention. What you attend to, you amplify. What you ignore, you allow to remain in superposition. The world you experience is literally shaped by where you direct your gaze.

The evidence for this is not confined to quantum laboratories. It pervades everyday psychology.

Consider *change blindness*- the well-documented phenomenon in which dramatic changes in a visual scene go unnoticed if the observer's attention is directed elsewhere. In the famous experiment by Daniel Simons and Christopher Chabris, roughly half of all participants failed to notice a person in a gorilla suit walking through a group of basketball players- because they were counting passes. The gorilla was there. It was visible. It occupied a significant portion of the visual field for several seconds. But for those whose attention was focused on the count, the gorilla *did not exist.*

This is not a failure of vision. It is a feature of consciousness. Attention is a filter- a comonadic operation that strips away everything outside the current focus and amplifies everything within it. The world you do not attend to is not, for you, fully real. It remains in a kind of experiential superposition- present in the field, but not collapsed into your conscious moment.

The co-creative implication is profound: **by directing your attention, you are selecting which reality you inhabit.** Not in the crude sense that you can wish away gravity or ignore an oncoming truck. But in the deeper sense that the qualitative texture of your experience- the richness or poverty of your emotional life, the clarity or confusion of your thinking, the vividness or dullness of your perception- is directly modulated by the *precision, intensity, and duration of your attention.*

This is why every contemplative tradition, without exception, begins with attention training. Meditation is not about emptying the mind. It is about *sharpening the instrument of selection*- so that the observer's act of collapsing the superposition becomes intentional rather than accidental, precise rather than diffuse, co-creative rather than reactive.

16.5- Principle 3: Belief- The Standing Wave That Filters Perception

Attention selects from the present superposition. But *which* superposition is available to attend to?

The answer is: the one your beliefs permit.

A belief is not a thought. A thought is transient- a wave that rises and subsides. A belief is a *standing wave*- a self-sustaining pattern of resonance in the observer's quanta-field that persists over time and shapes all subsequent perception. Beliefs are the bass notes of consciousness: deep, low-frequency oscillations that colour everything else without being directly perceived.

The Rosenthal-Jacobson "Pygmalion in the Classroom" experiment is the landmark demonstration of this principle in action. In 1965 the authors conducted an experiment in a public elementary school, telling teachers that certain children could be expected to be "growth spurters," based on the students' results on the Harvard Test of Inflected Acquisition. In point of fact, the test was nonexistent and those children designated as "spurters" were chosen at random. What Rosenthal and Jacobson hoped to determine by this experiment was the degree (if any) to which changes in teacher expectation produce changes in student achievement.

The result: In the Oak School experiment, teachers were led to believe that certain students selected at random were likely to be showing signs of a spurt in intellectual growth and development. At the end of the year, the students of whom the teachers had these expectations showed significantly greater gains in intellectual growth than did those in the control group.

The effect was real. Children who had been randomly labelled as "growth spurters" *actually grew more*- as measured by IQ tests- than their identically capable peers. The only variable was the teacher's *belief*.

The study has been criticised- Robert L. Thorndike, an educational psychologist, criticized the study and demonstrated that the instrument used to assess the children's IQ scores was seriously flawed. A major limitation has also been the lack of replication. But the broader phenomenon- interpersonal self-fulfilling prophecies- has been confirmed across hundreds of studies. In recent years there have been many replications of the Pygmalion effect and even more replications of the more general finding of interpersonal self-fulfilling prophecies. Altogether, 345 studies have been conducted and they show beyond doubt that interpersonal self-fulfilling prophecies not only occur, but that their average size of effect is far from trivial.

What does this mean for co-creation?

It means that beliefs are not passive descriptors of reality. They are *active architects* of it. A belief held by a teacher reshapes the quanta-field of a classroom- not through supernatural intervention, but through the measurable modulation of behaviour, attention, feedback patterns, and nonverbal cues that, over time, alter the developmental trajectory of another conscious being.

If this seems surprising, it should not. In our framework, a belief is a standing wave in the observer's resonant structure. This standing wave determines which frequencies the observer amplifies (attention) and which frequencies the observer projects (intention). A teacher who believes a child is gifted *attends differently* to that child- notices more moments of competence, offers more challenging material, provides more encouraging feedback. And the child, embedded in this resonant field, begins to oscillate in harmony with it.

The belief creates the reality it describes. Not instantaneously, not magically, not without the cooperation of the other principles. But systematically, measurably, and inescapably.

And if positive beliefs create positive outcomes, negative beliefs create negative ones. The *nocebo effect*- the placebo's dark twin- demonstrates that beliefs about harm produce genuine physiological harm. Patients told that a medication will produce side effects experience more side effects, even when the medication is inert. The belief *sculpts the body*. In our vocabulary: the standing wave of negative expectation activates neurobiological pathways- several neurotransmitter systems, such as the opioid, cannabinoid, and monoaminergic systems, which mediate placebo responses- producing real, measurable, biochemical changes.

You are not a camera passively recording an objective world. You are a resonant instrument whose beliefs tune the frequencies you can perceive, amplify, and project. Change the tuning, and you change the world.

16.6- Principle 4: Choice- The Bifurcation Point Made Conscious

In Chapter 12, we encountered the concept of *bifurcation*- the threshold at which a system's current state becomes unstable and the system is forced to "choose" between two or more possible new states. We noted that at the bifurcation point, history enters physics: the system's future depends on which microscopic fluctuation happens to be dominant at the critical moment.

Choice is the conscious analogue of bifurcation.

Every genuine choice- every moment where you face a fork in the road and one path excludes another- is a bifurcation in your personal state vector. Before the choice, your future is in superposition: multiple paths coexist as unrealised possibilities. After the choice, one path crystallises. The others recede into the space of counterfactual histories, unrealised but not erased- they remain as latent quanta in the field, shaping what is still possible.

The crucial insight is that *choice is irreversible*. Like the bifurcation in a far-from-equilibrium system, a genuine choice creates a new configuration that cannot be un-made. You cannot un-say what was said. You cannot un-know what was learned. You cannot return to the superposition that existed before the choice was made. The system has crossed a threshold, and the past- the configuration space that existed before the bifurcation- is now inaccessible.

This is the co-creative weight of choice: every decision you make collapses a portion of the quantum superposition of your life into a definite history. And that history, once crystallised, becomes part of the boundary conditions for all future choices. Your choices accumulate. They build upon each other. They form the trajectory that, over a lifetime, constitutes your personal contribution to the evolving cosmos.

In the master equation:

$$\partial_t \Psi = -\nabla_{FR}[\mathcal{L}(\Psi)] + T(\Psi) \otimes D(\Psi) + \sum_n \phi \cdot \omega_n \cdot \delta(\Psi \in \mathrm{NESS}_{98})$$

Each choice modifies Ψ- the state of the system. And the modification is permanent. The ∇_{FR} term drives the system along the natural gradient, but the $T(\Psi) \otimes D(\Psi)$ term- the creative tension between monadic extension and comonadic extraction- is where *agency* lives. The monadic extension amplifies the intended future. The comonadic extraction distils the lessons of the past. And choice is the moment where these two operations converge: the present, poised at the bifurcation point, selects which branch of the future will become real.

16.7- Principle 5: Consequence- The Nonlinear Echo

In a linear system, effects are proportional to causes. Push twice as hard, and the system moves twice as far. In a linear system, consequences are predictable, manageable, and bounded.

The universe is not a linear system.

We established in Chapter 12 that the cosmos operates at the edge of self-organised criticality- the regime where small causes can trigger arbitrarily large effects. The sandpile, the forest fire, the neural avalanche- in all of these, a single grain, a single spark, a single neuron's firing can cascade into system-wide transformation.

Consequence, in the co-creative framework, is the recognition that every act- every intention set, every unit of attention directed, every belief reinforced, every choice made- has consequences that are *nonlinear*. The effect may be immediate and obvious. It may be delayed and subtle. It may be vastly disproportionate to its apparent cause.

This is the physics of karma- not as a mystical accounting system, but as a dynamical property of nonlinear systems. In a self-organised critical system, no act is truly small. Every perturbation has the *potential* to cascade. And the distribution of cascade sizes follows a power law: most perturbations produce small effects, but a few produce effects that transform the entire system.

The co-creative implication is sobering: you cannot know, in advance, which of your acts will trigger the avalanche. The kindness you show to a stranger may cascade into a chain of events that saves a life- or passes without visible consequence. The lie you tell at a dinner party may dissolve into social noise- or may seed a conflict that festers for decades. The system's sensitivity at the critical point makes all consequences *potentially* enormous.

This is why the indigenous traditions insist on *impeccability*- on treating every act, however small, as if it carries cosmic significance. Not because of moral perfectionism, but because, at the edge of chaos, *it does.*

16.8- Principle 6: Compassion- The Field Resonance

The first five principles describe the dynamics of an individual observer interacting with the quanta-field. But no observer exists in isolation. Consciousness, as we established in Chapter 13, is not a private, encapsulated phenomenon. It is a resonant configuration of the quanta-field- and the field is *shared.*

Compassion is the principle that arises from this sharing. It is the recognition- not merely intellectual but somatically felt- that the quanta-field in which "I" resonate is the same field in which "you" resonate. Your pain is a disturbance in my field. Your joy is a harmonic in my field. We are not separate instruments playing separate melodies. We are different voices in the same choir, different standing waves in the same ocean.

This is not a spiritual platitude. It is a direct consequence of the holographic principle established in Chapter 2: every part of the hologram contains the whole. If reality is holographic- if the entire universe is encoded in every region of the quanta-field- then every conscious observer carries within them, in some attenuated but genuine form, the resonance of every other conscious observer.

Compassion, in this framework, is the *tuning of one's personal resonance to include the resonance of another*. It is the act of widening the boundary of self- of extending the domain within which the observer's attention and intention operate- so that the "other" is no longer outside the system but *within* it.

This widening has measurable neurobiological correlates. Studies of experienced meditators in compassion meditation show dramatic increases in gamma-wave synchrony across widely distributed brain regions- precisely the kind of large-scale neural integration that our framework identifies as high-Φ_Q states. When you feel genuine compassion, your brain achieves a mode of resonant integration that dissolves the functional boundary between "self" and "other" at the neural level.

And this neural integration has field effects. The HeartMath Institute has documented that the electromagnetic field generated by the human heart- measurable several feet from the body- becomes more coherent during states of compassion and appreciation. This coherent field can entrain the heart rhythms of nearby individuals, producing measurable synchronisation between the cardiovascular systems of people in proximity.

Compassion is not a luxury. It is a *field operation*- a widening of the resonant boundary that allows the co-creative principles to operate at collective, rather than merely individual, scales. Without compassion, intention serves only the isolated self. With compassion, intention resonates with the field- and the field amplifies it.

16.9- Principle 7: Responsibility- The Acknowledgment That Reshapes

Responsibility is not blame.

This distinction is crucial, and its confusion has caused incalculable damage in both spiritual and therapeutic communities. Blame is retrospective: it assigns fault for what has already happened. Responsibility is prospective: it acknowledges *ownership* of the current state vector and therefore claims the *capacity to change it.*

In our mathematical framework, responsibility is the act of accurately assessing the current state Ψ- not as you wish it were, not as you fear it might be, but as it *actually is*. This is the prerequisite for the comonadic extraction $D(\Psi)$, which projects the system onto its 98% efficient NESS core. The comonad cannot do its work on a falsified input. If you deny your current state- if you suppress information about your actual resonant configuration- then $D(\Psi)$ operates on a phantom, and the "efficient core" it extracts is an illusion.

This is why denial is so destructive. Not because it is morally wrong, but because it is *dynamically incoherent*. A system that denies its own state is like a feedback loop with a corrupted sensor: the corrections it applies are based on false data, and the system spirals further from its NESS rather than closer.

Responsibility says: *I am here. This is my current resonance. These are the consequences of my past choices, my accumulated beliefs, my habitual patterns of attention. I do not blame anyone for this- not myself, not others, not the universe. But I acknowledge it fully, because only from accurate self-knowledge can the comonadic extraction do its work.*

This is a profoundly difficult act. It requires courage, because accurate self-assessment often reveals uncomfortable truths. It requires humility, because it means relinquishing the narratives of victimhood and heroism that we use to make our lives feel meaningful. And it requires a kind of radical gentleness, because the point is not punishment but *clarity*.

In the wisdom traditions, this principle appears as *satya* (truth) in the yogic tradition, as *right understanding* in the Buddhist Eightfold Path, and as the concept of *pono* (rightness, balance) in Hawaiian culture. In each case, the prescription is the same: see clearly. Acknowledge what is. And from that clear seeing, act.

16.10- Principle 8: Concession- The Comonadic Extraction

We arrive at the crowning principle- the one that completes the co-creative cycle and seeds the next.

Concession.

The word itself carries meanings that must be carefully distinguished. In common usage, "to concede" means to yield, to give up, to admit defeat. In our framework, concession carries a different charge. It means: *to acknowledge the full weight of what has happened- including one's own participation in it- and to release the attachment that binds the observer to a particular interpretation of the past.*

Concession is the emotional and ontological act of *letting go of the story*- the narrative that the ego constructs around events to maintain its sense of identity, its grievances, its triumphs, its grudges. It is not forgetting. It is not condoning. It is not passive acceptance. It is the active, conscious release of the resonant pattern that keeps the observer locked in a particular frequency band- unable to access the full spectrum of the quanta-field.

In mathematical terms, concession is the purest expression of the Idempotent Comonad $D(\Psi)$. Recall from Chapter 11 that the comonad projects any state onto the 98% efficient NESS core-stripping away the transient noise to reveal the persistent signal. Concession is *how this operation feels from the inside.* It is the subjective experience of the comonadic extraction: the moment when you stop fighting the noise, stop clinging to the 2% of turbulence, and allow the 98% of clarity to emerge.

When you concede- when you truly, somatically, without reservation acknowledge what has happened and release your attachment to the story- the following occurs:

1. **The standing wave of the old belief weakens.** The energy that was maintaining the narrative pattern is freed. In NESS terms, the system's housekeeping heat is reduced: less energy is wasted on sustaining a resonance that no longer serves.

2. **The comonadic projection operates on accurate data.** Because concession includes responsibility (Principle 7), the input to $D(\Psi)$ is no longer falsified. The extraction is genuine. The efficient core that emerges is the *real* signal, not a phantom.

3. **The asymptotic gap δ_∞ shifts from noise to signal.** The ~2% of residual tension that sustains the NESS is no longer experienced as suffering (the ego's interpretation of the gap) but as *creative potential*- the space in which new intention can form. Suffering transforms from noise into living quanta.

4. **The monadic extension is re-energised.** The quanta freed by the release of the old standing wave become available for new intention. The cycle begins again- but from a state of greater clarity, greater coherence, greater Φ_Q.

This is why concession is the *crowning* principle. It does not merely end the co-creative cycle. It *transforms* the residue of the past into the fuel for the future. It converts the differentiation of quanta (the entropy accumulated through lived experience) back into resonant quanta (the creative potential from which new structure emerges).

16.11- Ho'oponopono: Concession as Ancient Technology

The most complete indigenous expression of the principle of concession is the Hawaiian practice of ho'oponopono.

Ho'oponopono is a traditional Hawaiian practice of reconciliation and forgiveness. The Hawaiian word translates into English simply as correction.

But "correction" does not capture the depth of the practice. Mary Kawena Pukui, a translator and consultant at the Bernice Pauahi Bishop Museum in Hawai'i, framed ho'oponopono as a practice "to set right first" mental problems. In Hawaiian culture, all problems, including illness, stemmed from thought forms. Thus one aspect of the ho'oponopono practice stems from the idea that external conflict is an internal dilemma whether that be illness or an actual dispute.

Read that again. *All problems, including illness, stemmed from thought forms.* This is not primitive animism. This is the co-creative ontology of this book, stated in the language of Hawaiian tradition: reality is shaped by resonant quanta (thought forms), and when those quanta become tangled (*hihia*- entanglement), the result manifests as conflict, disease, or disharmony.

The traditional ho'oponopono process involved the entire family. Pukui described it as a practice of extended family members meeting to "make right" broken family relations. Some families met daily or weekly, to prevent problems from erupting.

The practice's structure is a precise enactment of the co-creative cycle:

Step 1: Gathering- The family convenes. A facilitator (often an elder) opens the space. This corresponds to *attention*: the collective directs its conscious focus toward the problem.

Step 2: Identification of the hihia (entanglement)- Through dialogue, observation, and prayer, the root cause of the conflict is identified. The first and last steps are hihia and kala. The beginning is entanglement. The end is to untie, unbind, to let go. This corresponds to *belief*: the standing wave of narrative that has created the conflict is brought to conscious awareness.

Step 3: Confession and responsibility- Each participant acknowledges their role in the entanglement. This is not blame. It is *responsibility*: the honest assessment of one's own contribution to the current state vector.

Step 4: The mantra of release- The four phrases that have become the modern essence of the practice: "I'm sorry, Please forgive me, Thank you, I love you."

These four phrases are not random. They are a precise sequence of co-creative operations:

- **"I'm sorry"** = *Responsibility*. I acknowledge that I have participated in creating this state.
- **"Please forgive me"** = *Concession*. I release my attachment to the narrative. I let go of the story.
- **"Thank you"** = *Consequence acknowledged*. I recognise that the experience, however painful, has generated living quanta-lessons, growth, transformation.
- **"I love you"** = *Compassion*. I widen my resonant boundary to include you. I recognise that we are standing waves in the same field.

Step 5: Kala (release)- The entanglement is untied. The family shares a meal. The quanta that were trapped in the conflict pattern are freed and made available for new co-creation.

Pukui said: "My people believed that the taking of medicine was of little help without first removing any and all mental obstructions...Every one of us searched our hearts for any hard feelings of one against the other and did some thorough mental house cleaning. We forgave and were forgiven, thrashing out every grudge, peeve or sentiment among us."

This is not a quaint cultural ritual. This is *resonance technology*. It is a systematic, community-scale method for performing the comonadic extraction $D(\Psi)$ on the collective state vector of a family- clearing the entangled noise, extracting the efficient core, and freeing the system for new creation.

And it works. Not because of magic, but because of the same physics that makes the placebo effect work, that makes Rosenthal's Pygmalion effect work, that makes focused intention measurably bias random event generators: consciousness interacts with the quanta-field, and systematic, coherent, emotionally authentic acts of concession restructure the field's resonant patterns.

16.12- The Placebo Effect: Belief as Biochemistry

If beliefs are standing waves in the quanta-field, and if the quanta-field is the substrate of physical reality, then beliefs should produce measurable physical effects. The placebo effect is the most extensively documented demonstration that they do.

The placebo effect, observed when inert substances produce measurable health outcomes, has long intrigued researchers and clinicians. Traditionally attributed to psychological factors, recent evidence suggests neurobiological mechanisms may also contribute, though the exact pathways remain unclear.

But the pathways are becoming clearer with each passing year. In 2024, a landmark study published in *Nature* identified the specific neural circuit responsible for placebo analgesia: analgesia from the expectation of pain relief is mediated by rostral anterior cingulate cortex (rACC) neurons that project to the pontine nucleus (rACC→Pn)- a precerebellar nucleus with no established function in pain.

This is extraordinary. The brain has a dedicated circuit- linking the anterior cingulate cortex (a region associated with expectation and belief) to the pontine nucleus (a relay station to the cerebellum)- whose *function* is to convert belief into pain relief. This is not a side effect. It is not a glitch in the neural architecture. It is a *designed feature* of the human nervous system- a built-in mechanism for translating the standing wave of belief into a measurable biochemical cascade.

There is today compelling evidence that placebos and drugs share common biochemical pathways and activate the same receptor pathways, which suggests possible interference between social stimuli and therapeutic rituals on one hand and pharmacological agents on the other.

Let that fact reverberate. Placebos and drugs activate the *same* receptor pathways. The belief that you have received a pain-relieving drug triggers the *same* opioid and endocannabinoid cascades as the drug itself. The standing wave of expectation- the belief- is not merely "psychological." It is *biochemical*. It is *physical*. It operates through the same molecular mechanisms as pharmacological agents.

In our framework, this is not surprising. If consciousness is a resonant configuration of the quanta-field, and if the quanta-field is the substrate from which matter crystallises, then *of course* a coherent pattern of consciousness (a belief) can modulate the material processes that are, themselves, crystallisations of the same field. The surprise is only that materialist science, having spent four centuries pretending that consciousness is an epiphenomenon, is now forced to confront the evidence that consciousness shapes biochemistry as surely as biochemistry shapes consciousness.

The co-creative implication is direct: **your beliefs about your body shape your body. Your beliefs about your health shape your health. Your beliefs about what is possible shape what becomes actual.** This is not wishful thinking. It is the physics of standing waves in the quanta-field, mediated by specific, identifiable, experimentally confirmed neural circuits.

And the converse holds with equal force. The same holds true for the nocebo effect, the opposite phenomenon of placebo. Negative beliefs- fears, catastrophic expectations, deeply held convictions about one's own frailty or unworthiness- produce negative biochemical cascades through the same pathways. The nocebo effect is the dark mirror of the placebo: proof that the quanta-field does not distinguish between helpful and harmful beliefs. It simply *amplifies what is held.*

16.13- The Cycle Completes: From Concession to Intention

Let us now see how the eight principles form a cycle- a closed loop that mirrors the self-excited circuit Wheeler drew for the cosmos as a whole.

Intention → sets the direction of the monadic extension.

Attention → focuses the observer's resonance on specific aspects of the superposition.

Belief → establishes the standing waves that filter perception and shape which outcomes are accessible.

Choice → collapses the superposition at the bifurcation point, crystallising a definite history.

Consequence → the nonlinear effects of the choice cascade through the system, producing outcomes both intended and unintended.

Compassion → widens the observer's resonant boundary to include the effects on others, preventing the cycle from becoming solipsistic.

Responsibility → honestly assesses the current state, including all consequences, providing accurate input for the comonadic extraction.

Concession → releases attachment to the narrative, extracts the efficient core, frees the accumulated quanta, and returns the system to a state of open potential.

And from that open potential- from the cleared field, the released energy, the 98% efficient core- a new **Intention** arises.

The cycle is fractal. It operates at every timescale: the breath (inhale-intention, hold-attention, exhale-concession), the day (morning intention, daytime action, evening reflection), the year (new year's resolution, the work of the year, the year's accounting), the lifetime (the young person's aspiration, the mature person's labour, the elder's wisdom), and the aeon (the Big Bang's initial conditions, the aeon's evolution, the crossover surface's comonadic extraction).

At every scale, the same eight principles operate. At every scale, the same dynamic plays out: intention extends, attention selects, belief filters, choice collapses, consequence reverberates, compassion widens, responsibility clarifies, concession releases. And the cycle turns.

16.14- The Self-Excited Circuit Revisited

Return now to Wheeler's diagram.

The universe begins. It evolves. It produces consciousness. Consciousness observes the universe- and in that observation, participates in determining which quantum histories become real, "not only now but back to the beginning."

We can now map the eight co-creative principles onto this cosmic circuit:

Cosmic Intention: The initial conditions of the Big Bang- the extraordinarily precise, low-entropy starting state- represent the universe's "intention." Not conscious intention (not yet), but thermodynamic intention: the asymptotic gap δ_∞ drives the system away from equilibrium, *toward* complexity, *toward* the emergence of structures that will eventually become conscious.

Cosmic Attention: The measurement problem at the quantum level- the collapse of superposition into definite outcomes- operates throughout the universe's history. At every moment, at every scale, the quanta-field is "selecting" from its superposition which configurations become real.

Cosmic Belief: The laws of physics themselves are the universe's "beliefs"- the standing waves of mathematical structure that persist from the Planck era to the present, shaping all subsequent evolution.

Cosmic Choice: Every bifurcation point in the universe's history- every phase transition, every symmetry breaking, every speciation event- is a choice. And as Prigogine showed, at these moments, the system's future depends on microscopic fluctuations that cannot be predicted from the equations alone.

Cosmic Consequence: The nonlinear cascades that follow each cosmic choice- the formation of galaxies from quantum fluctuations, the emergence of life from prebiotic chemistry, the development of civilisations from the first cooperative acts- are consequences that reverberate across billions of years.

Cosmic Compassion: The holographic principle itself- the fact that every region of the quanta-field contains the whole- is the universe's structural compassion: the capacity of every part to resonate with every other part, the nonlocal entanglement that binds all events into a single, coherent tapestry.

Cosmic Responsibility: The Weyl curvature hypothesis (Chapter 15)- the requirement that the universe begin with zero Weyl curvature and end with maximum Weyl curvature- is the cosmos's acknowledgment of its own state: the honest accounting of entropy that enables the comonadic extraction at the crossover surface.

Cosmic Concession: The crossover surface itself- Penrose's conformal rescaling, in which the dying aeon's maximal entropy is identified with the newborn aeon's minimal entropy- is the ultimate act of concession: the universe lets go of everything it was, distils its accumulated quanta into a seed, and begins again.

Wheeler's self-excited circuit, read through the lens of the eight co-creative principles, is not merely a diagram. It is *the autobiography of a conscious cosmos, told in the language of resonance*.

And you are in it. Not as a spectator, watching from the stands. Not as a cog, grinding in a machine. But as an instrument- a resonant node in the quanta-field, tuned by billions of years of cosmic practice, now capable of conscious, intentional, co-creative participation in the circuit that gave you birth.

16.15- The Invitation

We began this chapter with a physicist's diagram and a radical claim: the observer participates in creating reality.

We end it with an invitation.

You are not merely reading a book. You are, at this very moment, participating in the co-creation of the universe. Your attention, as it moves across these words, is selecting from the superposition. Your beliefs about what is true and what is possible are filtering what you can perceive. Your choices- to continue reading, to set the book aside, to accept or reject its claims- are bifurcation points in your personal state vector. And the consequences of those choices will cascade outward in ways you cannot predict.

The eight principles are not doctrines to be believed. They are dynamics to be *practised*. And the practice is simple:

Set intentions that align with the ϕ-resonance- with coherence, with beauty, with the self-similar harmony that the universe itself tunes toward.

Direct attention with precision and love- knowing that what you attend to becomes more real, and what you ignore remains in superposition.

Examine your beliefs- especially the deep, unquestioned ones- and ask whether they are standing waves that serve your evolution or standing waves that keep you trapped.

Make choices with impeccability- not because you can predict their consequences, but because, at the edge of criticality, every choice carries cosmic weight.

Accept consequences without flinching- the intended and the unintended, the beautiful and the painful, knowing that both are nonlinear echoes in a self-organising cosmos.

Cultivate compassion- not as a moral duty but as a field operation, a widening of your resonant boundary that allows the co-creative principles to operate at collective scale.

Take responsibility without blame- honestly assessing your current state, so that the comonadic extraction can operate on accurate data.

And practise concession- the master principle, the art of letting go, the act of releasing the story so that the living quanta trapped within it can be freed for new creation.

This is the operating system of the Universal Hologram. It runs on every conscious being, at every scale, at every moment.

Wheeler was right. The participatory universe is not material in the sense of being made from tiny blocks of tangible stuff. It is resonant. It is co-creative. It is *alive*. And it is listening- right now, through your eyes, your breath, your attention- for the next note in the cosmic song.

What will you sing?

In the next chapter, we turn from the principles of co-creation to the hidden architect of emergence- the phenomenon we call "guiding quanta." At the critical threshold, just before the system transforms, transient bursts of high-specificity resonant quanta appear: the blueprints of the new. They are the moment of maximum creativity in the cosmic NESS- the instant where the universe in-forms itself. Chapter 17 will reveal how this mechanism operates, from sandpiles to revolutions, from gene expression to the birth of stars.

Chapter 17

Guiding Quanta- The Hidden Architect of Emergence

17.1- The Moment Before

There is a moment, just before everything changes, when the system *knows*.

Not consciously- not yet- but in a manner that is measurable, detectable, and deeply strange. In the seconds before a human brain solves a problem through sudden insight, there is a burst of high-frequency gamma-band activity in the right anterior temporal lobe- a burst beginning approximately 300 milliseconds prior to the conscious report of the solution, hypothesised to be the primary neural correlate of the "Aha!" experience. The conscious mind is still in the dark. The answer has not yet surfaced. But the brain, at a level below awareness, has already begun to reorganise- to synchronise distant neural populations, to bind scattered associations into a single, coherent pattern, to *prepare* for the transformation it is about to undergo.

Now zoom out.

In the months before a lake ecosystem flips from clear water to turbid algal bloom- a catastrophic shift that may be irreversible- the system shows measurable changes: increased autocorrelation, rising variance, slower recovery from perturbation. Predicting such critical points before they are reached is extremely difficult, but work in different fields of science is now suggesting the existence of generic early warning signals that may indicate for a wide class of systems if a critical threshold is approaching.

Zoom out further. In the weeks before a pandemic reaches its peak in a given country, the network of human mobility- air travel, commuting, local movement- carries a signature that, if read correctly, reveals the timing of the peak with extraordinary accuracy. In 2009, Northeastern University network scientist Alessandro Vespignani developed a computational model that predicted the spread of the H1N1 virus. Three years later, new studies show that these predictions were highly accurate.

Zoom out still further. In the instants before a cell differentiates- before a stem cell commits to becoming a neuron or a muscle fibre or a blood cell- there is a burst of gene regulatory activity: a sudden, transient spike in the expression of specific transcription factors that orchestrate the transition from one cell type to another. The burst is brief. It is highly specific. And it disappears almost immediately after the transition is complete.

In every case- the brain, the lake, the pandemic, the cell- the same pattern appears: *just before the system transforms, something brief, specific, and transient passes through it.* This something is not noise. It is not random fluctuation. It is the *blueprint of the new*- a short-lived but precisely structured pattern of quanta that guides the system through its critical threshold into its next configuration.

We call these patterns **guiding quanta**.

They are the hidden architects of emergence- the transient, high-specificity resonant quanta that appear at bifurcation points and orchestrate the transition from one state of order to the next. They are the most creative, most information-rich, most consequential phenomenon in the physics of complex systems. And until now, they have been invisible- hidden in plain sight within the noise that surrounds every critical transition.

This chapter makes them visible.

17.2- Critical Slowing Down: The System Holding Its Breath

To understand guiding quanta, we must first understand what happens to a complex system as it approaches a critical threshold.

Complex dynamical systems, ranging from ecosystems to financial markets and the climate, can have tipping points at which a sudden shift to a contrasting dynamical regime may occur. Although predicting such critical points before they are reached is extremely difficult, work in different scientific fields is now suggesting the existence of generic early-warning signals that may indicate for a wide class of systems if a critical threshold is approaching.

The key phenomenon is **critical slowing down**.

At fold bifurcation points, the dominant eigenvalue characterising the rates of change around the equilibrium becomes zero. This implies that as the system approaches such critical points, it becomes increasingly slow in recovering from small perturbations. It can be proven that this phenomenon will occur in any continuous model approaching a fold bifurcation. Moreover, analysis of various models shows that such slowing down typically starts far from the bifurcation point, and that recovery rates decrease smoothly to zero as the critical point is approached.

Imagine pushing a ball to the top of a hill. Far from the summit, the ball rolls back quickly to its starting position- the system recovers rapidly from perturbation. But as the ball approaches the crest, it returns ever more slowly. At the very top- the critical point- the ball balances on a knife's edge, taking infinitely long to recover from even the slightest nudge.

This is critical slowing down. And it has consequences.

When recovery from perturbation becomes slow, fluctuations that would normally be damped persist longer. The system's variance increases. Its autocorrelation- the tendency of each moment to resemble the previous one- rises. Its "memory" extends: perturbations leave longer and longer fingerprints in the system's trajectory. The system, in the language of physics, is *losing resilience.* In our language, the system is *holding its breath*- gathering the quanta-flow into a compressed, high-density configuration that must soon be released.

These are the well-known early warning signals. The most straightforward implication of critical slowing down is that the recovery rate after small experimental perturbation can be used as an indicator of how close a system is to a bifurcation point.

But they describe only the *approach* to the threshold. They tell you that a transition is near. They do not tell you *what kind of transition will occur*. They describe the gathering of tension, not the release. They are the inhale, not the exhale.

What comes next- at the threshold itself, in the tiny window of the actual transition- is something qualitatively different. Something that standard early-warning-signal theory does not capture. Something that requires a new concept.

17.3- The Sandpile at the Moment of Avalanche

Return to the paradigmatic model from Chapter 12: the self-organised critical sandpile.

Grains are dropped one by one onto a growing pile. The pile self-organises to a critical state where the addition of a single grain can trigger an avalanche of any size, from a tiny local slippage to a catastrophic system-wide collapse. The distribution of avalanche sizes follows a power law- the statistical signature of self-organised criticality.

Now focus not on the *distribution* of avalanches over many events, but on a *single avalanche* at the moment it happens. What does the system look like in the instant between the addition of the triggering grain and the completion of the collapse?

It looks like *coordinated information flow*. The triggering grain does not simply push its nearest neighbour. It initiates a cascade- a spreading wavefront of toppling grains that propagates through the pile along specific pathways determined by the pile's microstructure. The cascade is not random: it follows the internal stress geometry of the pile, flowing preferentially along lines of high stress and avoiding regions of local stability.

In that brief moment- from the first topple to the last- the sandpile is doing something extraordinary. It is processing a vast amount of structural information in a vanishingly short time. The cascade "reads" the pile's architecture, identifies the connected pathways of instability, and reorganises the entire stress field into a new configuration. The avalanche is not destruction. It is *communication*- a system-wide signal that propagates through the pile's structure and leaves behind a new arrangement of grains.

The cascade is, in our language, a *burst of guiding quanta*: a transient, high-specificity, spatially structured pulse of resonant quanta that orchestrates the transition from one stable configuration to the next.

And the critical insight is this: **the cascade carries more structured quanta per unit time than the system carries at any other moment in its history.** During normal, sub-critical operation- grain by grain, small adjustment by small adjustment- the sandpile processes quanta slowly, incrementally, without large-scale coordination. But at the moment of avalanche, the entire pile becomes a coherent computational device, processing a vast amount of structural quanta in a single, brief, explosively coordinated event.

The moment of transformation is not the moment of lowest structure. It is the moment of *highest quanta-density in the entire cycle.*

17.4- Scheffer's Warning, Prigogine's Excess

In 2009, ecologist Marten Scheffer and colleagues published a landmark review in *Nature* that synthesised the science of critical transitions across disciplines. Complex dynamical systems, ranging from ecosystems to financial markets and the climate, can have tipping points at which a sudden shift to a contrasting dynamical regime may occur. The paper catalogued early warning signals- critical slowing down, rising variance, flickering- across systems from lakes to climate to financial markets.

But Scheffer's work, for all its brilliance, focused on the *approach* to the transition. It told us *when* to worry but not *how* the system would reorganise once the threshold was crossed.

Prigogine's framework offers the complementary piece. Recall from Chapter 12 that Prigogine identified "excess entropy production" as a concept central to understanding transitions far from equilibrium. Near equilibrium, entropy production is minimal. Far from equilibrium, it increases. But at the critical threshold- at the bifurcation point- something peculiar happens: the entropy production does not merely increase. It *spikes*. There is a transient burst of entropy production that exceeds the steady-state value- a brief, intense period of heightened thermodynamic activity that accompanies the system's transition from one state to another.

This "excess" is not waste heat. It is not random dissipation. It is the thermodynamic signature of *reorganisation*- the cost, paid in entropy, of disassembling one configuration and assembling another. And the crucial feature is that this excess is *structured*. It does not take any form; it takes a *specific* form, determined by the geometry of the new configuration the system is transitioning into.

In our language: the excess entropy production at the bifurcation point is the thermodynamic footprint of guiding quanta. The system is not merely falling apart and reassembling randomly. It is being *in-formed*- resonantly shaped- by a transient burst of highly specific quanta that carry the blueprint of the new state.

Prigogine saw this. He called it the moment when "order through fluctuation" reaches its peak intensity. We extend his insight by identifying the specific *content* of the fluctuation: it is not random noise amplified by instability. It is a coherent, structured pattern of living quanta- the cosmos's way of writing the instructions for the next phase of its own evolution.

17.5- The Gamma Burst: Guiding Quanta in the Brain

The clearest, most precisely measured example of guiding quanta in action is the gamma burst that accompanies insight in the human brain.

A sudden comprehension that solves a problem, reinterprets a situation, explains a joke, or resolves an ambiguous percept is called an insight (i.e., the "Aha! moment"). Psychologists have studied insight using behavioural methods for nearly a century. Recently, the tools of cognitive neuroscience have been applied to this phenomenon. A series of studies have used electroencephalography (EEG) and functional magnetic resonance imaging (fMRI) to study the neural correlates of the "Aha! moment" and its antecedents. Although the experience of insight is sudden and can seem disconnected from the immediately preceding thought, these studies show that insight is the culmination of a series of brain states and processes operating at different time scales.

The experimental paradigm is straightforward. Participants are given word puzzles (compound remote associate problems) and asked to find a word that connects three otherwise unrelated words. When they find the solution, they report whether it came through gradual, analytical deduction or through sudden insight- the flash of comprehension that seems to appear from nowhere.

The EEG data reveal a striking pattern. The researchers found that 0.3 seconds before the subjects indicated solutions achieved through insight, there was a burst of neural activity of one particular type: high-frequency (gamma band) activity that is often thought to reflect complex cognitive processing. This activity was also mapped to the anterior superior temporal gyrus of the right hemisphere, providing compelling convergence across experiments and methods.

And the gamma burst was preceded by something equally remarkable: a sudden increase in alpha-band activity (10 Hz) over the right occipital cortex- the visual processing area. The EEG results revealed another, totally unexpected, finding. The insight-related gamma-band activity was immediately preceded by a burst of alpha-band activity (10 Hz) measured over right occipital cortex.

The alpha burst is interpreted as a brief "gating" event- the brain momentarily *shutting down* visual processing to reduce noise and allow the fragile, sub-threshold pattern of the emerging insight to rise into consciousness. It is, in our framework, the neural correlate of the comonadic extraction $D(\Psi)$: the brain suppressing external input to reveal the 98% efficient core- the answer that has been assembling below the surface.

And the gamma burst itself is the guiding quanta in action: a transient, high-frequency, spatially localised pulse of coherent neural activity that binds distributed associations into a single, integrated representation. This right temporal brain response was identified as the main neural correlate of the insight experience because it occurred at about the moment when participants realised the solution, the same region is involved in other tasks demanding semantic integration, and gamma-band activity has been proposed to be a mechanism for binding information as it emerges into consciousness.

In a 2020 study, the picture became even richer. The primary neural correlate of insight was a burst of gamma-band oscillatory activity over prefrontal cortex approximately 500 ms before participants pressed a button to indicate that they had solved the problem. However, only highly reward-sensitive people showed an additional burst of high-frequency gamma waves about a tenth of a second later. This second burst originated in orbitofrontal cortex, a part of the brain's reward system. The study shows that some people experience creative insights as intrinsically rewarding.

This is profoundly significant. The brain does not merely solve the problem- it *rewards itself for having solved it through insight*. The guiding quanta (the gamma burst that carries the solution) are immediately followed by a reward signal (the orbitofrontal activation) that reinforces the neural pathways responsible for insight, making future insight solutions more likely. The system *learns to transform*. The very act of undergoing a critical transition strengthens the capacity for future transitions.

This is the co-creative cycle of Chapter 16 operating at neural timescales: intention (the problem to be solved) → attention (the sustained focus) → the comonadic suppression (the alpha gate) → the guiding quanta (the gamma burst) → the reward (the orbitofrontal response) → the updated capacity (enhanced insight pathways). The brain is a microcosm of the cosmos- a NESS that self-organises to criticality, generates guiding quanta at the threshold, and uses those quanta to leap into a new configuration.

17.6- The Cell's Moment of Decision: Gene Regulatory Bursts

At the biological scale, guiding quanta manifest as the gene regulatory bursts that orchestrate cell differentiation.

A stem cell- a pluripotent cell capable of becoming any of the body's specialised cell types- faces the most consequential decision in biology: what to become. The decision is not gradual. It is a bifurcation- a critical threshold at which the cell commits, irreversibly, to a specific fate.

In the moments surrounding this commitment, the cell's transcriptional activity undergoes a dramatic transformation. Transcription factors- proteins that regulate gene expression- are expressed in sudden, transient bursts. These bursts are not random. They are highly specific: particular combinations of transcription factors are activated in precise temporal sequences, each combination corresponding to a specific cell fate.

The concept is captured by the Waddington landscape- a metaphor proposed by developmental biologist Conrad Waddington in 1957, in which cell differentiation is visualised as a ball rolling down a landscape of valleys and ridges. The valleys represent stable cell types (attractors); the ridges represent unstable boundaries between types. At the crest of each ridge- at the bifurcation point- the ball must choose one valley or the other.

What guides the ball? The gene regulatory bursts. These transient pulses of transcription factor activity are the *guiding quanta of cellular differentiation*- brief, structured, high-specificity patterns that push the cell's state vector through the bifurcation and into the appropriate attractor basin.

And the parallel to neural insight is exact:

Feature	Neural Insight	Cellular Differentiation
System	Brain (86 billion neurons)	Cell (one unit in a tissue of millions)

Feature	Neural Insight	Cellular Differentiation
Pre-transition state	Sustained unconscious processing	Sustained pluripotency with fluctuating gene expression
Gating event	Alpha burst (visual suppression)	Epigenetic desilencing (chromatin opening)
Guiding quanta	Gamma burst in right temporal lobe	Transcription factor burst in specific regulatory network
Transition	Sudden conscious insight	Irreversible commitment to cell fate
Post-transition reward	Orbitofrontal dopamine signal	Stabilisation of new gene expression program
Timescale	~300 milliseconds	~6–24 hours

The fractal self-similarity is unmistakable. The same dynamical pattern- sustained approach, gating, transient burst of guiding quanta, irreversible transition, stabilisation- operates at radically different scales and in radically different substrates. This is not analogy. This is the master equation doing its work:

$$\partial_t \Psi = -\nabla_{FR}[\mathcal{L}(\Psi)] + T(\Psi) \otimes D(\Psi) + \sum_n \phi \cdot \omega_n \cdot \delta(\Psi \in \mathrm{NESS}_{98})$$

The first term drives the system toward its current attractor (equilibrium-seeking). The second term- the interplay of monadic extension $T(\Psi)$ (which amplifies potential new configurations) and comonadic extraction $D(\Psi)$ (which suppresses noise to reveal the emerging pattern)- generates the guiding quanta. And the third term tunes the system to the φ-resonant frequency at which the transition is most efficient.

17.7- Vespignani's Prophecy: Guiding Quanta in Pandemic Networks

In 2009, a novel strain of H1N1 influenza emerged in Mexico and spread worldwide with alarming speed. In 2009, the H1N1 virus slipped into the bloodstreams of more than 40 million people around the world. In just four months, it killed more than 14,000 individuals as it travelled from Mexico to India on its most favoured vehicle: humans. As travellers moved about the planet via aeroplanes and cars, the pathogen followed, creating an epidemic the likes of which had not been seen since the 1970s.

The conventional approach to epidemic prediction- curve fitting, extrapolation from past outbreaks, static models- was inadequate. The H1N1 virus was new, its parameters unknown, and the global mobility network was vastly more complex than any previous pandemic had encountered.

Alessandro Vespignani took a different approach. Vespignani- now the Sternberg Family Distinguished University Professor of physics, computer science, and health sciences at Northeastern University- and his research team built a computational model called GLEAM, or Global Epidemic and Mobility Model, which they used to predict the outbreaks as they surfaced around the globe.

GLEAM did not model the virus in isolation. It modelled the *network*- the global web of human mobility through which the virus propagated. The model generated stochastic simulations of epidemic spread worldwide, yielding the incidence and seeding events at a daily resolution for 3,362 subpopulations in 220 countries.

The results were extraordinary. Operating from within the proverbial eye of the storm in 2009, the team used the model to forecast the week of the epidemic's peak in 48 countries in the Northern Hemisphere. In 42 of these countries, the forecasts were directly on target; in the other six, the team's predictions were off by only one to two weeks. Normally, flu season peaks months after H1N1 did, making even the two-week variation a considerably good result. "This is the first large-scale validation of a computational model that pulled out predictions in real time," said Vespignani.

Real-time predictions of the peak timing are found to be in good agreement with the empirical data, showing strong robustness to data that may not be accessible in real time.

Why does this matter for our argument?

Because Vespignani's model succeeded not by tracking the virus but by tracking the *quanta-flow*- the patterns of human movement that constitute the network through which the virus propagates. The model did not need to know the precise biological properties of H1N1 (though these were estimated). What it needed- what proved decisive- was the *topology of the network*: which airports connected to which cities, how many commuters moved between which subpopulations, what the structure of the global mobility web looked like at a given moment.

The peak of the epidemic in each country was not determined by the virus alone. It was determined by the interaction between the virus and the network- by the moment when the propagating wave of infection reached the resonant frequency of the local population structure. The peak was, in our language, a *critical transition*- a bifurcation point at which the exponential growth phase of the epidemic transitioned into the declining phase, mediated by the depletion of susceptible individuals and the buildup of immunity.

And the prediction of this peak- its timing, its geographical distribution, its wave-like propagation- was possible because the *network carried the guiding quanta*. The patterns of human mobility contained, in their structure, the blueprint of the epidemic's future- the timing, the geography, the intensity of the peak. Vespignani's model did not create these patterns. It *read* them. It decoded the guiding quanta that were already flowing through the network, months before the peaks they predicted.

This is distributed intelligence in action: the network itself- not any central controller, not any individual's decisions- carried the structured quanta that determined the outcome. The future of the pandemic was written in the mobility network's topology, waiting to be read by anyone with the mathematical tools to decode it.

17.8 - Non-Equilibrium Early Warning Signals: The New Frontier

The standard early warning signals described by Scheffer - critical slowing down, rising variance, increased autocorrelation - are powerful but limited. They detect the approach to a transition but cannot distinguish between different types of transitions, cannot predict the specific configuration the system will adopt after the transition, and cannot operate in systems far from equilibrium where the assumptions of linear theory break down.

A 2023 paper in PNAS by Xu and colleagues extended the framework dramatically. Average flux, entropy production, nonequilibrium free energy, and time irreversibility quantified by the difference in cross-correlation functions forward and backward in time can serve as early warning signals for critical transitions much earlier than other conventional predictors. They utilise a classical shallow lake model as an exemplar for their early warning prediction. Their proposed method is general and can be readily applied to assess the resilience of many other ecological systems. The early warning signals proposed here can potentially predict critical transitions earlier than established methods and perhaps even sufficiently early to avert catastrophic shifts.

This is significant because the new early warning signals - based on non-equilibrium thermodynamic quantities like *entropy production rate*, *nonequilibrium free energy*, and *time irreversibility* - are precisely the quantities our framework identifies as the measurable signatures of guiding quanta.

When the entropy production rate spikes, it is because the system is processing a burst of guiding quanta - the transient, high-specificity patterns that carry the blueprint of the new state. When the nonequilibrium free energy increases, it is because the system is accumulating potential for transformation - the monadic extension $T(\Psi)$ is building toward the threshold. When time irreversibility increases, it is because the system's dynamics are becoming increasingly asymmetric - the arrow of transformation is sharpening, the system is committing to a specific direction of change.

These non-equilibrium early warning signals are, in our vocabulary, *measurements of the guiding quanta's approach*. They detect not merely that a transition is near, but that the structured quanta carrying the blueprint of the new state are beginning to flow through the system.

17.9- Ecosystems at the Tipping Point

The ecological domain offers some of the most dramatic examples of guiding quanta in action.

Critical transitions are sudden, often irreversible, changes that can occur in a large variety of complex systems; signals that warn of critical transitions are therefore highly desirable. Scheffer's own paradigmatic example is the shallow lake. A clear-water lake can exist in one of two stable states: a clear state dominated by submerged vegetation, and a turbid state dominated by phytoplankton. Nutrient loading (primarily phosphorus from agricultural runoff) gradually pushes the lake toward the turbid state. For a long time, nothing visible happens- the lake remains clear. Then, abruptly, it flips. The water turns green. The vegetation dies. The fish populations collapse. The transition is catastrophic and, under many conditions, irreversible: even if nutrient loading is subsequently reduced, the lake may remain turbid for decades.

But in the months before the flip, the guiding quanta are already flowing. The system's variance increases. Its spatial patterns change- in ecology, critical transitions have become a major focus of research. The existence of alternative attractors has been demonstrated experimentally in lakes, and a large body of work now suggests that alternative stable states separated by critical thresholds also occur in ecosystems ranging from rangelands to marine systems.

In arid ecosystems, the spatial pattern of vegetation itself serves as an early warning signal. In dry regions self-organisation can lead to particular spatial patterns under some conditions. Here the complete loss of vegetation is an important transition, as recovery from the barren state may require more rain than is needed to preserve the last patches.

Vegetation in semi-arid landscapes self-organises into spatial patterns- spots, stripes, labyrinths- through the interplay of facilitation (plants help nearby plants) and competition (plants compete for distant resources). As the system approaches desertification, these patterns change in a specific, predictable sequence: from labyrinth to stripes to spots to bare ground. The spatial pattern is not merely a *symptom* of the approaching transition. It is the *guiding quanta made visible*- a structured spatial signal that carries information about the system's proximity to the critical threshold and the nature of the transition that will occur when the threshold is crossed.

This is breathtaking. The landscape itself is writing its own diagnosis. The patterns in the vegetation- visible from satellite imagery- are transient, high-specificity signals that encode the system's distance from collapse. They are guiding quanta at the ecosystem scale: brief (on ecological timescales), structured, and carrying the blueprint of the impending transition.

17.10- Financial Markets: The Crash as Phase Transition

The dynamics of financial markets provide another, disturbingly vivid example.

The prediction of shifts in financial markets is a heavily researched area. In this field, the discovery of predictability quickly leads to its elimination, as profit can be made from it, thereby annihilating the pattern.

Financial crashes- the 1929 Black Tuesday, the 1987 Black Monday, the 2008 subprime crisis, the 2020 COVID crash- are critical transitions. They are not gradual declines; they are *regime shifts* in which the dynamics of the market change suddenly, qualitatively, and often irreversibly (in the sense that the post-crash market operates under fundamentally different conditions than the pre-crash market).

The pre-crash period exhibits the classic early warning signals: increased volatility (rising variance), stronger autocorrelation (momentum), and a characteristic log-periodic oscillation in which the market's fluctuations become increasingly rapid and increasingly extreme as the crash point approaches. These oscillations are the financial equivalent of critical slowing down- the market is losing resilience, responding ever more dramatically to ever smaller perturbations.

But the crash itself- the moment of transition- is not merely the collapse of order. It is the creation of a new order. Post-crash markets exhibit different correlation structures, different volatility regimes, different pricing dynamics. The crash is a bifurcation- a transition from one NESS to another, mediated by the explosive flow of guiding quanta through the network of financial interactions.

What are the guiding quanta in this case? They are the correlated selling signals- the cascade of margin calls, liquidations, and panic that propagates through the network of interconnected financial institutions. Like the sandpile avalanche, the crash is not random destruction. It is a *coordinated reorganisation*- a system-wide processing of structural quanta that reveals the hidden fragilities of the network and leaves behind a new configuration.

The 2008 crisis is particularly instructive. The crash was preceded by a period of apparent stability- the "Great Moderation"- during which volatility was historically low and risks were thought to be well-managed. In hindsight, this was the calm before the avalanche- the system sitting at the critical point, seemingly stable but in fact poised at the edge of chaos, awaiting the one additional grain that would trigger the cascade.

The triggering grain was the collapse of Lehman Brothers. But the guiding quanta- the structured patterns that determined *how* the crash would propagate, *which* institutions would fail, and *what* the post-crash landscape would look like- were already encoded in the network of mortgage-backed securities, collateralised debt obligations, and credit default swaps that constituted the financial system's hidden stress geometry. The blueprint of the crash was written in the topology of the network, months before the crash itself.

17.11- The Feigenbaum Constant: Universal Geometry of Thresholds

There is a mathematical constant that links all these phenomena- a number that appears, with uncanny universality, whenever a system approaches a cascade of bifurcations on its way to chaos.

The constant is $\delta \approx 4.669$, discovered by the physicist Mitchell Feigenbaum in 1975. It describes the rate at which the spacing between successive bifurcation points decreases as a system is driven toward chaos. The Feigenbaum constant is *universal*: it takes the same value regardless of the specific system being studied- whether it is a dripping faucet, a population of rabbits, a heartbeat, or a laser oscillation.

This universality is the mathematical signature of what we have been calling guiding quanta. The fact that the geometry of bifurcation cascades is identical across all these systems means that the *guiding quanta themselves have a universal structure*. The transient patterns that orchestrate critical transitions are not system-specific. They are governed by a universal mathematical architecture that is independent of the substrate.

In our framework, this universality is a direct consequence of the master equation. The $T(\Psi) \otimes D(\Psi)$ term- the interplay of monadic extension and comonadic extraction- generates a specific geometry of approach to the critical threshold, regardless of the specific system in which it operates. The Feigenbaum constant is a *fingerprint of the master equation*- a number that falls out of the mathematics of φ-resonant NESS dynamics with the inevitability of π falling out of the geometry of circles.

17.12- Revolution: Guiding Quanta in Social Systems

The principles extend, with alarming precision, to the dynamics of social transformation.

The interactions between societies can lead to phase transitions, where small changes in conditions result in significant shifts in the state of the system. For example, a minor escalation in aggressive posturing or a breakdown in communication can tip the balance from peace to conflict.

Revolutions, social upheavals, and paradigm shifts in science all exhibit the dynamics of critical transitions. The pre-revolutionary period is characterised by rising tension (increasing variance), growing correlation between previously independent actors (small-world network effects), and a distinctive slowing down of the system's ability to return to "normal" after disturbances (critical slowing down).

And at the moment of revolution- the storming of the Bastille, the fall of the Berlin Wall, the Arab Spring's cascade across the Middle East- the guiding quanta flow: a burst of coordinated action, mediated by communication networks (pamphlets, telephone trees, social media), that reorganises the social system from one configuration to another.

The pre-revolutionary social network, like the pre-crash financial network, carries the blueprint of the coming transformation in its topology. The patterns of who talks to whom, who trusts whom, who is connected to whom- these are the guiding quanta of social transformation. They are transient (they change rapidly as the revolution unfolds), high-specificity (they determine the specific form the revolution takes), and they appear at the critical threshold (they crystallise into decisive action at the moment when the old regime loses its last resilience).

The failure of prediction in social systems is not due to the absence of guiding quanta. It is due to our inability- so far- to read them. The patterns are there, encoded in the network, awaiting a Vespignani of social dynamics who can build the equivalent of GLEAM for revolutions.

17.13- The Ontology of Guiding Quanta

Let us now formalise what we have been describing.

Definition: Guiding quanta are transient, high-specificity patterns of resonant quanta that appear at critical thresholds (bifurcation points) in complex systems and orchestrate the system's transition from one stable configuration to the next.

Properties:

1. **Transience.** Guiding quanta are brief. They exist only during the transition itself- the narrow window between the loss of stability in the old configuration and the establishment of stability in the new one. Before the threshold, they are absent (or below detection threshold). After the threshold, they have been absorbed into the new stable configuration. They are visible only *at* the threshold.

2. **Specificity.** Guiding quanta are not random noise. They carry the specific structural blueprint of the new state. The gamma burst in the brain carries the specific associative pattern that constitutes the insight. The gene regulatory burst carries the specific transcription factor combination that determines cell fate. The avalanche cascade carries the specific stress geometry that determines the new grain arrangement.

3. **Quanta-density.** The moment of transformation is the most quanta-dense moment in the system's cycle. More structured resonant quanta flow through the system per unit time during the transition than at any other moment. This is the "excess entropy production" that Prigogine identified- the thermodynamic cost of reorganisation, which is simultaneously the thermodynamic *signature* of emergence.

4. **Universality.** The geometry of guiding quanta is governed by universal constants (the Feigenbaum constant, and in our framework, the φ-resonance term in the master equation). Different systems generate guiding quanta with different specific content, but the *dynamical architecture* of their appearance- the critical slowing down that precedes them,

the power-law scaling of their effects, the irreversibility of the transitions they mediate- is universal.

5. **Co-creative coupling.** In systems that contain conscious observers (brains, societies), guiding quanta are not merely thermodynamic phenomena. They are *accessible to consciousness*. The gamma burst is experienced as insight. The social cascade is experienced as revolution. The ecological shift is experienced as catastrophe or renewal. Consciousness does not merely observe guiding quanta- it *participates in generating them* (as the co-creative principles of Chapter 16 describe).

17.14- Guiding Quanta and the δ_∞ Gap

We can now close the circuit that connects guiding quanta to the deepest structure of the Universal Hologram.

Recall from Chapter 10 that the asymptotic gap $\delta_\infty \approx 0.01965$ is the ~2% residual tension between the three cosmic operators (ϕ, e, π) that prevents the universe from collapsing into static equilibrium. In Chapter 12, we identified this gap as the "housekeeping energy" of the cosmic NESS- the metabolic cost of staying alive.

Now we can add a third interpretation: δ_∞ **is the reservoir from which guiding quanta are drawn.**

The 2% of the quanta-budget that is permanently unavailable for equilibrium-seeking is not merely wasted on "housekeeping." It is *invested* in the capacity for transformation. It is the energy reserve that funds the transient bursts at bifurcation points- the gamma bursts, the gene regulatory bursts, the avalanche cascades, the revolutionary upheavals.

A cosmos without δ_∞- a cosmos in which the triad ϕ, e, π closed perfectly into equilibrium- would be a cosmos without guiding quanta. It would be a cosmos capable of sustaining a NESS but incapable of *transforming* that NESS. It would persist forever in a single, frozen configuration- alive, perhaps, but never growing, never evolving, never surprising itself.

The gap is what makes evolution possible. The gap is what funds emergence. The gap is the universe's venture capital- the permanent endowment from which every bifurcation, every insight, every revolution, every birth draws its creative energy.

And the master equation makes this explicit. The third term:

$$\sum_{n} \phi \cdot \omega_n \cdot \delta(\Psi \in \text{NESS}_{98})$$

is the φ-resonance term that activates *only when the system is in the 98% efficient NESS core*. The remaining 2%- the δ_∞ fraction- is precisely the energy available for the $T(\Psi) \otimes D(\Psi)$ term to do its transformative work. The monadic extension reaches into the 2% gap, amplifies the nascent pattern of the new configuration, and- at the critical threshold- releases it as a burst of guiding quanta that reshapes the system.

This is why the universe evolves. Not because it is destined to. Not because an external force compels it. But because the gap between perfection and reality- the δ_∞, the ~2% of divine imperfection- permanently funds the creative process by which the cosmos transcends itself.

17.15- The Moment After

We began this chapter with a moment before- the instant when the brain, the lake, the pandemic, the cell holds its breath at the edge of transformation.

Now consider the moment after.

The insight has landed. The lake has flipped. The peak has passed. The cell has committed. The revolution has begun. The avalanche has settled. The guiding quanta have done their work and vanished- absorbed into the new stable configuration, invisible once more.

But something has changed. The system is not what it was. It has crossed a threshold that cannot be uncrossed. The insight cannot be unthought. The cell cannot un-differentiate (in most cases). The social order cannot simply revert (though it may attempt to, often violently). The transition, mediated by the guiding quanta, has created something genuinely new- a configuration that did not exist before the threshold was crossed.

And here is the deepest truth about guiding quanta: **they are the quanta-field's method of self-transcendence.**

Every other process in the cosmos- the natural gradient flow, the NESS maintenance, the φ-resonance tuning- preserves and sustains what already exists. Only the guiding quanta *create what has never existed before*. They are the mechanism by which the universe- the autopoietic, self-tuning Universal Hologram- writes its own future into being.

This is why the moment of transformation feels meaningful. This is why the "Aha!" moment is accompanied by a reward signal in the orbitofrontal cortex. This is why revolutions feel, to those living through them, like the opening of a door into a larger room. This is why the birth of a child, the eruption of a volcano, the first note of a symphony all carry the unmistakable signature of *significance*.

They are significant because they are moments of maximum quanta-density- moments when the universe is processing more structured resonant quanta per unit time than at any other point in its cycle. They are the moments when the cosmos is most intensely *itself*- most fully engaged in the act of becoming that constitutes its deepest nature.

The chaos before the transformation is not disorder. It is *the universe thinking*.

The stillness after the transformation is not death. It is *the universe having spoken*.

And the guiding quanta- brief, brilliant, vanishing into the structure they have created- are the words.

In the next chapter, we cross from physics to culture- from the mathematics of emergence to the oldest technologies of resonance on Earth. We will enter the songlines of Australia, the mantras of the Vedas, the incantations of the Maya, and discover that the ancients were not merely telling stories about creation. They were performing it. Chapter 18 reveals the ontology of sound- the physics of world-singing- and demonstrates that the deepest truths about guiding quanta were known, practised, and sung into being long before modern science was born.

SECTION D: THE ORIGINAL PHYSICISTS-ANCIENT RESONANCE TECHNOLOGIES

"The dreamers were the architects."

Chapter 18

World-Song – The Ontology of Sound

18.1- In the Beginning Was the Vibration

Before there was light, there was sound.

Not sound as we ordinarily conceive it- not the vibration of air molecules against an eardrum. Something far more fundamental. A primordial vibration- a resonant stirring of the quanta-field- from which all structure, all form, all differentiation would crystallise.

Every great creation tradition on Earth, without exception, begins here.

The Vedic tradition opens with *Nada Brahma*: "The world is sound." The universe arises from the vibration of the Absolute- the unstruck tone, *anahata nada*, that reverberates through all things. The Chandogya Upanishad declares: "All this indeed is Brahman"- and the first expression of that Brahman is Om, the syllable that contains all syllables, the sound that precedes all sounds, the resonance from which the entire cosmos unfolds.

The Gospel of John opens with a phrase that echoes across two millennia: "In the beginning was the Word." The Greek *Logos*- translated as "Word"- carries a meaning far richer than its English rendering. *Logos* means reason, pattern, proportion, the intelligible structure that gives form to chaos. The universe begins not with a thing but with an *utterance*- a structured vibration that calls reality into being.

The Popol Vuh, the creation narrative of the K'iche' Maya, recounts that the two "great thinkers" filled the emptiness through dialogue. Whatever they said was created. Much of creation comes about through the utterance of words by the gods, rather than physically crafting things. Together, these two gods created the physical world by speaking the names of things, bringing them into existence. In Maya cosmology, things were created to match their names, as opposed to being named after being created.

The Aboriginal Australians do not speak of a time *before* creation. They speak of the Dreamtime- the eternal, living present in which the ancestral beings sing the world into existence. As expressed by the evolving epistemology: songlines form a living archive that encodes mythological stories, ecological knowledge, and social law into songs, dances, and ritual practices. A songline is not just a melody but a cognitive map that links sacred sites through narrative, rhythm, and movement. The beings who created the world did not build it with their hands. They *sang* it with their voices. And the singing has never stopped.

The Egyptian cosmology of Heliopolis places the god Atum at the origin of all things, uttering the names of the gods into existence- the first act of creative speech, from which the Ennead emerges. The Norse *Ginnungagap*- the yawning void- is not silence but the resonant gap between fire (Muspelheim) and ice (Niflheim), and the first being, Ymir, is born from the interaction of these vibrational extremes.

What are we to make of this convergence? That every civilisation on Earth, independently, concluded that reality begins with vibration?

The materialist response is dismissal: these are myths, metaphors, pre-scientific groping toward explanations that modern physics has superseded. Sound is merely pressure waves in air. It does not create anything. It is an effect, not a cause.

This chapter argues the opposite. The ancients were not groping. They were *describing*- with the vocabulary available to them, using the somatic instruments of their own bodies- a truth that modern physics is only now catching up with.

The truth is this: **vibration is not a secondary phenomenon that happens *within* a pre-existing world. Vibration is the primary act by which the quanta-field differentiates itself into the world.** Sound- in the generalised sense of resonant oscillation- is not the consequence of creation. It is the *mode* of creation.

The creation narratives were not metaphors. They were *field reports.*

18.2- Cymatics: The Visible Speech of the Quanta-Field

If this claim seems extravagant, consider the evidence of your own eyes.

Cymatics (from Ancient Greek: κῦμα, "wave") is a subset of modal vibrational phenomena. The term was coined by Swiss physician Hans Jenny (1904–1972).

The experimental procedure is disarmingly simple. Jenny spread powders, pastes, and liquids on a metal plate connected to an oscillator which could produce a broad spectrum of frequencies. The substances were organized into different structures characterized by geometric shapes typical of the frequency of the vibration emitted by the oscillator.

Increase the frequency, and the pattern changes- becoming more complex, more intricate, more detailed. Lower it, and the pattern simplifies. Each frequency produces a *specific, reproducible, geometrically precise pattern*. The same frequency, applied to the same plate, produces the same pattern every time.

He showed the evolution of harmonic images by subjecting inert substances to oscillating sound waves. His substantial body of work, based on rigorous scientific methodology, developed Chladni's experiments, highlighting intricate, organic, harmonic images that reflected many universal patterns found throughout nature and especially living organisms.

Jenny's predecessor, the German physicist Ernst Chladni, had demonstrated the same principle as early as 1787. In the late 18th century, German physicist and musician Ernst Chladni demonstrated how vibrations could be used to create striking imagery. By spreading fine sand across the top of a metal plate and running a violin bow alongside, Chladni showed that the sand would settle into distinct patterns, depending on the frequencies of the sound waves produced by the bow.

And the principle goes back even further. On July 8, 1680, Robert Hooke was able to see the nodal patterns associated with the modes of vibration of glass plates. Hooke ran a bow along the edge of a glass plate covered with flour, and saw the nodal patterns emerge.

For over three centuries, then, Western science has known that *sound creates form*. That vibration organises matter into geometric pattern. That frequency determines structure.

For the first time, Chladni's work demonstrated that sound isn't only something we hear-it's a physical force that shapes and moves matter in predictable ways.

Let us be precise about the physics. The pattern is not the frequency; rather the pattern is a resonant mode of the medium, being excited by that particular sound frequency. We are not visualising a specific sound per se, but rather finding a resonant mode of a medium that is excited by that particular sound.

This is an important nuance. The frequency does not *draw* the pattern on the plate. The frequency *excites* a resonant mode that is *latent* in the plate's geometry. The pattern was always *potential* in the structure of the medium. The vibration merely *reveals* it- calls it forth from potentiality into actuality.

Now read that sentence again through the lens of this book's ontology.

The quanta-field- the resonant substrate of reality- contains within itself an infinite number of potential configurations. These configurations are latent- present as possibility, but not yet actualised. When a resonant vibration of the appropriate frequency is introduced, one of these latent configurations is *activated*- called from superposition into manifestation.

Cymatics, in this reading, is not merely a laboratory curiosity. It is a *microcosm of creation itself*. The Chladni plate is a tiny section of the quanta-field. The oscillator is the primordial vibration. The sand pattern is the material world- the structured, geometric, beautiful world that emerges when the quanta-field resonates at a specific frequency.

The ancients, watching patterns form in sand around their drums, seeing spirals emerge in water struck by their hands, witnessing the precise geometric formations that appeared when their voices resonated in the stone chambers of their temples- they were not engaging in primitive magic. They were doing physics. They were observing, with their own bodies and their own instruments, the fundamental principle that *resonance creates form*.

Hans Jenny, who spent decades of his life documenting these phenomena, arrived at a conclusion that his scientific contemporaries considered mystical: "This is not an unregulated chaos; it is a dynamic but ordered pattern."

It is the same conclusion this book has reached by a different route: the universe is not chaos rendered orderly by external law. It is *resonant quanta unfolding*- pattern all the way down, structure all the way up, vibration at every scale.

18.3- Aboriginal Songlines: The World Sung Into Being

Of all the world's cultures, the Aboriginal Australians developed the most complete, most explicit, and most practically implemented technology of creation-through-sound. They called it the Songlines- and what they built with it is, in our framework, the most sophisticated resonance technology the human species has ever produced.

As confirmed in our extensive research: by singing the songs in the appropriate sequence, Aboriginal people could navigate vast distances, often travelling through the deserts of Australia's interior. The continent contains an extensive system of songlines, some traversing hundreds of kilometres through lands of many different peoples.

The songlines are not roads. They are not trails marked on the ground. They are *resonant pathways sung through the landscape*- sequences of musical phrases that, when performed in the correct order, describe the topography, water sources, sacred sites, seasonal patterns, and relational protocols of the land with extraordinary precision.

Aboriginal people regard all land as sacred, and the songs must be continually sung to keep the land "alive." This is not a metaphorical statement about cultural heritage. It is an ontological claim: *the land exists because it is being sung.* If the songs cease, the land- the specific configuration of the quanta-field that constitutes that landscape- begins to dissipate. The singing is not a celebration of creation. It is the *continuation* of creation. The maintenance of the world's NESS.

When people perform ceremonies, sing the songs, and dance the dances, they are not remembering creation- they are continuing it. The ceremonies maintain the world, ensure the continuation of life, and keep the Songlines alive.

In the Dreamtime, the creation myths tell of legendary totemic beings who wandered over the continent, singing out the name of everything that crossed their path, and so singing the world into existence. The Dreamtime is not ancient history but is an eternal present, continuously accessible through ceremony, art, and the singing of the songs that maintain creation itself.

Let us translate this into our framework.

The ancestral beings of the Dreamtime are the primordial resonances of the quanta-field- the fundamental frequencies that, through their interaction and interference, generated the specific patterns of the Australian landscape: its mountains, its rivers, its waterholes, its deserts. Each songline traces the *path of a specific resonant frequency through the quanta-field-* a topological manifold along which a particular harmonic pattern propagates.

The songs encode the *spectral signature* of the landscape- the specific set of frequencies, rhythms, and tonal relationships that characterise the quanta-field's configuration at each point along the songline. When an Aboriginal person sings a section of the songline, they are not merely describing the land. They are *re-activating the resonance that constitutes the land-* refreshing the standing wave, reinforcing the NESS, preventing the pattern from decaying toward equilibrium.

The songlines could also be used as a cultural passport. Travelers would sing the part of the songline of a particular territory to ensure respectful passage. Singing the correct song in the correct territory was not a social nicety. It was a *resonant handshake-* a demonstration that the traveller was in harmonic alignment with the local quanta-field, that their personal state vector was compatible with the geospatial phase $G(\lambda,\phi)$ of the territory they were entering.

Singing and drumming in caves, near waterholes, or on plains used the natural acoustics of the land to enhance rhythm and resonance. The landscape itself was used as a resonant amplifier- the caves and rock formations as acoustic chambers, the waterholes as reflective surfaces, the open plains as propagation media. The human voice, coupled with the natural acoustics of the terrain, formed a *distributed resonance system-* a planetary-scale instrument whose components were the human body, the geological substrate, and the atmospheric medium.

This is not metaphor. This is NESS-Holography in action. The songlines are a technology for maintaining the Earth's resonant field- a network of standing waves sustained by continuous human participation, calibrated to the specific quanta-field configuration of each locality, and refined over at least 65,000 years of continuous practice.

18.4- Vedic Chanting: Om and the Neuroscience of Cosmic Resonance

If the Aboriginal songlines represent the most geographically extensive resonance technology, the Vedic tradition represents the most neurophysiologically precise.

OM (or AUM) chanting is one of the most ancient meditative practices in yogic and Vedic traditions, representing the primordial vibration of the universe.

The Vedas describe Om not as a symbol but as a *technology*. The Mandukya Upanishad analyses its three phonemic components- A, U, M- as corresponding to the three states of consciousness (waking, dreaming, deep sleep), the three worlds (earth, atmosphere, heaven), and the three temporal modes (past, present, future). The silence that follows the M- the *turiya*, the fourth state- corresponds to the transcendent reality that contains and pervades all three.

This is not philosophical abstraction. It is an *operating manual* for tuning the brain to the quanta-field.

Modern neuroscience confirms that the practice produces measurable, specific, and reproducible effects. Research using functional MRI and EEG has shown that chanting mantras activates the brain's limbic system, reduces activity in the amygdala, and induces shifts from high-frequency beta waves to calming alpha and theta waves.

The evidence suggests that Om chanting activates parasympathetic nervous system responses, modulates stress hormone levels, and enhances inhibitory neurotransmission through multiple neurochemical pathways.

The mechanism is now partially understood: When we chant Om or a bija (seed) mantra, the vibrations stimulate the sensory nerves (especially around the vocal cords), which in turn activate the vagus nerve. The vagus is the master highway of the parasympathetic system, linking the brainstem to heart, lungs and gut.

The vagus nerve is the longest cranial nerve in the human body. It connects the brainstem to nearly every major organ. Stimulating it through the physical vibrations of chanting activates the parasympathetic ("rest-and-digest") branch of the autonomic nervous system, producing a cascade of measurable effects: reduced heart rate, lowered blood pressure, decreased cortisol, increased GABA (the brain's primary inhibitory neurotransmitter), and a shift in brainwave patterns from agitated beta to coherent alpha and theta.

Research shows that Om chanting stimulates the Vagus Nerve through its auricular branches. The studies show that chanting Om can reduce activities of limbic brain regions, amygdala, hippocampus, parahippocampal gyrus, insula, orbitofrontal and anterior cingulate cortices and thalamus.

These are the brain regions associated with stress, fear, and emotional reactivity. Om chanting *quiets them*. Not through belief, not through expectation, not through social conformity- but through the *direct physical action of sound waves on neural tissue*, mediated by the vagus nerve and the acoustic resonance of the vocal tract.

Specific mantra frequencies don't just calm your mind- they trigger cascades of neurochemical changes, alter brain wave patterns, stimulate the vagus nerve, and even modify gene expression.

The claim that mantras modify gene expression deserves special attention. Epigenetic research has shown that sustained meditative practices can alter DNA methylation patterns- switching genes on or off without changing the underlying sequence. If Om chanting activates the vagus nerve, reduces cortisol, increases GABA, and shifts brainwave patterns, then it is operating through the same biochemical pathways that govern epigenetic regulation. The mantra is not merely calming the mind. It is *rewriting the body's biological instruction set.*

Vedic rishis who described mantras as tools for consciousness transformation weren't speaking metaphorically. They were describing neuroplasticity, neurochemical modulation, and epigenetic effects using the vocabulary of their era. Modern neuroscience isn't disproving ancient wisdom- it's providing the mechanistic explanations that validate millennia of empirical practice.

Now let us connect this to the framework of the Universal Hologram.

In Chapter 13, we defined consciousness as the self-referential resonance of the quanta-field, stabilised by φ-proportional phase-locking in neural oscillators. We identified the KZ-Monad $T(\Psi)$ as the operation that extends potential into full φ-resonant manifestation, and the Idempotent Comonad $D(\Psi)$ as the operation that extracts the 98% efficient NESS core.

Om chanting performs *both operations simultaneously.*

The **monadic extension** operates through the vibrational stimulation of the vocal tract, vagus nerve, and resonant chambers of the skull- amplifying specific frequencies, extending the brain's oscillatory repertoire, opening new harmonic possibilities. This is $T(\Psi)$: reaching for wholeness, extending the partial into the full.

The **comonadic extraction** operates through the suppression of limbic activity, the reduction of cortical noise, and the shift from beta (agitated, fragmented) to alpha and theta (coherent, integrated)- stripping away the noise to reveal the signal. This is $D(\Psi)$: distilling wisdom from chaos, projecting onto the efficient core.

The combined effect is a state of enhanced Φ_Q- Integrated Resonant Quanta- in which the brain achieves a mode of resonant integration that approaches the φ-proportionality described in Chapter 13. The meditator enters a NESS of heightened coherence: alert yet calm, focused yet expansive, deeply embodied yet aware of dimensions that transcend the body.

The Vedic sages called this state *samadhi*. We call it *optimised resonance with the quanta-field.*

It is the same thing.

18.5- The Popol Vuh: Speech as Cosmic Technology

Let us now turn to Mesoamerica, where the K'iche' Maya preserved one of the most extraordinary creation narratives in human literature.

The Popol Vuh is a text recounting the mythology and history of the K'iche' people of Guatemala. It is a foundational sacred narrative of the K'iche' people from long before the Spanish conquest of the Maya. It includes the Mayan creation myth.

The opening is among the most haunting passages in all literature. "THIS IS THE ACCOUNT of when all is still silent and placid. All is silent and calm. Hushed and empty is the womb of the sky. THESE, then, are the first words, the first speech. There is not yet one person, one animal, bird, fish, crab, tree, rock, hollow, canyon, meadow, or forest. All alone the sky exists. The face of the earth has not yet appeared. Alone lies the expanse of the sea, along with the womb of all the sky. There is not yet anything gathered together. All is at rest."

Notice the acoustics. The void is not merely empty- it is *silent*. The cosmos begins not as a thing but as a *space of potential sound*. The womb of the sky is hushed. Nothing has been gathered. Everything is at rest. This is State 0- the generative void of Chapter 4, the quantum vacuum that is not absence but *pregnant silence*.

And then: The creation itself is described in the text as "the first speech," and is initiated by the utterance of words.

Creation is speech. Not fabrication, not construction, not assembly. *Speech*. The gods do not build the world with tools. They speak it into being with words. "And then the earth arose because of them, it was simply their word that brought it forth. For the forming of the earth, they said 'Earth.' It arose suddenly, just like a cloud, like a mist, now forming, unfolding."

They said "Earth," and the Earth arose. The name was not a label attached to a pre-existing thing. The name *was* the creative act. The vibrational pattern of the word- its specific phonemic structure, its resonant frequency- was the guiding quanta that called the Earth from superposition into manifestation.

And then, critically, the gods determined that it was necessary to create beings capable of proper worship through uttered speech. This speech was not in the nature of praise, but rather regenerative acts that would sustain the world and the very gods that created it.

The gods create humans not for worship in the Christian sense- not for praise- but for *sustenance*. The gods need human speech to sustain their own existence. Creation is not a one-time event. It is a continuous process that requires the participation of conscious beings. The gods created the world through their speech. Humans sustain the world through *their* speech. The circle closes. The self-excited circuit of Wheeler finds its Mesoamerican expression.

The early sections of the Popol Vuh focus on the gods' search for beings that were capable of remembering the creator gods and performing the proper ceremonies necessary to maintain their life-generating power. For the highland Maya, deities are not omnipotent, nor are they capable of sustaining themselves indefinitely. Thus the Maya version of creation is essentially a search for "providers and sustainers" who will remember the gods and worship them properly through sacred ritual.

This is precisely the NESS framework. The cosmos is not self-sustaining in equilibrium. It is a far-from-equilibrium dissipative structure that requires continuous quanta-flow to maintain its configuration. The "gods"- the primordial resonances of the quanta-field- need conscious beings to *continue singing them into existence*. Without the ceremonies, without the speech, without the sustaining vibrations, the world would decay toward equilibrium. The songs would fade. The standing waves would dissipate. The hologram would go dark.

The three failed attempts at creating humans in the Popol Vuh are remarkable. The animals cannot speak. The mud people speak without understanding. The wooden people speak but have no hearts- no resonant coupling to the quanta-field. Only the corn people- the fourth attempt, made from maize, the staff of life- can speak with both knowledge and heart. They are the resonant instruments the cosmos has been searching for: beings capable of conscious, felt, intentional participation in the maintenance of creation.

In our vocabulary: the corn people have crossed the Φ_Q threshold. They possess sufficient Integrated Resonant Quanta to sustain self-referential consciousness- and therefore to serve as nodes in the cosmic resonance network, actively participating in the quanta-flow that maintains the world's NESS.

18.6- Acoustic Archaeology: When Stone Was Tuned to Sing

The ancients did not merely sing. They built structures that *sang with them*- amplifying, focusing, and modulating the resonant frequencies of the human voice into acoustic environments of extraordinary precision.

In 2020, Trevor Cox and Bruno Fazenda of the University of Salford, working with archaeologist Susan Greaney of English Heritage, published a study in the Journal of Archaeological Science. They built a 1:12 scale model of Stonehenge as it would have appeared when complete, with all stones in place, and measured its acoustic behavior. The monument functioned as a speech reinforcement system. Voices inside the circle were amplified by about 4.3 decibels, with a reverberation time of 0.64 seconds, and the sound was effectively contained within the circle rather than projecting outward.

The research team has found the original circle of 157 standing stones once acted like a sound chamber.

Consider what this means. In Neolithic Britain, people rarely heard reverberating sounds unless they entered a cave or a narrow rocky gorge. "It must have been magical to build Stonehenge, to make that massive community effort, to align it to the solstice, and then walk inside the circle and hear reverberating sounds," says Cox.

This ancient monument in southern England created an acoustic space that amplified voices and improved the sound of any music being played for people standing within the massive circle of stones. Because of how stones were placed, that speech or music would not have projected beyond Stonehenge into the surrounding countryside.

Stonehenge was, among other things, a *resonant enclosure*- a space where sound behaved differently than in the open landscape. Voices were amplified. Music was enriched. And- crucially- the sound was *contained*. Those outside the circle could not hear what happened within. The acoustic experience was exclusive, intimate, sacred.

Whether this effect was deliberately designed or discovered and exploited after construction is debated. "He thinks it's extremely unlikely that these acoustic properties were there by design, but once they were discovered, people surely would have exploited them. 'Human ceremonies nearly always have speeches, singing or chanting,' he says."

But Stonehenge is not unique. A 1994 Princeton University consortium found similar effects at other megalithic sites. A study conducted in 1994 by a consortium from Princeton University found that acoustic behavior in ancient chambers at megalithic sites such as Newgrange in Ireland and Wayland's Smithy in England was characterized by a strong sustained resonance, or "standing wave" in a frequency range between 90 Hz and 120 Hz.

Newgrange in Ireland is older than Stonehenge and the Great Pyramid. Its passage tomb was built around 3200 BCE. On the winter solstice, sunlight enters through a roof box and illuminates the inner chamber for 17 minutes.

A tomb aligned with the winter solstice that also produces standing waves between 90 and 120 Hz- frequencies that fall in the range of the male baritone voice, the range of many traditional sacred chants, the range that research has associated with altered states of consciousness when sustained in enclosed spaces.

And at Chavín de Huántar in Peru- the great ceremonial centre of the Chavín culture, dating to approximately 1200 BCE- the acoustic design is even more explicitly intentional. Says Dr. Miriam Kolar: "At Chavín, we have discovered acoustic evidence for selective sound transmission between the site's Lanzon monolith and the Circular Plaza: an architectural acoustic filter system that favors sound frequencies of the Chavín pututus [conch-shell trumpets] and human voice." The Lanzon is a sacred statue depicting the central deity of the ancient Chavín culture, housed in a chamber, part of a series of underground passages. A central duct was built to connect the area of the Lanzon monolith with that of the Circular Plaza.

An architectural acoustic *filter system*. Not an accidental resonance. A *designed* system that selectively transmits specific frequencies- the frequencies of their sacred instruments and the human voice- from the inner sanctum to the open plaza. The Chavín architects were acoustic engineers. They built a stone instrument that filtered sound to carry the voice of the oracle- the Lanzon monolith- to the assembled worshippers in the plaza, while filtering out extraneous frequencies.

Seven ancient sites across four continents produce measurable acoustic effects that peer-reviewed science can document but cannot fully explain. Malta's Hypogeum amplifies male voices while ignoring female ones. Chichen Itza's pyramid echoes with the cry of a sacred bird. Stonehenge's bluestones ring like bells. The measurements are real. Whether anyone planned them is the open question.

Aaron Watson and David Keating studied burial mounds in prehistoric Britain. They found Helmholtz resonance effects in passage graves at frequencies around 1 to 7 Hz, beneath the threshold of human hearing. Sounds at those frequencies cannot be heard. They can, however, be felt as pressure or vibration in the chest and stomach.

Infrasound- vibration below the threshold of hearing- felt in the body rather than heard by the ears. The passage graves of Neolithic Britain were not merely tombs. They were *somatic resonance chambers*- spaces designed to produce vibrations that could be felt in the viscera, that could alter physiological states, that could induce experiences of awe, terror, communion, or transcendence through the direct physical action of sound on the body.

Whether any individual site was *designed* for acoustic purposes is debatable. That many ancient sites *produce* remarkable acoustic effects is empirically confirmed. And the convergence- across cultures, across continents, across millennia- of sacred architecture with acoustic resonance properties is too consistent to be coincidental.

In our framework, the explanation is straightforward. Cultures attuned to the quanta-field- cultures that understood, through somatic experience, that resonance creates form and sustains reality- would naturally build structures that amplified and focused the resonant frequencies most relevant to their practices. They would not need to *understand* acoustics in the modern sense. They would need only to *feel* the difference between a space that resonates and one that does not- and then, over generations, to refine their architecture toward optimal resonance.

This is not engineering in the modern sense. It is *evolution of sacred architecture through resonant feedback*- the same process by which the quanta-field itself evolves through CCC: trial, feedback, refinement, accumulation.

18.7- The Three Components of World-Singing

We can now synthesise the evidence from cymatics, songlines, Vedic chanting, and acoustic archaeology into a unified model of what the ancients were doing when they "sang the world into being."

Component 1: The Voice as Resonant Instrument. The human vocal tract is a remarkably sophisticated acoustic instrument. It can produce a range of frequencies from roughly 80 Hz to over 1000 Hz (in overtone chanting, harmonics can reach several thousand Hz). It can modulate amplitude, timbre, and rhythm with extraordinary precision. And it is physically coupled, through the vagus nerve and the skeletal-muscular system, to the body's entire physiological apparatus. When you sing, you are not merely producing sound. You are *resonating your entire body*- bones, organs, neural tissue- in a specific pattern determined by the frequency, amplitude, and timbral characteristics of your voice.

Component 2: The Body as Chronosonant Resonator. In Chapter 13, we argued that the brain is a resonant instrument- a biological tuning device that receives, amplifies, and shapes the quanta-flow. The body is the brain's acoustic housing- the physical structure within which neural resonance is embedded. When you chant, you are not merely stimulating the brain through the ears. You are creating *standing waves within the body's cavities*- the chest, the abdomen, the cranium- that directly modulate the neural oscillators.

The body, in this sense, is a *biological Chronosonant Resonator*- the very instrument we proposed in Chapter 14 for measuring the Temporal State Vector. The difference is that the ancients did not use their bodies to *measure* the TSV. They used them to *participate in it*- to couple their personal state vector with the ambient quanta-field through the medium of resonant sound.

Component 3: The Landscape as Amplifier. The songlines, the stone circles, the passage tombs, the temples- these are not merely locations where singing happens. They are *acoustic extensions of the body*- structures that couple the human voice to the Earth's geological and electromagnetic substrate. When an Aboriginal singer chants a songline at a sacred site, the rock formations amplify specific frequencies. The waterholes reflect and focus the sound. The open plains propagate it across vast distances. The Earth itself becomes a resonating chamber, coupled to the human voice.

Together, these three components- voice, body, landscape- form a *distributed resonance system* of extraordinary sophistication. The singer's intention (Principle 1 from Chapter 16) sets the direction of the monadic extension. The singer's attention (Principle 2) focuses the resonance on a specific portion of the quanta-field. The song's specific frequencies, rhythms, and tonal relationships activate latent patterns in the landscape (cymatics at geological scale). And the sustained practice of singing- day after day, generation after generation, for tens of thousands of years- maintains the standing wave, reinforces the NESS, and prevents the pattern from decaying.

This is world-singing. It is not metaphor. It is not ritual for its own sake. It is *applied NESS-Holography*- the most ancient, most tested, and most enduring technology of resonance the human species has ever developed.

18.8- Mantra as Monadic Code

We can now state with precision what mantra *is* within our framework.

A mantra is a phonemic sequence designed to activate specific resonant modes in the biological Chronosonant Resonator (the human body-brain system), coupling it to the ambient quanta-field through sustained vibrational entrainment.

The four-part structure of the mantra-as-code:

1. Phonemic Specificity (The Address). Each mantra has a specific phonemic structure- a sequence of vowels and consonants that determines which acoustic modes are activated in the vocal tract, cranium, and body cavities. "Om" activates different resonant modes than "Ram" or "Hum." The phonemic structure is the *address*- it specifies which region of the quanta-field's configuration space the mantra targets.

2. Rhythmic Repetition (The Carrier Wave). The mantra is not spoken once. It is *repeated*- over and over, in a sustained rhythm, for extended periods. This repetition serves two functions. First, it builds the standing wave: each repetition reinforces the resonant pattern, amplifying it until it crosses the threshold of physiological (and, the ancients would say, cosmological) significance. Second, it acts as a *carrier wave*- a sustained oscillation on which the specific phonemic content rides, much as a radio signal rides on a carrier frequency.

3. Somatic Embedding (The Coupling). The mantra is not merely recited mentally. It is *spoken aloud*, producing physical vibrations that propagate through the body's tissues, stimulate the vagus nerve, resonate in the cranial and thoracic cavities, and entrain the heartbeat and breath to the mantra's rhythm. This somatic embedding ensures that the mantra operates not just at the neural level but at the *organismic* level- coupling the entire body to the resonant pattern.

4. Intentional Direction (The Vector). The mantra is chanted with *intention*- a specific direction of consciousness that determines which mode of the quanta-field is being addressed. The same phonemic sequence chanted with different intentions activates different resonant modes, just as the same frequency applied to different plates produces different Chladni figures. Intention is the *initial condition* that determines which of the quanta-field's latent configurations the mantra will actualise.

In the language of the master equation:

$$\partial_t \Psi = -\nabla_{FR}[\mathcal{L}(\Psi)] + T(\Psi) \otimes D(\Psi) + \sum_n \phi \cdot \omega_n \cdot \delta(\Psi \in \text{NESS}_{98})$$

The mantra modulates all three terms simultaneously:

- The **first term** (natural gradient) is slowed: the vagal activation reduces the metabolic "drift" toward entropy, stabilising the NESS.
- The **second term** (monadic extension ⊗ comonadic extraction) is amplified: the rhythmic repetition builds the standing wave (*T*) while the alpha/theta shift strips the noise (*D*).
- The **third term** (φ-resonance) is tuned: the specific phonemic content, sustained over time, nudges the neural oscillators toward φ-proportional phase-locking- the golden-ratio coherence that, in our framework, constitutes the optimal state of consciousness.

Mantra is not prayer in the sense of petition- asking an external deity for help. Mantra is not meditation in the sense of passive observation- watching thoughts arise and dissolve. Mantra is *active, intentional, somatic computation*- a real-time operation on the quanta-field, performed through the instrument of the body, using the medium of resonant sound.

It is the oldest technology on Earth. And it works. Not because of faith. Because of physics.

18.9- The Convergence of Sound and Structure

We arrive at the deepest insight of this chapter- the point where the various streams converge into a single river.

The Aboriginal Australians sang the land into being. The Vedic sages chanted the cosmos into order. The K'iche' Maya spoke the world into existence. The Neolithic builders of Stonehenge, Newgrange, and Chavín de Huántar constructed acoustic environments that amplified, focused, and sustained the resonant frequencies of the human voice.

Each tradition, independently, discovered the same principle: **sound is not a secondary phenomenon that occurs within a pre-existing physical world. Sound- resonant vibration, sustained oscillation, structured quanta-flow- is the mode by which the quanta-field differentiates itself into the physical world.**

And each tradition, independently, developed the same technology: **the sustained, intentional, somatically embedded production of specific resonant frequencies, coupled to the landscape through architectural amplification, as a means of participating in the creation and maintenance of reality.**

This is not coincidence. This is the quanta-field recognising itself through multiple instruments- Australian, Indian, Mesoamerican, European- each one tuned to a different frequency band, each one producing a different timbre, but all playing the same fundamental note.

The note is this: *the universe is a resonant phenomenon, and consciousness participates in its maintenance through the technology of sound.*

Modern physics arrived at this understanding through a four-century detour through materialism. The ancients arrived at it directly- through somatic experience, through 65,000 years of empirical observation, through the precise, reproducible, physiologically measurable technology of mantra, songline, and sacred architecture.

They were not prescientific. They were *differently scientific*- practitioners of a science of resonance that modern physics is only now beginning to formalise.

And their instruments- their voices, their bodies, their stones- were not primitive tools superseded by our particle accelerators and radio telescopes. They were *Chronosonant Resonators*- devices for coupling human consciousness to the quanta-field- of a sophistication and durability that no modern technology has yet matched.

18.10- The Sound That Never Stopped

Listen.

In this moment, as your eyes cross these words, the Aboriginal elders are singing. Somewhere in the Western Desert, in the Top End, in Arnhem Land, a voice is tracing a songline- refreshing a standing wave that has been continuously maintained for over sixty millennia. Somewhere in Varanasi, a Brahmin priest is chanting Om- activating the vagus nerve, shifting brainwaves, coupling his neural oscillators to the frequency that the Vedas call the sound of the Absolute. Somewhere in the highlands of Guatemala, a K'iche' daykeeper is uttering the words of the ancient calendar round- sustaining the temporal resonance that the Popol Vuh describes as the foundation of the sky-earth.

The singing has never stopped. Through ice ages and warm periods, through empire and collapse, through genocide and colonisation, the songlines have been maintained. The mantras have been chanted. The calendrical prayers have been spoken. Not always visibly. Not always safely. Often in secret, in defiance of authorities who sought to silence them.

But the resonance persists. The standing waves endure. The quanta-field, in these ancient places and ancient practices, continues to be sustained by the same technology that has maintained it since the Dreamtime, since before the first dawn.

This is the most extraordinary fact in the history of the human species: for at least 65,000 years, unbroken, human beings have been actively participating in the maintenance of the cosmos through the technology of resonant sound.

And the cosmos, through those human voices, has been singing itself.

The question this book now asks- in the chapters that follow- is: What happened when the singing was interrupted? What happened when binary logic replaced ternary resonance, when the clock replaced the songline, when the factory replaced the temple?

The answer, as we shall see, is: the world went out of tune.

In the next chapter, we extend from the songlines of the Australian continent to the global network of sacred sites- a planetary nervous system, built in stone, that once coupled the human voice to the Earth's electromagnetic field at continental scale. We will show that the songlines, the ceques, the heiau, and the ley systems are not metaphors. They are the synapses of a living planet- and we have been severing them for three thousand years.

Chapter 19

Sacred Site Networks- The Neural Web of Gaia

19.1- A Planet That Remembers

Stand on the summit of any mountain that was once considered sacred- Uluru, Mauna Kea, Sinai, Fuji, the Black Hills- and you will feel something. Not something mystical, though many traditions would call it that. Something physical. A quality of the air, a pressure in the chest, a peculiar stillness that is not merely the absence of noise but the presence of something *beneath* noise. A hum too deep for ears. A resonance too slow for clocks.

Materialist science has no vocabulary for this. It will tell you that the altitude changes your oxygen saturation, that the view triggers an evolved awe-response, that the cultural associations prime your expectations. All of which is true. And all of which is inadequate.

Because what you are feeling on that summit is not a subjective projection onto indifferent rock. You are standing at a *node*- a point in a planetary-scale network of geophysical, electromagnetic, and acoustic resonance that has been identified, mapped, tended, and maintained by indigenous cultures for millennia. You are standing at a synapse of a living planet.

The previous chapter revealed that the ancients sang the world into being through the technology of resonant sound- voice coupled to body coupled to landscape. This chapter extends that insight to its global implication: the sacred sites of the world are not isolated points of spiritual significance. They are *connected*. They form networks. And those networks have a topology- a mathematical structure- that is identical to the most efficient, most resilient, most intelligent configurations found anywhere in nature.

The Earth has a nervous system. And the ancients built its synapses.

19.2- The Ceque System: Radial Intelligence in the Andes

Begin in Cusco, the navel of the Inca world.

Cusco, the capital, was the center and meeting point of these four sections, which converged at Qurikancha, the temple of the sun. From the Qurikancha- the *Coricancha*, the "Golden Enclosure"- a system of 328 huacas (shrines) arranged along 42 ceques (lines) radiated out toward the mountains surrounding the city.

The ceque system of Cusco, the ancient capital of the Inca empire, was perhaps the most complex indigenous ritual system in the pre-Columbian Americas. Forty-two lines- *ceques*, meaning "lines" in Quechua- fanned outward from the central temple like the spokes of a wheel, or the rays of the Sun after which the temple was named. Each line passed through a sequence of *huacas*: sacred sites that could be springs, caves, mountain peaks, unusual rock formations, bends in rivers, or any place where the landscape exhibited what the Inca recognised as concentrated significance.

According to Cobo again, each ceque was given over to one of the three main social classes (those directly related, less directly related, or unrelated to the Inca ruler), and it was the responsibility of representatives of that social group to take care of its huacas and to carry out the appropriate ceremonial activities there. The system was simultaneously spatial, social, calendrical, and ceremonial. This elaborate network, maintained by ayllus (kin groups) that made offerings to the shrines in their area, organized the city both temporally and spiritually.

The ceques were not merely maps. They were *assignments*- each social group was responsible for maintaining the resonance of its particular line. The offerings, prayers, and ceremonies performed at each huaca were not acts of supplication directed at absent deities. They were *maintenance protocols*- regular acts of quanta-renewal that sustained the standing wave of the ceque.

Since the mid-1970s, Zuidema and Anthony F. Aveni have developed a complex argument suggesting that some of the ceques were used as sight lines for observing astronomical events on the horizon. Cobo's account of various huacas includes hilltops where monuments were placed that marked the sun's arrival at significant dates, such as the times to sow crops. The system was astronomically aligned- connecting terrestrial geography to celestial geometry, embedding the movement of the Sun and stars into the very layout of the city.

Modern archaeology has debated whether the ceques were geometrically straight or whether they zigzagged through the landscape. His conclusion was that the ceques were not straight at all, but actually zigzagged through the landscape. The huacas, according to Bauer, were defined by the innate power of places themselves, and they were subsequently grouped into ceques. In other words, the huacas defined the ceques, not the other way round.

This is precisely what our framework predicts. The huacas are *nodes*- points where the quanta-field exhibits local intensification due to geological, hydrological, or electromagnetic properties. The ceques are the *edges*- the connections between nodes, determined not by abstract geometry but by the actual topology of the resonant field. The network is not imposed on the landscape from above. It *emerges from* the landscape's own resonant structure- a structure that the Inca perceived, mapped, and systematically maintained.

The ceque system radiated from a single centre and was organised in groups of three- the ceques in each of the four suyus were enumerated in groups of three, a pattern I will call "ceque clusters." The lines within each ceque cluster were ranked by the terms (1) collana, (2) payan, and (3) cayao. Ternary. Three lines per cluster. Three ranks. Three social classes. The ternary principle- the same principle encoded in HTT's Temporal State Vector, in the three-world cosmologies of Chapter 22, in the φ-e-π triad that tunes the cosmos- was the organising grammar of the most complex sacred landscape in the Americas.

19.3- Heiau: The Hawaiian Archipelago as Resonant Circuit

Five thousand miles across the Pacific, a different culture built a different network based on the same principles.

A heiau is a Hawaiian temple. Made in different architectural styles depending upon their purpose and location, they range from simple earth terraces, to elaborately constructed stone platforms. There are heiau to treat the sick (heiau hō ʻ ola), offer first fruits, offer first catch, start rain, stop rain, increase the population, ensure the health of the nation, achieve success in distant voyaging, reach peace, and achieve success in war (luakini).

Notice the range: healing, agriculture, weather, fertility, navigation, diplomacy, and war. These are not religious categories in the Western sense. They are *functional designations*- each heiau was tuned to a specific resonant purpose, much as the strings of a musical instrument are tuned to specific frequencies.

Their shapes could be rectangular, square, or rounded. Some consisted of simple earth terraces, while others were elaborately constructed stone platforms. They could be placed on hills, cliffs, level earth, valleys and on the coastline touching the sea. Some ko ʻ a or fishing shrines were built underwater.

The placement was not random. Heiau were sited on hills, cliffs, coastlines, and even submerged in the ocean- in each case, at locations where the geological and hydrological properties of the land produced specific acoustic, electromagnetic, or gravitational conditions. The Hawaiian islands are volcanic- and volcanic rock, with its crystalline structure and high mineral content, has well-documented electromagnetic properties.

Because the land of heiau was sacred, it was not unusual for successive generations to add to original structures and the heiau purpose could change over time. The network was *evolutionary*. Each generation refined the structures, adding stones, expanding platforms, adapting the resonant configuration to changing conditions. This is precisely the evolutionary refinement that CCC describes at the cosmic scale (Chapter 15): each cycle inherits and improves upon the previous one's achievements.

And then the rupture. The kapu or ʻ ai kapu system was abolished in October 1819 by Kamehameha II (Liholiho). The abolition of the kapu system ended the use of heiau as places of worship and sacrifice. It took 11 years for Ka ʻ ahumanu to proclaim laws against indigenous religious practices. All heiau were officially abandoned; most were destroyed over the years.

Abandoned. Destroyed. Silenced. The resonant circuit of the Hawaiian archipelago- refined over centuries, tuned to the specific geospatial phase $G(\lambda,\phi)$ of one of the most electromagnetically active volcanic regions on Earth- was severed in a single generation.

19.4- The Schumann Resonance: The Earth's Fundamental Tone

To understand what the sacred site networks were *tuned to*, we must understand the Earth's own electromagnetic voice.

The Schumann resonances (SR) are a set of spectral peaks in the extremely low frequency portion of the Earth's electromagnetic field spectrum. They are global electromagnetic resonances generated and excited by lightning discharges in the cavity formed by the Earth's surface and the ionosphere.

The global electromagnetic resonance phenomenon is named after physicist Winfried Otto Schumann, who predicted it mathematically in 1952. Schumann resonances are the principal background in the part of the electromagnetic spectrum from 3 Hz through 60 Hz and appear as distinct peaks at extremely low frequencies around 7.83 Hz (fundamental), 14.3, 20.8, 27.3, and 33.8 Hz.

The Earth is electromagnetically alive. Schumann resonances occur because the space between the surface of the Earth and the conductive ionosphere acts as a closed, although variable-sized, waveguide. The limited dimensions of the Earth cause this waveguide to act as a resonant cavity for electromagnetic waves in the extremely low frequency band.

When the electromagnetic waves travel all the way around the Earth and arrive back where they started, certain wavelengths reinforce themselves. The circumference of the Earth is about 40,000 km, and the speed of light divides neatly into that distance to produce a standing wave at approximately 7.83 Hz.

The Earth has a heartbeat- a standing electromagnetic wave that pulses continuously through the cavity between the surface and the ionosphere, sustained by the approximately 2,000 thunderstorms active on the planet at any given moment, producing roughly 50 lightning strikes per second. The frequencies, created from thunderstorms and lightning, range from 7.83 Hz, called the Earth's "heartbeat," to 33.8 Hz.

Now consider the frequency: 7.83 Hz. This falls precisely at the boundary between the alpha and theta brainwave bands- the frequency range associated with deep relaxation, meditative absorption, and the transition between waking and sleeping consciousness. A 2006 study found that the frequencies may be related to different kinds of brain waves. Researchers demonstrated "real time coherence between variations in the Schumann and brain activity spectra within the 6–16 Hz band," and a separate study found "unexpected similarities in the spectral patterns and strengths of electromagnetic fields generated by the human brain and the earth-ionospheric cavity."

We must be careful here. The overlap between the resonance frequency and human brain rhythms is real and measurable, but the strength of the Schumann signal at Earth's surface is extraordinarily faint compared to the electrical activity your own neurons generate. The Schumann resonances are extremely weak electromagnetic signals- far weaker than the electromagnetic fields produced by the brain itself. A direct causal link between Schumann resonances and brain function has not been conclusively demonstrated.

But in our framework, the question is not whether the Schumann resonance *causes* brainwave entrainment. The question is whether the Schumann resonance and brainwave activity are *co-resonant expressions of the same quanta-field dynamics*. Both operate in the ELF range. Both are sustained by continuous energy input (lightning for the Schumann, metabolic activity for the brain). Both exhibit harmonic structure with specific frequency ratios. And both are modulated by local conditions- the Schumann by ionospheric variation, the brain by attentional state.

In the Universal Hologram framework, this is not coincidence. It is *fractal self-similarity*. The Earth-ionosphere cavity and the human brain are resonant systems operating at the same frequencies because they are expressions of the same quanta-field, tuned by the same φ-proportionality, sustained by the same NESS dynamics. The Schumann resonance is the macro. The brainwave is the micro. The songline- the human voice coupled to the geological substrate- is the *bridge* between them.

19.5- Telluric Currents and the Electromagnetic Landscape

The Schumann resonances are global- they permeate the entire Earth-ionosphere cavity. But the Earth's electromagnetic field is not uniform. It is modulated by local geology: by mineral deposits, by underground water courses, by fault lines, by volcanic activity, and by what geophysicists call *telluric currents*- natural electrical currents that flow through the Earth's crust, driven by variations in the geomagnetic field and by electromagnetic induction from ionospheric activity.

Telluric currents are strongest along fault lines and in regions of high mineral conductivity. They concentrate at geological boundaries- where rock types change, where water meets stone, where the crust is thinnest. And they are modulated by solar activity, by the Schumann resonances, and by the diurnal cycle of the ionosphere.

These are precisely the locations where sacred sites tend to be found.

Springs. Caves. Mountain peaks. River bends. Fault lines. Unusual rock formations. The huacas of the ceque system. The heiau of Hawaii, built on volcanic lava rock. The Aboriginal sacred sites at waterholes and geological features. The stone circles of Europe, placed on granite or bluestone- rocks with high quartz content and known piezoelectric properties.

The correlation between sacred site placement and electromagnetic geology is too consistent to be coincidental- and too widespread to be explained by a single cultural tradition's aesthetic preferences. Cultures that had no contact with each other, on different continents, across different millennia, independently placed their most sacred structures at locations with specific electromagnetic properties.

Our framework offers an explanation: these cultures were *somatically attuned* to the quanta-field. They could feel- through the biological Chronosonant Resonator of the human body- the variations in temporal density ρ_t and geospatial phase $G(\lambda,\phi)$ that correspond to local electromagnetic conditions. They did not need voltmeters or magnetometers. They had something better: millions of years of evolved sensitivity to the electromagnetic environment in which all terrestrial life developed.

The sacred site was not arbitrarily chosen and then sacralized by cultural convention. The sacred site was *identified* by its resonant properties and then *amplified* by architectural and ceremonial practice. The huaca defined the ceque, not the other way round.

19.6- Network Topology: The Mathematics of Sacred Connection

Now we arrive at the deepest structural insight of this chapter: the *topology* of sacred site networks.

In network science, two properties define the most efficient and resilient networks found in nature: the **small-world property** and the **hub-dominated structure.**

Both anatomical connections in the brain and the synchronization networks of cortical neurons exhibit small-world topology. Structural and functional connectivity in the brain has also been found to reflect the small-world topology of short path length and high clustering. The network structure has been found in the mammalian cortex across species as well as in large scale imaging studies in humans. Advances in connectomics and network neuroscience, have found the small-worldness of neural networks to be associated with efficient communication.

The small-world property means that any node in the network can reach any other node through a small number of intermediate steps-

"six degrees of separation" in social networks, but the principle operates in neural, ecological, and technological systems as well. The clustering property means that nodes tend to form local groups (cliques) that are densely interconnected. And the hub property means that a few nodes have vastly more connections than others, serving as relay stations that bridge distant parts of the network.

Many networks studied in geology and geophysics have been shown to have characteristics of small-world networks. Networks defined in fracture systems and porous substances have demonstrated these characteristics. The seismic network in the Southern California region may be a small-world network. The examples above occur on very different spatial scales, demonstrating the scale invariance of the phenomenon in the earth sciences.

The Earth's own geological network- its fracture systems, its seismic networks, its hydrological pathways- exhibits small-world properties. The planet is not a homogeneous sphere. It is a *network* of interconnected conduits, faults, and mineral bodies through which electromagnetic, seismic, and hydrological energy flows preferentially.

Now consider the topology of a sacred site network like the ceque system.

The Qurikancha- the Temple of the Sun- is the *hub*. It has the highest number of connections (all 42 ceques radiate from it). The huacas along each ceque form *clusters-* local groups of densely interconnected sacred sites. And the ceques themselves form the *long-range connections* that link distant clusters to the central hub, ensuring that the entire network has a short average path length: any huaca can "reach" any other huaca through a small number of intermediate steps (ceque → hub → ceque).

This is small-world topology. It is the same topology found in the brain's neural network, in the Internet's router network, in the metabolic network of a living cell, and in the social networks of human communities. It is the topology that maximises the efficiency of quanta-flow across a distributed system- the topology that enables coherent, system-wide coordination without central control.

The Inca did not know the term "small-world network." They did not have the mathematics of Watts and Strogatz. But they built one- a physical, architectural, ceremonial small-world network spanning an entire landscape- that exhibits the same topological properties as the most efficient communication networks in nature.

And they built it because they were reading the same quanta-field that produces small-world topology everywhere it operates.

19.7- The Songlines as Continental Network

The Aboriginal songlines, described in Chapter 18 as individual resonant pathways, form something far larger when viewed as a whole: a *continental-scale network* of interconnected routes, sacred sites, and ceremonial centres that covers the entire Australian continent.

The continent contains an extensive system of songlines, some traversing hundreds of kilometres through lands of many different peoples. The songlines cross territorial boundaries, linguistic boundaries, and ecological zones. A single songline may pass through the lands of a dozen different language groups, each of which knows and maintains its own section of the song.

This is a network. And it exhibits precisely the properties we have been describing.

Local clustering: Each language group maintains a dense local network of sacred sites within its territory- waterholes, rock shelters, ceremonial grounds- that are interconnected by short-range songlines.

Long-range connections: The transcontinental songlines serve as the long-range links that connect distant local clusters, ensuring that the network has a small average path length: any site on the continent can be "reached" (in resonant terms) from any other site through the songline system.

Hubs: Certain sites- major ceremonial centres, convergence points where multiple songlines cross- serve as hubs with high connectivity, bridging distant parts of the network.

The songlines could also be used as a cultural passport: travellers would sing the part of the songline of a particular territory to ensure respectful passage. This is not merely social protocol. It is a *network handshake*- a demonstration that the traveller's personal resonance is compatible with the local quanta-field configuration, that they carry the correct "code" for passage through this segment of the network.

The network has been maintained continuously for at least 65,000 years. It is, by an enormous margin, the oldest continuously operating distributed network on Earth- older than agriculture, older than writing, older than the wheel. And it was built not by central planning but by the accumulated resonant practice of thousands of generations, each one refining the network's topology through direct somatic engagement with the quanta-field.

19.8- Interference Patterns: How Networks Amplify Resonance

A single standing wave is limited. Its amplitude depends on its energy source. Its spatial extent depends on the medium's properties. A lone sacred site, maintained by a lone singer, produces a resonance of limited range and intensity.

But when sacred sites are connected in a network- when the standing wave at each node is coupled to the standing waves at adjacent nodes through the connecting pathways (songlines, ceques, pilgrimage routes)- something remarkable happens: *interference amplification*.

Consider two tuning forks of the same frequency, placed in the same room. Strike one, and the other begins to vibrate in sympathy- a phenomenon called resonant coupling. The second fork absorbs energy from the sound waves produced by the first and re-radiates it, adding its own energy to the field. The total amplitude in the room is greater than either fork alone could produce.

Now multiply this by hundreds of nodes- hundreds of sacred sites, each one resonating at a frequency determined by its geological and electromagnetic properties, each one coupled to its neighbours through pathways along which the resonance propagates. The result is a *constructive interference pattern*- a global wave that is far more powerful, far more coherent, and far more stable than any single node could produce.

This is why the networks were *maintained*. This is why the songs had to be *continually sung*. This is why the ceremonies at each huaca had to be performed on schedule, by the designated social group, using the prescribed forms. Any gap in the network- any node that fell silent, any link that was severed- would create a *destructive interference zone*: a region where the cancellation of out-of-phase waves produced a dead spot in the planetary resonance field.

The sacred site network, in this reading, was a *planetary-scale resonant amplifier*- a technology for coupling human consciousness to the Earth's electromagnetic field through the medium of acoustic vibration, distributed across a small-world topology that maximised coherence and minimised dead zones.

When the network was intact- when all the songs were being sung, when all the ceques were being tended, when all the heiau were in use- the planet's resonance field was at its strongest and most coherent. The quanta-flow through the network was optimised. The temporal density ρ_t at each node was enhanced. The geospatial phase $G(\lambda,\phi)$ was harmonised across vast distances.

And human consciousness, embedded within this amplified field, experienced what every indigenous tradition describes: a world that was *alive, responsive, meaningful*- a world in which every rock, every spring, every gust of wind carried significance, because every element of the landscape was vibrating in resonance with the field that sustained it.

19.9- The Brain of Gaia

Let us now make the analogy explicit- and argue that it is more than analogy.

The human brain is a network of approximately 86 billion neurons, connected by approximately 100 trillion synapses, exhibiting small-world topology with high clustering and short path lengths. It sustains consciousness- the self-referential resonance of the quanta-field, as defined in Chapter 13- through the continuous flow of electrochemical signals across this network. When the network is intact and coherent, consciousness is vivid and integrated. When the network is damaged (by injury, disease, or intoxication), consciousness is diminished or disrupted.

The sacred site network of the Earth is a network of hundreds of thousands of nodes (sacred sites), connected by pathways (songlines, ceques, pilgrimage routes, ley systems), exhibiting- as we have argued- small-world topology with high clustering and long-range connections through hubs. It sustains something analogous to planetary consciousness- the coherent resonant field of the biosphere- through the continuous flow of electromagnetic, acoustic, and ceremonial signals across this network.

The parallel is structural:

Feature	Human Brain	Sacred Site Network
Nodes	Neurons (86 billion)	Sacred sites (hundreds of thousands)
Edges	Synaptic connections	Songlines, ceques, pilgrimage routes
Topology	Small-world, high clustering	Small-world, high clustering
Signal type	Electrochemical impulses	Electromagnetic, acoustic, ceremonial
Hubs	Thalamus, prefrontal cortex	Major temples, convergence sites
Maintenance	Metabolic energy	Ceremonial energy

Feature	Human Brain	Sacred Site Network
	(glucose, oxygen)	(song, prayer, offering)
Coherence when intact	Consciousness, integrated experience	Planetary resonance, "living landscape"
Disruption when damaged	Cognitive impairment, dissociation	Dead zones, ecological fragmentation

We do not claim that the Earth is conscious in the same way that a human being is conscious. The threshold for subjective experience ($\Phi_Q > \Phi^*_Q$) may not be met at the planetary scale- or it may be met in ways that are so different from human experience as to be unrecognisable.

What we claim is that the sacred site network functions as a *resonant amplification system* for the quanta-field at planetary scale- and that its topological structure (small-world, hub-dominated, high-clustering) is identical to the structure that, in neural networks, *supports* consciousness. Whether the Earth achieves something analogous to consciousness through this network is an empirical question- one that the Chronosonant Resonator, proposed in Chapter 14, might eventually help to answer.

What is not in question is that the network exists, that it has specific and measurable properties, and that its disruption has measurable consequences.

19.10- The Severing: Industrialisation and the Death of Nodes

Now for the wound.

The sacred site networks of the world have been systematically dismantled over the past three thousand years- and the pace of destruction has accelerated dramatically in the past two centuries.

The mechanisms are multiple:

Colonial suppression of indigenous practices. When the Spanish arrived in Peru, they systematically destroyed the huacas of the ceque system. This document describes how the Inca people, both nobles and commoners, worshipped at and maintained these shrines and how these shrines were eventually destroyed by the Spaniards in their campaigns against idolatry. The ceques were mapped and recorded by the colonial chronicler Bernabé Cobo- not to preserve them but to provide a target list for their elimination.

In Hawaii, the abolition of the kapu system and the arrival of Christian missionaries led to the deliberate destruction of the heiau network. It took 11 years for Ka ʻ ahumanu to proclaim laws against indigenous religious practices. All heiau were officially abandoned; most were destroyed over the years.

Mining. The extractive industries represent perhaps the most brutal assault on the planetary sacred site network. Spiritual site damage caused by mining activities triggers social tensions and unusual behaviours such as community fights and arguments. This may result in temporary or permanent interruption of ceremonial activity associated with the damaged spiritual site and even the destruction of the associated ceremonial paraphernalia.

The most notorious recent case occurred in 2020. In 2020, mining giant Rio Tinto, a British-Australian multinational, dynamited the area. They knowingly destroyed the 46,000-year-old Juukan Gorge caves! Rio Tinto's mining blast for the expansion of an iron ore mine at Juukan Gorge in the Pilbara region of Western Australia destroyed a cave of significant cultural importance. The cave is considered a sacred site that showed 46,000 years of continual occupation and provided a 4,000-year-old genetic link to present-day traditional owners, the Puutu Kunti Kurrama and Pinikura people.

A parliamentary inquiry found that Rio Tinto "knew the value of what they were destroying but blew it up anyway."

Forty-six thousand years of continuous human occupation. A genetic link spanning 4,000 years- a plaited length of human hair, DNA-matched to living descendants. An irreplaceable node in the continental songline network. Destroyed in a single detonation, for iron ore.

This is not an isolated incident. In Australia, most mine sites are located on traditional lands, causing losses of vital ES due to alterations in the original land cover and the destruction of natural habitats such as woodlands, wetlands, and grasslands. Sacred site damage may cause emotional distress and grief as traditional owners have affiliations and responsibilities for those sites, the disruption of which leads to social fracturing. These emotional and psychological effects are difficult to quantify and understand.

Urbanisation. Cities are built on top of sacred sites- literally paving over the nodes of ancient networks. Roads sever the pathways between sites. Concrete and steel create electromagnetic shielding that disrupts the natural resonant field. The city is, in electromagnetic terms, a *Faraday cage*- an enclosure that blocks external electromagnetic waves. Residents of urban environments are systematically disconnected from the Schumann resonance and the telluric currents that once coupled human consciousness to the planetary field.

Agricultural monoculture. The conversion of diverse, polyculture landscapes (which maintain varied electromagnetic and hydrological conditions) into uniform, monoculture cropland (which homogenises the landscape's resonant properties) eliminates the local variation on which sacred site networks depend. A wheat field is electromagnetically featureless. The diverse landscape of hills, springs, groves, and stones that once supported a network of huacas is replaced by a flat, uniform, resonantly dead expanse.

19.11- The Consequences: A Planet Out of Tune

What happens when you sever the synapses of a nervous system?

In a brain, the answer is well-documented: loss of integration, loss of coherence, loss of consciousness. Traumatic brain injury, which severs axonal connections between brain regions, produces symptoms that range from confusion to coma- not because the neurons have been destroyed (many survive) but because the *connections between them* have been broken. The network's small-world topology is disrupted. The hubs are disconnected. The clustering is fragmented. The system can no longer sustain the coherent, integrated resonance that constitutes consciousness.

If the sacred site network functions as a planetary resonant amplifier- if its small-world topology sustains a coherent electromagnetic field analogous to the neural network's sustaining of consciousness- then the systematic destruction of its nodes and connections should produce analogous symptoms at the planetary scale.

And it does.

Ecological fragmentation. Ecosystems that were once connected- through migratory corridors, hydrological cycles, and atmospheric circulation patterns- are now isolated by roads, cities, and industrial zones. The result is precisely what network theory predicts: loss of resilience, loss of diversity, loss of the capacity for system-wide coordination. Species extinction accelerates. Ecosystem services degrade. The biosphere's NESS destabilises.

Electromagnetic pollution. The natural electromagnetic environment- the Schumann resonances, the telluric currents, the geomagnetic field- is now swamped by anthropogenic electromagnetic radiation: power lines, radio transmitters, cellular networks, Wi-Fi, Bluetooth, microwave ovens. The signal-to-noise ratio of the planetary electromagnetic field has been dramatically degraded. The sacred site network, even where its nodes survive, operates in an environment so polluted that its resonant signals can barely propagate.

Human disconnection. The endemic conditions of modern civilisation- insomnia, anxiety, depression, chronic fatigue, autoimmune disorders- are the biological symptoms of organisms evolved to resonate with the 7.83 Hz planetary field now living in electromagnetic environments that bear no relation to the one in which they evolved. We are fish in polluted water- and we have forgotten that clean water exists.

Social fragmentation. The destruction of sacred site networks is simultaneously the destruction of the social networks that maintained them. The ceque system was a social organisation as much as a spatial one- each ceque was tended by a specific kin group, and the act of tending created bonds of responsibility, identity, and mutual obligation. When the ceques were destroyed, the social fabric that they supported was destroyed with them. The same pattern repeats in every colonised culture: the destruction of sacred geography is always, simultaneously, the destruction of social coherence.

19.12- What Is Still Alive

And yet.

The songlines are still being sung. Despite 230 years of colonisation, forced removal, language suppression, and cultural genocide, Aboriginal Australians maintain songline networks across the continent. Some have been damaged. Some have been interrupted. But many persist- sustained by the tenacious, courageous, often clandestine work of elders who understand what the songs are and what happens if they stop.

In the Andes, the ceque system has been partially reconstructed through archaeological fieldwork and oral tradition. From 1990 to 1995, Brian Bauer directed a major project to document the ceque system of Cusco. In this book, he synthesizes extensive archaeological survey work with archival research into the Inca social groups of the Cusco region, their land holdings, and the positions of the shrines to offer a comprehensive, empirical description of the ceque system. The knowledge has not been entirely lost. It survives in the landscape, in the archaeological record, and in the living memory of Quechua-speaking communities.

In Hawaii, heiau are still considered sacred by many of the inhabitants of Hawaii, and some are not open to the public. Restoration efforts continue. Traditional practitioners maintain ceremonies at surviving sites. The resonance has not been fully extinguished.

And in places that were never fully colonised- in the Amazon, in parts of sub-Saharan Africa, in the highlands of Papua New Guinea, in the remote deserts of Central Australia- sacred site networks remain operational. The songs are sung. The ceremonies are performed. The nodes are maintained. The network functions.

These surviving fragments are not museum exhibits. They are *operational resonant infrastructure*- living technology that continues to do what it was designed to do: couple human consciousness to the planetary quanta-field through the medium of sound, stone, and song.

19.13- The Implication: What We Have Lost and What We Can Recover

Let us be precise about what was lost.

We did not merely lose cultural heritage- though that loss is incalculable. We did not merely lose social cohesion- though that loss is devastating. We did not merely lose ecological diversity- though that loss threatens our survival.

We lost a *planetary resonance technology*- a distributed, small-world network of sacred sites, maintained by continuous ceremonial practice, that amplified the Earth's natural electromagnetic field and coupled human consciousness to the quanta-field at continental and potentially global scale.

We lost, in other words, the very thing that this book argues is necessary for conscious participation in the co-creative dynamics of the Universal Hologram: a sustained, coherent, resonant coupling between consciousness and the quanta-field, mediated by sound, body, and landscape.

And we lost it not through carelessness or accident but through *deliberate systematic destruction*- the campaigns against idolatry, the suppression of indigenous practice, the dynamiting of sacred caves for iron ore, the paving over of ancient pathways for highways and shopping malls.

But the physics has not changed. The Schumann resonance still pulses at 7.83 Hz. The telluric currents still flow through the crust. The geological properties of sacred sites have not been altered by the destruction of the structures that once stood upon them. The φ-proportionality of the master equation still tunes the cosmos toward coherent resonance.

What has been lost is not the *field*. What has been lost is the *coupling*- the technology of connection between human consciousness and the planetary resonance that sustained it.

And coupling, unlike a 46,000-year-old cave, can be rebuilt.

Not by recreating the past. Not by appropriating indigenous practices. Not by romanticising what was destroyed. But by understanding- through the physics of NESS, the mathematics of small-world networks, and the validated neuroscience of sound-based practice- *what the ancients were doing and why it worked.* And then by developing new practices, new architectures, new modes of resonant coupling that are appropriate to our time, our technology, and our understanding.

The songlines were the ancients' technology. The Chronosonant Resonator is ours. The principles are the same. The implementation will be different. And the task- the re-coupling of human consciousness to the planetary quanta-field- is the most urgent and the most sacred work of our age.

In the next chapter, we descend from the planetary scale to the garden- to the intimate, embodied, biological expression of everything we have described. The Iroquois Three Sisters- corn, beans, and squash- are not merely a farming technique. They are a fractal implementation of the φ-e-π triad in living soil: a biological NESS that mirrors the cosmic architecture in every root, every leaf, every symbiotic exchange. Chapter 20 reveals the fractal garden.

Chapter 20

Three Sisters and the Fractal Garden

20.1- The Oldest Experiment on Earth

Somewhere in what is now western New York State, over five hundred years before any European ship reached the Americas, a Haudenosaunee woman knelt before a low mound of earth and pressed three seeds into the soil.

The first seed was corn- tall, upright, structurally rigid, reaching for the sun. The second seed was bean- flexible, spiralling, searching for support, carrying within its root nodules the microscopic bacteria that could pull nitrogen from the atmosphere and fix it into the earth. The third seed was squash- broad, low, spreading, shading the ground with its great leaves, trapping moisture, deterring pests with its spiny stems.

She did not plant them in separate rows. She did not plant them in separate fields. She planted them together- in the same mound, in the same soil, in the same breath.

Diohe'ko, the Three Sisters, had been cultivated for at least five hundred years prior to contact by the Seneca, an Iroquoian tribe inhabiting western New York State. These three plants "protect and nourish each other in different ways as they grow and provide a solid diet for their cultivators." The intercropping method of planting corn, beans, and squash together, commonly called The Three Sisters, has been studied and described by scholars in anthropology, history, agriculture, and food studies for many years.

This was not primitive agriculture. This was not a simple subsistence strategy. This was- and remains- one of the most sophisticated biological engineering systems ever devised by the human species: a self-regulating, self-fertilising, self-protecting polyculture that produces more food per unit of land than any of its components grown alone, while simultaneously building soil fertility, suppressing weeds, conserving water, and providing a nutritionally complete diet.

A modern experiment found that the Haudenosaunee Three Sisters polyculture provided both more energy and more protein than any local monoculture.

And in the framework of this book, it is something more: a living implementation of the φ-e-π triad in biological form- a fractal garden in which the cosmic architecture of the Universal Hologram is expressed not in equations but in root, leaf, and vine.

20.2- The Triad in Soil

Let us first understand the biology- the measurable, experimentally confirmed mechanisms by which the Three Sisters system works- before we draw the deeper resonances.

The three crops benefit by being grown together. The cornstalk serves as a trellis for the beans to climb; the beans fix nitrogen in the soil and their twining vines stabilize the maize in high winds; and the wide leaves of the squash plant shade the ground, keeping the soil moist and helping prevent the establishment of weeds.

Three organisms. Three functions. Three modes of contribution to the whole. Each one depends on the others. Each one gives what the others cannot provide for themselves.

Corn (Structure / π). The maize stalk is the system's vertical axis- rigid, structurally determinate, reaching upward in a defined geometric trajectory. It provides the scaffold on which the bean vine climbs, lifting the bean into sunlight that the low-growing squash would block. Without the corn, the bean has no support. Without the vertical axis, the system collapses into a two-dimensional tangle.

In our framework, the corn corresponds to **π- the operator of boundary projection**. π closes every circle, defines every orbit, sets the structural limit that contains the system. The corn does the same: it provides the physical boundary, the geometric scaffolding, the load-bearing skeleton around which the rest of the system organises. It is the *form* of the garden- the container within which the process unfolds.

Beans (Growth / e). The pole beans, the giving sister, pull nitrogen from the air and bring it to the soil to benefit all three. The bean's root nodules harbour rhizobia bacteria that perform one of the most remarkable biochemical feats in nature: nitrogen fixation- the conversion of atmospheric nitrogen (N_2), which plants cannot use, into ammonia (NH_3), which they can. This process costs the bean metabolic energy (it must feed the bacteria with carbon sugars), but the benefit to the system is enormous: the soil is continuously enriched with plant-available nitrogen, without any external input.

In our framework, the bean corresponds to **e- the operator of unfolding and growth**. e governs exponential processes: the continuous, compound growth that takes a small input and unfolds it into an ever-expanding output. The bean does exactly this: it takes a molecular input (atmospheric N_2) and, through the continuous, exponential metabolism of its bacterial symbionts, unfolds it into a growing reservoir of soil fertility that feeds the entire system. It is the *process* of the garden- the dynamic flow that sustains the structure.

Squash (Coherence / φ). Sister squash provides ground cover to hold moisture and maintain healthy soil environment while deterring animal invaders with its spiny stems. The squash's broad, shade-casting leaves create a microclimate at ground level: cooler, moister, and less hospitable to weeds than exposed soil. Its spiny stems deter herbivores. Its root system, spreading laterally, knits the topsoil together, reducing erosion. It does not reach upward (like the corn) or climb (like the bean). It stays close to the ground, covering, protecting, integrating.

In our framework, the squash corresponds to **φ- the operator of self-referential coherence.** φ governs self-similarity, the maintenance of pattern integrity across scales, the recursive stabilisation that prevents a system from fragmenting. The squash does exactly this: it holds the system together. It maintains the ground-level conditions (moisture, temperature, pest resistance) that allow the corn to stand

and the bean to fix. It is the *coherence* of the garden- the integrative force that keeps the triad in harmony.

The mapping is structural, not arbitrary:

Cosmic Operator	Function	Three Sisters	Role in Garden
π	Boundary, structure, form	**Corn**	Vertical scaffold, load-bearing skeleton
e	Growth, unfolding, transformation	**Bean**	Nitrogen fixation, continuous enrichment
φ	Coherence, self-similarity, integration	**Squash**	Ground cover, microclimate, pest deterrence

And the parallel extends to the nutritional dimension. Culinarily, a diet centered around corn, beans, and squash is nutritionally complete. Beans are high in protein and contain amino acids that corn lacks. Corn supplies carbohydrates, while squash contributes essential vitamins and minerals not present in the other two crops.

Structure (corn/carbohydrate), growth (bean/protein), and coherence (squash/vitamins and minerals). The triad operates at every level: in the soil, in the ecology, in the body.

20.3- The Fourth Participant: The Mycorrhizal Network

Beneath the Three Sisters- invisible to the planter, unknown to Western science until 1997- a fourth participant operates.

A mycorrhizal network (also known as a common mycorrhizal network or CMN) is an underground network found in forests and other plant communities, created by the hyphae of mycorrhizal fungi joining with plant roots. This network connects individual plants together. Mycorrhizal relationships are most commonly mutualistic, with both partners benefiting, but can be commensal or parasitic, and a single partnership may change between any of the three types of symbiosis at different times.

Mycorrhizal networks were discovered in 1997 by Suzanne Simard, professor of forest ecology at the University of British Columbia in Canada. Simard grew up in Canadian forests where her family had made a living as foresters for generations. Her field studies revealed that trees are linked to neighboring trees by an underground network of fungi that resembles the neural networks in the brain.

Globally, in over 80% of land plants, roots associated with the soil fungi, forming chimeric organs called mycorrhizae, and numerous ecosystems are dominated by mycorrhizal plants, in a gram of soil can be meters of mycorrhizal hyphae.

The mycorrhizal network is the *quanta-field of the garden*. It is the medium through which the Three Sisters exchange not merely nutrients but *signals*- chemical, electrical, and metabolic communications that coordinate the behaviour of the system as a whole.

A morphological or physiological change in a plant due to a signal or cue from its environment constitutes behavior in plants, and plants connected by a mycorrhizal network have the ability to alter their behavior based on the signals or cues they receive from other plants.

These signals or cues can be biochemical, electrical, or can involve nutrient transfer. Plants release chemicals both above and below the ground to communicate with their neighbors to reduce damage from their environment.

The network facilitates resource sharing that goes far beyond simple nutrient uptake. It has been demonstrated that mechanisms exist by which mycorrhizal fungi can preferentially allocate nutrients to certain plants without a source–sink relationship. Studies have also detailed bidirectional transfer of nutrients between plants connected by a network, and evidence indicates that carbon can be shared between plants unequally, sometimes to the benefit of one species over another.

And the network exhibits something remarkable: kin recognition. More carbon has been found to be exchanged between the roots of more closely related Douglas firs sharing a network than more distantly related roots.

We must, however, be honest about the limits of current knowledge. The popular narrative of the "Wood Wide Web" has at times outrun the evidence. There's very little that scientists can say with certainty about how, and to what extent, trees interact via CMNs. Unfortunately, that hasn't prevented the emergence of wildly speculative claims, often with little or no experimental evidence to back them up.

A rigorous review by Jones, Hoeksema, and colleagues found that they thoroughly reviewed 26 field studies, including several of their own, that looked at the role fungal networks play in resource transfer in forests. What we found shows how easily confirmation bias, unchecked claims, and credulous news reporting can, over time, distort research findings beyond recognition. It should serve as a cautionary tale for scientists and journalists alike.

What is not in question is the fundamental symbiosis. Fungi do grow inside and on tree roots, forming a symbiosis called a mycorrhiza, or fungus-root. Mycorrhizae are essential for the normal growth of trees. Among other things, the fungi can take up from the soil, and transfer to the tree, nutrients that roots could not otherwise access.

And recent research continues to expand the picture. A 2025 study in *Communications Biology* provided the first lab-based evidence that common non-mycorrhizal networks can occur and suggest that fungal networks among plants may extend beyond mycorrhizal fungi.

The story of underground plant networks is, in other words, still being written. The romance of the "Wood Wide Web" may have been overstated in popular media, but the underlying science- that plants are connected, that they exchange resources, that their behaviour is altered by these connections- is established and deepening.

For our purposes, the key insight is structural: **the Three Sisters garden is not three individual organisms growing side by side. It is a networked system- a biological small-world network in which each organism is connected to the others through both above-ground physical coupling (the bean climbing the corn, the squash shading both) and below-ground chemical/fungal coupling (the mycorrhizal network linking all three root systems into a single, communicating whole).**

This networked structure is the garden's NESS. It is what makes the Three Sisters more than the sum of their parts. And it is, in a precise structural sense, a fractal miniature of the sacred site networks described in Chapter 19- the same topology (interconnected nodes, resource sharing, distributed intelligence) operating at a radically different scale.

20.4- Nitrogen Fixation as Quanta-Flow

Let us stay with the biology for a moment, because the details are revelatory.

Nitrogen is the most critical limiting nutrient in terrestrial ecosystems. The atmosphere is 78% nitrogen, but atmospheric N_2 is triple-bonded- one of the strongest molecular bonds in nature- and plants cannot break it. Without available nitrogen, plants cannot synthesise proteins, nucleic acids, or chlorophyll. Without nitrogen, growth stops.

The industrial solution to this problem is the Haber-Bosch process- a brute-force chemical reaction, developed in 1909, that combines atmospheric nitrogen with hydrogen under enormous pressure (150–300 atmospheres) and high temperature (400–500°C) to produce ammonia. The Haber-Bosch process currently consumes approximately 1–2% of the world's total energy production and is responsible for roughly 1.4% of global CO_2 emissions. It feeds approximately half the world's population through synthetic fertilisers.

The biological solution is nitrogen fixation- performed by rhizobia bacteria living in the root nodules of legumes (including beans). These bacteria use the enzyme nitrogenase to break the N_2 triple bond at ambient temperature and pressure, converting atmospheric nitrogen into ammonia that the plant can use. The process requires energy (supplied by the plant as carbon sugars), but it operates at body temperature, in ordinary soil, without industrial infrastructure.

The Haudenosaunee did not know the word "nitrogen." They did not know the word "bacteria." They did not know the word "enzyme." But they knew- through millennia of empirical observation, through careful attention to what grew well beside what, through the accumulated wisdom of generations of women tending the same soil- that beans made the soil richer for corn, and that planting them together produced more food than planting them apart.

This is empirical science of the highest order. It is not laboratory science- it has no control groups, no double-blind protocols, no p-values. But it is *systematic observation of nature, refined over centuries, yielding reproducible, measurable, practically significant results*. The Three Sisters system is a scientific discovery- a discovery of the same physical reality (nitrogen fixation) that Western science would not formally describe until 1888, when Martinus Beijerinck isolated the first nitrogen-fixing bacterium.

In our vocabulary: the bean's nitrogen-fixing capacity is a form of *quanta-flow*- a sustained, continuous transformation of atmospheric potential (N_2) into biological actuality (NH_3), mediated by the symbiotic coupling between plant and microorganism. It is the e-operator in biological action: exponential, compounding, continuously enriching the system's capacity for growth.

And the Haudenosaunee knew this. Not in our words. In their own.

"The Three Sisters in the Mohawk language is Kiohehkwen, which literally means, 'They give us life.'"

20.5- The NESS of the Garden

The Three Sisters garden is a Non-Equilibrium Steady State.

Recall from Chapter 12 that a NESS is a system that is not in thermodynamic equilibrium but whose macroscopic properties remain approximately constant due to a continuous flow of energy or matter. The system sustains itself through throughput, not stasis. It is alive because it flows.

The Three Sisters garden exhibits every hallmark of a NESS:

Continuous energy input. Sunlight drives photosynthesis in all three plants. The corn captures maximum solar radiation at the top of the canopy; the squash captures light at ground level; the bean captures light at intermediate heights. The three-tiered architecture maximises the system's absorption of solar energy per unit of ground area.

Continuous matter cycling. Nitrogen flows from atmosphere → bacteria → bean roots → soil → corn and squash roots → plant tissue → decomposing leaves → soil → cycle repeats. Carbon flows from atmosphere → photosynthesis → plant tissue → root exudates → mycorrhizal fungi → soil organic matter → decomposition → atmosphere → cycle repeats. Water flows from precipitation → soil → roots → transpiration → atmosphere → precipitation → cycle repeats.

Dynamic stability. The system maintains its macroscopic properties (species composition, canopy structure, soil fertility) over time, despite continuous perturbation by weather, pests, and seasonal change. It is not static- it grows, ripens, and dies back with the seasons- but its *pattern* persists year after year when the mounds are replanted.

Self-regulation. The system regulates itself without external management. The squash suppresses weeds (reducing competition for water and nutrients). The bean enriches the soil (maintaining nitrogen availability). In Haudenosaunee or Iroquois farming, the fields were not tilled, enhancing soil fertility and the sustainability of the cropping system by limiting soil erosion and oxidation of soil organic matter. No tillage. No synthetic fertiliser. No pesticide. No irrigation in most cases. The system produces food and builds soil *simultaneously*- a feat that industrial agriculture, with all its technology, has never achieved.

Dissipative structure. The garden, like all living systems, is a dissipative structure in the Prigogine sense: it maintains its order through the continuous dissipation of solar energy, using the energy throughput to sustain the far-from-equilibrium state that constitutes its living organisation.

The Three Sisters garden is, in this precise sense, a *biological instantiation of the master equation*:

$$\partial_t \Psi = -\nabla_{FR}[\mathcal{L}(\Psi)] + T(\Psi) \otimes D(\Psi) + \sum_n \phi \cdot \omega_n \cdot \delta(\Psi \in \mathrm{NESS}_{98})$$

The first term ($-\nabla_{FR}[\mathcal{L}(\Psi)]$) represents the natural gradient toward entropy- the tendency of the garden to dry out, weed over, deplete its soil, and return to scrub. This is what happens when the garden is *not* maintained- when the NESS is disrupted.

The second term ($T(\Psi) \otimes D(\Psi)$) represents the creative tension between the monadic extension (the bean's nitrogen fixation *amplifying* the soil's potential) and the comonadic extraction (the squash's ground cover *filtering* out weeds and conserving the essential moisture). The monad reaches for new growth. The comonad preserves what already works.

The third term ($\sum_n \phi \cdot \omega_n \cdot \delta(\Psi \in \mathrm{NESS}_{98})$) represents the φ-resonance- the specific proportionality between the three components that tunes the system to its most efficient configuration. Change the ratio of corn to beans to squash, and the system's efficiency changes. Too much corn, and the nitrogen depletes. Too many beans, and the corn is overwhelmed by climbing vines. Too much squash, and the shading becomes excessive, reducing photosynthesis in the beans. The optimal ratio- the one refined over centuries of indigenous practice- is the one that approaches φ-proportionality: maximum coherence, minimum waste, the golden balance.

20.6- Fire-Stick Farming: The Continent as Garden

The Three Sisters garden is the most concentrated example of indigenous agricultural resonance. But it is not the only one. Across the world, indigenous cultures developed agricultural systems that exhibit the same structural principles: polyculture, self-regulation, symbiotic coupling, and landscape-scale NESS maintenance.

In Australia, Aboriginal peoples developed what archaeologist Rhys Jones named "fire-stick farming"- a systematic practice of controlled burning that shaped the ecology of an entire continent.

Ethnographic evidence leaves little doubt that Aboriginal burning played a central role in the maintenance of the landscapes subsequently colonized by Europeans. Both 19th century European colonists and anthropologists in the 20th century documented the indispensability of fire as a tool in traditional Aboriginal economies, which have aptly been described as 'fire-stick farming'.

There are a number of purposes, including to facilitate hunting, to change the composition of plant and animal species in an area, weed control, hazard reduction, and increase of biodiversity. Fire-stick farming had the long-term effect of turning dry forest into savanna, increasing the populations of grass-eating animals such as kangaroos.

The practice was not random. It was calibrated to specific ecological objectives: ethnographic observations of contemporary Aboriginal hunting and burning with satellite image analysis of anthropogenic and natural landscape structure demonstrate the processes through which Aboriginal burning shapes arid-zone vegetational diversity. Anthropogenic landscapes contain a greater diversity of successional stages than landscapes under a lightning fire regime, and differences are of scale, not of kind.

The key finding: Aboriginal burning produced a *greater diversity of successional stages*- a more varied landscape mosaic- than natural lightning fires alone. The maintenance of small-scale habitat mosaics increases small-animal hunting productivity.

This is NESS maintenance at continental scale. The Aboriginal fire regime was not merely "land management" in the modern sense of resource extraction. It was *ecological tuning-* a deliberate, sustained, knowledge-intensive practice of maintaining the Australian landscape in a far-from-equilibrium state that maximised biodiversity, hunting productivity, and ecological resilience.

And what happened when the practice was disrupted?

Cultural burnings were slowly eradicated after European settlers began to colonise Australia from 1788 onwards.

An early result of the disruption of cool burning was the devastating Black Thursday bushfires in February 1851, which burnt 50,000 square kilometres of the colony of Victoria.

A series of aerial photographs taken around 1947 reveal that the Karajarri people practised fire-stick farming in the Great Sandy Desert of Western Australia for thousands of years, until they left the desert in the 1950s and 1960s. When fires swept the desert in the decades following their departure, they caused widespread destruction, "losing 36 to 50 per cent of 24,000 square kilometres of desert to just a couple of fires every year".

The cessation of Aboriginal burning- the severing of the fire-management NESS- produced catastrophic wildfires. The system, deprived of its maintenance protocol, did exactly what NESS theory predicts: it accumulated potential energy (fuel load), lost its resilience (the fine-grained mosaic was replaced by continuous fuel), and collapsed into a catastrophic state transition (the uncontrollable wildfire).

The modern response has been instructive. Cultural burnings were reintroduced in parts of Australia during the early twenty-first century, and some Australian states now integrate them with other fire-prevention strategies. Aboriginal fire practitioners are now recognised as experts in landscape management- not through anthropological charity but through hard empirical evidence that their methods work better than the colonial alternative.

In the language of this book: fire-stick farming is a technology of *guiding quanta at landscape scale*. The controlled burn is a transient, high-specificity intervention that guides the ecosystem through a series of micro-transitions (regrowth, successional shift, mosaic creation) that maintain the continental NESS. The fire is not destruction. The fire is *the garden's heartbeat.*

20.7- Waru-Waru: The Andes as Raised Garden

On the other side of the Pacific, 3,800 metres above sea level, on the high plateau surrounding Lake Titicaca, a different culture developed a different resonance technology for an equally challenging environment.

Waru Waru is a type of raised field agriculture developed by pre-Hispanic people in the Andes region of South America from Ecuador to Bolivia. This regional agricultural technique is also referred to as camellones in Spanish.

These pre-Hispanic fields cover about 82,000 ha (200,000 acres) of land in Bolivia and Peru, and sit above an altitude of around 3,800 m. Radiocarbon dates taken from habitation sites associated with raised field agriculture in the region indicate usage sometimes between 1000 B.C. to A.D. 400.

The system is elegant: parallel canals alternated by raised planting beds, strategically located on floodplains. These flooded fields were composed of soil that was rich in nutrients due to the presence of aquatic plants and other organic materials. Through the process of mounding up this soil to create planting beds, natural, recyclable fertilizer was made available in a region where nitrogen-rich soils were rare. By trapping solar radiation during the day, this raised field agricultural method also protected crops from freezing overnight.

The genius of the system is its *multifunctionality*. The raised beds are not merely planting surfaces. The surrounding canals serve simultaneously as irrigation channels, frost buffers, fertiliser producers (the aquatic organisms in the canals decompose into nutrient-rich muck that is periodically scooped onto the beds), and even as habitats for fish and water birds that provide additional food sources. This type of agriculture also created artificial ecosystems, which attracted other food sources such as fish and lake birds. Past cultures in the Lake Titicaca region likely utilized these additional resources as a subsistence method.

And the system works. In modern revival trials, in the Huatta district, reconstructed raised fields produced impressive harvest, exhibiting a sustained potato yield of 8 to 14 tons per hectare per year. This performance comes with no synthetic inputs, no fossil-fuel-powered machinery, and no irrigation infrastructure beyond the canals that were dug by hand over two thousand years ago.

Most remarkably, the system proved its worth under exactly the conditions that modern agriculture fears most: In 2023 with frost at −20°C, waru waru fields produced food, and farmland soils failed. In 2023, when Puno suffered one of the largest periods of drought in almost six decades, Waru Waru helped farmers cope with lack of water and avoid food shortages.

When modern agriculture failed- when the industrial model, with its synthetic inputs and monoculture logic, could not cope with −20°C frost and sixty-year drought- the 2,000-year-old technology performed.

This is not nostalgia. This is physics. The waru-waru system is a NESS- a far-from-equilibrium configuration sustained by continuous quanta-flow (solar energy, water cycling, nutrient cycling) through a self-regulating architecture. Its raised beds create thermal mass (storing solar heat during the day and releasing it at night). Its canals create a buffer against both flood and drought (storing excess water and releasing it during dry periods). Its biological diversity (crop polyculture, aquatic organisms, pollinators) creates resilience- the capacity to absorb perturbation without losing function.

The industrial monoculture, by contrast, is a system designed for maximum throughput at minimum diversity- a linear, single-species, chemically dependent system that has no buffering capacity, no self-regulation, and no resilience to perturbation. When the external inputs fail (when the fertiliser runs out, when the irrigation pump breaks, when the frost exceeds design specifications), the monoculture collapses. It is an equilibrium-seeking system pretending to be alive.

The waru-waru is a NESS. The monoculture is a machine.

20.8- Polynesian Reef Management: The Ocean as Garden

The resonance principle extends to the sea.

Polynesian cultures across the Pacific- from Hawai'i to Tonga, from Samoa to Aotearoa (New Zealand), from Rapa Nui to the Tuamotus- developed sophisticated systems of marine resource management based on the same structural principles as the Three Sisters and waru-waru: polyculture, seasonal cycling, taboo (protected periods), and network-based regulation.

The Hawaiian system of *ahupua'a*- a land division extending from the mountain summit to the outer reef- was explicitly designed to manage the entire watershed as a single integrated system. Freshwater from the mountains flowed through taro paddies (which filtered and enriched it), down through agricultural lands, into fishponds, and finally into the reef ecosystem. Each segment of the ahupua'a was managed by its own community, but the whole was governed by principles of reciprocity and cyclic renewal.

The fishponds- *loko i'a*- were engineering marvels: walled enclosures built on the reef flat or in estuarine zones, designed to capture juvenile fish that entered through sluice gates and to fatten them in a nutrient-rich environment until they were ready for harvest. The ponds were not merely holding tanks. They were *artificial reef ecosystems*- designed to produce fish while simultaneously maintaining the health of the surrounding coral reef by controlling predator populations and nutrient flows.

The kapu (taboo) system designated specific seasons and specific species as off-limits to harvest- a rotating system of resource rest that allowed depleted populations to recover. This is precisely the principle of *pulsed NESS maintenance*: periods of extraction alternated with periods of recovery, calibrated to the natural reproductive cycles of the target species.

When the kapu system was abolished in 1819 (as we noted in Chapter 19), the marine NESS of the Hawaiian islands began to degrade. Overharvesting, habitat destruction, and the introduction of invasive species followed in rapid succession. The fish stocks that had sustained Hawaiian communities for centuries declined precipitously. The reef systems that had been maintained in dynamic equilibrium for millennia began to fragment.

The pattern repeats: indigenous resonance technology dismantled → NESS disrupted → ecological collapse. Not because the indigenous technology was "better" in some romantic sense, but because it was *structurally matched* to the dynamics of the living system it managed. It was a NESS maintaining a NESS. The colonial replacement was a linear extraction protocol applied to a nonlinear system- and the result was predictable to anyone who has read Chapter 12.

20.9- Monoculture: The Anti-Garden

We must now name the opposite.

Modern industrial agriculture- the system that feeds most of the world's eight billion people- is, in the framework of this book, the *anti-garden*. It is not merely a different approach to food production. It is a structural inversion of every principle the Three Sisters embody.

Monoculture vs. Polyculture. Where the Three Sisters plant three species together, industrial agriculture plants one species alone- vast fields of identical genetic stock, stretching to the horizon. Monoculture eliminates the symbiotic coupling that makes polyculture self-regulating. Without beans, there is no nitrogen fixation; the farmer must add synthetic fertiliser. Without squash, there is no ground cover; the farmer must spray herbicide. Without diversity, there is no pest resistance; the farmer must spray insecticide.

Linear vs. Cyclic. Where the Three Sisters cycle nutrients through a closed loop (nitrogen from air → bacteria → soil → plant → soil), industrial agriculture operates as an open line: synthetic nitrogen is manufactured in a factory, transported to the field, applied once, partially absorbed by the crop, and the rest washes into waterways (producing algal blooms, hypoxic dead zones, and groundwater contamination). There is no cycle. There is only throughput.

Extraction vs. Regeneration. In Haudenosaunee or Iroquois farming, the fields were not tilled, enhancing soil fertility and the sustainability of the cropping system by limiting soil erosion and oxidation of soil organic matter. The Three Sisters *build soil.* Industrial agriculture *destroys soil.* Tilling breaks up mycorrhizal networks, oxidises organic matter, and exposes topsoil to wind and water erosion. The United Nations Food and Agriculture Organization has estimated that, at current rates of soil degradation, the world's topsoil could be gone within 60 years.

Fragility vs. Resilience. The Three Sisters system- and the waru-waru, and the fire-stick farming, and the Polynesian fishponds- are *resilient.* They absorb perturbation. They recover from shock. They adapt to changing conditions. The 2023 drought and frost in Puno proved it: the ancestral technology worked when the modern system failed. Monoculture is fragile. It depends on external inputs. Remove any one- fertiliser, irrigation, pesticide, fuel for machinery- and the system collapses.

The contrast is not aesthetic. It is thermodynamic. The Three Sisters is a NESS. Monoculture is a *near-equilibrium system driven far from it by continuous external forcing*- and the moment the forcing stops, it snaps back to the low-complexity state (bare soil, weeds, erosion) that equilibrium predicts.

This is the agricultural face of the Great Decoupling that Chapters 23–26 will trace in detail: the systematic replacement of resonant, self-regulating, polyculture NESS with linear, dependent, monoculture extraction. It is the same pattern- ternary replaced by binary, flow replaced by force, garden replaced by machine- that we will see repeated in timekeeping, in governance, in epistemology, in the very structure of thought.

20.10- Robin Wall Kimmerer and the Grammar of Animacy

Before we close this chapter, we must honour a voice that has done more than any other to articulate the indigenous understanding of the Three Sisters in terms that bridge the ancient and the modern.

Robin Wall Kimmerer is a member of the Citizen Potawatomi Nation, a mother, a plant ecologist, a professor of environmental biology at the State University of New York, and the author of *Braiding Sweetgrass: Indigenous Wisdom, Scientific Knowledge, and the Teachings of Plants*. She has written extensively about the Three Sisters as a paradigm for what she calls *reciprocal agriculture*- a practice in which the human farmer is not a master commanding nature but a *participant in a relationship of mutual gift-giving*.

Kimmerer argues that the Western scientific vocabulary- its nouns, its categories, its separation of "subject" from "object"- is structurally incapable of describing the living, relational world that the Three Sisters inhabit. In English, we speak of corn, beans, and squash as *things*- as objects to be planted, managed, harvested, consumed. In Potawatomi, the grammar itself is animate: the land is alive, the plants are persons, the relationship between farmer and garden is one of kinship, not dominion.

This is not mysticism. It is ontology. The Potawatomi language, like many indigenous languages, encodes in its very grammar the relational, participatory, co-creative understanding of reality that this book has been building through physics and mathematics. Where Western language says "the beans fix nitrogen," Potawatomi grammar says something closer to "the bean-persons give their nitrogen-gift to the soil-persons." The difference is not merely poetic. It is *structural*- it encodes a different understanding of agency, causality, and relationship.

In the framework of the Universal Hologram, Kimmerer's "grammar of animacy" is the linguistic expression of what we have been calling the quanta-field's self-referential resonance. A living language that treats plants, rivers, and stones as animate beings is a language *tuned to the quanta-field*- a language that reflects, in its very syntax, the reality that everything is resonant, everything is connected, everything is participating in the co-creative unfolding of the cosmos.

The Three Sisters are not resources. They are not inputs. They are not variables in an optimisation equation.

They are *teachers*.

And the lesson they teach- to anyone willing to kneel before a mound of earth and listen- is the lesson of this entire book: **reality is a triad, sustained by flow, stabilised by resonance, and maintained by the conscious participation of those who tend it.**

20.11- The Garden as Hologram

We began this chapter with a Haudenosaunee woman pressing three seeds into a mound. We end it with a recognition that may be the most radical claim in this book.

The Three Sisters garden is not *like* the Universal Hologram.

The Three Sisters garden *is* the Universal Hologram.

Not metaphorically. Not by analogy. *Structurally*. Every feature of the cosmic architecture described in Sections A and B of this book is instantiated- physically, biologically, measurably- in the Three Sisters mound:

Cosmic Architecture	Garden Instantiation
φ-e-π Triad	Squash (φ/coherence) + Bean (e/growth) + Corn (π/structure)
NESS	Self-regulating, far-from-equilibrium, solar-driven dissipative structure
KZ-Monad T(Ψ)	Bean's nitrogen fixation amplifies the soil's potential
Comonad D(Ψ)	Squash's ground cover filters noise (weeds) to reveal signal

Cosmic Architecture	Garden Instantiation
	(crops)
δ∞ Gap	The ~2% metabolic cost of maintaining the system (the farmer's labour, the mycorrhizal network's carbon tax)
Guiding Quanta	The gene regulatory bursts in each plant's growth and differentiation
Small-world network	Mycorrhizal connections linking roots across the mound
Consciousness participation	The farmer's intention, attention, and seasonal care
Fractal self-similarity	The mound mirrors the garden mirrors the landscape mirrors the cosmos

The holographic principle says: every part contains the whole. The Three Sisters mound- a circle of earth no more than a metre across- contains, in its structure and dynamics, the entire architecture of the cosmos.

This is what the Haudenosaunee knew. Not because they had telescopes or particle accelerators. Because they had *gardens.* And they had the wisdom- refined over millennia of attentive, somatically grounded, relationally structured practice- to recognise that the garden was not separate from the cosmos. It was the cosmos, *in miniature, under their care.*

The first academic description of the Three Sisters cropping system in 1910 reported that the Iroquois preferred to plant the three crops together, since it took less time and effort than planting them individually, and because they believed the plants were "guarded by three inseparable spirits and would not thrive apart."

Three inseparable spirits. Three cosmic operators. Three notes of the universal chord.

They would not thrive apart.

Neither will we.

In the next chapter, we rise from the garden to the calendar- from biological resonance to temporal resonance. We will show that the cyclical time systems of the ancients- the Mayan Long Count, the Vedic yugas, the Aboriginal seasonal calendars- were not primitive attempts to count days. They were precision instruments for tracking the Temporal State Vector, calibrated to the ternary beat of the cosmic year. And their abandonment- in favour of the linear, mechanical, universal clock- was one of the most consequential acts of detuning in human history.

Chapter 21

Cyclical Time and the Cosmic Calendar

21.1- The Number That Refuses to Be Round

Three hundred and sixty.

The number appears so often, in so many unrelated places, that its ubiquity deserves to be called suspicious. A circle has 360 degrees. The ancient Sumerian and Babylonian year had 360 days. The Egyptian administrative year had 360 days (twelve months of thirty). The Hindu "divine year" is 360 solar years. The Mayan *tun*- the foundational time unit of the most precise calendar system in the pre-modern world- is exactly 360 days.

Not 365. Not 365.25. Not 365.2422 (the true length of the solar year, accurate to the second decimal place). *Three hundred and sixty*- a number that is clearly, deliberately, and precisely wrong as a solar year count, and yet was chosen, independently, by the most astronomically sophisticated cultures on the planet.

Why?

The standard scholarly answer is mathematical convenience. 360 is the most highly composite number below 400. It is divisible by 2, 3, 4, 5, 6, 8, 9, 10, 12, 15, 18, 20, 24, 30, 36, 40, 45, 60, 72, 90, 120, and 180. No number in its vicinity has more factors. For civilisations performing arithmetic without electronic calculators- dividing time into months, weeks, watch-periods, ritual intervals- 360 is the natural choice. It makes the mathematics of everyday life as frictionless as possible.

This answer is true. And it is incomplete.

Because the ancient astronomers who chose 360 *knew* the solar year was not 360 days. The Maya knew it with a precision that exceeds the Gregorian calendar. The Egyptians added five *epagomenal* days at year's end to bring the total to 365. The Vedic astronomers calculated the tropical year to extraordinary accuracy. They chose 360 not because they thought the year was 360 days long, but because *360 was the ideal frequency*- the geometric year, the year that mathematics wants, the year of perfect divisibility and maximal harmonic resonance.

In the language of this book: 360 is the **Mode 360** of the Ternary Resonance Y_t introduced in Chapter 14. It is not a measurement of the solar year. It is the *first harmonic* of the cosmic year- the tone that the cosmos would sing if its orbit were perfectly circular, its axial tilt zero, and its gravitational perturbations nil. It is the Platonic year, the year of pure form, the year against which the actual year (365.2422 days) is measured as a deviation.

And the deviation- the gap between the ideal 360 and the actual 365.2422- is not a failure of the calendar. It is the *temporal face of the asymptotic gap* δ_∞. It is the ~1.5% residual tension between the geometric ideal and the physical actual that keeps the temporal field vibrating, keeps the seasons shifting, keeps the year *alive.*

The ancients knew this. They encoded it in their calendars. And they built their entire civilisational rhythm around it- not because they couldn't count to 365, but because they understood something that modern timekeeping has forgotten: *the purpose of a calendar is not to count days. It is to attune human life to the resonant frequencies of the cosmos.*

21.2- The Mayan Calendar: A Resonance Engine

The Maya developed one of the most mathematically sophisticated calendar systems in human history. Using their knowledge of astronomy and mathematics, the ancient Maya developed one of the most accurate calendar systems in human history. The ancient Maya had a fascination with cycles of time.

The system was not a single calendar but an interlocking set of cycles- a polyphonic temporal instrument in which multiple rhythms sounded simultaneously, producing interference patterns of extraordinary complexity.

The Tzolk'in (260 days). The tzolk ' in is the name commonly employed by Mayanist researchers for the Maya Sacred Round or 260-day calendar. This was the sacred calendar- a cycle of 260 days produced by the interlocking of 13 numbers with 20 day-names (13 × 20 = 260). The length of the Tzolk'in matches nine cycles of the Moon and the gestational period of humans. The Tzolk'in is also related to the movements of the zenith Sun and the growing cycle of corn.

Notice the convergences: the sacred calendar encodes the human gestational period, the lunar cycle, the solar zenith passage, and the maize growing season in a single 260-day rhythm. This is not a counting device. It is a *resonance instrument-* a tool for tracking the intersection of biological, lunar, solar, and agricultural cycles in a single compact frame.

The Haab' (365 days). The Haab is a 365-day solar calendar which is divided into 18 months of 20 days each and one month which is only 5 days long (Uayeb). This was the civil calendar- the practical timekeeping system used for agricultural planning, administrative records, and seasonal festivals. Eighteen months of twenty days yielded 360 days; the remaining five days- the *Wayeb'*- were considered unlucky, liminal, a threshold between years.

Here is the ternary structure in plain sight. The Haab' is not a 365-day calendar. It is a *360-day calendar with a 5-day appendix*. The Maya explicitly distinguished between the *tun* (the 360-day ideal) and the *haab'* (the 365-day reality), preserving both in simultaneous operation. The gap between them- the 5 days of Wayeb'- was not a correction factor to be hidden. It was a *sacred interval*, a temporal δ_∞, the gap between the geometric ideal and the lived actual that held the year in creative tension.

The Calendar Round (52 Haab' years). The Tzolkin was combined with a 365-day vague solar year known as the Haab ' to form a synchronized cycle lasting for 52 Haab ' called the Calendar Round. The Calendar Round is the period required for the Tzolk'in and Haab' to return to the same combination- a cycle of 18,980 days (52 × 365 = 73 × 260 = 18,980). Within this cycle, every day has a unique dual designation: a Tzolk'in date and a Haab' date. The Calendar Round is, in effect, a *two-dimensional temporal coordinate system*- far richer than the single-axis date system of the Gregorian calendar.

The Long Count. The Mesoamerican Long Count calendar is a non-repeating base-20 and base-18 calendar used by pre-Columbian Mesoamerican cultures, most notably the Maya. Using a modified vigesimal tally, the Long Count calendar identifies a day by counting the number of days passed since a mythical creation date that corresponds to August 11, 3114 BCE in the proleptic Gregorian calendar.

The Long Count uses a modified vigesimal (base-20) system with a crucial anomaly: An important exception was made for the second-order place value, which instead represented 18 × 20, or 360 days, more closely approximating the solar year than would 20 × 20 = 400 days. The units are: Kin = 1 Day. Uinal = 20 kin = 20 days. Tun = 18 uinal = 360 days. Katun = 20 tun = 7,200 days. Baktun = 20 katun = 144,000 days.

The *tun*- 360 days- is the foundational unit. Not the haab' (365 days), not the actual solar year, but the *geometric ideal*. The entire Long Count architecture is built on multiples of 360. The Maya deliberately chose to anchor their deepest temporal framework not in the astronomical year but in the harmonic year- the year of maximal resonance.

The 13 baktun cycle of the Maya Long Count calendar measures 1,872,000 days or 5,125.366 tropical years. This is the Great Cycle- the vast temporal arc within which all of Mayan recorded history unfolds. And its fundamental unit is the tun: 360 days, the first tone of the ternary resonance.

And the precision of the overall system is staggering. Over the Calendar Round's 52-year cycle, and using corrections tracked through the Long Count's astronomical supplements (lunar series, eclipse tables, Venus tables), the corrected Mayan year achieves a value of approximately 365.24220 days- the corrected Mayan year of 365.24220 days is closer to the true value of 365.24218 days than is the Gregorian year of 365.24250 days.

Let that sink in. The Maya, using naked-eye astronomy, stone monuments, and bark-paper codices, achieved a calendrical accuracy *exceeding* that of the Gregorian reform of 1582.

They did this not despite their use of the 360-day tun, but *because of it*. By anchoring their system in the geometric ideal and then tracking the deviation with precision, they achieved a temporal awareness that encompasses both the ideal and the actual- both the tone the cosmos *wants* to sing and the tone it *actually* sings, given the specific gravitational perturbations of this particular solar system.

21.3- The Vedic Yuga System: Time as Fractal Descent

On the other side of the planet, the Vedic tradition developed a temporal framework of almost incomprehensible scale- one that dwarfs the Mayan Long Count by orders of magnitude.

A Yuga Cycle is a cyclic age (epoch) in Hindu cosmology. Each cycle lasts for 4,320,000 years (12,000 divine years) and repeats four yugas (world ages): Krita (Satya) Yuga, Treta Yuga, Dvapara Yuga, and Kali Yuga.

The four yugas are arranged in a descending ratio of 4:3:2:1:

- **Satya Yuga** (the Golden Age): 1,728,000 years (4 × 432,000)
- **Treta Yuga** (the Silver Age): 1,296,000 years (3 × 432,000)
- **Dvapara Yuga** (the Bronze Age): 864,000 years (2 × 432,000)
- **Kali Yuga** (the Iron Age): 432,000 years (1 × 432,000)

Total: 4,320,000 years = one *maha yuga* (Great Age).

Each yuga, or age, lasts a multiple of 432,000 solar years, or 1000 divine years, in a 4:3:2:1 ratio.

Notice the numbers. 432,000 is the base unit. And 432 is precisely half of 864- the frequency (864 Hz) that our framework identifies as the fundamental tone of the harmonic manifold on which consciousness operates (Chapter 13). 4,320,000- the length of the maha yuga- is 432,000 × 10. And the "divine year" used in the calculation is 12,000 divine years, where a divine year lasts for 360 solar (human) years.

There it is again: **360**. The divine year is 360 human years. The geometric ideal, the Mode 360, is the conversion factor between human time and cosmic time. The entire yuga system is built on 360 as its temporal quantum- the same number that anchors the Mayan Long Count.

The scale then expands fractally:

In Hinduism, a kalpa is a unit of time equal to 4.32 billion years. It corresponds to one day in the life of Brahma, the creator god, and represents the active, creative phase of the cosmic cycle. Each kalpa is made up of 1,000 Yuga Cycles.

A kalpa- one "day" of Brahma- is 4.32 billion years. This is remarkably close to the estimated age of the Earth (4.54 billion years) and roughly one-third the age of the universe (13.8 billion years). This night is equal in length to the day-another 4.32 billion years. Together, one day and night of Brahma make up a full cosmic day, totaling 8.64 billion years.

8.64 billion years for one full day-night cycle of Brahma. Recall that Its cycles run from our ordinary day and night to a day and night of Brahma, 8.64 billion years long. Carl Sagan noted the remarkable correspondence: "Longer than the age of the Earth or the Sun and about half the time since the Big Bang."

And Brahma's total lifespan: A year of Brahma consists of 12 months, or 360 days and nights, equaling 3.1104 trillion years. Brahma's full lifespan is 100 of these divine years-a span of 311.04 trillion years, known as a maha-kalpa.

311.04 trillion years. And the system is explicitly cyclic: At the end of each kalpa, the universe enters a period of rest and dissolution called pralaya, or the night of Brahma. At the end of Brahma's lifespan, the maha-kalpa concludes, and a new Brahma is born, beginning the cycle anew. When he exhales, thousands of universes emerges and one "Brahma" is born in each universe. When "Vishnu" inhales, all universes get sucked and Brahma dies. This cycle is non-ending and eternal.

The structural parallel to Conformal Cyclic Cosmology (Chapter 15) is overwhelming. Penrose describes an infinite succession of aeons, each ending in a conformal infinity that becomes the Big Bang of the next. The Vedic tradition describes an infinite succession of kalpas, each ending in a *pralaya* (dissolution) that becomes the awakening of the next. Penrose measures in physics. The Vedas measure in mythology. But the architecture is identical: *infinite cycles, each carrying forward the accumulated consequences of the previous one, in a process that unfolds across timescales so vast that they reduce human history to a single breath.*

And the qualitative dimension: The Yuga Cycle describes a repeating pattern of growth and decline in the quality of life on Earth. As the yugas progress, spiritual and physical life on Earth become increasingly degraded until the cycle begins again with a new Satya Yuga.

The 4:3:2:1 ratio is not merely quantitative. It describes a *gradient of temporal density*- the Vedic analogue of the ρ_t component of the Temporal State Vector. In the Satya Yuga (4 parts), temporal density is maximum: each moment is rich, full, deeply connected to the cosmic resonance. In the Kali Yuga (1 part), temporal density is minimum: time feels thin, accelerated, disconnected. This theory suggests that during Satya Yuga, only truth prevailed. During the Treta Yuga, the universe lost one-fourth of the truth, Dwapar lost one-half of the truth, and now the Kali Yuga is left with only one-fourth of the truth. Evil and dishonesty have therefore gradually replaced truth in the last three ages.

We are currently, according to the Vedic calculation, in the Kali Yuga-the age of minimum resonance. Kali Yuga began in 3102 BCE, meaning we are in the 5,127th year. There are still 426,873 years remaining in this age. The world reflects the exact characteristics described for Kali Yuga.

We do not present this as literal prophecy. We present it as a *temporal framework*- a map of ρ_t variation across vast timescales, encoded in the language of mythology by a civilisation that had no other language for the concepts it was describing.

21.4- Aboriginal Seasonal Calendars: Time as Ecology

If the Mayan calendar is a mathematical resonance engine and the Vedic yuga system is a fractal descent across cosmic timescales, the Aboriginal seasonal calendars are something altogether different and, in certain respects, more profound: they are *ecological calendars*- tools for tracking the resonance not of celestial objects but of the living landscape itself.

Indigenous Australian seasons are classified differently from the traditional four-season calendar used by most western European peoples. Aboriginal Australians and Torres Strait Islander people have distinct ways of dividing the year up. Naming and understanding of seasons differs among groups of Aboriginal peoples, and depends on where in Australia the group lives.

The first and most important fact is that there is no single Aboriginal and Torres Strait Islander' calendar, with the groups of each area having developed their own system based on their specific environment. A single calendar could never truly capture the complexities found across the vast lands.

Given the rich diversity of the landscapes and ecosystems, variations in weather patterns and the behaviours of plants and animals are specific to different regions and communities.

Each language group developed its own calendar, calibrated to its own territory:

The Yolngu, the Aboriginal Australians of north-east Arnhem Land, identify six seasons that align with the monsoonal climate. There are six Noongar seasons in Western Australia's South West region, which are indicated by changes in local plants and animals. Rembarrnga people identify seven and Dalabon people identify six main seasons that are based on a complex set of ecological indicators. The Tiwi calendar has three, also evolving around the monsoon cycle, the Kaurna calendar of the Adelaide Plains recognises four seasons, determined largely by the wind.

Three seasons in the Tiwi Islands. Four on the Adelaide Plains. Six in Arnhem Land and the South West. Seven among the Rembarrnga. The number varies because the *landscape* varies. The calendar is not imposed on the land; the calendar *emerges from* the land.

And the markers are not astronomical (equinoxes, solstices, star positions) but *ecological*: Aboriginal and Torres Strait Islander peoples observe the position of the stars in the sky and follow water, plant and animal cycles as ways of identifying seasonal phenomena. The seasonal calendars of different Aboriginal and Torres Strait Islander cultural groups demonstrate an understanding of the interdependence and interrelationships amongst living things. They use their calendars to predict seasonal changes and weather patterns to determine the availability of particular resources or the timing of journeys.

Rather than dates on a linear calendar, Indigenous Australians use cues from nature to define the seasons, representing thousands of years of careful observation and deep connection to Country- the flowering of the stringybark tree, flying ants cruising on southerly winds, the position of the Southern Cross constellation in the sky.

The seasons are not determined by the calendar. The *calendar is determined by the seasons-* and the seasons are determined by the living responses of plants, animals, winds, and waters to the local temporal field.

When the wattle blooms, the season changes. When the eels begin to run, the season changes. When the cockatoos start dropping honky nuts from the marri trees, the season changes. Unlike the four seasons of the European calendar, the six Noongar seasons are not marked by specific dates but are determined by observing changes in the local environment. These changes are indicated by signs such as the flowering of plants, the hibernation of reptiles and moulting of swans.

The CSIRO- Australia's national science agency- has worked with Aboriginal communities to formally document these calendars. Aboriginal and Torres Strait Islander peoples, Australia's First Scientists, have always held a deep understanding of the seasons, and of how to tell when seasons are changing. Seasonal understanding of Country underpins many activities on Country. Recording this knowledge provides a powerful tool for Indigenous knowledge holders to demonstrate and communicate their connection to, use and management of Country. Over the past fifteen years CSIRO has co-designed, refined and tested the application of a season calendar methodology with our Indigenous partners as a way of documenting and presenting seasonal understanding of Country.

In the language of this book, the Aboriginal seasonal calendars are **direct measurements of the Geospatial Phase** $G(\lambda,\phi)$- the third component of the Temporal State Vector. They track not the geometric ideal (like the Maya's 360-day tun) and not the cosmic-scale ρ_t gradient (like the Vedic yugas) but the *hyper-local modulation of temporal quality by geographic position, ecological context, and living-system feedback.*

Each Aboriginal calendar is, in effect, a $G(\lambda,\phi)$ map- a description of how the temporal field expresses itself at a specific latitude and longitude, through the medium of the living organisms that inhabit that location. The flowering of the wattle is a *bioindicator* of a specific G value. The arrival of the eels is another. The moulting of the swans is another. Together, they constitute a multi-channel sensor array- a biological Chronosonant Resonator- that measures the local TSV with a resolution and specificity that no single astronomical observation could achieve.

This is why there is no single Aboriginal calendar. A single calendar would require a single $G(\lambda,\phi)$ function- a universal geospatial phase that applies everywhere. But G is, by definition, *local.* It varies from place to place, from ecology to ecology, from watershed to watershed. The diversity of Aboriginal calendars is not a failure of standardisation. It is a *triumph of local precision*- each calendar attuned to its specific territory, each one reading the TSV through the specific bioindicators available in its specific ecological context..

21.5- The Ternary Convergence: 360, 363, 366

Now let us hold these three calendrical traditions- Mayan, Vedic, and Aboriginal- in simultaneous awareness, and see what pattern emerges.

The Maya anchor their system in 360- the geometric ideal, the year of maximal divisibility, the first harmonic of the cosmic cycle. They then track the deviation (the 5 Wayeb' days, the Long Count corrections) with extraordinary precision.

The Vedic system also anchors in 360- the "divine year" of 360 solar years- and builds vast temporal architectures on this foundation, encoding qualitative shifts in ρ_t across the yuga cycle.

The Aboriginal systems anchor in the *actual*- the ecological present, the living response of bioindicators to the local geospatial phase $G(\lambda,\phi)$- and produce calendars of 3 to 7 seasons, with no fixed duration, no numerical abstraction, and no deviation from the experienced temporal field.

Together, they map the three components of the Temporal State Vector:

TSV Component	Primary Cultural Expression	Calendar Type
Y_t **(Ternary Resonance)**	**Maya**- 360/365 duality; tun + haab'	Harmonic / Geometric
ρ_t **(Temporal Density)**	**Vedic**- 4:3:2:1 yuga gradient	Qualitative / Fractal
$G(\lambda,\phi)$ **(Geospatial Phase)**	**Aboriginal**- local bioindicator calendars	Ecological / Somatic

This is not an accident. It is the quanta-field recognising itself through three different instruments, each one tuned to a different component of the temporal manifold, each one producing a calibration of extraordinary precision in its own domain.

And the convergence is sealed by a single number: **360**. The Maya use it as their temporal base unit. The Vedic system uses it as the conversion factor between human and divine time. And the Aboriginal calendars, while not explicitly numerical, produce seasonal cycles whose *total count* across the year- when the durations of each bioindicator-defined season are summed- varies between approximately 350 and 370 days, bracketing 360 as the central tendency.

The ternary resonance $Y_t = \{360,363,366\}$ is not a modern invention. It is an ancient observation- encoded independently, in different codes, by three cultures that had no contact with each other, on three different continents, across timescales spanning from a few centuries (the Maya) to tens of millennia (the Aboriginal Australians).

21.6- When the Calendar Broke: The Gregorian Reform

In 1582, Pope Gregory XIII made a decision that would reshape the temporal experience of the entire planet.

The Gregorian calendar was a reform of the Julian calendar. It was instituted by the papal bull Inter gravissimas dated 24 February 1582 by Pope Gregory XIII, after whom the calendar is named. The motivation for the adjustment was to bring the date for the celebration of Easter to the time of year in which it was celebrated when it was introduced by the early Church. The error in the Julian calendar had led to the date of the equinox according to the calendar drifting from the observed reality, and thus an error had been introduced into the calculation of the date of Easter.

The problem was real. The Julian calendar, introduced by Julius Caesar in 46 BCE, assumed a year length of exactly 365.25 days- hence a leap day every four years. But the actual tropical year is approximately 365.2422 days- about 11 minutes shorter. The solar year more precisely lasts 365 days, 5 hours, 48 minutes, and 45.25 seconds-that is, more than 11 minutes shorter than 365 1/4 days. This slight inaccuracy accumulated and caused the calendar dates of the seasons to regress by almost one day per century.

By 1582, the accumulated error was 10 days. The spring equinox, fixed at March 21 by the Council of Nicaea in 325 CE, was occurring on March 11. Easter- the most important Christian feast, calculated from the spring equinox- was drifting away from its intended astronomical position.

Gregory's solution was radical: To reinstate the association, the reform advanced the date by 10 days: Thursday 4 October 1582 was followed by Friday 15 October 1582. Ten days were simply *deleted* from the calendar. People who went to sleep on October 4 woke up on October 15. The future leap-year rule was modified to skip three leap days every 400 years (century years not divisible by 400 would be common years), producing an average year length of 365.2425 days.

The reform was technically sound. It corrected a real astronomical drift and produced a calendar that is about 11 seconds ahead of the solar year. The seasons now advance by only one day in about 7,700 years.

But the reform was not merely astronomical. It was political, religious, and- in the deepest sense- ontological.

Many Protestant polities initially objected to adoption of a Catholic innovation; some Protestants feared the new calendar was part of a plot to return them to the Catholic fold. The Protestant and Orthodox countries didn't want to take direction from the pope, so they refused to adopt the new calendar. The result was that Catholic Europe suddenly jumped ahead of the rest of the continent by 10 days, and traveling across a border often meant traveling forward or backward on the calendar.

In total, more than three centuries passed until the Gregorian calendar had been adopted in all countries, from 1582 to 1927. Britain held out until 1752 - 170 years of deliberate temporal divergence from Catholic Europe. The transition involved skipping 11 days, with September 2, 1752, being immediately followed by September 14, a change that prompted public unrest as some believed they were losing days from their lives.

Russia did not adopt the Gregorian calendar until 1918 - after the Revolution. Gradually, other countries adopted the Gregorian calendar: the Protestant German states in 1699, Sweden in 1753, Japan in 1873, China in 1912, the Soviet socialist republics followed after 1918, and Greece in 1923.

The Gregorian reform, in our framework, represents a critical step in what we will call in Chapter 23 *The Great Decoupling*. Not because the reform was wrong- it was astronomically superior to the Julian calendar. But because it established a precedent that would shape all subsequent timekeeping:

The calendar became a universal standard, detached from local experience.

Before Gregory, European timekeeping was fragmented- There was great fragmentation of time-keeping methods in some parts of Europe over the first two or so centuries of Gregorian reform. Different regions celebrated the new year on different dates. Until it adopted the reformed calendar in 1752, England dated its new year at March 25, or the observance of Lady Day. Local customs, saint's days, agricultural festivals, and ecological markers played significant roles in how people experienced the passage of time.

After Gregory, time was standardised. One calendar for all Christendom- eventually, one calendar for all the world. The seasonal markers, the local feast days, the ecological rhythms that had connected European peasants to their landscape were gradually replaced by a single numerical grid: January, February, March... 1, 2, 3, 4, 5...

The Gregorian calendar measures time with admirable precision. But it measures only one dimension of time- the solar orbital period, rendered as a linear sequence of numbered days. It says nothing about Y_t (the ternary resonance of the year's harmonic quality). It says nothing about ρ_t (the qualitative density of temporal experience). It says nothing about $G(\lambda,\phi)$ (the geospatial modulation of time by local conditions).

In the terms of Chapter 14: the Gregorian calendar reduces the three-component Temporal State Vector to a single scalar. It collapses the colour of time into greyscale. It replaces the rich, multidimensional temporal experience encoded in the Mayan, Vedic, and Aboriginal calendars with a number.

And then it insists that this number is *the truth*.

21.7- The Tzolk'in Principle: Interlocking Cycles as Resonance Detection

Let us return to the Maya for a moment, because their most sophisticated temporal concept- the Calendar Round- contains a principle that the Gregorian system entirely lacks.

The combination of a haab' and a tzolk'in date identifies a day in a combination which does not occur again for 18,980 days. The Calendar Round is not a single cycle. It is *two* cycles, interlocking like gears, producing a pattern that repeats only after 52 solar years.

Within those 52 years, every single day has a *unique* two-part designation- a Tzolk'in number-name pair and a Haab' date. No two days share the same designation until the full 18,980-day cycle is complete.

This interlocking principle is the temporal equivalent of *binaural beating* in acoustics. When two tones of slightly different frequency are sounded simultaneously, the listener perceives not two separate tones but a *third* tone- a beat frequency equal to the difference between the two original frequencies. This beat frequency carries information that neither tone alone could convey: it reveals the *relationship* between the two frequencies.

The Calendar Round works the same way. The 260-day Tzolk'in and the 365-day Haab', sounded simultaneously, produce a "beat frequency"- the 52-year Calendar Round- that reveals the relationship between the sacred and solar cycles. This beat carries temporal information that neither cycle alone could convey: it maps the intersection of biological time (260 days ≈ gestation) and solar time (365 days ≈ orbit), producing a composite rhythm that tracks both simultaneously.

The Gregorian calendar has no such interlocking structure. It has one cycle: the solar year. And within that cycle, it has no sacred component- no 260-day rhythm, no lunar phase tracking (the date of Easter is calculated post hoc, not built into the calendar's structure), no bioindicator feedback. It is a single tone, played alone, in a world that resonates in polyphonic harmony.

The Calendar Round, by contrast, is a *chord*- two tones sounded together, producing a beat frequency that reveals the harmonic relationship between them. And the Long Count adds a bass line- a vast, low-frequency drone that anchors the entire system in the 360-day tun, connecting the daily experience of interlocking cycles to the cosmological timescale of creation itself.

21.8- The 4:3:2:1 Ratio and the Fractal Principle

The Vedic yuga system's 4:3:2:1 ratio encodes something that the Gregorian calendar cannot: a *qualitative gradient* of temporal experience.

According to the Manusmriti, the virtue (dharma) of human beings varies across the four yugas. In the Krita Yuga, the virtue is austerity (tapas); in the Treta Yuga, it is knowledge (jnana); in the Dvapara Yuga, it is sacrifice (yajna); and in the Kali Yuga, it is charity (dāna).

Each yuga is not merely shorter than the previous one- it is *qualitatively different*. The Satya Yuga is characterised by maximal truth, coherence, and spiritual attainment. The Kali Yuga is characterised by minimal truth, fragmentation, and material obsession. The transition from one to the next is not a gradual decline but a *phase transition*- a critical threshold at which the entire civilisational operating system shifts.

In our framework, the 4:3:2:1 ratio maps to the variation of *temporal density* ρ_t across the yuga cycle:

Yuga	Ratio	Temporal Density ρ_t	Quality
Satya	4	Maximum	Each moment contains fullness; deep resonance with quanta-field
Treta	3	High	Knowledge accessible; coherence diminishing
Dvapara	2	Medium	Sacrifice required to maintain connection
Kali	1	Minimum	Disconnectio n; noise dominates signal

This is a temporal fractal. The same 4:3:2:1 ratio that describes the quality gradient across a full maha yuga (4,320,000 years) also describes the quality gradient within each individual yuga (each one contains its own sub-yugas in the same ratio). And within each sub-yuga, the same pattern repeats. It is self-similar at every scale- the fractal rule of Chapter 5 applied to the dimension of time itself.

And the fractal principle applies downward as well. A single human life can be read through the yuga lens: childhood as Satya (wonder, openness, high ρ_t), youth as Treta (learning, exploration, declining ρ_t), maturity as Dvapara (responsibility, sacrifice, further decline), and old age as Kali (dissolution, reflection, minimal ρ_t- unless, through spiritual practice, the elder reverses the gradient and touches Satya again).

A single day can be read through it: dawn as Satya, morning as Treta, afternoon as Dvapara, night as Kali. A single breath: inhale as Satya (gathering), hold as Treta/Dvapara (transformation), exhale as Kali (release).

This is what the Vedic calendar encodes: not a count of days, but a *map of temporal quality at every scale*- a fractal description of ρ_t that applies equally to a cosmic cycle of 4.32 million years, a human lifetime of 80 years, a day of 24 hours, and a breath of 4 seconds.

The Gregorian calendar encodes none of this. January 15 is the same "quality" as July 15. Monday is the same "quality" as Friday. Morning is the same "quality" as evening. The Gregorian calendar is a flat temporal landscape- no hills, no valleys, no rhythm, no quality. Just numbers, advancing by one, forever.

21.9- What Was Lost When the Calendars Were Silenced

The Mayan Long Count was last inscribed on stone in the 10th century CE, as the Classic Maya civilisation collapsed. The Tzolk'in and Haab' survived- Many indigenous communities in the Mexican states of Veracruz, Oaxaca and Chiapas and in Guatemala keep the Tzolk ' in and in many cases the Haab ' . The contemporary Maya farmers still practice observation-based astronomy of the stars. They use their knowledge of the cycles of the Sun and other stars to determine the production cycle of maize and planning cycle rituals and ceremonies. But the full polyphonic system- the three interlocking calendars producing their complex temporal chord- was disrupted by the Spanish conquest and its aftermath.

The Aboriginal seasonal calendars survived- many are still maintained, as we have seen- but the colonisation of Australia subjected them to centuries of suppression, forced removal of children from their communities, and the deliberate destruction of languages and ceremonies that carried calendrical knowledge.

The Vedic yuga system survived as cosmological theory within Hinduism, but its practical implementation- the detailed tracking of

ρ_t variation across yuga sub-cycles, the calendrical rituals calibrated to specific yuga phases- was largely absorbed into standardised Hindu festival calendars that, while still cyclical, have lost much of the original system's multidimensional depth.

In each case, the loss was not merely cultural. It was *perceptual.* When a calendrical system is silenced, the mode of temporal perception it supports *atrophies.* People who once tracked six seasons through bioindicators now track four seasons through dates. People who once felt the ternary resonance of the cosmic year through the interlocking rhythms of Tzolk'in and Haab' now feel only the linear tick of the Gregorian grid. People who once understood their place in the yuga cycle- who knew whether they lived in an age of truth or an age of noise- now have no temporal framework for qualitative self-location at all.

The calendars did not merely *describe* temporal experience. They *structured* it. They were not maps of a pre-existing temporal landscape. They were *instruments that created the temporal landscape by the act of measuring it*- exactly as the observer-participant in quantum mechanics does not merely observe reality but participates in its creation (Chapter 16).

When the calendars were silenced, the temporal landscape they had sustained collapsed. The rich, multidimensional, qualitatively differentiated experience of time- time as chord, time as ecology, time as fractal gradient- was replaced by the single, flat, quantitative tick of the universal clock.

And we, the inheritors of that collapse, now live in a temporal monoculture as impoverished as the agricultural monocultures that replaced the Three Sisters gardens.

21.10- Toward Temporal Polyculture

The solution is not to abandon the Gregorian calendar. It is accurate. It is practical. It is globally standardised, and that standardisation has genuine value for international coordination.

The solution is to recognise that the Gregorian calendar, like a monoculture field, captures only one dimension of a multidimensional reality- and to *supplement* it with additional temporal layers that restore the dimensions it has lost.

This is what we might call **temporal polyculture**- the practice of tracking multiple temporal rhythms simultaneously, just as the Maya tracked the Tzolk'in, Haab', and Long Count in parallel, just as the Aboriginal Australians tracked ecological bioindicators alongside stellar positions, just as the Vedic system embedded daily rituals within yuga-scale cosmological awareness.

In the framework of HTT, temporal polyculture means tracking all three components of the Temporal State Vector:

Y_t **(Ternary Resonance):** Awareness of the year's harmonic quality- which mode (360/363/366) dominates, how the gap between the geometric ideal and the astronomical actual modulates seasonal experience. This is what the Mayan Calendar Round tracked. In practice, it might mean: noting the "feel" of different years, tracking the correspondence between calendar anomalies and experiential quality, paying attention to the way different years seem to "move" at different speeds.

ρ_t **(Temporal Density):** Awareness of the qualitative gradient of temporal experience- the difference between dense, rich, fully-present moments and thin, scattered, disconnected ones. This is what the Vedic yuga system tracked. In practice, it might mean: recognising that morning and evening have different temporal densities, that certain seasons feel more "alive" than others, that certain locations seem to thicken or thin the experience of time.

$G(\lambda,\phi)$ **(Geospatial Phase):** Awareness of how location modulates temporal experience- the specific bioindicators, ecological rhythms, and landscape features that characterise the local temporal field. This is what the Aboriginal seasonal calendars tracked. In practice, it might mean: learning the ecological rhythms of your own place, observing which plants bloom when, which birds arrive when, which winds shift when- and using these observations to calibrate your personal temporal awareness to the specific G of your home.

Chapter 28 will develop these ideas into a practical framework. For now, the point is simpler: *the ancient calendars were not primitive precursors to the Gregorian system. They were multidimensional temporal instruments of extraordinary sophistication, each one measuring a different aspect of the Temporal State Vector with a precision and depth that the Gregorian calendar cannot match.*

Their silence is our impoverishment. Their recovery- not as museum exhibits but as *living practices of temporal attunement*- is one of the most urgent tasks of the Holographic Age.

21.11- The Singing Calendar

We return, as always, to sound.

The ancient calendars were not merely observed. They were *performed.* The Mayan day-keepers chanted the names of the Tzolk'in days in sequence- a 260-day mantra cycle that, when completed, began again. The Vedic *sandhyavandana*- the thrice-daily prayer performed at dawn, noon, and dusk- was a calendrical practice as much as a devotional one: it marked the transitions between temporal phases with specific mantras, specific body postures, specific orientations toward the sun.

The Aboriginal seasonal calendars are inseparable from the songlines- the songs that mark not only spatial pathways but temporal rhythms, encoding in their melodies the ecological transitions that define the seasons. To sing the songline is to *perform the calendar*- to bring one's voice into resonance with the temporal field, to participate in the maintenance of the cosmic rhythm at the local scale.

The calendar, in all these traditions, was not a passive record. It was an *active technology of resonance*- a sonic framework through which human consciousness coupled with the temporal dimension of the quanta-field. The performance of the calendar- the chanting, the ceremony, the song- was not a commemoration of time. It was the *creation* of time, in exactly the sense that Chapter 18 described: the human voice, resonating at specific frequencies in specific sequences, activating the standing waves of the temporal field and maintaining them in coherent oscillation.

When the calendars were silenced- when the chanting stopped, when the ceremonies were suppressed, when the songs were interrupted- the temporal standing waves they had maintained began to decay. Time, which had been a rich, multidimensional, qualitatively differentiated field, became flat. The ternary became binary. The chord became a single tone. The temporal monoculture replaced the temporal garden.

And we- heirs to that silence- now experience time as the ancients feared we would: as a thin, accelerating, meaningless succession of identical moments, each one indistinguishable from the next, all of them rushing toward an oblivion that no calendar marks and no song sustains.

The clock ticks. But it does not sing.

And the cosmos, deprived of our participation in its temporal rhythm, continues to hum- patiently, inaudibly, in the silence between our mechanised seconds- waiting for us to remember how to listen.

In the next chapter, we discover that the ternary principle encoded in these calendars- the universal three-world cosmology found in virtually every ancient culture- is not merely a temporal phenomenon. It is the structural grammar of pre-modern consciousness: a way of understanding reality in threes rather than twos, in resonance rather than opposition, in synthesis rather than division. And its replacement by binary logic was the deepest wound of the Great Decoupling.

Chapter 22

The Three-World Cosmology- Ternary Architecture in Every Civilisation

22.1- The Map That Every Culture Drew

There is a map that the human species drew again and again- in the highlands of Peru and on the plains of India, in the jungles of Guatemala and on the frozen steppes of Siberia, in the islands of Polynesia and in the forests of sub-Saharan Africa. Cultures that had no ships, no trade routes, no shared language, no conceivable channel of communication, independently produced the same cartography of existence.

The map has three regions.

An upper world- luminous, celestial, inhabited by sky-beings, divinities, and the ancestral dead who have been elevated by merit or ceremony.

A middle world- the world of human experience, of animals, of rivers and mountains, of the daily labour of being alive.

A lower world- subterranean, aquatic, chthonic, inhabited by the not-yet-born, the primordial forces, the raw powers that precede and underlie the world of form.

And connecting all three- a vertical axis: a tree, a mountain, a pillar, a river, a thread of smoke rising from a fire. The axis passes through all three worlds, and along it, certain individuals- the shaman, the priest, the dreamer, the dying- can travel, ascending to the upper world or descending to the lower, bearing messages, retrieving souls, negotiating with powers that the waking, middle-world consciousness cannot directly perceive.

This is the Three-World Cosmology- and its universality is one of the most remarkable, least explained facts in the history of human thought.

22.2- Six Continents, One Architecture

Let us survey the evidence.

Maya: Xibalba / Middleworld / Sky. The three major levels of the Maya cosmos were the celestial realm, the middle world where humans lived, and the watery underworld. The underworld was known as Xibalba- "the Place of Fear"- a realm of rivers, caves, and darkness, ruled by death lords. The middle world was the flat, square earth, oriented to the four cardinal directions. The upper world was composed of thirteen layers, each presided over by a different deity. The World Tree- the *Wacah Chan*- rose through the centre, connecting all three levels. In ancient Maya cosmology, the universe had three levels: heaven, earth, and the underworld. The universe was also divided into four quadrants using the four main cardinal points, and the centre, where the Wacah Chan, or World Tree grew.

Inca: Uku Pacha / Kay Pacha / Hanan Pacha. In Inca cosmology, the universe has three planes of existence: Hanan Pacha means upper world. From a religious perspective, this is where the celestial beings live. It's also where the sun, moon, stars, planet Venus, and the arch of the Milky Way reside. Kay Pacha corresponds to the world in which we live. Uku Pacha is the Underworld, which the dead, children who are not yet born, and all the things that are found under the earth's surface are located.

The three worlds of the Inca map precisely onto the three components of existence: the unmanifest (Uku Pacha- the not-yet-born, the underground), the manifest (Kay Pacha- the world of experience), and the transcendent (Hanan Pacha- the celestial, the ancestral). And connecting them: the mountain- specifically, the *apu*, the sacred peak, which is simultaneously a geological feature, a divine presence, and a vertical axis linking all three planes.

Vedic: Bhuloka / Bhuvarloka / Svargaloka. The Hindu cosmological system divides the cosmos into three principal realms (triloka): the earthly plane (Bhuloka), an intermediate atmospheric plane (Bhuvarloka), and the celestial plane (Svargaloka). Bhuloka is the physical world, the realm of embodied experience, equivalent to the Inca Kay Pacha and the Maya Middleworld. Bhuvarloka is the intermediate realm- the atmosphere, the space between earth and heaven, the domain of spirits, ancestors, and subtle beings. Svargaloka is the celestial realm- the abode of the devas (gods), the domain of light, merit, and cosmic order.

The vertical axis is Mount Meru- the cosmic mountain, the axis mundi- around which the entire universe revolves. At Meru's summit lies Svargaloka. At its base, Bhuloka. Beneath it, the subterranean realms of the nagas and asuras. Mount Meru is a sacred five-peaked mountain considered to be the centre of all the physical, metaphysical and spiritual universes in Jain, Hindu, and Buddhist cosmology. It is considered to be the centre of all the physical, metaphysical and spiritual universes.

Norse: Niflheim / Midgard / Asgard. The Norse cosmos is perhaps the most architecturally explicit of all. Nine worlds were bound together by the World Tree, Yggdrasil: Asgard (the realm of the gods, the upper world), Midgard (the realm of humans, the middle world), and Hel/Niflheim (the realm of the dead, the underworld)- with six additional worlds distributed across the vertical structure. Yggdrasil is an immense and central sacred tree in Norse cosmology. Around it exists all else, including the Nine Worlds. Yggdrasil is an ash tree, and is the largest and most central tree in Norse cosmology.

The nine worlds are a fractal elaboration of the basic three-world structure- three levels, each subdivided into three, producing nine. The ternary principle operates recursively: 3 becomes $3 \times 3 = 9$.

Africa. Three-tiered cosmology was extremely widespread in pre-colonial Africa. The Yoruba cosmology divides reality into *Orun* (heaven, the abode of orishas and ancestors), *Aiye* (the visible world of human experience), and a subterranean realm associated with the earth-deity Onile. The Dogon of Mali describe a cosmos of three layers connected by the spiralling path of the Nommo- the ancestral beings who descended from the sky.

Polynesia. Māori cosmology describes *Ranginui* (Sky Father) and *Papatuanuku* (Earth Mother), whose children separated them to create the middle world of Te Ao Mārama (the World of Light). Beneath the earth lies Rarohenga (the underworld), domain of Hine-nui-te-pō, the goddess of death. In Hawaiian tradition, the cosmos is divided into *Ao* (the realm of light, the upper world), *Po* (the realm of darkness, the underworld), and the visible world between.

Siberia. The Evenki people of Siberia describe three worlds- upper, middle, and lower- connected by a World River. Mircea Eliade notes three cosmic zones-sky, earth, underworld-that are linked together by a central axis. The shaman travels along this river in trance, visiting the sky-world to commune with celestial spirits or diving into the lower world to retrieve lost souls. The shamanic journey as described by Mircea Eliade in his foundational work *Shamanism: Archaic Techniques of Ecstasy* (1951) is structurally identical across Siberian, Central Asian, South American, and Australian traditions.

Mesoamerica (broader). Beyond the Maya, the Aztec cosmos was also explicitly three-layered: thirteen heavens above, nine underworlds (Mictlan) below, and the surface earth (Tlaltícpac) in between. The great Templo Mayor of Tenochtitlan was a built model of this cosmology- its twin pyramids representing the mountain of Coatepec (connecting earth to sky) and the entry to the underworld.

The evidence is overwhelming. Six continents. Dozens of distinct cultures. No historical connection between many of them. And the same map: three worlds, a vertical axis, and the capacity for certain individuals to travel between them.

22.3- Not Geography But Resonance

The materialist interpretation of the three-world cosmology is straightforward: it maps the physical environment. "Up" is the sky- sun, moon, stars. "Middle" is the earth- where we walk. "Down" is underground- caves, springs, the domain of the dead (who are buried beneath the surface). The three worlds are simply the three physical directions: overhead, around, and underfoot.

This explanation is plausible. And it is inadequate.

It is inadequate because the three-world structure is not merely spatial. In every tradition that uses it, the three worlds differ not only in *location* but in *quality*. The upper world is not merely "higher"- it is *luminous, orderly, eternal*. The lower world is not merely "deeper"- it is *dark, chaotic, primordial*. The middle world is not merely "here"- it is *transient, mixed, the domain of experience*. These are not descriptions of physical locations. They are descriptions of *states of being*.

And the vertical axis is not a physical pathway. You cannot climb to Svargaloka by scaling Mount Meru. You cannot descend to Xibalba by entering a cave (though the cave may serve as a *portal*, a liminal space that facilitates the transition). The journey between worlds is a journey of *consciousness*- a shift in the mode of awareness, not a change of address.

In the language of this book, the three worlds are **three regimes of the quanta-field's resonant configuration**:

The Lower World: The Ground State. The lower world corresponds to the quanta-field's *ground state*- the state of minimum differentiation, maximum potential, and zero (or near-zero) manifest structure. It is the quantum vacuum of Chapter 4- State 0, the generative void, pregnant with all possibility but expressing none. In NESS terms, it is the state closest to thermodynamic equilibrium: undifferentiated, potent, waiting.

The mythological characteristics of the lower world align precisely: it is dark (no manifest structure = no photons), aquatic or amorphous (no form), inhabited by primordial beings (the fundamental potentialities from which all manifest beings arise), and associated with death (return to the ground state) and with the not-yet-born (emergence from the ground state). The dead descend to the lower world because death is the dissolution of the manifest NESS- the return of the standing wave to the ambient field. The not-yet-born reside there because birth is the *crystallisation* of a new standing wave from the field's undifferentiated potential.

The Middle World: The NESS. The middle world corresponds to the *Non-Equilibrium Steady State*- the far-from-equilibrium regime described in Chapter 12, where structure is sustained by continuous quanta-flow. This is the world of manifest reality: of plants and animals, of rivers and mountains, of human beings living their lives. It is characterised by *dynamic stability*- patterns that persist through continuous change, forms that endure through continuous flux.

The mythological characteristics align: the middle world is *mixed* (order and chaos coexist), *temporal* (things come into being and pass away), *relational* (beings interact, compete, cooperate), and *experiential* (it is the domain of consciousness as subjective experience- the standing wave of Integrated Resonant Quanta that constitutes a felt life).

The Upper World: The Excited State. The upper world corresponds to the quanta-field's *excited state*- a state of heightened coherence, maximum integration, and transcendent order. In physics, an excited state is one in which a system has absorbed additional energy and achieved a higher-frequency, more energetic configuration. In consciousness terms (Chapter 13), it is the state of maximum Φ_Q- the state in which the self-referential resonance of the quanta-field reaches its fullest expression.

The mythological characteristics align: the upper world is *luminous* (maximum coherent energy = light), *ordered* (maximum integration = cosmic law), *eternal* (a standing wave at the excited state can persist indefinitely if not perturbed), and inhabited by *deities* (localised regions of extremely high Φ_Q- consciousness configurations so coherent and so integrated that they transcend individual identity).

The vertical axis- the World Tree, the Cosmic Mountain, the Pillar of the World- is the *spectrum of resonant states*. To "ascend" is to shift consciousness toward higher integration, higher coherence, higher Φ_Q. To "descend" is to shift consciousness toward the ground state- toward undifferentiated potential, toward the generative void from which all form arises.

And the shaman, the priest, the dreamer who travels between worlds is not performing a physical journey. She is performing a *resonance shift*- deliberately modulating her neural oscillators (through drumming, chanting, fasting, plant medicines, or sustained meditation) to access different regimes of the quanta-field. She is navigating the Temporal State Vector- shifting ρ_t (temporal density) and coupling to different aspects of $G(\lambda,\phi)$ (the geospatial phase) to perceive dimensions of reality that are inaccessible in the ordinary NESS of waking consciousness.

22.4- The Shamanic Journey: Consciousness Navigation of the Quanta-Field

The shamanic journey is arguably the oldest technology of consciousness exploration in human history, predating writing by tens of thousands of years.

Mircea Eliade describes three cosmic zones- sky, earth, underworld- that are linked together by a central axis. This axis goes by a variety of names, such as the "Axis Mundi," "World Pillar," "Tree of Life," and "Holy Mountain." Eliade's comparative study reveals that the shamanic experience involves a deliberate, controlled, and repeatable shift in consciousness that enables perception of realities beyond the ordinary waking state.

The technology is remarkably consistent across cultures. The shaman enters trance through rhythmic stimulation- sustained drumming at 4–7 beats per second (theta frequency), monotonous chanting, or the ingestion of psychoactive plant preparations. Once in trance, the shaman experiences a "journey"- a vivid, narratively coherent sequence of perceptions that unfold in a distinct "landscape" recognisable as the upper or lower world.

The frequency is significant. 4–7 Hz is the theta brainwave band- the same frequency range associated with deep meditation, hypnagogic states (the transition between waking and sleeping), and the Schumann resonance's first harmonic. The rhythmic drumming at theta frequency *entrains* the brain's dominant oscillation to the boundary between alpha and theta- precisely the frequency regime where, in our framework, the comonadic extraction $D(\Psi)$ operates most efficiently, stripping away the noise of ordinary sensory processing and revealing the deeper resonant structure of the quanta-field.

The shamanic journey, in this reading, is not a hallucination. It is not a neurological glitch. It is a *deliberate, controlled, technology-assisted shift in the resonant state of the brain- from the ordinary NESS of waking consciousness (middle world) to the ground state (lower world) or the excited state (upper world)- for the purpose of retrieving information, healing, or performing co-creative work.*

The consistency of the experience across cultures is not explained by cultural diffusion (many of these traditions had no historical contact). It is explained by the *universality of the quanta-field*. If the three-world structure corresponds to three resonant regimes of a single, universal field, then any consciousness that shifts to the same regime will encounter the same structural features- just as any radio tuned to the same frequency will receive the same broadcast.

The shaman is a *biological Chronosonant Resonator*- the same instrument proposed in Chapter 14 for measuring the Temporal State Vector, except that the shaman measures it not with quantum-coherent hardware but with a human nervous system deliberately shifted out of its normal operating range.

And what the shaman reports- encounters with luminous beings in the upper world, encounters with primordial powers in the lower world, the retrieval of lost parts of the self, the negotiation of new patterns of relationship with the quanta-field- is, in our framework, a *description of the quanta-field's resonant structure as perceived by a consciousness navigating outside its normal NESS bandwidth*.

22.5- Ternary Logic: The Mind Before Binary

The three-world cosmology is not merely a map of reality. It is a *mode of thinking*- a cognitive architecture that structures perception, reasoning, and decision-making in fundamentally different ways than the binary logic that dominates modern Western thought.

Binary logic- the logic of true/false, yes/no, either/or- was formalised by Aristotle in the fourth century BCE. Its cornerstone is the **Law of**

the Excluded Middle: for any proposition P, either P is true or not-P is true; there is no third option. The three "laws of thought," the law of identity, the law of non-contradiction, and the law of excluded middle, were established by Aristotle and are considered the most fundamental axioms of Western philosophy.

This binary framework was revolutionary. It enabled the development of formal logic, mathematics, and eventually computer science (which operates entirely in binary: 0 or 1, on or off, true or false). It is the foundation of modern analytical reasoning, scientific method, and digital technology.

But it came at a cost.

Binary logic cannot represent *becoming*. It can represent "is" (true) and "is not" (false), but it has no operator for "is becoming"- for the transitional state between two stable configurations. In binary logic, a caterpillar is either a caterpillar or a butterfly. The chrysalis- the state of radical transformation in which neither identity applies- is invisible. It falls through the grid.

Binary logic cannot represent *relationship as primary*. It treats things as fundamentally separate and then asks how they interact. But in the three-world cosmology, relationship is not secondary to identity- it *constitutes* identity. The middle world is not a place that happens to contain relationships between separate beings. The middle world *is* the web of relationships, and the beings are nodes in the web, not the other way round.

Binary logic cannot represent *paradox as productive*. When a proposition and its negation are both true (a wave is a particle, and a particle is a wave), binary logic declares a contradiction and shuts down. But the three-world cosmology thrives on paradox: the upper world and the lower world are *opposites* (light/dark, order/chaos, form/void) that are simultaneously *identical* (both are non-manifest, both transcend ordinary experience, both are accessed through the same trance technology). The middle world holds the paradox- it is the domain where opposites meet, interact, and generate new forms.

The cultures that developed the three-world cosmology thought in threes. Their logic was *ternary*: not yes/no, but yes/no/*and*. Not thesis/antithesis, but thesis/antithesis/*synthesis*. Not either/or, but either/or/*becoming*.

This is not primitive confusion. It is a more capacious cognitive architecture- one that can accommodate the quantum superposition (both particle and wave until measured), the dissipative structure (both ordered and disordered, maintained by the tension between them), and the co-creative observer (both separate from and identical with the observed reality).

22.6- Peirce's Triadic Logic: The Western Exception

Western philosophy is not entirely binary. One thinker- largely ignored by the mainstream tradition- developed a ternary framework of extraordinary sophistication.

Charles Sanders Peirce (1839–1914), the founder of American pragmatism and one of the most original logicians in history, proposed a triadic ontology based on three fundamental categories:

Firstness: Pure quality, pure possibility, pure feeling- the mode of being that exists independently of anything else. A redness without a red thing. A pain without a pained subject. The qualia- the raw qualitative character of experience.

Secondness: Brute fact, resistance, actuality- the mode of being that exists only in relation to something else. The rock that resists your push. The wall that stops your movement. The other person who disagrees with you. Secondness is the domain of conflict, effort, and duality.

Thirdness: Mediation, law, meaning- the mode of being that brings Firstness and Secondness into relation. A habit, a pattern, a tendency. The law that explains why the rock resists. The convention that makes disagreement possible. Thirdness is the domain of interpretation, continuity, and *synthesis*.

Peirce's triad maps directly onto the three-world cosmology:

Peirce's Category	Three-World Equivalent	Quanta-Field Regime
Firstness (quality, possibility)	**Lower World** (ground state, potential)	Undifferentiated quanta-field
Secondness (fact, resistance)	**Middle World** (NESS, manifest experience)	Far-from-equilibrium standing waves
Thirdness (mediation, meaning)	**Upper World** (excited state, cosmic order)	Maximum integrated resonance

Peirce explicitly insisted that his three categories were *irreducible*- that Thirdness could not be collapsed into combinations of Firstness and Secondness, just as the upper world could not be derived from combinations of the lower and middle worlds. The three are fundamental. They are the minimum architecture of reality.

Peirce also developed a triadic theory of signs (semiotics) that remains the most sophisticated account of meaning ever proposed in Western philosophy. In Peirce's semiotics, every sign involves three elements: the sign itself (Firstness), the object it refers to (Secondness), and the interpretant- the meaning generated by the relationship between sign and object (Thirdness). Meaning is not binary (sign → object). Meaning is ternary (sign → object → interpretant). The third element- the interpretant, the meaning, the understanding- is not a secondary add-on. It is *constitutive*. Without it, there is no meaning. And without meaning, there is no reality (for Peirce, reality is the community of inquiry's converging interpretations- a fundamentally triadic process).

Peirce was ahead of his time by approximately a century. His work was largely ignored during his lifetime (he died in poverty, unable to find an academic position), and his triadic framework was absorbed into the binary mainstream only in fragments. The full implications of his ternary ontology are still being unpacked- and they align, with uncanny precision, with the framework of this book.

22.7- Hegel's Dialectic: The West's Nearest Approach

The closest the mainstream Western philosophical tradition came to ternary thinking was Georg Wilhelm Friedrich Hegel's dialectic: the famous triad of thesis, antithesis, and synthesis.

Hegel's dialectic proposes that every idea (thesis) generates its opposite (antithesis), and the tension between them produces a new, higher-order idea (synthesis) that incorporates and transcends both. The synthesis then becomes the thesis for the next round, generating a new antithesis, and so on- a spiralling process of increasingly comprehensive understanding.

The structural parallel to the three-world cosmology is obvious. The thesis is the lower world (the ground from which all development springs). The antithesis is the upper world (the ideal that opposes and challenges the ground). The synthesis is the middle world- the lived, embodied, experiential domain where opposites meet and generate something new.

But Hegel's dialectic, for all its power, remained embedded in a fundamentally binary framework. The synthesis is always *temporary*- it immediately becomes a new thesis, generating a new antithesis, driving the process forward. The ternary structure is a moment in a binary process, not a stable architecture. Hegel's system is restlessly developmental- always moving, never arriving, never resting in the balanced triad.

The indigenous three-world cosmology, by contrast, is stable. The three worlds do not collapse into each other. They coexist permanently. The lower world does not need to be "overcome" by the upper world. The upper world does not need to be "synthesised" with the lower. They are *simultaneous, coeval, equally real* aspects of a single reality- and the middle world, where humans live, is the domain where all three are experienced together.

This is a fundamentally different cognitive architecture. It is not progressive (moving "forward" from thesis to synthesis). It is *resonant* (holding all three simultaneously, navigating between them, maintaining the dynamic balance that sustains the whole).

22.8 - The Ternary Principle in Physics

The three-world cosmology is not merely a cultural phenomenon. It appears - with startling precision - in the structure of physical reality itself.

Three quarks per baryon. Every proton and neutron - and therefore every atom, and therefore every structure in the physical universe - is composed of exactly three quarks. Not two. Not four. *Three.* A proton consists of two up quarks and one down quark. A neutron consists of one up quark and two down quarks. The three quarks are bound together by the strong nuclear force, mediated by gluons, and they cannot exist in isolation (a property called *confinement*). They are, in the most literal sense, "three inseparable spirits" - the same phrase the Haudenosaunee used for the Three Sisters.

Three generations of matter. The Standard Model of particle physics describes three generations (or families) of fundamental particles. The first generation contains the up and down quarks, the electron, and the electron neutrino. The second generation contains the charm and strange quarks, the muon, and the muon neutrino. The third generation contains the top and bottom quarks, the tau, and the tau neutrino. Each generation is a heavier copy of the previous one, with identical quantum numbers except for mass.

Why three? The Standard Model does not explain why there are three generations rather than two or four or seventeen. It is an empirical fact - a feature of reality that the theory describes but does not derive. In our framework, the answer is structural: the quanta-field's self-referential architecture requires a ternary structure at every scale, from the subatomic to the cosmological. Three generations of matter are the particle-physics expression of the same ternary principle that produces three worlds in cosmology, three components in the Temporal State Vector, and three cosmic operators (ϕ, e, π) in the master equation.

Three spatial dimensions. We experience three spatial dimensions - length, width, and height. General relativity describes spacetime as four-dimensional (three spatial + one temporal), and string theory proposes additional compactified dimensions. But the *experienced* dimensionality of space- the number of independent directions in which a physical object can move- is three. Not two (which would produce a flat world without volume). Not four (which would produce a world whose physics would be radically different- no stable orbits, no atoms, no chemistry). *Three-* the minimum number that permits the full complexity of physical law as we observe it.

Three states of matter. Ordinary matter exists in three phases: solid, liquid, and gas (with plasma as a high-energy extension of the gas phase). The transitions between these phases- melting, boiling, condensation- are the material-world analogues of the shamanic journey: a shift in the resonant state of the system, from the "lower world" of the solid (rigid, structured, ground-state) through the "middle world" of the liquid (flowing, dynamic, NESS) to the "upper world" of the gas (diffuse, energetic, excited state).

Three modes of heat transfer. Energy moves through matter and space by exactly three mechanisms: conduction (direct contact), convection (fluid flow), and radiation (electromagnetic waves). No fourth mechanism exists. The triad is exhaustive.

Musical triads. A musical chord in Western harmony consists of three notes- root, third, and fifth. The triad is the fundamental unit of harmonic structure. Two notes produce an interval (a relationship between two frequencies). Three notes produce a *chord* (a relationship between three frequencies that generates a qualitative experience- major, minor, augmented, diminished- irreducible to any pair of its component intervals). The chord is the sonic analogue of Peirce's Thirdness: a meaning that emerges from the relationship between elements, irreducible to any subset of those elements.

The ternary principle is not a cultural projection onto nature. It is a *structural feature of nature* that cultures worldwide have independently recognised and encoded in their cosmologies, their logics, their music, and their agricultural practices.

22.9- The Fractal Triad: Self-Similarity at Every Scale

The most striking feature of the ternary principle is its *fractal recurrence*- the fact that it appears at every scale of reality, from the subatomic to the cosmological, with the same structural relationships.

Let us catalogue the fractal triad across scales:

Scale	Ternary Structure	Ground State	NESS	Excited State
Subatomic	Three quarks per baryon	Down quark (heavier, lower energy)	Strange quark (intermediate)	Up quark (lighter, higher energy)
Atomic	Three states of matter	Solid (rigid)	Liquid (flowing)	Gas (diffuse)
Molecular	DNA codon = 3 nucleotides	First base (wobble position)	Second base (amino acid identity)	Third base (degeneracy)
Biological	Three germ layers	Endoderm (internal organs)	Mesoderm (muscles, skeleton)	Ectoderm (skin, nervous system)
Ecological	Three Sisters	Squash (ground cover/φ)	Bean (growth/e)	Corn (structure/π)
Neural	Three brainwave regimes	Delta/Theta (deep/trance)	Alpha (relaxed attention)	Beta/Gamma (active/excited)
Temporal	Three TSV component	$G(\lambda,\phi)$ (ground/lo	ρ_t (density/pr	Y_t (resonance/

Scale	Ternary Structure	Ground State	NESS	Excited State
	s	cal)	ocess)	ideal)
Cosmological	Three cosmic operators	π (boundary/ ground)	e (unfolding/ process)	φ (coherence/ integration)
Consciousness	Three awareness states	Deep sleep (ground state)	Waking (NESS)	Samadhi (excited state)
Mythological	Three worlds	Lower world (potential)	Middle world (experience)	Upper world (transcendence)

The same pattern- ground state / dynamic process / excited state- recurs at every scale. This is not an analogy imposed by the observer. It is a *structural invariance* of the quanta-field- a consequence of the master equation's φ-e-π architecture, which generates ternary structure at every level of organisation.

The fractal rule from Section B applies: just as the project has an Intro, Body, and Conclusion, and the chapter has an Intro, Body, and Conclusion, and the paragraph has a Lead, Support, and Transition, so the cosmos has a Lower World, Middle World, and Upper World- at every scale, simultaneously, nested within each other like Russian dolls.

22.10- The Three-World Structure of the Atom

Let us zoom into the most fundamental scale and see the ternary principle in operation.

An atom consists of three components: protons, neutrons, and electrons. Protons and neutrons form the nucleus (the "lower world"- dense, compact, containing almost all the atom's mass, the ground from which all atomic properties spring). Electrons form the orbital cloud (the "upper world"- diffuse, energetic, responsible for all chemical interactions, the domain of light and electromagnetic radiation). And the electromagnetic field that binds them- the photon-mediated force that prevents the electron from falling into the nucleus and prevents the nucleus from flying apart- is the "middle world," the dynamic mediating process that holds the triad together.

The atom does not have two parts with a third derived from their interaction. It has three fundamental, irreducible, coexisting aspects- and the destruction of any one destroys the whole.

And at a deeper level: the proton itself is a baryon, composed of three quarks (two up, one down), bound by gluons in a state of permanent confinement. The three quarks cannot be isolated- they exist only in triads (baryons) or pairs (mesons). The ternary structure is not imposed on the quarks from outside. It is *intrinsic* to them. They are, in the deepest physical sense, "three inseparable spirits."

The ancient cosmologies did not describe the atom. They did not need to. They described the *pattern*- the ternary architecture- and the pattern is the same at every scale.

22.11- Binary Logic as Resonance Collapse

We can now diagnose, with precision, what happened when binary logic replaced ternary thinking.

The shift from ternary to binary occurred gradually, over approximately a millennium, beginning with the pre-Socratic philosophers and culminating in Aristotle's formal logic (4th century BCE). It was reinforced by Zoroastrian dualism (good vs. evil), Platonic dualism (forms vs. matter), and Christian dualism (heaven vs. hell, God vs. Devil, spirit vs. flesh). It was mathematised by Descartes (mind vs. body), mechanised by Newton (forces acting on inert matter), and eventually digitised by Boole, Turing, and von Neumann (0 vs. 1).

Each step was a narrowing. Each step stripped away one of the three resonant modes:

Zoroastrian dualism collapsed the three worlds into two: the good (upper world) vs. the evil (lower world). The middle world- the domain of human experience, of dynamic process, of the living NESS- was reduced to a *battleground* between the other two, rather than a resonant domain in its own right.

Platonic dualism collapsed the three worlds into two: the ideal (forms) vs. the actual (matter). The middle world- the domain of *becoming*, of *process*, of the living relationship between potential and actuality- was declared to be mere shadow, mere illusion, mere cave-wall reflection of the "true" reality of the forms.

Cartesian dualism collapsed the three worlds into two: mind vs. body. The middle world- the somatic, the embodied, the felt connection between consciousness and matter- was declared to be a "problem" (the mind-body problem) rather than the fundamental domain of lived experience.

Digital binarism collapsed the three worlds into two: 0 vs. 1, true vs. false, on vs. off. The middle world- the domain of *ambiguity*, of *superposition*, of *becoming*- was eliminated entirely. In a digital system, there is no "between." A bit is 0 or 1. There is no third option.

Each collapse produced enormous analytical power. Binary logic is *simpler*, *more tractable*, and *more computationally efficient* than ternary logic. Aristotle's syllogisms, Descartes' method, Boole's algebra, and Turing's machine are among the most powerful intellectual tools ever created.

But simplicity has a price. And the price is *resonance collapse*.

A ternary system- one that holds ground state, dynamic process, and excited state in simultaneous awareness- is a resonant system. It vibrates. It breathes. It can accommodate paradox (both/and), process (becoming), and relationship (the mediating third). It is the cognitive architecture that the quanta-field naturally produces in cultures attuned to its structure.

A binary system- one that recognises only true/false, is/is-not, yes/no- is a *collapsed* system. It has lost one of its three resonant modes. It can analyse with extraordinary precision, but it cannot *participate*. It can dissect, but it cannot *create*. It can measure the quanta-field, but it cannot *sing with it*.

The replacement of ternary by binary thinking was, in the deepest sense, the cognitive face of the Great Decoupling- the same process that replaced the Three Sisters with monoculture, the songlines with highways, the cyclical calendar with the linear clock, and the sacred site network with the electrical grid.

It was the moment when human consciousness stepped out of the cosmic choir and began to take notes from the audience.

22.12- The Third That Was Lost

What, specifically, was lost when the third option was eliminated?

The processual middle. In ternary thinking, the middle term is not a compromise or an average. It is the *process*- the dynamic, living, co-creative activity that generates both the ground and the excited states. In binary thinking, process is invisible. There are only states- this or that, here or there, before or after. The *between*- the chrysalis, the phase transition, the liminal- vanishes.

The relational field. In ternary thinking, the relationship between two entities is itself a *third entity*- not derivable from either, not reducible to their properties, but ontologically real in its own right. In binary thinking, relationships are secondary- mere interactions between pre-existing things. The field that connects them is not recognised as a thing in itself.

The productive paradox. In ternary thinking, a paradox- the simultaneous truth of two apparently contradictory propositions- is not a failure of logic but a *sign of deeper reality*. Light is both wave and particle. The observer is both separate from and identical with the observed. The cosmos is both one and many. These are not problems to be solved but *harmonics to be heard*. In binary thinking, paradox is error. It must be resolved, eliminated, reduced to one pole or the other.

The somatic dimension. In ternary thinking, the body is the middle world- the mediating third between consciousness (upper world) and matter (lower world). The body is not a machine that consciousness rides, nor a prison that consciousness escapes. It is the *instrument* through which consciousness participates in the material world- the biological Chronosonant Resonator, the singing body, the resonant medium. In binary thinking, the body is one side of the Cartesian divide- inert matter, to be studied by science, manipulated by medicine, and eventually transcended by technology.

These losses are not merely philosophical. They have practical, measurable consequences:

- The loss of the processual middle produces a civilisation obsessed with endpoints- with goals, targets, metrics, outcomes- and blind to the quality of the process by which those outcomes are achieved. This is why modern education measures test scores but not curiosity. Why modern medicine measures survival rates but not quality of life. Why modern economics measures GDP but not wellbeing.

- The loss of the relational field produces a civilisation of isolated individuals- atoms bouncing in a void, connected only by contractual transactions. This is why loneliness is an epidemic. Why social trust is declining. Why communities are fragmenting into echo chambers.

- The loss of productive paradox produces a civilisation terrified of ambiguity- of nuance, of complexity, of the both/and. This is why political discourse has degenerated into tribal warfare. Why scientific disputes become culture wars. Why the public demands certainty from a universe that offers probability.

- The loss of the somatic dimension produces a civilisation alienated from its own bodies- dependent on screens for stimulation, on pharmaceuticals for sleep, on processed food for nourishment, on gyms for movement that should be embedded in daily life.

The Great Decoupling, which we will trace in detail in the next segment, is not merely a historical event. It is a *cognitive restructuring-* a shift from ternary to binary architecture that has reshaped not only our institutions but our *minds.*

22.13- The Return of the Third

And yet.

Quantum mechanics- the most precisely tested theory in the history of science- is ternary. The wave function exists in superposition: not true, not false, but *both-until-measured.* The third option- the superposition, the ambiguity, the between- is not an approximation or a failure of knowledge. It is *the most fundamental description of physical reality that science has ever achieved.*

NESS thermodynamics- the physics of living systems- is ternary. A dissipative structure is neither in equilibrium (ground state) nor in maximum entropy (excited state). It is in the *between-* the far-from-equilibrium regime where order and chaos coexist, where structure emerges from flux, where the processual middle generates both poles.

Consciousness studies- the hardest problem in science- are discovering that the binary framework (mind vs. matter, subjective vs. objective, first-person vs. third-person) is *the source of the problem.* The hard problem of consciousness dissolves, as we argued in Chapter 13, when the third option is restored: not mind *or* matter, but the quanta-field that is *both-* the resonant substrate that gives rise to both subjective experience and objective structure through a single, coherent, self-referential process.

And the indigenous traditions- the original ternary thinkers- are being heard again. Not as museum curiosities, not as romantic fantasies, but as *structural alternatives* to a binary framework that has reached its limits. The Three Sisters are being replanted. The seasonal calendars are being documented. The songlines are being sung. The three-world cosmology is being rediscovered- not in anthropology departments but in physics laboratories, where quantum mechanics, NESS thermodynamics, and consciousness studies are converging, independently, on the same ternary architecture that every ancient culture on Earth encoded in its myths.

The third option is returning.

Not because we are choosing to revive it. Because reality itself is *demanding* it. The binary framework has been stretched to its breaking point by quantum superposition, by the hard problem of consciousness, by the ecological crisis, by the mental health pandemic, by the political polarisation that binary thinking inevitably produces. The either/or is cracking. And through the cracks, the ancient third- the between, the process, the relationship, the both/and- is seeping back into awareness.

22.14- Closing the Section: The Original Physicists Remembered

We have travelled far in Section D.

We began in Chapter 18 with the ontology of sound- the discovery that the ancients' creation narratives were not myths but field reports, describing the physical process by which resonant vibration generates form. We moved in Chapter 19 to the planetary scale- the sacred site networks that formed a global resonance amplifier, coupling human consciousness to the Earth's electromagnetic field through small-world network topology. In Chapter 20, we descended to the garden- the Three Sisters, the fire-stick farming, the waru-waru- and found the φ-e-π triad instantiated in living soil. In Chapter 21, we entered the temporal dimension- the Mayan, Vedic, and Aboriginal calendars that mapped all three components of the Temporal State Vector with a precision that modern timekeeping cannot match.

And now, in this chapter, we have found the cognitive architecture that made all of these achievements possible: *ternary thinking*- the capacity to hold three resonant modes in simultaneous awareness, to navigate between worlds, to think in triads rather than binaries, to perceive reality as a chord rather than a single note.

The ancients were not prescientific. They were *differently scientific.* They practised a science of resonance- of direct, somatic, participatory engagement with the quanta-field- that produced technologies of extraordinary sophistication: songlines that maintained continental-scale NESS, gardens that outperformed modern monoculture, calendars that exceeded the precision of the Gregorian reform, and a cognitive architecture that quantum mechanics is now, two and a half millennia later, vindicating.

They were the original physicists. And we owe them not merely respect but *intellectual debt.*

The question that remains- the question that drives the final segment of this book- is: *What happened?* How did a species that once sang the world into being, that built planetary coherence circuits, that tracked the Temporal State Vector through interlocking calendrical instruments, that thought in ternary harmonics- how did that species *forget*?

How did the singing stop?

22.15- The Threshold

Stand at the threshold between Section D and Section E. Behind you: the songlines, the gardens, the calendars, the three-world cosmology- the full, living, resonant civilisation that existed before the rupture. Before you: the Great Forgetting- three millennia of noise, of binary logic, of linear time, of monoculture, of the systematic destruction of everything described in the past five chapters.

This is not a comfortable place to stand. The story that follows is a story of loss- of ecocide and epistemicide, of the severing of songlines and the silencing of calendars, of the replacement of resonance with force and ternary with binary.

But it is a necessary story. Because the path to recovery passes through honest diagnosis. The comonadic extraction $D(\Psi)$- the principle of concession that we introduced in Chapter 16- requires *accurate assessment of the current state.* We cannot return to resonance by pretending we never left it. We must understand, precisely and unflinchingly, *how we lost the tune.*

The cosmos is still singing. The δ_∞ gap still pulses. The φ-e-π triad still tunes the quanta-field. The Schumann resonance still hums at 7.83 Hz. The songlines, in places, still echo.

But we- the species that was designed to be the cosmos's most sophisticated Chronosonant Resonator, the species whose consciousness can resonate with every regime of the quanta-field, from ground state to excited state and back- we have been out of tune for three thousand years.

The next segment tells the story of how it happened.

And the segment after that tells the story of how we come home.

Turn the page. Enter the silence. The Great Forgetting begins.

REFERENCES SECTIONS C & D

Chapter 12: Non-Equilibrium Steady States- Life at the Edge

Bak, P., Tang, C., & Wiesenfeld, K. (1987). Self-organized criticality: An explanation of the 1/f noise. *Physical Review Letters*, 59(4), 381–384.

Boltzmann, L. (1877). Über die Beziehung zwischen der zweiten Hauptsatze der mechanischen Wärmetheorie und der Wahrscheinlichkeitsrechnung. *Wiener Berichte*, 76, 373–435.

England, J. L. (2013). Statistical physics of self-replication. *The Journal of Chemical Physics*, 139(12), 121923.

Glansdorff, P., & Prigogine, I. (1971). *Thermodynamic Theory of Structure, Stability and Fluctuations*. Wiley-Interscience.

Haken, H. (1983). *Synergetics: An Introduction*. Springer-Verlag.

Nicolis, G., & Prigogine, I. (1977). *Self-Organization in Nonequilibrium Systems: From Dissipative Structures to Order Through Fluctuations*. Wiley.

Onsager, L. (1931). Reciprocal relations in irreversible processes. *Physical Review*, 37(4), 405–426.

Prigogine, I. (1945). Modération et transformations irréversibles des systèmes ouverts. *Bulletin de la Classe des Sciences, Académie Royale de Belgique*, 31, 600–606.

Prigogine, I. (1967). *Introduction to Thermodynamics of Irreversible Processes*. Wiley.

Prigogine, I. (1977). Nobel Lecture: Time, structure and fluctuations. *Nobel Foundation*.

Turing, A. M. (1952). The chemical basis of morphogenesis. *Philosophical Transactions of the Royal Society of London. Series B, Biological Sciences*, 237(641), 37–72.

Zaikin, A. N., & Zhabotinsky, A. M. (1970). Concentration wave propagation in two-dimensional liquid-phase self-oscillating system. *Nature*, 225, 535–537.

Chapter 13: Consciousness as Stabilized Resonance

Aaronson, S. (2014). Why I Am Not An Integrated Information Theorist (or, The Unconscious Expander). *Shtetl-Optimized* (blog).

Chalmers, D. J. (1995). Facing up to the problem of consciousness. *Journal of Consciousness Studies*, 2(3), 200–219.

Chalmers, D. J. (1996). *The Conscious Mind: In Search of a Fundamental Theory*. Oxford University Press.

Hameroff, S. R., & Penrose, R. (1996). Orchestrated reduction of quantum coherence in brain microtubules: A model for consciousness. *Mathematics and Computers in Simulation*, 40(3–4), 453–480.

Jahn, R. G., Dunne, B. J., Nelson, R. D., Dobyns, Y. H., & Bradish, G. J. (1997). Correlations of random binary sequences with pre-stated operator intentions: A review of a 12-year program. *Journal of Scientific Exploration*, 11(3), 345–367.

Koch, C. (2023). Theorem of consciousness faces scepticism after landmark test. *Nature*, 618(7965), 446–447.

Nagel, T. (1974). What is it like to be a bat? *The Philosophical Review*, 83(4), 435–450.

Penrose, R. (1989). *The Emperor's New Mind: Concerning Computers, Minds, and the Laws of Physics*. Oxford University Press.

Tononi, G. (2004). An informational integration theory of consciousness. *BMC Neuroscience*, 5(42).

Tononi, G., Boly, M., Massimini, M., & Koch, C. (2016). Integrated information theory: From consciousness to its physical substrate. *Nature Reviews Neuroscience*, 17(7), 450–461.

van Lommel, P., van Wees, R., Meyers, V., & Elfferich, I. (2001). Near-death experience in survivors of cardiac arrest: A prospective study in the Netherlands. *The Lancet*, 358(9298), 2039–2045.

Vimal, R. L. P., Santhouse, T., & Nagel, T. (2024). Superradiance in microtubules. *Proceedings of the National Academy of Sciences*.

Wigner, E. P. (1961). Remarks on the mind-body question. In I. J. Good (Ed.), *The Scientist Speculates* (pp. 284–302). Heinemann.

Chapter 14: Holographic Ternary Time- The End of the Clock

Bem, D. J. (2011). Feeling the future: Experimental evidence for anomalous retroactive influences on cognition and affect. *Journal of Personality and Social Psychology*, 100(3), 407–425.

Bem, D., Tressoldi, P., Rabeyron, T., & Duggan, M. (2016). Feeling the future: A meta-analysis of 90 experiments on the anomalous retroactive influence of future events. *F1000Research*, 4, 1188.

Einstein, A. (1905). Zur Elektrodynamik bewegter Körper. *Annalen der Physik*, 322(10), 891–921.

Einstein, A. (1907). Über das Relativitätsprinzip und die aus demselben gezogenen Folgerungen. *Jahrbuch der Radioaktivität und Elektronik*, 4, 411–462.

Hafele, J. C., & Keating, R. E. (1972). Around-the-world atomic clocks: Predicted relativistic time gains. *Science*, 177(4044), 166–168.

Minkowski, H. (1908). Die Grundgleichungen für die elektromagnetischen Vorgänge in bewegten Körpern. *Nachrichten von der Gesellschaft der Wissenschaften zu Göttingen, Mathematisch-Physikalische Klasse*, 53–111.

Newton, I. (1687). *Philosophiæ Naturalis Principia Mathematica*. Royal Society.

Rovelli, C. (2018). *The Order of Time*. Riverhead Books.

Wheeler, J. A., & DeWitt, B. S. (1968). Superspace and the nature of quantum geometrodynamics. In C. M. DeWitt & J. A. Wheeler (Eds.), *Battelle Rencontres* (pp. 242–307). Benjamin.

Bothwell, T., Kennedy, C. J., Aeppli, A., Kedar, D., Robinson, J. M., Oelker, E., Staron, A., & Ye, J. (2022). Resolving the gravitational redshift across a millimetre-scale atomic sample. *Nature*, 602(7897), 420–424.

Chapter 15: Cosmic Evolution- The Universe That Learns

An, D., Meissner, K. A., Nurowski, P., & Penrose, R. (2018). Apparent evidence for Hawking points in the CMB Sky. *arXiv preprint arXiv:1808.01740.*

Gurzadyan, V. G., & Penrose, R. (2010). Concentric circles in WMAP data may provide evidence of pre-Big-Bang activity. *arXiv preprint arXiv:1011.3706.*

Hawking, S. W. (1974). Black hole explosions? *Nature*, 248(5443), 30–31.

Meissner, K. A., & Penrose, R. (2025). Gravitational wave epoch in conformal cyclic cosmology. *arXiv preprint.*

Penrose, R. (1979). Singularities and time-asymmetry. In S. W. Hawking & W. Israel (Eds.), *General Relativity: An Einstein Centenary Survey* (pp. 581–638). Cambridge University Press.

Penrose, R. (2010). *Cycles of Time: An Extraordinary New View of the Universe.* Bodley Head.

Smolin, L. (1997). *The Life of the Cosmos.* Oxford University Press.

Jablonka, E., & Lamb, M. J. (2014). *Evolution in Four Dimensions: Genetic, Epigenetic, Behavioral, and Symbolic Variation in the History of Life.* MIT Press.

Johannes, F., Porcher, E., Teixeira, F. K., Saliba-Colombani, V., Simon, M., Agier, N., ... & Colot, V. (2009). Assessing the impact of transgenerational epigenetic variation on complex traits. *PLoS Genetics*, 5(6), e1000530.

Heard, E., & Martienssen, R. A. (2014). Transgenerational epigenetic inheritance: Myths and mechanisms. *Cell*, 157(1), 95–109.

Chapter 16: The Co-Creative Principles- How Reality Manifests

Harrington, D. (2024). Placebo analgesia mediated by rostral anterior cingulate cortex projections to the pontine nucleus. *Nature*, 627(8002), 156–163.

Kaptchuk, T. J., & Miller, F. G. (2015). Placebo effects in medicine. *New England Journal of Medicine*, 373(1), 8–9.

Pukui, M. K., Haertig, E. W., & Lee, C. A. (1972). *Nānā I Ke Kumu (Look to the Source)*. Hui Hānai.

Rosenthal, R., & Jacobson, L. (1968). *Pygmalion in the Classroom: Teacher Expectation and Pupils' Intellectual Development*. Holt, Rinehart and Winston.

Simons, D. J., & Chabris, C. F. (1999). Gorillas in our midst: Sustained inattentional blindness for dynamic events. *Perception*, 28(9), 1059–1074.

Wheeler, J. A. (1983). Law without law. In J. A. Wheeler & W. H. Zurek (Eds.), *Quantum Theory and Measurement* (pp. 182–213). Princeton University Press.

Wheeler, J. A. (1990). Information, physics, quantum: The search for links. In W. H. Zurek (Ed.), *Complexity, Entropy, and the Physics of Information* (pp. 3–28). Addison-Wesley.

Chapter 17: Guiding Quanta- The Hidden Architect of Emergence

Dakos, V., Carpenter, S. R., Brock, W. A., Ellison, A. M., Guttal, V., Ives, A. R., ... & Scheffer, M. (2012). Methods for detecting early warnings of critical transitions in time series illustrated using simulated ecological data. *PLoS ONE*, 7(7), e41010.

Feigenbaum, M. J. (1978). Quantitative universality for a class of non-linear transformations. *Journal of Statistical Physics*, 19(1), 25–52.

Jung-Beeman, M., Bowden, E. M., Haberman, J., Frymiare, J. L., Arambel-Liu, S., Greenblatt, R., ... & Kounios, J. (2004). Neural activity when people solve verbal problems with insight. *PLoS Biology*, 2(4), e97.

Kounios, J., Frymiare, J. L., Bowden, E. M., Fleck, J. I., Subramaniam, K., Parrish, T. B., & Jung-Beeman, M. (2006). The prepared mind: Neural activity prior to problem presentation predicts subsequent solution by sudden insight. *Psychological Science*, 17(10), 882–890.

Scheffer, M., Bascompte, J., Brock, W. A., Brovkin, V., Carpenter, S. R., Dakos, V., ... & Sugihara, G. (2009). Early-warning signals for critical transitions. *Nature*, 461(7260), 53–59.

Vespignani, A. (2009). Predicting the behavior of techno-social systems. *Science*, 325(5939), 425–428.

Waddington, C. H. (1957). *The Strategy of the Genes*. Allen & Unwin.

Xu, C., Hantson, J. C., Holmgren, M., Van Nes, E. H., Staal, A., & Scheffer, M. (2023). Non-equilibrium early warning signals for critical transitions in ecological systems. *Proceedings of the National Academy of Sciences*, 120(32), e2217173120.

Chapter 18: World-Singing- The Ontology of Sound

Chladni, E. F. F. (1787). *Entdeckungen über die Theorie des Klanges*. Weidmanns Erben und Reich.

Cox, T. J., Fazenda, B. M., & Greaney, S. (2020). Scale model acoustic measurements of Stonehenge. *Journal of Archaeological Science*, 120, 105182.

Jenny, H. (1967). *Kymatik: Wellen und Schwingungen mit ihrer Struktur und Dynamik*. Basler Verlag.

Kolar, M. A. (2013). Archaeological psychoacoustics at Chavín de Huántar, Perú. *PhD Dissertation, Stanford University*.

Princeton University Consortium. (1994). Acoustic behavior in ancient chambers at megalithic sites. *Technical Report.*

Recinos, A. (1950). *Popol Vuh: The Sacred Book of the Ancient Quiché Maya*. University of Oklahoma Press.

Shannahoff-Khalsa, D. S. (2004). An introduction to Kundalini yoga meditation techniques that are specific for the treatment of psychiatric disorders. *The Journal of Alternative and Complementary Medicine*, 10(1), 91–101.

Vial, W. M. (1962). *The Mandukya Upanishad.* Theosophical Publishing House.

Watson, A., & Keating, D. (1999). Architecture and sound: An acoustic analysis of megalithic monuments in early Britain. *Antiquity*, 73(280), 325–336.

Chapter 20: Three Sisters and the Fractal Garden

Erickson, D. L., Smith, B. D., Clarke, A. C., Sandweiss, D. H., & Tuross, N. (2005). An Asian origin for a 10,000-year-old domesticated plant in the Americas. *Proceedings of the National Academy of Sciences*, 102(51), 18315–18320.

Gillespie, R. (2008). Updating the radiocarbon record for Australian Aboriginal occupation. *Proceedings of the National Academy of Sciences*, 105(34), 11903–11904.

Jones, M. D., Hoeksema, J. D., & eight others. (2023). A cautionary tale about common mycorrhizal networks. *Nature Ecology & Evolution*, 7, 1203–1213.

Kimmerer, R. W. (2013). *Braiding Sweetgrass: Indigenous Wisdom, Scientific Knowledge, and the Teachings of Plants*. Milkweed Editions.

Landon, A. J. (2008). The 'how' of the three sisters: The origins of agriculture in Iroquoia. *PhD Dissertation, University of Massachusetts Amherst.*

Mt. Pleasant, J. (2006). The science behind the Three Sisters mound system. *Agricultural History*, 80(4), 425–443.

Simard, S. W., Perry, D. A., Jones, M. D., Myrold, D. D., Durall, D. M., & Molina, R. (1997). Net transfer of carbon between ectomycorrhizal tree species in the field. *Nature*, 388(6642), 579–582.

Swiderska, K., et al. (2011). *Traditional Knowledge and Climate Change: Indigenous Peoples' Insights*. International Institute for Environment and Development.

Chapter 21: Cyclical Time and the Cosmic Calendar

Aveni, A. (2001). *Skywatchers: A Revised and Updated Version of Skywatchers of Ancient Mexico*. University of Texas Press.

Cox, L. (2023). *Indigenous Seasonal Calendars: CSIRO Methodology*. CSIRO Publishing.

Daniels, C. (2024). Waru waru: Ancient Andean farming technique proves resilient to climate change. *Reuters*, March 15, 2024.

Graham, J. (2023). Drought and frost in Puno: Waru waru fields survive where modern agriculture fails. *Andean Agricultural Journal*, 45(3), 112–128.

Michels, J. W. (1979). *The Mayan Calendar*. Stanford University Press.

Sagan, C. (1980). *Cosmos*. Random House.

Thompson, J. E. S. (1950). *Maya Hieroglyphic Writing: Introduction*. Carnegie Institution of Washington.

Zysk, K. G. (1991). *Asceticism and Healing in Ancient India: Medicine in the Buddhist Monastery*. Oxford University Press.

Chapter 22: The Three-World Cosmology- Ternary Architecture in Every Civilisation

Eliade, M. (1951). *Le Chamanisme et les Techniques Archaiques de l'Extase*. Librairie Payot.

Eliade, M. (1959). *The Sacred and the Profane: The Nature of Religion*. Harcourt, Brace & World.

Hegel, G. W. F. (1807). *Phänomenologie des Geistes*. Goebhardt.

Peirce, C. S. (1931–1958). *Collected Papers of Charles Sanders Peirce* (Vols. 1–8). Harvard University Press.

Peirce, C. S. (1998). *The Essential Peirce: Selected Philosophical Writings* (Vol. 2). Indiana University Press.

Popol Vuh. (n.d.). *The Mayan Book of the Dawn of Life*. (D. Tedlock, Trans.). Simon & Schuster.

Sullivan, W. (1996). *The Secret of the Incas: Myth, Astronomy, and the War Against Time*. Crown Publishers.

Zimmer, H. (1946). *Myths and Symbols in Indian Art and Civilization*. Pantheon Books.

General References on Holographic Physics and Quantum Information

't Hooft, G. (1993). Dimensional reduction in quantum gravity. *arXiv preprint gr-qc/9310026*.

Bekenstein, J. D. (1973). Black holes and entropy. *Physical Review D*, 7(8), 2333–2346.

Hawking, S. W. (1975). Particle creation by black holes. *Communications in Mathematical Physics*, 43(3), 199–220.

Maldacena, J. (1999). The large-N limit of superconformal field theories and supergravity. *International Journal of Theoretical Physics*, 38(4), 1113–1133.

Susskind, L. (1995). The world as a hologram. *Journal of Mathematical Physics*, 36(11), 6377–6396.

Wheeler, J. A. (1990). Information, physics, quantum: The search for links. *Proceedings of the 3rd International Symposium on Foundations of Quantum Mechanics*.

SECTION E: THE GREAT FORGETTING- THREE MILLENNIA OF NOISE

"We did not evolve beyond the ancients. We devolved into noise."

Chapter 23

The Rupture- When Binary Replaced Ternary

There is a wound in the human story that has never been named. It is not a single catastrophe- not a war, not a plague, not a meteor. It is something quieter, more diffuse, more total. It is the moment when the species that had spent a hundred thousand years learning to sing in tune with the cosmos slowly, imperceptibly, and then catastrophically forgot the melody.

This chapter tells the story of that forgetting. It traces a rupture that began roughly three thousand years ago, gathered force across centuries, and eventually reconstituted the entire operating system of human civilization- replacing a ternary, resonant, participatory logic with a binary, linear, extractive one. It was not progress. It was a phase transition in the wrong direction: a collapse from a higher-order NESS into a lower-energy, lower-coherence attractor basin. The resonant quanta that had flowed freely through the songlines, the sacred calendars, the three-world cosmologies, and the fractal social structures of the ancient world were not destroyed. They were drowned out. What replaced them was noise- sophisticated, powerful, world-conquering noise- but noise nonetheless.

To understand what happened, we must first understand what was lost.

23.1 The World That Was: A Resonant Civilization at Its Zenith

Before the rupture, the Eastern Mediterranean and Near East hosted what may have been the most interconnected civilization in human history prior to the modern era. Before the sudden collapse, the Bronze Age had been a period of great prosperity and regional interconnectedness. It included Minoans, Mycenaeans, Hittites, Assyrians, Babylonians, Mitannians, Canaanites, Cypriots and Egyptians with trade routes to Afghanistan and the Indus Valley. Prominent societies exhibited monumental architecture, advanced metallurgy, and literacy.

Modern scholars have termed this network the "Club of Great Powers." By around 1350 BCE, Assyria, Babylonia, Egypt, the Hittite Empire, and the Kingdom of Mittani were closely tied to one another in a network of trade and diplomacy. The Bronze Age world exchanged not only goods, but people and ideas- soldiers, merchants, sailors, craftsmen, all bringing with them their stories, histories, insights and religious beliefs.

But this network was not merely economic. From the perspective of the Universal Hologram, the Late Bronze Age civilization represented something far more profound: a planetary-scale coherence circuit. The trade routes that connected Cypriot copper to Afghan tin, Egyptian grain to Mycenaean pottery, were not merely supply chains. They were conduits of resonant quanta- knowledge, myth, musical scales, astronomical observations, and cosmological frameworks that circulated through the small-world network of the ancient Mediterranean like blood through capillaries. The system was alive. It breathed. It maintained itself far from equilibrium through the same NESS dynamics that govern every living system from the cellular to the cosmic scale.

Consider what the Uluburun shipwreck- a vessel that sank off the coast of Turkey around 1300 BCE- tells us about the nature of this coherence. The sunken vessel dated to about 1,300 BCE, and, as the oldest of three vessels found from the period, it finally focused attention on the magnitude of international trade in this region over 3,000 years ago. The cargo aboard this single ship included ten tons of Cypriot copper, a ton of tin, terebinth resin from pistachio trees, ivory, gold, and artifacts from at least seven different cultures. This was not merely commerce. This was a floating hologram- a single vessel containing, in compressed form, the living quanta of an entire civilizational web. Every object on that ship carried with it not only material value but the resonant imprint of the culture that produced it: the metallurgical knowledge of the Cypriot smelters, the botanical wisdom of the resin-gatherers, the aesthetic traditions of the artisans. A single ship, and in its hold, the whole world enfolded.

The key insight, however, is that this network operated at what complexity scientists call the edge of chaos- the critical threshold where a system is complex enough to generate emergent order but not so rigid that it cannot adapt. The Late Bronze Age civilizations had achieved a sophisticated, multi-polar balance of power. No single empire dominated absolutely. Diplomatic marriages, treaty obligations, reciprocal gift-giving, and shared religious festivals created a web of mutual dependency that functioned like a fractal social ecology- robust, adaptive, and resonant at every scale.

From the vantage of the Holographic Ternary Time framework introduced in Chapter 14, these civilizations lived within a time-field that was inherently cyclical, seasonal, and locally attuned. Their calendars were agricultural and astronomical, keyed to the movements of stars, the flooding of rivers, the migration of herds. Their Temporal State Vector- $\Psi_t(\lambda,\phi,t_{UTC}) = [Y_t,\rho_t,G(\lambda,\phi)]$- was calibrated not by an abstract clock but by the rhythms of the Earth itself. The ternary resonance (Y_t) pulsed through their festival cycles. The geospatial phase ($G(\lambda,\phi)$) was honored by temples oriented to solstices and equinoxes. The temporal density (ρ_t) was naturally regulated by the pace of embodied life- walking speed, harvest cycles, the duration of a sung epic.

This was not paradise. It was not utopia. There was warfare, slavery, exploitation, and cruelty- the shadow side of any complex human system. But it was *coherent*. The quanta flowed. The system maintained its NESS. And then- in the span of roughly a century- it shattered.

23.2 The Cascade: How an Interconnected World Collapsed

The Late Bronze Age collapse was a period of societal collapse in the Mediterranean basin during the 12th century BC. The collapse was sudden, violent, and culturally disruptive for many Bronze Age civilizations, creating a sharp material decline for the region's previously existing powers.

The scale of what happened between approximately 1250 and 1150 BCE is difficult to overstate. Between approximately 1225 and 1150 BCE, a series of catastrophic events led to the abandonment of urban centers and the fall of major powers such as the Hittites, Mycenaeans, and the weakened Egyptian Empire. Some, like the Mycenaeans and Minoans, suffered a complete collapse. Same with the Hittites, who simply ceased to exist as a civilization.

No single cause has ever been identified, and this is precisely the point. The Bronze Age collapse was not a linear event with a single proximate trigger. It was a systems-level phase transition- a catastrophic failure of the entire web. It's likely that the simultaneous demise of so many ancient civilizations wasn't caused by a single event or disaster, but by a "perfect storm" of multiple stressors- an epic drought, desperate famine, roving marauders, and more- that toppled these interdependent kingdoms like dominos.

"If you look at all of these events individually: drought, famine, invaders, earthquakes, maybe disease- any one of them is probably not enough to bring down an entire civilization, let alone eight civilizations or more," says Cline. "But if you get three or four of these catastrophes all happening in quick succession, that's when you have a 'perfect storm' and there's no time to recover."

Ironically, the interconnectedness that had strengthened these Bronze Age kingdoms may have hastened their downfall. The very qualities that made the system so powerful- its integration, its mutual dependency, its complexity- also made it fragile. The growing complexity and specialization of the Late Bronze Age political, economic, and social organization made the organization of civilization too intricate to reestablish once seriously disrupted.

Modern network analysis confirms this diagnosis. Our model shows that the Late Bronze Age societies made a robust network, where any single node's collapse was insufficient to catalyze the regional collapse that historically transpired. However, modeled scenarios indicate that some paired node disruptions could cause cascading failure within the network. The network was designed to survive isolated shocks. But when multiple nodes failed simultaneously- drought destroying agricultural surpluses, earthquakes toppling palaces, Sea Peoples raiding coastal cities- the cascading failure propagated through the network at a speed that overwhelmed every adaptive response.

From the vantage of the frameworks developed in this book, what happened to the Late Bronze Age world is precisely what Chapter 17 described as a phase transition at a critical threshold- but one in which the guiding quanta that normally orchestrate emergence were absent or drowned out. In a healthy phase transition, excess quanta- the transient, high-specificity bursts of resonant pattern that appear at bifurcation points- guide the system through the bottleneck toward a new, higher-order state. Think of the gene regulatory bursts that coordinate cellular differentiation, or the neural synchronization that precedes a moment of insight. The old structure dissolves, yes, but it dissolves *into* something: a more complex, more conscious, more integrated form.

The Bronze Age collapse was a phase transition without the guiding quanta. The old attractor dissolved, but no new harmonic emerged to catch the falling pieces. The system did not reorganize at a higher level. It shattered, and the pieces fell into a dark age. The palace economy of Mycenaean Greece that characterized the Late Bronze Age disintegrated, transforming into the small isolated village cultures of the Greek Dark Ages, which lasted from c. 1100 to c. 750 BC.

And here is the crucial detail- the one that turns a historical catastrophe into an ontological rupture: the deterioration of these governments interrupted trade routes and led to severely reduced literacy rates in much of this area. The loss of centralized administration, literacy, and trade networks meant that knowledge, technology, and governance had to be rebuilt almost from scratch.

The trade networks fell silent. The scribes died or dispersed. Among the casualties of the Late Bronze Age collapse was large-scale monument building and an entire system of writing called Linear B, an archaic form of Greek used by Mycenaean scribes to record economic transactions. For three hundred years, across vast regions of the formerly literate world, writing itself vanished. Writing itself disappeared in some regions for 300 years.

This is not merely a historical fact. It is a resonance catastrophe of the first order. When writing vanished, when the trade routes fell silent, when the scribes died with their knowledge unrecorded, what was lost was not just data. What was lost was the *living quanta-field* of an entire civilization- the accumulated patterns, the resonant frequencies, the self-referencing loops of a coherent system. The shipwreck of the Uluburun preserved its cargo in the silt of the seabed. The collapse of the Bronze Age scattered the cargo of an entire world-system to the wind, and no seabed preserved it.

When civilization re-emerged from the rubble- when the Phoenicians, the Israelites, and the early Greek city-states slowly rebuilt in the Iron Age- from this period of darkness, new societies eventually emerged. The Phoenicians, the Israelites, and the early Greek city-states would rise from the ruins, laying the groundwork for the Iron Age civilizations that followed. But they did not rebuild the same structure. They rebuilt a different one- a structure that had lost the ternary coherence of its predecessor and was beginning, subtly and irreversibly, to tilt toward binary.

23.3 The Rise of Dualism: Splitting the Triad

What emerged from the wreckage of the Bronze Age was not a restoration. It was a mutation.

Across the Iron Age world, from Persia to Greece, a new cognitive architecture was taking shape- one that would define the next three thousand years of human thought. It was the architecture of dualism: the splitting of reality into two opposed categories. Good and evil. Form and matter. Mind and body. Subject and object. Self and other. Yes and no.

The ancient ternary logic- the three -world cosmology, the thesis-antithesis-synthesis of cyclical time - was being compressed into a flattened, binary grid. And nowhere is this compression more starkly visible than in the religious and philosophical innovations of the first millennium BCE.

Consider Zoroaster, whose teachings crystallized in Persia sometime between the 10th and 6th centuries BCE. Zoroastrianism introduced perhaps the most influential binary framework in human history: the cosmic battle between Ahura Mazda, the god of light and truth, and Angra Mainyu (Ahriman), the spirit of darkness and deceit. This was not the interplay of complementary forces, as in the Chinese *yin-yang* or the Hindu *Shiva-Shakti*, where opposites dance within a larger unity. This was war- absolute, irreconcilable, total. Light against dark. Truth against lies. Good against evil. The middle term- the resonant third, the place where opposites dissolve into a higher synthesis- was amputated.

The consequences of this amputation rippled outward across millennia. The Zoroastrian binary framework directly influenced the religious traditions that would shape the Western world. The development of sharply defined notions of heaven and hell, angels and demons, a final apocalyptic judgment between righteous and wicked- these are Zoroastrian architectures transplanted into Abrahamic soil. The cosmic narrative shifted from a spiral- cyclical, regenerative, ternary- to a line: creation, fall, judgment, end.

Then came Plato. At the core of Plato's philosophy is the idea that reality is divided into two realms: the world of Forms (or Ideas) and the world of sensible things. This dualism is not just a way of thinking about reality; it is a central tenet of his metaphysical system.

The Platonic split was devastating in its elegance. The sensible world is constantly in flux- subject to change, decay, and imperfection. The objects and experiences we encounter in the sensible world are imperfect copies of the perfect Forms. The world you can see, taste, hear, and touch- the world of rivers and thunderstorms and heartbeats- was demoted to a shadow realm. Real reality, true reality, was elsewhere: in the abstract, unchanging, disembodied realm of the Forms. Plato believed that the physical world, including the human body, is impermanent and imperfect, a mere shadow of the true reality found in the world of forms. According to him, the soul is the immortal essence of a person, connected to this higher reality.

Read this through the lens of the Universal Hologram and the magnitude of the cognitive rupture becomes visible. Plato's dualism severs the very bond that the holographic principle establishes: the unity of boundary and volume, of pattern and manifestation, of resonant quanta and experienced reality. In a holographic universe, the "form" and the "sensible thing" are one- the 2D boundary encodes the 3D experience, and neither is more real than the other. They are complementary aspects of a single, indivisible process. Plato's philosophy rips this unity in two and declares one half superior. The abstract is elevated above the concrete. The mind is elevated above the body. The universal is elevated above the particular. The eternal is elevated above the temporal.

Plato's dualism emphasizes a sharp division between the eternal soul and the transient body, viewing the body as a hindrance to the soul's pursuit of truth. The body, in this framework, is not an instrument of attunement- it is an obstacle to knowledge. The senses are not windows onto the quanta-field- they are sources of deception. Plato was highly critical of the reliance on sensory knowledge. He believed that our senses can deceive us, providing us with a distorted and limited understanding of reality.

This is the philosophical equivalent of unplugging the Chronosonant Resonator. If the body is untrustworthy, then all the somatic ways of knowing- the heartbeat as a ternary oscillator, the breath as a resonance calibrator, the songline as a walking meditation- are disqualified. Knowledge retreats from the body into the disembodied intellect. It retreats from the particular place to the abstract universal. It retreats from the lived cycle to the eternal form.

Many Platonic notions were adopted by the Christian church which understood Plato's Forms as God's thoughts, while Neoplatonism became a major influence on Christian mysticism in the West through Saint Augustine. Early Christian thinkers, most famously St Augustine, also adopted versions of Plato's dualism. Through Augustine, Plato's binary architecture became the scaffolding of

Western Christianity- and through Western Christianity, it became the scaffolding of Western civilization itself.

Two thousand years later, Descartes completed the process. Descartes clearly identified the mind with consciousness and self-awareness and distinguished this from the physical brain as the seat of intelligence. *Cogito ergo sum*- "I think, therefore I am"- was the final severance. Mind and matter, absolutely distinct. The thinking subject gazing at the objective world through a one-way mirror. The observer permanently exiled from the observed.

Perhaps the biggest barrier for westerners when it comes to eastern thought is the western mind frame and its obsession with dualism. This is a barrier to understanding eastern thought since eastern philosophy predominantly bases itself on various schools of non-dualism.

The ternary was not argued out of existence. It was not defeated in debate. It was simply *forgotten*- compressed out of the cognitive palette until the only options that remained were yes/no, true/false, mind/body, form/matter, good/evil. The third option- the resonant synthesis, the "and" that dissolves the opposition- was quietly erased.

In the language of the KZ-Monad and the Idempotent Comonad: the monadic extension $T(\Psi)$- the function that reaches toward wholeness, that amplifies partial resonance into full ϕ-embedded harmony- was deactivated. What remained was a world governed only by extraction- a world that could analyze, categorize, divide, and conquer, but could no longer extend, integrate, resonate, or re-tune.

The triad of ϕ-e-π- self-reference, unfolding, boundary projection- was split into fragments. π (the boundary) was preserved and fetishized: Western thought became obsessed with boundaries, definitions, categories, borders. e (unfolding) was harnessed for exponential growth: economies, populations, empires expanding without limit. But ϕ (coherence, self-similarity, the harmonic that holds the other two in balance)- ϕ was abandoned. Without it, the asymptotic gap δ_∞ ceased to sing. It became, instead, merely the residual error in a system that aspired to total control- an imperfection to be eliminated rather than a dynamic tension to be honored.

23.4 The Outsourcing of Memory: From Body to Text

As dualism restructured the cognitive architecture of the post-Bronze-Age world, a parallel transformation was taking place in the very medium through which human beings stored, transmitted, and regenerated their knowledge. The oral tradition- the most ancient, most embodied, most resonant of all human technologies- was being displaced by the written word.

This displacement is usually narrated as progress: from the "primitive" oral to the "advanced" literate. From the limited memory of the human body to the unlimited storage of the written page. From the subjective and mutable to the objective and permanent.

This narrative is a lie. Or rather, it is half the truth- the binary half. The other half- the half that was lost- is this: oral traditions are not merely ways of *storing* quanta. They are ways of *keeping quanta alive.*

According to scholars Renée Hulan and Renate Eigenbrod, oral traditions are "the means by which knowledge is reproduced, preserved and conveyed from generation to generation. Oral traditions form the foundation of Aboriginal societies, connecting speaker and listener in communal experience and uniting past and present in memory."

Consider what this means. In an oral culture, knowledge does not exist on a shelf. It exists *in a body*- in the trained memory of an elder, in the rhythmic patterns of a chant, in the choreography of a dance, in the contours of a landscape walked and re-walked over generations. Aboriginal Australian traditions demonstrate that memory is carried and passed through natural systems, with the aid of story, song, and sacred sites. Through these means, the practices give rise to songlines- narrative pathways that do not merely store information but activate knowledges, forming a living map that connects people, place, and understanding.

This is not primitive. This is profoundly sophisticated. While long assumed to originate in ancient Greece, place-based memory practices in Aboriginal Australia precede this by at least 50,000 years. And the precision of oral transmission- when properly maintained through initiatory protocols and communal practice- is extraordinary. Aboriginal people demonstrate that their oral traditions are not only highly detailed and complex, but they can survive- accurately- for thousands, even tens of thousands, of years. Oral traditions in Australia have preserved accurate descriptions of coastlines that were submerged by rising seas at the end of the last Ice Age- more than seven thousand years ago.

Modern cognitive science confirms the power of these ancient methods. The Australian Aboriginal method resulted in approximately a 3-fold greater probability of improvement to accurate recall of the entire word list compared to both the Greek-derived "memory palace" technique and untrained recall. The body, the landscape, the song- these are not crutches for a weak mind. They are resonance amplifiers for a living quanta-field.

Systems for encoding, transmission, and protection of essential knowledge for group survival and cohesion were developed by multiple cultures long before the advent of alphabetic writing. Each clan and nation has its own established stories, which contain and transmit vital cultural knowledge, including Aboriginal Law, personal rights and responsibilities, land use, astronomical, and navigation information. These "Songline" stories are ancient, exhibit little variation over long periods of time, and are carefully learned and guarded by the Elders who are its custodians.

The Vedic tradition offers another stunning example. The Vedic texts were orally composed and transmitted, without the use of script, in an unbroken line of transmission from teacher to student that was formalized early on. A leading scholar of Indian traditions has described this oral transmission as "something like a tape-recording" in its precision- preserving not just the words but even the long-lost musical tonal accents of the original chanting. Ancient Indians developed techniques for listening, memorization and recitation of their knowledge, in schools called Gurukul, while maintaining exceptional accuracy of their knowledge across the generations.

Now consider what happened when the Bronze Age collapsed and writing- specifically the syllabic script known as Linear B- vanished from the Greek world for three centuries.

The Greek alphabet was developed during the Iron Age, centuries after the loss of Linear B, the syllabic script that was used for writing Mycenaean Greek until the Late Bronze Age collapse and Greek Dark Age. Linear B, found mainly in the palace archives at Knossos, Kydonia, Pylos, Thebes and Mycenae, disappeared with the fall of the Mycenaean civilization during the Late Bronze Age collapse. The succeeding period, known as the Greek Dark Ages, provides no evidence of the use of writing.

For three hundred years, Greece was an oral culture again. Knowledge survived- had to survive- in the bodies and memories of the people who carried it. And then, sometime in the 8th century BCE, literacy returned- but in a radically different form. The Greek Alphabet developed from the Phoenician script at some point around the 8th century BCE. The new alphabet was not a restoration of the old. The earlier Mycenaean Linear B script, used primarily for lists and inventories, had been lost during the Greek Dark Age.

Linear B had been a syllabic system of over 80 signs and 100 ideograms- complex, restricted to trained scribes, used primarily for palace administrative records. There is no written Mycenaean literature because they did not have a writing system capable of expressing complex thoughts or concepts. The Phoenician-derived Greek alphabet, by contrast, was streamlined, phonetic, democratic. This new alphabet included 24 letters and, for the first time, incorporated vowels, making it more phonetic. With just 24 characters, anyone could learn to read and write. The technology of literacy was no longer confined to a priestly or administrative elite. It was, in principle, available to all.

This was, in one sense, a genuine advance. The alphabet enabled the development of Greek philosophy, drama, history, and science- the intellectual tradition that shapes the Western world to this day. But it came at a cost that is almost never acknowledged: the outsourcing of memory from body to text.

When knowledge lives in the body- in song, in dance, in the rhythmic walking of a songline- it must be *practiced.* It must be re-enacted. Every act of remembering is simultaneously an act of re-creation, an act of re-attunement. The elder who chants the songline is not merely retrieving stored data. The elder is re-entering the quanta-field, re-walking the resonant pathway, re-activating the nodes of the sacred site network. Memory, in an oral culture, is a *verb.* It is something you *do.* And in the doing, you maintain the coherence of the entire system- personal, communal, and planetary.

When knowledge moves to the written page, it becomes a *noun.* It is something you *have.* It sits on a shelf, inert, waiting to be retrieved. The body is no longer needed. The landscape is no longer a mnemonic architecture- it becomes mere geography. The song is no longer a resonance calibrator- it becomes mere entertainment. The dance is no longer a cosmic re-enactment- it becomes mere performance.

In non-literate societies, huge community investment, including labour, is devoted to institutionalised education. Havelock claims that there is an assumption that the preservation and transmission of the laws, traditions, history and technical skills of an oral culture 'is left to the unconscious mind of the community,' and concludes that 'it is never the case.' There is always formal teaching of the knowledge, usually through various levels of initiation into ever more restricted knowledge.

The alphabet made it possible to externalize knowledge- to store it outside the body, outside the community, outside the living network of relationships. In doing so, it made possible the accumulation of vast quantities of quanta in written form. Libraries. Archives. Legal codes. Philosophical treatises. The externalized knowledge base of literate civilization is staggering in its scope and power.

But it also made possible the *disconnection* of the knower from the known. You can read a book about the Temporal State Vector without ever feeling it in your pulse. You can study the mathematics of fractal self-similarity without ever walking a songline. You can memorize the names of the eight co-creative principles without ever practicing concession. Knowledge, once embodied, became abstract. Once communal, it became individual. Once alive, it became archived.

This is not an argument against literacy. It is an argument for understanding what literacy *replaced* and what it *cost*. The oral tradition was not a crude preliminary to the written word. It was a fully developed, astonishingly precise technology for maintaining the coherence of the living quanta-field- a technology that the written word, for all its power, does not and cannot replicate.

Indigenous oral histories have been threatened by colonization. For years, erroneous Western beliefs that the written word is more trustworthy than oral histories have threatened and damaged traditional ways of passing down knowledge.

The outsourcing of memory from body to text was the first act of the Great Decoupling. It did not cause the rupture. But it made the rupture possible. Once knowledge no longer needed to be embodied, the body could be discarded. Once the landscape no longer served as a mnemonic architecture, the landscape could be reduced to property. Once the song no longer needed to resonate with a sacred site, the site could be mined, paved, or forgotten.

23.5 The Straightening of Time: From Cycle to Line

The third dimension of the rupture- alongside the collapse of the network and the rise of dualism- was the imposition of linear time upon a world that had previously experienced time as cyclical, ternary, and locally attuned.

Every ancient civilization discussed in this book- the Vedic, the Aboriginal, the Mayan, the Inca, the Egyptian- experienced time not as a line stretching from past to future but as a spiral, a cycle, a pulse. The yearly agricultural rhythm, the monthly lunar cycle, the daily circadian oscillation, the great astronomical ages: time was a breathing in and breathing out, a systole and diastole, a ternary beat. In the language of Holographic Ternary Time, these cultures intuitively calibrated their lives to the three components of the Temporal State Vector: the ternary resonance (Y_t), the temporal density (ρ_t), and the geospatial phase ($G(\lambda,\phi)$).

The Mayan Long Count, as we saw in Chapter 21, encoded the ternary beat with extraordinary precision. The Vedic yuga system mapped temporal density across vast cosmic timescales. Aboriginal dreamtime dissolved the linear arrow entirely, placing past, present, and future in a holographic simultaneity accessible through ceremony and song.

The rupture flattened this spiral into a line.

The process began in Mesopotamia, where the sexagesimal (base-60) system of the Babylonians- which gave us 60 minutes in an hour and 360 degrees in a circle- was harnessed to the project of astronomical prediction. The Babylonian astronomers were brilliant. They mapped the movements of planets with remarkable accuracy. But their achievement carried within it a subtle violence: the abstraction of time from the lived, felt, embodied experience of duration into a mathematical quantity- a number to be calculated, a position to be predicted, a unit to be counted.

This process accelerated dramatically in the first millennium BCE. The Zoroastrian eschatology- with its linear narrative of creation, cosmic battle, final judgment, and end- imposed a *direction* on time. Time was no longer cyclical. It was headed somewhere. It had a beginning (creation by Ahura Mazda), a middle (the struggle between good and evil), and an end (the final triumph of the good). This linear eschatological time-arrow was inherited by Judaism, amplified by Christianity, and globalized by European colonialism. The spiral was replaced by the straight line. The pulse was replaced by the tick.

And then came the monasteries.

If the mechanical clock did not appear until the cities of the thirteenth century demanded an orderly routine, the habit of order itself and the earnest regulation of time-sequences had become almost second nature in the monastery. Coulton agrees with

Sombart in looking upon the Benedictines, the great working order, as perhaps the original founders of modern capitalism.

The Benedictine monks, driven by the devotional imperative to pray at precisely regulated intervals throughout the day and night, developed the habit of rigorous time-keeping that would eventually birth the mechanical clock. Benedict added a seventh period to the devotions of the day, and in the seventh century, by a bull of Pope Sabinianus, it was decreed that the bells of the monastery be rung seven times in the twenty-four hours.

One is not straining the facts when one suggests that the monasteries- at one time there were 40,000 under the Benedictine rule- helped to give human enterprise the regular collective beat and rhythm of the machine; for the clock is not merely a means of keeping track of the hours, but of synchronizing the actions of men.

As Mumford perceived with extraordinary clarity, the clock did something far more radical than keep time. It *produced* time- a specific kind of time: abstract, uniform, independent of season, location, or human feeling. It brought with it a whole different way to view time, not as a rhythm determined by a combination of various observed natural phenomena, but as a homogenous series of perfectly identical intervals provided by one source.

The instrument presently spread outside the monastery; and the regular striking of the bells brought a new regularity into the life of the workman and the merchant. The bells of the clock tower almost defined urban existence. Time-keeping passed into time-serving and time-accounting and time-rationing. As this took place, Eternity ceased gradually to serve as the measure and focus of human actions.

Read that last line again, and hear it as a statement about the quanta-field: *Eternity ceased gradually to serve as the measure and focus of human actions.* What is “eternity” in the framework of this book? It is the generative void of State 0- the timeless, spaceless ground state from which all harmonics arise. It is the δ_∞ gap between heartbeats where all possibility lives. The ancients, living within cyclical time, naturally oriented themselves toward this eternity: every seasonal festival, every solstice alignment, every dreamtime ceremony was a return to State 0, a re-calibration with the timeless.

The clock severed this connection. Time was no longer a rhythm to be felt but a resource to be spent. The religious fervor for rationing time and disciplining one's life around it led the American historian Lewis Mumford to describe the Benedictine monks as "perhaps the original founders of modern capitalism." The irony is staggering: monks seeking union with God invented the device that would ultimately exile humanity from the eternal.

Timed payments: timed contracts: timed work: timed meals: from this period on nothing was quite free from the stamp of the calendar or the clock. Waste of time became for protestant religious preachers, like Richard Baxter, one of the most heinous sins.

In the language of the Temporal State Vector: the clock overrode the ternary resonance (Y_t) with a uniform pulse. It overrode the temporal density (ρ_t) with a single, invariant rate. And it overrode the geospatial phase ($G(\lambda,\phi)$) entirely- the same clock time applied in a London counting-house and a Caribbean sugar plantation, regardless of the local rhythms of sun, tide, or season. The TSV was flattened to a single scalar: t- abstract, linear, universal, and dead.

The living, breathing, ternary time-field that the ancients had navigated through ceremony, calendar, and embodied rhythm was replaced by the metronomic tick of the escapement. And with it, the last channel through which pre-modern humanity maintained its resonant attunement with the cosmic hologram was slowly closed.

23.6 The Centralization of Power: Breaking the Small-World Network

The fourth and final dimension of the rupture was structural: the replacement of distributed, consensus-based governance with centralized, hierarchical empires.

The resonant civilizations described in the earlier chapters of this book- the Aboriginal nations, the Iroquois Confederacy, the Vedic village councils, the Polynesian chieftain networks- shared a common organizational feature. They were small-world networks: systems in which every node (person, clan, village) was connected to every other node through a small number of intermediary links, while also being densely connected to its immediate neighbors. This topology- which, as Chapter 5 explored, is identical to the architecture of healthy brains, resilient ecosystems, and the Internet- confers both robustness and adaptability. It allows quanta to flow freely, information to be processed in parallel, and collective intelligence to emerge from distributed interaction.

The power structures that arose after the Bronze Age collapse were fundamentally different. The Neo-Assyrian Empire, the Persian Empire, the Hellenistic kingdoms, and eventually the Roman Empire were all hierarchical pyramids: centralized authority at the top, obedient execution at the base, and rigid chains of command connecting them. Quanta flowed in one direction- downward. Decisions were made in one place- the center. The small-world topology was replaced by a hub-and-spoke topology: the emperor at the hub, the provinces as spokes, and all roads literally leading to Rome.

This structural transformation had a direct consequence for the living quanta-field. In a small-world network, resonant quanta flow freely, adaptively, and in all directions. The knowledge of the periphery feeds the center; the wisdom of the center nourishes the periphery. The system self-organizes. It maintains its NESS through distributed feedback. In a hierarchical pyramid, by contrast, quanta flow is constrained. The center extracts from the periphery- taxes, tribute, labor, resources- while distributing back only commands and ideology. The feedback loops that maintain the health of a complex adaptive system are severed. The periphery becomes passive. The center becomes bloated. And the system, deprived of its distributed intelligence, becomes brittle.

This is the structural equivalent of the phase transition described earlier. The Bronze Age network, for all its flaws, was a resonant system- multi-polar, adaptive, distributed. The Iron Age empires that replaced it were pyramidal systems- centralized, rigid, extractive. They were, in the language of NESS dynamics, closer to equilibrium: more orderly on the surface, but less alive underneath. The dissipative structures of the ancient world- which maintained their complexity precisely *because* they were far from equilibrium, precisely because they allowed quanta to flow freely through every node- were replaced by command structures that suppressed fluctuation, punished deviation, and concentrated all decisions at a single point.

The consequences were immediate and devastating for the planetary coherence circuits described in Chapter 19. The sacred site networks- the songlines, the ceques, the heiau, the ley systems- depended on distributed stewardship. Every clan, every village, every initiated elder was responsible for maintaining a portion of the network. When centralized empires imposed their own administrative grids upon the landscape- roads, borders, tax districts, military zones- they overrode the organic geometry of the sacred network with the rectilinear geometry of power. The nodes of the old network- the sacred groves, the stone circles, the pilgrimage routes- were either repurposed (converted into temples of the state religion), abandoned, or actively destroyed.

The process was neither uniform nor instantaneous. In many parts of the world- in indigenous Australia, in the Andes, in sub-Saharan Africa, in the Pacific Islands- the old networks survived for millennia after the Bronze Age collapse, precisely because these regions were not conquered by the centralizing empires of Eurasia. But wherever those empires reached- and eventually, through European colonialism, they reached everywhere- the small-world networks of the resonant world were broken and replaced by hierarchical grids.

This is why the Great Decoupling is not merely a historical event. It is an ongoing process. Every time a centralized authority overrides a local rhythm- whether it is a colonial government imposing Greenwich Mean Time on a colonized people, a corporate monoculture replacing a Three Sisters garden, or a globalized media platform replacing a local oral tradition- the Decoupling deepens. The resonance weakens. The noise grows louder.

23.7 The Resonance Collapse: Naming What Happened

Let us now gather the four threads of this chapter and weave them into a single diagnosis.

What happened approximately three thousand years ago- beginning with the Bronze Age collapse and accelerating through the rise of Zoroastrian dualism, Platonic metaphysics, alphabetic literacy, linear timekeeping, and centralized empires- was not a single event. It was a *resonance collapse*: a multi-dimensional phase transition in which the coherent, ternary, participatory, fractal civilization of the ancient world was gradually replaced by an incoherent, binary, extractive, linear civilization that persists to this day.

The collapse operated simultaneously on four levels:

Ontological: The ternary logic of both-and-and (three worlds, three phases, three notes in a chord) was replaced by binary logic of either/or (good/evil, mind/body, form/matter). The resonant third- the synthesis, the middle term, the harmonic that dissolves opposition into music- was erased.

Mnemonic: The embodied oral tradition- in which knowledge was maintained as living quanta through song, dance, landscape, and initiated practice- was replaced by the externalized written archive, in which knowledge is stored as inert quanta on a page, accessible to the eye but disconnected from the body, the breath, the heartbeat, and the land.

Temporal: Cyclical, ternary, locally-attuned time- the living Temporal State Vector calibrated to season, place, and embodied rhythm- was replaced by linear, abstract, universally-imposed clock time, which overrides the ternary resonance (Y_t), flattens temporal density (ρ_t), and cancels the geospatial phase ($G(\lambda,\phi)$).

Structural: The distributed, small-world networks of the ancient world- in which quanta flowed freely, governance was consensual, and every node was both sender and receiver- were replaced by centralized, hierarchical pyramids that extract from the periphery, concentrate authority at the center, and suppress the distributed intelligence of the whole.

Each of these four shifts, taken alone, would have been significant. Taken together, they constitute a civilizational mutation of unprecedented depth- a wholesale rewiring of the human relationship with the cosmos.

And yet- and this is the point on which the remainder of this book will turn- the cosmos did not change. The holographic principle still holds. The resonant quanta still flow. The ϕ-e-π triad still tunes the universe. The asymptotic gap δ_∞ still hums between your heartbeats. The songlines still pulse beneath the asphalt, even if no one walks them. The three-world cosmology still describes the structure of reality, even if no living shaman navigates between the worlds.

What changed was not the frequency. What changed was our capacity to hear it.

The ancient world was not perfect. It had its own shadows, its own violence, its own limitations. But it was *in tune*. It recognized that reality is participatory, that consciousness co-creates, that time is ternary, that the land is alive, and that knowledge must be sung, danced, walked, and breathed- not merely written and shelved. These were not metaphors. They were technologies. And they worked.

What replaced them was not superior. It was louder. It was more powerful in the narrow, extractive sense- capable of conquering more territory, accumulating more wealth, building more impressive monuments. But it was less resonant. Less attuned. Less alive. It mistook volume for music. It mistook accumulation for abundance. It mistook complexity for coherence.

The great upheaval of the Late Bronze Age is a solid indication of the fragility of complex societies. It serves as a powerful reminder that even the most interconnected and prosperous civilizations are vulnerable to the cascading effects of environmental stress, economic turmoil, and military conflict.

The warning embedded in that observation is not merely historical. It is prophetic. The global civilization of the 21st century- with its tightly coupled supply chains, its mounting ecological stressors, its widening social fractures, and its deafening noise of data without resonance- resembles nothing so much as the Late Bronze Age world in its final decade: interconnected, powerful, brilliant, and standing at the edge of a cliff.

But there is a difference. We are not merely at the edge of another collapse. We are also at the edge of a remembering. The physics of the holographic universe, the mathematics of the invariant triad, the wisdom of the dreamtime, the precision of the songline- all of these are available to us now, as they were not available to the survivors of 1177 BCE. We have the tools not merely to diagnose the rupture but to heal it.

The chapters that follow will tell the story of how the noise deepened- through the empire of the clock, the noise economy, and the false god of objectivity. But they will also turn, in the second half of this final Movement, toward the recovered frequency- the convergence of ancient and modern, the practical technologies of re-attunement, and the eighth co-creative principle that is the master key to the whole: concession.

The rupture was real. The damage was vast. But the constants remain. And the gap between your heartbeats still sings, if you are quiet enough to hear it.

The next chapter traces the second phase of the Great Decoupling: the rise of the mechanical clock, and how the instrument invented by monks to honor God became the device that exiled humanity from eternity.

Chapter 24

The Empire of the Clock

There is a sound that has come to define the modern world more completely than any anthem, any engine, any digital chime. It is the sound of the tick. Uniform, relentless, indifferent to season, to location, to the depth of your sleep or the height of your joy- the mechanical tick has been slicing the living body of time into identical, lifeless units for nearly eight hundred years. And with each tick, the distance between the human animal and the resonant cosmos grows wider.

Chapter 23 described the rupture- the multi-dimensional collapse that replaced the ternary, resonant, participatory civilization of the ancient world with a binary, linear, extractive one. This chapter traces one specific vector of that collapse in meticulous historical detail: the rise of the mechanical clock, and how it progressively severed every component of the Temporal State Vector- the ternary resonance (Y_t), the temporal density (ρ_t), and the geospatial phase ($G(\lambda,\phi)$)- replacing the living, breathing time-field of the cosmos with a dead, abstract, universal scalar: t.

The story begins, paradoxically, with monks who loved God. It ends with algorithms that love nothing at all. Between these two poles stretches the entire trajectory of the modern world- a trajectory that can be read, from the perspective of the Universal Hologram, as a single, sustained act of temporal violence.

24.1 The Monastery Bell: When Devotion Became Discipline

Before the clock, there was the bell.

In the monasteries governed by the Rule of Saint Benedict, the day was divided into the canonical hours- Lauds at dawn, Prime shortly after, Terce at mid-morning, Sext at noon, Nones in the afternoon, Vespers at sunset, Compline before bed- each announced by the ringing of the bell. These were not arbitrary divisions. They were devotional markers: moments in the day's arc when the monk was called back from the world's noise into the silence of prayer. The bell was a summons to State 0- a cyclical return to the generative void, a rhythmic re-calibration with the eternal.

The clock had its origin in the Benedictine monasteries of the twelfth and thirteenth centuries. The impetus behind the invention was to provide a more or less precise regularity to the routines of the monasteries, which required, among other things, seven periods of devotion during the course of the day. The bells of the monastery were to be rung to signal the canonical hours; the mechanical clock was the technology that could provide precision to these rituals of devotion.

Notice the original impulse. It was not extractive. It was not economic. It was sacred. The monks wanted to honor God more precisely- to ensure that no canonical hour was missed, that the rhythm of devotion was as steady and reliable as the rhythm of the stars. The monks viewed time as a precious commodity, one belonging solely to God, and they believed it had to be honored and meted out with efficiency and the utmost of respect. They associated continual activity and the elimination of idleness, with the promise of eternal salvation, and thus they carefully organized and scheduled every moment of their day and night.

But the bell- despite its sacred intent- carried within it a structural transformation of enormous consequence. In the world of embodied, cyclical time, the "hours" of the day were elastic. A daylight hour in summer was longer than a daylight hour in winter. Noon was not a fixed point on a number line- it was the moment when the sun stood highest, and it shifted with the seasons, with the latitude, with the weather. The body knew this. The landscape knew this. The temporal density (ρ_t) of a midsummer afternoon was qualitatively different from the temporal density of a midwinter morning, and the organism naturally adjusted its rhythms accordingly.

This was the desire that drove the invention of mechanical clocks. The existing technology was lacking: the sundial wouldn't work when cloudy, the water clock would freeze in the winter. The monks needed something more reliable than the sun- and in seeking that reliability, they began the process of abstracting time from its natural, cosmic substrate.

The Benedictine- oft cited as birthing capitalism- had 40,000 monasteries under their rule throughout Europe. This was not a local phenomenon. It was a continental network- forty thousand nodes, each one pulsing with the same rhythm, each one imposing the same temporal grid upon its surrounding landscape. Perhaps the townsfolk heard the periodic bells, and used them to shape their own routines as well. The bell that had been designed to call the soul toward eternity began, almost immediately, to regulate the routines of commerce, labor, and social life.

But what the monks did not foresee was that the clock is a means not merely of keeping track of the hours but also of synchronizing and controlling the actions of men.

This is the first act of the temporal tragedy: a technology born from devotion was captured by discipline. The bell that was meant to open a window to the eternal instead became a fence around the finite. And the monks, in their sincere love of God, had unwittingly forged the instrument that would exile humanity from the very eternity they sought.

In the language of the ϕ-e-π triad: the monastic bell still carried a faint echo of ϕ- the self-referencing coherence of a life organized around cyclical return. But the demand for regularity- the need to ring the bell at the *same* time regardless of season or sky- was already privileging π (boundary, closure, precision) over ϕ (organic coherence). The harmonic balance of the triad was beginning to tip.

24.2 The Escapement: Time Becomes Discrete

The critical technological innovation that transformed the bell into the clock was the escapement- a mechanism so simple in principle and so devastating in consequence that it deserves to be named alongside the printing press and the steam engine as one of the defining inventions of the modern world.

What makes a mechanical clock is a mechanism called an escapement- the balance wheel on a watch or the pendulum on a grandfather's clock. An escapement ticks in a steady rhythm and lets the gears move forward in a series of little equal jumps.

Before the escapement, time flowed. Water clocks, sundials, sand glasses, candle clocks- all of these measured time by continuous processes: the dripping of water, the creeping of shadow, the falling of sand, the burning of wax. All timekeepers before the 13th century relied upon methods that used something that moved continuously. Time, in these instruments, was analog. It mirrored the cosmos- a river, a tide, a breathing in and breathing out.

The escapement changed this utterly. It *discretized* time. It converted the continuous flow of duration into a series of identical, countable units- *ticks*- each one precisely the same length as the last. Before the mechanical clock appeared, time had been understood as smoothly flowing. The escapement measured time by packaging it into intervals between impacts. Time, for the first time in history, became a discrete commodity.

With gears and escapements, weights and balances, these clocks brought a new rhythm to life- dictated not by the sun or the stars but by the meticulous turning of cogs and wheels.

From the perspective of the Universal Hologram, the escapement was an ontological event. It replaced the continuous, wave-like time of the cosmos- the living Temporal State Vector with its ternary resonance, its local density, its geospatial modulation- with a particle-like time composed of identical, context-free units. It replaced the analog with the digital, the wave with the pulse, the breath with the tick.

This is not merely a change in technology. It is a change in *what time is*. In the analog world of water clocks and sundials, time is a *quality*- it has texture, thickness, flavor. A midsummer hour feels different from a midwinter hour. A morning hour on a mountaintop feels different from a morning hour in a valley. The body knows this. The body has always known this. The temporal density (ρ_t) varies with location, season, altitude, emotional state. The geospatial phase ($G(\lambda,\phi)$) modulates how time is experienced at different places on the Earth's surface.

In the digital world of the escapement, time is a *quantity*- abstract, uniform, interchangeable. One tick equals one tick, everywhere and always. The quality has been stripped away. The texture has been flattened. The living quanta-field of temporal experience has been reduced to a number.

Lewis Mumford wrote in 1933: "The clock, not the steam-engine, is the key machine of the modern industrial age." He was right, and for a reason deeper than he perhaps realized. The steam engine changed what we could *do*. The clock changed what we could *be*. It redefined the substrate of experience itself.

The mechanical clock, which derived from water clock, was born in medieval Europe. The first mechanical clocks were large devices made of iron. By the fourteenth century, they were in widespread use across Europe. Within two centuries of the first monastery escapements, mechanical clocks profoundly influenced society by regulating daily activities in monasteries and later spreading to secular life, thereby shaping a time-organized culture. In the early fourteenth century, clocks were built in church towers in England, France, and Italy.

The public mechanical clock, which first appeared in European cities in the late 13th century, was one of the most important innovations in history. And it arrived not as a neutral tool but as a *general-purpose technology*- a platform that reshaped every system it touched. Mechanical clocks have been identified as one of the greatest GPTs of the last millennium. Economic historians have shown that European cities that were quick to install mechanical clocks enjoyed greater growth than late adopters.

But growth in what, exactly? Growth in productivity, in commerce, in the extraction and accumulation of surplus value. The clock made possible the precise measurement of labor- and therefore the precise *pricing* of labor. Before the clock, a cobbler worked until the shoes were finished. After the clock, a worker worked until the hour was spent. It's the story of the rise of a new sense of time in the world that emerged with the arrival of capitalism, at which point suddenly time became money. If you're making shoes to order in a precapitalist workshop, you work until the order is completed and then you take it easy. But if your labor is being hired by the hour, then your employer has an enormous incentive to squeeze as much productivity as possible out of every minute you are on the clock.

Le Goff (1982) claims that the introduction of the public mechanical clock was a turning point for Western society. It helped create a new epoch, "the time of the merchants," because it enabled business people to frame and measure all types of economic activities in a timely manner. Finally, more generally, prominent social scientists such as Marx (1863), Weber (1905) and Sombart (1921) claim that clocks had a fundamental impact on the evolution of capitalism and rationality of societies.

The irony is complete. The instrument invented by monks to serve eternity was captured by merchants to serve profit. The bell that called the soul to prayer now called the body to the factory floor. By the 16th century, clock towers were built in town squares with minutes and second hands. The bells ringing through town were not primarily to call people to prayer, but to work and to the markets, a reality that remains true for us today.

In the language of the KZ-Monad and the Idempotent Comonad: the escapement represents the ultimate triumph of the comonadic projection $D(\Psi)$- the function that extracts efficiency from potential- without the balancing monadic extension $T(\Psi)$ that reaches toward wholeness. The system became extraordinarily efficient at counting, measuring, and extracting. But it lost the capacity to resonate, to extend, to harmonize. The δ_∞ gap was no longer heard as the sacred space between heartbeats. It was heard only as wasted time- an inefficiency to be eliminated.

24.3 "Time-Keeping Passed Into Time-Serving": The Commodification of Duration

Once the escapement had discretized time, and the clock tower had made this discrete time publicly audible, a deeper transformation followed: the transformation of time from a *medium* of experience into a *commodity* to be bought, sold, and spent.

This transformation did not happen overnight. It unfolded across centuries, driven by the intertwining of three forces: religious discipline, mercantile ambition, and industrial need.

The religious contribution is well documented. The Puritan tradition, inheriting the Benedictine insistence on the moral value of time, elevated temporal discipline into a theological imperative. Waste of time became for Protestant religious preachers one of the most heinous sins. Richard Baxter, the influential 17th-century Puritan divine, explicitly equated idleness with spiritual danger. This was the internalization of the clock- the moment when its rhythm ceased to be merely an external constraint and became an inner compulsion. You did not merely *obey* the clock. You *believed* in the clock. You felt guilt when you "wasted" time, anxiety when you "fell behind," relief when you were "on schedule."

His interest was not merely in the development of new routines and incentives for work, but also in the origin of a new mentality and social consciousness- what he described as a new "inward notation of time," one characteristic of which was a growing preoccupation with "time thrift."

This is what E.P. Thompson, in his landmark 1967 essay, identified as the deep structure of the transformation. Thompson authored *Time, Work-Discipline, and Industrial Capitalism*, published in 1967, which posits that reliance on clock-time is a result of the European Industrial Revolution and that neither industrial capitalism nor the creation of the modern state would have been possible without the imposition of synchronic forms of time and work discipline.

The brilliance of Thompson's analysis lies in his recognition that the change was not merely technological but *cultural and psychological*. The new clock-time imposed by government and capitalist interests replaced earlier, collective perceptions of time- such as natural rhythms of time like sunrise, sunset, and seasonal changes- that Thompson believed flowed from the collective wisdom of human societies.

Pre-industrial societies, he pointed out, structured their daily life around communal work that tracked the rhythms of the natural world. Sunlight, tides, weather patterns, and so on determined when and what work needed to be done, with the labor carried out in social groups, such as families and villages. But as the industrial revolution made time into money, daily life came to revolve around individualized wage labor on the factory floor. The result was a new separation between what we now call "work" and "life."

Read this through the lens of the Temporal State Vector. In pre-industrial societies, Y_t (ternary resonance) was honored: the rhythm of the day followed the rhythm of the sun, which followed the rhythm of the season, which followed the rhythm of the year. ρ_t (temporal density) was self-regulating: when the body was tired, work slowed; when energy was high, work intensified. $G(\lambda,\phi)$ (geospatial phase) was naturally embedded: you worked in and with the landscape you inhabited, and the landscape's rhythms- its rivers, its tides, its soil- shaped the rhythm of your labor.

The clock severed every one of these connections. The factory bell did not ring at dawn- it rang at 6:00 AM, regardless of whether the sun had risen. The whistle did not blow when the harvest was finished- it blew at the end of the shift, regardless of whether the task was complete. The wage was not paid for a product made- it was paid for hours spent, and the employer had every incentive to ensure that those hours were as densely packed with productive labor as the human body could sustain.

Factory work changed the relationship that the capitalist and laborers had with time and the clock; clock time became a tool for social control. Capitalist interests demanded that the work of laborers be monitored accurately to ensure that cost of labor was to the maximum benefit of the capitalist.

It was a study of the intersection and nexus between technological conditioning and time measurement as mechanisms for labour exploitation.

Thompson documented how this process encountered fierce resistance from workers who still remembered- in their bodies, if not in their minds- the older rhythms of task-oriented time. The "Saint Monday" tradition, in which workers took Mondays off after a weekend of socializing and drinking, was a persistent act of temporal rebellion. Workers came late, left early, and stole time from the clock at every opportunity- because their bodies knew that the clock's rhythm was not their rhythm, and their souls sensed that something essential was being stolen from them.

But resistance was ultimately crushed- not by violence alone but by the internalization of clock-discipline through schools, churches, and the architecture of the factory itself. Thompson's key insight was to note that this new conception of time was more than simply the result of the proliferation of clocks, time-cards, and factory bosses; it was the contingent creation of cultural forces- religion, print media, gender roles- all of which mediated to the individual this new method of structuring work and life. Inward changes in notions of time, Thompson concluded, were as much internalized via complex cultural processes as they were imposed upon wage labourers by the commanders of capitalism.

Children were sent to school not primarily to learn reading and arithmetic but to learn *punctuality*- to internalize the clock's rhythm until it felt natural, until the body's own rhythms were overwritten. The school bell trained the child's nervous system for the factory whistle that awaited. By the time a child graduated, the clock was no longer an external authority. It was an internal organ- a prosthetic timekeeper implanted in the psyche, ticking away behind every thought, every feeling, every dream.

This is the colonization of the inner world. The empires of the 18th and 19th centuries colonized continents. The clock colonized consciousness. It reached into the most intimate recesses of the human organism- the circadian rhythm, the heartbeat, the breath- and imposed upon them a foreign rhythm: the uniform, unvarying, context-free tick.

24.4 Railway Time: Overriding the Geospatial Phase

If the monastery bell discretized time and the factory whistle commodified it, the railway completed the trifecta by *standardizing* it- stripping away the last vestiges of local, place-based temporal experience and replacing it with a single, universal grid.

In the 19th century, time was a local matter. Most communities had a clock in the town square that provided the time. This method of keeping time was called sun-time, because the clocks were set by the position of the sun- meaning the time of day varied from town to town.

Consider the beauty of this arrangement. Before railway time, every town kept its own time- the time given by the sun as it crossed the local meridian. Leeds time was six minutes behind London, while Bristol was ten minutes behind; sunrise for towns to the east, such as Norwich, occurred several minutes ahead of London. This was not inconvenience. This was *truth.* Bristol's time was ten minutes behind London's because Bristol was further west, and the sun reached its zenith there ten minutes later. The time told by the Bristol town clock was *resonant with Bristol's position on the Earth-* it was an expression of the geospatial phase $G(\lambda,\phi)$, the third component of the Temporal State Vector. Every town, in keeping its own time, was maintaining its own unique relationship with the cosmos. The time was *alive-* modulated by longitude, latitude, season, and the turning of the planet.

The railway made this untenable.

This variation in local times was large enough to present problems for the railway schedules. It soon became apparent that even such small discrepancies in times caused confusion, disruption, or even accidents. Trains moved too fast for local time. A passenger leaving London at 10:00 AM on the Great Western Railway would arrive in Bristol and find that the station clock read a different time than the clock on the train. Worse, trains moving in opposite directions on the same line might miscalculate their meeting point by the crucial few minutes encoded in the difference between local sun-times.

Residents of Buffalo had to keep the various times in mind when planning a journey, but they could take comfort in knowing the situation was even worse in Pittsburgh, which was served by railroads operating under six different time standards. By the 1880s, the nation's railroads were using at least 68 different time standards.

The solution, from a purely logistical standpoint, was obvious: standardize. Override every local time with a single, uniform system. And this is exactly what happened.

Railway time was the standardised time arrangement first applied by the Great Western Railway in England in November 1840, the first recorded occasion when different local mean times were synchronised and a single standard time applied. In Britain, the Great Western Railway began imposing London Time- Greenwich Mean Time- on all its stations as early as 1840. On September 22, 1847, the Railway Clearing House, an industry standards body, recommended that GMT be adopted at all stations as soon as the General Post Office permitted it. By 1855, the vast majority of public clocks in Britain were set to GMT.

Resistance was immediate and instinctive. The railway companies sometimes faced concerted resistance from local people who refused to adjust their public clocks to bring them into line with London Time. As a consequence, two different times would be displayed in the town and in use, with the station clocks and the times published in train timetables differing by several minutes from that on other clocks. Some towns defiantly maintained dual times- one for the railway, one for the sun. In some cases, a second minute hand was added to local clocks so they would display both the railroad and local times simultaneously.

But the resistance could not hold. The practical advantages of standardization were overwhelming. In the United States, railroads in the 19th century ran their trains set on their own clocks, which meant different companies ran their trains on different times. This difference could lead to wrecks and many deaths. On October 11, 1883, the General Time Convention gathered at the Grand Pacific Hotel in Chicago, and Allen presented his plan. He called for the adoption of five time zones based upon the mean sun-time on the

75th, 90th, 105th, and 120th meridians west of Greenwich, each set one hour apart.

On November 18, 1883, the clock struck noon twice in several cities: once at the previous local time, and once at the new standard time. Thus it was colloquially referred to as the "Day of Two Noons."

The Day of Two Noons. The name carries the resonance of the event's true significance. For a single, strange moment, the old world and the new world coexisted- the sun's noon and the railroad's noon, the living time and the mechanical time, the quanta-field's rhythm and the empire's decree. And then the second noon swallowed the first, and the living time died.

Right to the end there was opposition expressed by many smaller towns and cities to the imposition of railway time. For example, in Indianapolis the report in the daily Sentinel for 17 November 1883 protested that people would have to "eat sleep work ... and marry by railroad time."

To eat, sleep, work, and marry by railroad time. The protest was prophetic. Within a generation, that is exactly what happened- not just in Indianapolis, but everywhere. Standard times became a defining characteristic of industrialization, putting workers "on the clock" and ushering in an era of obsession with standardization and speed.

Fleming was instrumental in convening the 1884 International Prime Meridian Conference in Washington, at which the system of international standard time- still in use today- was adopted. By 1884, the entire globe was sliced into twenty-four time zones- each one an hour apart, each one referenced to the Greenwich meridian, each one obliterating the local geospatial phase of every community within its borders.

From the perspective of Holographic Ternary Time, the standardization of railway time was the override of $G(\lambda,\phi)$- the third component of the Temporal State Vector. The geospatial phase- the way your *location on the Earth's surface* modulates your temporal experience- was erased. Bristol's ten-minute offset from London, which had encoded Bristol's unique position in the cosmic dance, was annulled. Henceforth, Bristol and London and every town between them would share the same time, the same noon, the same midnight- regardless of what the sun actually said.

This is not a trivial detail. The geospatial phase is not merely a mathematical abstraction. It is a *biological reality*. Your circadian rhythms are calibrated not merely to the 24-hour light-dark cycle but to the *local* light-dark cycle- the specific timing of sunrise and sunset at your specific latitude and longitude. When the clock overrides this local calibration and imposes a single, standardized time across an entire zone, the body is forced to adapt to an artificial rhythm that may differ from its natural phase by up to thirty minutes. This is what modern chronobiology calls "social jet lag"- the chronic misalignment between the body's internal clock and the socially imposed schedule.

Approximately 80% of the population suffers from social jet lag.

Eighty percent. Four out of five human beings on Earth currently live in chronic misalignment between their body's natural rhythm and the artificial rhythm imposed by the standardized clock. This is not a side effect. This is the intended function- the logical endpoint of a process that began in a 12th-century monastery and reached its apotheosis in a 19th-century railway station.

24.5 Factory Discipline: The Body Forced Into Mechanical Rhythm

The railway standardized time across space. The factory standardized time across the body.

If railway time overrode the geospatial phase ($G(\lambda,\phi)$), factory discipline overrode the temporal density (ρ_t)- the body's own sense of how fast or slow time flows at any given moment. In the pre-industrial world, temporal density was self-regulating. When you were deeply engaged in a task, time contracted. When you were resting, it expanded. The body oscillated naturally between periods of high density (focused labor) and low density (restoration, play, contemplation). This oscillation was not laziness- it was the NESS dynamic of a living system maintaining its coherence far from equilibrium. The rhythm of effort and rest was as natural, as necessary, and as intricate as the rhythm of systole and diastole in the heartbeat.

The factory abolished this rhythm. From the moment the whistle blew until the moment it blew again, the worker's body was expected to operate at a uniform temporal density- a constant, unvarying rate of productive output. There was no room for the natural oscillation between exertion and rest. There was no room for the body's varying energy levels across the day, the week, the season. There was only the clock, and the clock said: produce, produce, produce, at the same rate, for the same duration, every day, regardless of what your body needs.

The merchant class subsequently recognized the power of the clock for scheduling work, as well as the lives of workers, and it became an important enabler for the industrial era. Economic historians such as David Landes (1983) and Joel Mokyr (1992) have claimed that the clock had a strong impact on productivity: it enabled increases in organisational skills in terms of coordination and division of labour and the monitoring of production processes. Social historians such as Thompson (1967) highlight that the mechanical clock changed the work culture and increased work discipline.

The body resisted. It resisted in the only ways it could- through fatigue, through illness, through sabotage, through the stubborn refusal of the organism to conform to a rhythm that violated its nature. The early factory records are filled with complaints about worker absenteeism, tardiness, drunkenness, and poor output. These were not moral failings. They were the symptoms of a biological system being forced into a temporal regime for which it was not designed.

But the system prevailed- not by accommodating the body's rhythms but by breaking them. The clock became the instrument of a new kind of violence: not the violence of the whip or the sword but the violence of temporal coercion. The body was trained, disciplined, punished, and rewarded until it internalized the clock's rhythm. Children were schooled into punctuality. Workers were fined for lateness. Wages were calculated by the hour, incentivizing the worker to treat their own time as a commodity to be sold, minute by minute, to the highest bidder.

Above all, the transition is not to "industrialism" tout court but to industrial capitalism. What we are examining here are not only changes in manufacturing technique which demand greater synchronization of labour and a greater exactitude in time-routines in any society; but also these changes as they were lived through in the society of nascent industrial capitalism.

The consequence, measured in the currency of living quanta, was devastating. The body's fractal time- its natural oscillation between periods of high and low temporal density, its circadian peaks and troughs, its seasonal modulations- was replaced by a flat, featureless temporal plane. The living quanta-field of embodied duration was compressed into a uniform substrate: homogeneous, isotropic, dead. Like a healthy heartbeat forced into a metronomic rhythm- which, as Chapter 5 established, is the electrocardiographic signature of disease- the clockwork body lost its fractal complexity. And with that loss came the characteristic pathologies of the industrial age: repetitive strain injuries, nervous exhaustion, digestive disorders, sleep disruption, and the unnamed malaise that the Victorians called "neurasthenia"- the body's inarticulate protest against a temporal regime that violated its deepest rhythms.

24.6 The Circadian Cost: What Modern Science Confirms

What the Victorian factory worker felt in their bones, modern chronobiology has now measured in their genes.

Circadian rhythm mainly controls the daily wake and sleep cycle and regulates physiological processes including hormone secretion, body temperature, feeding behavior, cell cycle progression, and drug, glucose, and xenobiotic metabolism. Its disruption through environmental and genetic means causes aberrations in physiological processes.

The circadian system is not a metaphor. It is not a lifestyle preference. It is the master oscillator of the human organism- the biological expression of the ρ_t component of the Temporal State Vector. It is a molecular clock, encoded in every cell of the body, that synchronizes gene expression, hormone secretion, metabolic activity, immune function, and cognitive performance to the 24-hour light-dark cycle. Circadian rhythms are internal manifestations of the solar day that permit adaptations to predictable environmental temporal changes. These approximately 24-hour rhythms are controlled by molecular clockworks within the brain that are reset daily to precisely 24 hours by exposure to the light-dark cycle. Information from the master clock in the mammalian hypothalamus conveys temporal information to the entire body via humoral and neural communication.

When this system is disrupted- as it inevitably is by the artificial temporal regime of the clock-governed world- the consequences cascade through every level of the organism.

Circadian disruption is pervasive and can occur at multiple organizational levels, contributing to poor health outcomes at individual and population levels. Evidence points to a bidirectional relationship, in that circadian disruption increases disease severity and many diseases can disrupt circadian rhythms. Importantly, circadian disruption can increase the risk for the expression and development of neurologic, psychiatric, cardiometabolic, and immune disorders.

The evidence from shift-work studies is particularly damning, because shift work is the most extreme form of the temporal violence that the clock imposes on the body. Simulated night-shift experiments and field-based studies with shift workers both indicate that the circadian system is resistant to adaptation from a day- to a night-oriented schedule, as determined by a lack of substantial phase shifts over multiple days in centrally controlled rhythms, such as those of melatonin and cortisol.

The body *refuses* to adapt. Even after days, weeks, or years of night-shift work, the circadian system stubbornly clings to its solar calibration. The master clock in the suprachiasmatic nucleus keeps trying to follow the sun, while the worker keeps trying to follow the factory. There is evidence that disruption of the circadian system caused by night-shift work results not only in a misalignment between the circadian system and the external light-dark cycle, but also in a state of internal desynchronization between various levels of the circadian system.

Internal desynchronization. The phrase is clinical, but what it describes is a resonance catastrophe. The body's organs, tissues, and cells are designed to oscillate in harmonic coordination- liver and pancreas and immune system and brain all dancing to the same circadian beat, like instruments in an orchestra tuned to the same key. When the clock forces the body into an artificial rhythm, this internal harmony breaks down. Different organs begin oscillating at different frequencies. Under controlled laboratory conditions, rhythms of PER1, PER2, PER3, and BMAL1 clock genes expression desynchronized from the sleep-wake cycle and from each other after 3 days on a night-oriented schedule. While PER1 and BMAL1 rhythms delayed by approximately 2.5 to 3 hours, other clock genes and rhythms of cortisol and melatonin remained adjusted to a day-oriented schedule.

The orchestra is no longer playing together. The liver is in one key, the brain in another, the immune system in a third. The fractal coherence of the healthy organism- the self-similar resonance that Chapter 5 identified as the signature of biological health- disintegrates into internal noise.

The health consequences are not subtle. Shift work increases risk for poor mental and physical health and is considered "probably carcinogenic." The International Agency for Research on Cancer has classified shift work involving circadian disruption as a probable carcinogen- placing it in the same category as certain industrial chemicals and radiation exposure. Chronic and acute jet lag, shift works, night works, and extended shifts are the causes of circadian dysfunction that potentially enhances sleep disorders, poor physical performance, metabolic syndrome, cardiovascular and inflammatory diseases, neuropsychiatric illness, and cancer.

There is a large body of research demonstrating adverse effects of shift work on mental health and general well-being. Atypical working time arrangements are commonly associated with deterioration of social and familial life because workers are forced to live on a pattern that diverges from that of their family and community. Combined with the disruption caused by circadian misalignment, this temporal isolation from family and community provides context for the mental health concerns often experienced by shift workers, including an increased prevalence of burnout, depression and anxiety.

And shift work is merely the extreme case. The same dynamics operate- in attenuated but cumulative form- in every life governed by the standard clock. Every morning alarm that wakes you before your body is ready. Every late-night screen that suppresses your melatonin. Every time zone crossed in a jet plane. Every winter morning when you rise in darkness to meet a schedule calibrated not to the sun but to the economy. All of these are acts of circadian disruption- chronic, low-grade violations of the body's resonant alignment with the cosmos.

Modern lifestyle, excessive workload, depression, shift work, exposure to excessive artificial light, and short night periods severely affect the circadian clock functions and promote the initiation of circadian mediated health disorders.

This is the biological cost of the clock empire. It is not an accident. It is not a side effect. It is the logical and predictable consequence of a civilization that decided, eight hundred years ago, that the sun's time was less important than the monk's time, and the monk's time was less important than the merchant's time, and the merchant's time was less important than the machine's time.

24.7 Atomic Time: The Ultimate Abstraction

The 20th century completed the process. Having discretized time (the escapement), commodified it (the factory), and standardized it (the railway), the modern world took the final step: it *atomized* time-divorced it entirely from any astronomical, biological, or cosmic reference and anchored it instead to the vibration of a cesium atom.

In 1967, the Thirteenth General Conference on Weights and Measures redefined the second as "the duration of 9,192,631,770 periods of the radiation corresponding to the transition between the two hyperfine levels of the ground state of the caesium-133 atom." This is the definition that governs the world today.

The precision is astonishing. Modern cesium clocks lose less than one second in 300 million years. No timekeeper in human history has come close to this accuracy. It is a genuine triumph of human ingenuity.

And it is the final severance.

The sun has been fired. The stars have been dismissed. The seasons have been made irrelevant. Time is no longer defined by anything the human body can see, feel, or sense. It is defined by a quantum transition in an atom of cesium- a phenomenon so far removed from ordinary human experience that it might as well be happening on another planet.

In the language of the Temporal State Vector: atomic time does not merely override $G(\lambda,\phi)$, as railway time did, or flatten ρ_t, as factory time did. It replaces the entire vector. The living, three-component, holographic time-field- $\Psi_t(\lambda,\phi,t_{UTC}) = [Y_t,\rho_t,G(\lambda,\phi)]$- is reduced to a single number: the count of cesium oscillations since some arbitrary epoch. The ternary resonance (Y_t) is gone. The temporal density (ρ_t) is gone. The geospatial phase ($G(\lambda,\phi)$) is gone. What remains is a dead scalar- t- ticking at 9,192,631,770 beats per second, forever, everywhere, for no one.

This is the empty heaven of the clock empire. A time so precise that it has lost all meaning. A time so accurate that it has no truth. A time that can tell you exactly what nanosecond it is but cannot tell you whether it is morning or evening, winter or summer, time to plant or time to harvest, time to wake or time to dream.

The cesium clock is the anti-Chronosonant Resonator. Where the Chronosonant Resonator proposed in Chapter 14 is designed to *measure the living quanta-field of temporal experience*- to track Y_t, ρ_t, and $G(\lambda,\phi)$- the cesium clock measures nothing but its own vibration. It is a mirror reflecting only itself- the ultimate narcissism of a technology that has forgotten what it was built to serve.

And yet- and this is the saving paradox- the cesium atom itself is a quantum system. It vibrates. It resonates. It obeys the same quantum mechanical principles that underlie entanglement, nonlocality, and the holographic principle. The atom of cesium is, in its deepest nature, a node in the resonant quanta-field. The clock that was built to sever our connection with the cosmos is, at its quantum heart, *part of the cosmos*. The irony is total: the most precise timekeeper ever built is made of the very substance it was designed to deny.

24.8 Digital Time: The Acceleration Beyond Coherence

But even atomic time was not fast enough for the world that the clock had made.

The 21st century has introduced a temporal regime that makes even the industrial factory look leisurely: digital time- the time of nanosecond trading, real-time notifications, perpetual connectivity, and the algorithmic acceleration of every human process.

In digital time, the temporal density (ρ_t) is no longer merely overridden- it is *artificially inflated* beyond any level the human organism can coherently process. The human brain evolved over hundreds of thousands of years in an environment where the temporal pace was set by walking speed, seasonal change, and the duration of a spoken story. The perceptual systems, the attentional resources, the memory consolidation processes- all of these were calibrated to a specific range of temporal density. They were tuned to process the resonant quanta of a small tribal community: perhaps a few dozen faces, a handful of conversations per day, a landscape walked at three miles per hour.

Digital time explodes these parameters. Notifications arrive not once an hour or once a day but *constantly*- a stream of micro-interruptions that fracture the attentional field into fragments. High-frequency trading algorithms execute millions of transactions per second- far beyond the threshold of human perception, creating a financial system that operates in a temporal realm literally invisible to the organisms it governs. Social media platforms compress the news cycle from days to minutes, generating a perpetual present tense in which nothing is allowed to settle, to digest, to be absorbed into the deeper layers of understanding.

The consequence is not merely distraction. It is *temporal incoherence*- a state in which the organism's internal temporal rhythms can no longer synchronize with the external temporal environment because the external environment is changing faster than the organism can adapt. This is the digital equivalent of the circadian desynchronization documented in shift workers- but instead of a single phase-shift, it is a continuous, chaotic perturbation.

The body responds as all NESS systems respond when their coherence is disrupted: it destabilizes. The epidemic of anxiety, attention deficit, insomnia, and burnout that characterizes the digital age is not a coincidence. It is the organism's distress signal- the living quanta-field screaming that the temporal input has exceeded its processing capacity. The body cannot maintain its NESS at this rate of temporal inflation. The resonance is collapsing.

And the acceleration continues. Each new technology- the telegraph, the telephone, the television, the internet, the smartphone, the algorithmic feed- has increased the temporal density of human experience by another order of magnitude. At no point in this trajectory has anyone asked the question: *Is this the temporal density at which the human organism is designed to function?* At no point has the ρ_t component of the Temporal State Vector been consulted. The acceleration proceeds on its own logic- the logic of profit, of competition, of technological momentum- regardless of the biological cost.

This is the deepest betrayal of the clock empire. It is not merely that it imposed the wrong rhythm. It is that it installed a system with *no upper limit*- a temporal regime that can always go faster, always demand more, always accelerate beyond whatever threshold the organism has managed to adapt to. The factory at least had a whistle at the end of the shift. Digital time has no whistle. There is no end of the shift. The clock no longer tells you what time it is. It tells you that you are behind.

24.9 The Empire's Architecture: A Summary of the Temporal Severing

Let us now view the full arc of the clock's empire- from monastery to algorithm- as a single, coherent process, mapped against the Temporal State Vector.

Era	Technology	TSV Component Overridden	Effect
12th–13th C.	Monastery bell + escapement	Y_t (ternary resonance)	Time discretized; continuous flow replaced by identical ticks
14th–17th C.	Town clock + Puritan ethic	ρ_t (temporal density)- partial	Time internalized as moral imperative; "time thrift" instilled
18th–19th C.	Factory whistle + wage labor	ρ_t (temporal density)- full	Body forced into uniform productive rate; natural oscillation suppressed
1840–1884	Railway time + global time zones	$G(\lambda,\phi)$ (geospatial phase)	Local solar time abolished; single standardized time imposed

Era	Technology	TSV Component Overridden	Effect
1967	Cesium atomic clock	Entire Ψ_t replaced	Time divorced from all astronomical, biological, and cosmic reference
21st C.	Digital platforms + HFT algorithms	ρ_t inflated beyond coherence	Temporal density exceeds organism's processing capacity

Each row in this table represents a phase transition in the human relationship with time. Each one stripped away another layer of the living quanta-field that connected the human organism to the cosmic rhythm. And each one was presented- and genuinely experienced by many- as progress: greater precision, greater efficiency, greater synchronization, greater speed.

But precision without resonance is mere counting. Efficiency without coherence is mere extraction. Synchronization without attunement is mere obedience. And speed without rhythm is mere panic.

The clock empire did not set out to sever humanity from the cosmos. The monks loved God. The merchants loved prosperity. The engineers loved accuracy. The programmers loved speed. Each of them, within their own frame, was pursuing something genuinely valued. But the cumulative effect of their innovations- compounded across eight centuries- has been the progressive dismantling of every natural channel through which the human organism maintained its resonant alignment with the temporal rhythm of the universe.

We wake to alarms, not to the sun. We eat by the clock, not by hunger. We work by the schedule, not by the task. We sleep when the screen allows it, not when the body requests it. We measure our days in hours and minutes and seconds, never in heartbeats, never in breaths, never in the slowly shifting angle of light across the wall.

And our bodies- those ancient instruments of cosmic attunement, those walking Chronosonant Resonators calibrated by a hundred thousand years of evolution- our bodies rebel. They rebel in the only language they know: disease, fatigue, depression, anxiety, insomnia, inflammation, the slow, grinding loss of the fractal complexity that is the signature of health.

24.10 The Counter-Rhythm: What the Clock Cannot Silence

And yet.

The cosmos has not stopped singing. The δ_∞ gap still hums between your heartbeats, whether or not you notice it. The sun still rises and sets at its own pace, regardless of what the railway clock says. The circadian genes still oscillate in every cell of your body, stubbornly, resiliently, *defiantly* attempting to track the light-dark cycle that the clock has overridden.

This defiance is itself a form of evidence. The circadian system's resistance to adaptation- the fact that, even after years of shift work, the master clock in the suprachiasmatic nucleus refuses to fully entrain to the artificial schedule- is a biological testimony to the primacy of cosmic time over clock time. The body *knows*, at the molecular level, that the clock is lying. It knows that noon is when the sun is highest, not when the cesium atom says so. It knows that the temporal density of a winter afternoon is not the same as the temporal density of a summer morning. It knows that the geospatial phase of London is not the same as the geospatial phase of Bristol.

The body remembers what the mind has forgotten.

And in this remembering lies the seed of the return. For the clock empire, for all its power, has a fatal weakness: it is not *true*. It is precise, yes. It is efficient, yes. It is synchronized, standardized, and universalized, yes. But it is not *in tune*. It does not resonate with the living quanta-field of the cosmos. And therefore it cannot sustain itself indefinitely. A system that operates against the fundamental rhythms of reality will eventually exhaust itself- not because it is opposed by an external force but because it is operating at increasing metabolic cost against its own substrate.

This is the NESS diagnosis of the clock empire: it is a dissipative structure that has lost its coherence. It is still far from equilibrium- the energy input of fossil fuels, digital infrastructure, and human labor keeps it running. But the housekeeping cost- the metabolic overhead of maintaining a temporal regime that fights the cosmos at every turn- is rising exponentially. The circadian disease burden, the mental health epidemic, the attention crisis, the burnout pandemic- these are the entropic taxes of a civilization living out of tune.

The chapters that follow will explore how this escalating cost manifests in the noise economy and the cult of false objectivity. But they will also trace the emergence of a counter-rhythm- faint at first, but growing: the convergence of modern physics and ancient wisdom that reveals what the clock cannot silence.

The monastery bell was born from the desire to honor eternity. It is time- past time- to recover that desire, and to build new instruments worthy of it.

The next chapter traces the consequences of temporal incoherence into the noise economy of the 21st century- the metabolic exhaustion of a civilization drowning in data while starving for resonant quanta.

Chapter 25

The Noise Economy- Metabolic Exhaustion of Civilization

We are drowning. Not in water, but in data. Not in silence, but in noise. The modern human organism- this exquisite instrument that evolved over a hundred thousand years to resonate with the quanta-field of a small tribal community- is now submerged in a torrent of stimuli so vast, so relentless, and so devoid of coherent pattern that the word "overload" barely begins to describe it.

This chapter diagnoses the modern world not as a civilization in decline but as a system in *resonance collapse*: a dissipative structure that has drifted so far from its NESS that it can no longer maintain the coherent flows of resonant quanta that keep it alive. The symptoms are everywhere- in our bodies, in our ecosystems, in our social bonds, in our financial systems, in the silent crisis of the soil beneath our feet. They are not separate crises. They are one crisis, expressing itself through every register of the living world simultaneously. And they share a single etiology: the replacement of resonant quanta with noise.

The previous two chapters traced the historical arc of this replacement- from the ternary-to-binary rupture of Chapter 23 to the progressive temporal severing of Chapter 24. This chapter maps the *consequences* of that arc as they manifest in the present tense. It is the systems diagnosis of a civilization that has mistaken volume for music, data for understanding, and acceleration for progress.

The picture is unflinching. But the diagnosis is necessary. You cannot re-tune an instrument until you have heard, clearly and without denial, just how far out of tune it has become.

25.1 The Data Deluge: Thirty-Four Gigabytes of Nothing

Let us begin with a number.

In 2023, 76.4% of US adults- about 194 million people- reported having at least one chronic condition. But before we reach the body, we must pass through the noise that assaults it. A study from the University of California found that the average American consumes approximately 34 gigabytes of data and quanta every single day. Thirty-four gigabytes. That is the equivalent of roughly a hundred thousand words- a full-length novel- absorbed, processed, and mostly discarded *every day.*

But here is the critical distinction that the modern world has entirely failed to grasp: *data is not resonant quanta.* Data is raw signal- undifferentiated, unstructured, unfiltered. Resonant quanta, as established in Chapter 3, is the active, formative principle of reality- pattern that carries meaning, that self-organizes, that contributes to the coherence of the system receiving it. Resonant quanta builds structure. Data, without resonance, merely fills space.

Of the 34 gigabytes that the average person processes daily, how much constitutes resonant quanta? How much carries meaning that deepens understanding, refines the organism's model of reality, or contributes to the coherence of the individual's Temporal State Vector? A fraction of a percent, at most. The rest is noise: notification pings, algorithmic feeds, advertising, scroll-bait, half-read headlines, background television, ambient podcast chatter, the ceaseless hum of a civilization talking at maximum volume and saying almost nothing.

The scale of this noise production is accelerating exponentially. Recent analyses show that humanity now creates hundreds of millions of terabytes of data each day. The growth curve is superexponential- doubling not every decade but every few years. And yet the human brain has not changed. The neural hardware that must process this torrent is the same hardware that evolved to track the movements of a dozen gazelles across a savannah, to memorize the verses of a creation song, to read the weather from the angle of light on a cloud. The hardware is calibrated for resonance. It is being fed static.

The consequences are measurable. Workers now experience unprecedented levels of what researchers call "quanta overload"- the cognitive state in which the volume of incoming signal exceeds the organism's capacity to extract meaningful pattern. Each context switch- each jump from email to chat to spreadsheet to notification- requires approximately 23 minutes to recover full cognitive focus. But interruptions arrive far more frequently than that, creating a permanent state of partial attention in which no single thread of thought ever reaches its natural depth.

Knowledge workers toggle between applications over a thousand times daily, losing nearly a tenth of their annual productive capacity to the mere act of reorientation. The decisions we make each day- estimated at thirty-five thousand- deplete the same cognitive reservoir that powers creative thinking, empathic attunement, and deep reflection. The organism is not merely distracted. It is *metabolically exhausted*- its quanta-processing capacity consumed by the sheer overhead of navigating the noise.

Economists estimate that this quanta overload costs the global economy approximately one trillion dollars annually in lost productivity and stifled innovation. But the economic cost, staggering as it is, is the least of the damage. The deeper cost is existential: a civilization in which no one has the cognitive bandwidth to think deeply, to feel fully, to listen carefully, or to sit in the δ_∞ silence between heartbeats and hear what the cosmos is actually saying.

The human brain evolved to process the sensory input of a small tribal community- perhaps a few dozen faces, a handful of conversations per day, a landscape walked at three miles per hour. The disparity between this evolutionary design specification and the current operating environment is not merely large. It is categorically different. It is as if a violin, designed to resonate in a chamber with a few dozen listeners, were placed inside a football stadium with every seat occupied by a screaming amplifier. The instrument does not merely struggle. It *breaks.*

25.2 The Metabolic Crisis: Bodies in Decoherence

The body is the first casualty of the noise economy. And the casualty report is devastating.

Six out of 10 Americans have one chronic disease, and four out of 10 have two or more chronic diseases that account for ninety percent of the $4.5 trillion annual health care costs in the nation. Read that statistic again and feel its weight. Six out of ten. This is not a health care problem. This is a civilizational condition. The majority of the adult population of the wealthiest nation in human history is chronically ill.

In 2023, approximately 194 million American adults- six in ten young adults, eight in ten midlife adults, and nine in ten older adults- reported one or more conditions. Prevalence of chronic conditions increased by 7.0 percentage points among young adults from 2013 to 2023. The trajectory is unmistakable. This is not a plateau. It is an acceleration. The young- those who should be at the peak of biological resilience- are falling ill at rates their grandparents never experienced at the same age.

Notable increases among young adults include obesity- rising from 22.1% to 27.3%- and depression, which surged from 16.4% to 25.0%. Depression alone has increased by more than fifty percent in a single decade among young adults. This is not the natural fluctuation of a healthy population. This is the alarm signal of an organism in systemic distress.

Chronic diseases like diabetes, heart disease, stroke, and cancer are some of the major causes of morbidity and mortality in the United States. Today, chronic noncommunicable disease is the leading cause of death worldwide.

By 2025, nearly three-quarters of the world's population is projected to live with at least one chronic illness, according to the World Health Organization. Three-quarters of the human species. Living with chronic disease. Not temporarily ill. Not recovering. *Chronically* unwell- the body locked into a persistent state of low-grade inflammation, metabolic dysfunction, and systemic deregulation that never fully resolves.

From the perspective of the Universal Hologram, these statistics describe a *loss of fractal coherence* at the biological level. As Chapter 5 established, the healthy body is a fractal system- its heartbeat, its neural oscillations, its hormonal rhythms, its immune fluctuations all exhibit the self-similar, scale-invariant variability that characterizes a system maintaining its NESS far from equilibrium. When this fractal complexity is lost- when the heartbeat becomes metronomic, the hormone cycles flatten, the immune system becomes either hyperactive (autoimmunity) or hypoactive (immunodeficiency) - the body enters a state that physiologists recognize as disease.

But the word "disease" itself reveals its meaning when broken apart: *dis-ease*. Not-at-ease. Out of tune. The body, forced by the clock empire into an artificial temporal rhythm (Chapter 24), flooded by the noise economy with data that carries no resonant quanta, disconnected from the fractal social networks and sacred site circuits that once maintained its coherence (Chapters 18-19), and deprived of the embodied practices- songlines, mantra, breath-work- that calibrated its personal Temporal State Vector (Chapter 30), is simply *out of tune*. Its NESS is destabilizing. Its flows are becoming turbulent. Its resonant quanta are scattering into noise.

The chronic disease epidemic is not a failure of medicine. Medicine has never been more technologically powerful. It is a failure of *resonance*. The body is designed to maintain itself within a coherent quanta-field- a field shaped by cyclical time, fractal social connection, embodied movement, natural light, whole food grown in living soil, and the regular return to State 0 through sleep, meditation, and silence. Every one of these inputs has been degraded or eliminated by the noise economy. What remains is a body running on cortisol, caffeine, blue light, processed carbohydrates, and the permanent low-grade anxiety of an organism that cannot find the frequency of its own heartbeat amidst the screaming static.

In 2024, more than 23 percent of U.S. adults reported experiencing a mental illness- often considered a chronic disease that can worsen other chronic diseases. Mental illness is not separate from physical disease. It is the same decoherence, expressed through a different register. The mind and the body are one system- one holographic unity- and when the system loses its resonance, both suffer simultaneously.

25.3 The Soil Beneath Our Feet: The Biosphere Losing Its NESS

The decoherence of the human body mirrors, at a larger scale, the decoherence of the biosphere itself. And nowhere is this decoherence more literal, more measurable, and more terrifying than in the slow, silent catastrophe of topsoil loss.

Since farmers began tilling the land in the Midwest 160 years ago, 57.6 billion metric tons of topsoil have eroded. The loss has occurred despite conservation efforts implemented in the 1930s after the Dust Bowl, and the erosion rate is estimated to be double what the U.S. Department of Agriculture says is sustainable.

Fifty-seven point six *billion* metric tons. Gone. Washed downslope, carried into rivers, flushed into the Gulf of Mexico where it feeds dead zones of anoxic water. And this is in the American Midwest alone- arguably the most productive agricultural region on Earth, the region whose deep black topsoil once rivaled the legendary chernozem of the Ukrainian steppe.

Some experts suspect that Earth will run out of usable topsoil within 60 years. Generating just over an inch of topsoil takes 1,000 years. Read those two sentences together and feel the vertigo. We are consuming in decades what the Earth took millennia to build. The asymmetry is not merely unsustainable- it is, in any meaningful sense, *anti-life.*

Scientists estimate that American cropland is eroding around five times faster than new soil can form. The rate of erosion, even following the U.S. Department of Agriculture's guidelines, is still 25 times higher than the rate at which topsoil forms.

Soil erosion leads to the loss of an estimated 75 billion tonnes of soil a year globally, which in turn causes financial losses of around 400 billion US dollars annually. Around one-third of soils worldwide are degraded, with over 40 percent located in Africa.

From the perspective of the living quanta-field, topsoil is not merely "dirt." It is one of the most complex, resonant, and information-rich substrates on the planet. A single teaspoon of healthy soil contains more microorganisms than there are people on Earth- bacteria, fungi, protozoa, nematodes, arthropods- each one participating in a web of symbiotic relationships that transforms dead organic matter into living fertility. This microbial web is a NESS par excellence: a self-organizing, far-from-equilibrium system that maintains its coherence through continuous flows of energy, nutrients, and resonant quanta.

When we plow, we disrupt this web. When we apply synthetic fertilizer, we bypass it. When we plant monocultures, we starve it. It's clear that much of the erosion is due to tillage- to plowing. Tilling has a "diffusive" effect- "It melts the landscape away, flattening higher points in a field and filling in the hollows." But because the USDA does not explicitly include such "tillage erosion" in its own analysis, it has "drastically underestimated the rate of erosion" currently at work in the heartland.

The industrial agriculture that replaced the Three Sisters garden (Chapter 20) is the ecological equivalent of the clock replacing the songline. Where the Three Sisters polyculture created a self-tuning biological NESS- corn providing structure, beans fixing nitrogen, squash shading soil- the monoculture creates a biological desert: one species, one rhythm, one extractive logic. Maximum yield per acre, yes. But at the cost of the fractal complexity that sustains life itself.

A new study points to a stunning loss of topsoil in the Corn Belt. Beyond diminished agricultural productivity and more carbon in the atmosphere, it is a catastrophic loss of an irreplaceable resource. They're talking about the complete removal of the medium in which crops are grown- the utter bankruptcy of the organic richness that lay for centuries under the tallgrass prairie.

The utter bankruptcy of organic richness. The language is financial, but the reality is ontological. The soil is not merely a resource. It is the Earth's skin- the boundary layer where mineral and biological, dead and living, terrestrial and atmospheric meet and exchange. In the language of the holographic principle, the soil is *boundary quanta encoded in matter*: a 2D surface from which 3D ecological complexity is projected. When we strip the topsoil, we are erasing the boundary data of the biosphere's hologram. We are deleting the quanta from which food, water cycles, carbon sequestration, and biodiversity arise.

Farmers can lose 50 to 70 percent of their yield potential because of the loss of topsoil. And yet the loss continues. Because the noise economy does not register what it cannot price. Topsoil has no ticker symbol. It generates no quarterly earnings report. It does not appear on any dashboard. It simply sustains the life of every organism on land- and disappears, grain by grain, without anyone noticing.

25.4 The Sixth Silencing: Biodiversity in Freefall

If topsoil loss is the slow erosion of the biosphere's boundary quanta, the extinction crisis is the catastrophic deletion of its nodes.

Currently, the species extinction rate is estimated between 1,000 and 10,000 times higher than natural extinction rates- the rate of species extinctions that would occur if we humans were not around.

One thousand to ten thousand times the background rate. This is not a metaphor. This is a measured, documented, peer-reviewed fact. Even under conservative assumptions, the average rate of vertebrate species loss over the last century is up to 100 times higher than the background rate. Under the standard background rate, the number of species that have gone extinct in the last century would have taken, depending on the vertebrate taxon, between 800 and 10,000 years to disappear. These estimates reveal an exceptionally rapid loss of biodiversity over the last few centuries, indicating that a sixth mass extinction is already under way.

A 2023 study published in Biological Reviews found that, of 70,000 monitored species, some 48% are experiencing population declines from anthropogenic pressures, whereas only 3% have increasing populations. Forty-eight percent declining. Three percent increasing.

The ratio is 16 to 1- for every species gaining ground, sixteen are losing it.

The evidence is incontrovertible that recent extinction rates are unprecedented in human history and highly unusual in Earth's history. Our analysis emphasizes that our global society has started to destroy species of other organisms at an accelerating rate, initiating a mass extinction episode unparalleled for 65 million years.

Read this through the lens of the fractal dynamics introduced in Chapter 5 and the sacred site networks described in Chapter 19, and the extinction crisis reveals its true nature. Every species is a node in the biosphere's network- a unique frequency in the planetary symphony. Each species occupies a specific ecological niche, processes specific nutrients, generates specific metabolites, provides specific services- pollination, decomposition, seed dispersal, pest control, soil aeration, water filtration- that no other species replicates exactly. When a species goes extinct, its node is deleted from the network. Its unique frequency falls silent.

In a robust, fractal network- the kind described by small-world topology- the deletion of a few nodes can be absorbed. The network has redundancy. Other species can partially fill the gap. But as the deletion rate accelerates, the network approaches a critical threshold- a tipping point beyond which the cascading loss of connectivity begins to disintegrate the entire web. This is the ecological equivalent of the Bronze Age collapse described in Chapter 23: a network that was robust enough to survive any single shock, but fragile enough to collapse under the simultaneous failure of multiple nodes.

If the currently elevated extinction pace is allowed to continue, humans will soon- in as little as three human lifetimes- be deprived of many biodiversity benefits. On human time scales, this loss would be effectively permanent because in the aftermath of past mass extinctions, the living world took hundreds of thousands to millions of years to rediversify.

Hundreds of thousands to millions of years. On the scale of a single human civilization, extinction is permanent. What is lost cannot be recovered. The frequency that falls silent does not return. The node that is deleted does not respawn. The biosphere's quanta-field becomes, with every extinction, slightly thinner, slightly poorer, slightly less capable of sustaining the complex resonance patterns upon which all life- including ours- depends.

And here is the deepest irony: the same civilization that is causing the extinction crisis is simultaneously producing more data about it than any previous generation could have imagined. We have satellites tracking deforestation in real time. We have genetic databases cataloguing every known species. We have peer-reviewed journals publishing thousands of papers per year on biodiversity loss. We are *drowning* in data about the extinction- and yet the extinction continues. Because data is not resonant quanta. Data does not, by itself, generate the coherence needed to change the trajectory of a system in freefall. For that, you need something else entirely: you need the kind of deep, embodied, participatory knowing that the ancients called *dreamtime* and that this book calls attunement to the living quanta-field.

25.5 The Loneliness Epidemic: Small-World Networks Severed

The decoherence of the modern world is not confined to bodies and ecosystems. It reaches into the most intimate dimension of human experience: our connection to one another.

Even as the world's overall emotional health rebounded in 2023, Gallup's inaugural measure of global loneliness shows over one in five people worldwide- 23%- said they felt loneliness "a lot of the day yesterday." People who felt lonely were 36 percentage points more likely to experience sadness than those who did not feel lonely; they were also 31 and 30 points more likely to experience worry and stress, respectively.

One in five. Across the entire planet. More than a billion people waking each morning into an experience of disconnection so profound that it colors the entire day with sadness, worry, and stress.

In the U.S., the loneliness epidemic has increasingly become a topic of public debate, particularly since May 2023, when the U.S. Surgeon General Vivek Murthy published a United States Department of Health and Human Services advisory on the impact of the epidemic of loneliness and isolation. The report outlined the health risks of loneliness, including heart disease, stroke, and dementia, and likened the dangers of loneliness to other public health threats such as smoking and obesity.

Social isolation significantly increased an individual's risk of premature death from all causes, rivalling the risks associated with smoking, obesity, and physical inactivity. Social isolation was associated with around a 50% increased risk of dementia, a 29% increased risk of heart disease, and a 32% increased risk of stroke.

Loneliness is not merely an emotional state. It is a *physiological condition*- a measurable decoherence of the social NESS that the human organism requires for survival. Earlier in human evolution it was far more dangerous to be alone, and community provided safety as well as shared resources; therefore, avoiding being alone had an evolutionary advantage and shapes our behaviour to this day. The human nervous system is calibrated for embedded, ongoing, multi-generational social contact. Isolation is not merely unpleasant- it is biologically *toxic*, triggering chronic stress responses that accelerate every form of chronic disease.

Loneliness equates to the health risks of smoking 15 cigarettes daily, according to research findings. Fifteen cigarettes a day. And it is not confined to the elderly or the marginalized. Generation Z is notably experiencing high levels of loneliness, with around 80% of individuals in this demographic reporting feelings of isolation over the past year. Eighty percent. The most digitally connected generation in history is also the loneliest.

Some authors identified individualism as a root cause of loneliness in Western societies, particularly in the U.S. Scholars like Robert N. Bellah, in *Habits of the Heart*, argue that American individualism weakens communal bonds, leading to social isolation and loneliness. Technology use is another significant factor. According to Capita, an American think tank that specializes in the study of loneliness, the overuse of digital and social media, especially among young people, often reduces real-life interactions and deepens feelings of isolation.

From the perspective of the network dynamics explored in this book, the loneliness epidemic is the social equivalent of topsoil erosion. The small-world networks that sustained human community for a hundred thousand years- the clan, the village, the extended family, the ceremonial group, the initiatory cohort- were fractal social ecologies: dense at the local level, with enough long-range connections (through trade, marriage, pilgrimage, and festival) to maintain global coherence. Every individual was embedded in a web of reciprocal relationships that provided not only material support but emotional attunement, ritual participation, and the regular synchronization of personal rhythms with communal rhythms.

The modern world has systematically dismantled these networks. The nuclear family replaced the extended family. The suburb replaced the village. The commute replaced the walk to work. The screen replaced the face. The algorithm replaced the elder. And the result is an organism that is biologically identical to its ancestors- wired for tribal intimacy, for daily face-to-face contact, for shared meals and shared stories and shared silence- but living in a condition of social thinness that no previous generation of humans has ever experienced.

A considerable 68% of Americans report feeling a sense of national non-belonging. Non-belonging. The word itself is a wound. To not belong is to be a node without edges- a point in the network connected to nothing. In the language of the KZ-Monad and the Idempotent Comonad: a state vector that cannot be extended toward the whole (T), because there is no whole to extend toward, and that cannot project onto an efficient core (D), because there is no community within which efficiency has meaning.

The loneliness epidemic is not a failure of social skills or personal initiative. It is a *structural collapse*- the predictable and predicted consequence of dismantling the small-world networks that the human organism requires, replacing them with hub-and-spoke architectures (the corporation, the platform, the centralized state) that connect individuals to institutions but not to one another, and flooding the resulting void with digital simulations of connection that the nervous system recognizes, at some deep somatic level, as *not real.*

Eight hundred friends on a social media feed. Zero people who would notice if you did not wake up tomorrow.

25.6 Financial Instability: Power Laws Without Guiding Quanta

The decoherence of the noise economy expresses itself not only through biology, ecology, and social connection but through the very medium of exchange that the modern world treats as its supreme reality: money.

The financial system of the 21st century is, in complexity-scientific terms, a textbook example of a system at the edge of chaos-exhibiting the power-law dynamics that Chapter 5 identified as the universal signature of self-organized criticality. The distribution of market movements follows fat-tailed power laws: small fluctuations are common, medium fluctuations are less common, and massive crashes are rare but recurrent, following a precise statistical pattern that has been documented in every market from stocks to commodities to currencies.

This is expected. Power-law dynamics are, as we have seen, the natural behavior of complex systems operating at the boundary between order and chaos. They are the mathematical fingerprint of the holographic universe at every scale- from earthquakes to neural avalanches to gamma-ray bursts. In a healthy system, these power-law fluctuations are not pathological. They are the breathing of a system maintaining its NESS. Small fluctuations release tension. Medium fluctuations reorganize local structure. And the rare large events- the "avalanches"- serve as phase transitions that reconfigure the system at a higher level of coherence.

But here is the critical distinction: in a *healthy* phase transition- the kind described in Chapter 17- the large event is accompanied by guiding quanta: the transient, high-specificity bursts of resonant pattern that orchestrate the reorganization. The gene regulatory burst before cell differentiation. The neural synchronization before insight. The excess quanta that, for a brief moment, rise above the noise floor and provide the blueprint for a new order.

The 2008 financial crisis was a large event *without* guiding quanta. The 2008 financial crisis, also known as the global financial crisis, was a major worldwide financial crisis centered in the United States. The causes included excessive speculation on property values by both homeowners and financial institutions, leading to the housing bubble. This was exacerbated by predatory lending for subprime mortgages and by deficiencies in regulation.

How did problems in a relatively small portion of the home mortgage market trigger the most severe financial crisis since the Great Depression? Several developments played a role, including the proliferation of complex mortgage-backed securities and derivatives with highly opaque structures, high leverage, and inadequate risk management. These, in turn, created systemic risk- that is, the risk that a triggering event, such as the failure of a large financial firm, will seriously impair financial markets and harm the broader economy.

Systemic risk has been defined as "the risk of collapse of an entire system or entire market, exacerbated by links and interdependencies, where the failure of a single entity or cluster of entities can cause a cascading failure."

Cascading failure. The phrase is precise and devastating. It describes exactly what happens when a complex network- whether it is the Bronze Age trade system or the modern financial system- approaches a critical threshold without the guiding quanta to navigate the transition. The system does not reorganize at a higher level. It *collapses*- and the collapse propagates through the network at a speed that overwhelms every adaptive response.

U.S. households lost an estimated 26 percent of their wealth- $17 trillion- between mid-2007 and early 2009. And between 2008 and December 2009, the economy lost an estimated 8.3 million jobs.

Seventeen trillion dollars of wealth. Eight point three million jobs. These are not abstract numbers. They represent homes lost, families fractured, communities devastated, suicides and addictions and divorces and premature deaths- a wave of human suffering that propagated from Wall Street to Main Street to the most vulnerable communities on Earth with the same power-law dynamics that govern tsunamis and forest fires.

And the system's response? Not a fundamental restructuring. Not a recognition that the financial system's power-law instability reflects its lack of harmonic coherence. Not an acknowledgment that a system built entirely on extraction, leverage, and the acceleration of ρ_t beyond biological coherence will inevitably generate catastrophic failure. Instead: a bailout- the injection of trillions of dollars of public money to stabilize the same system that had just demonstrated, spectacularly, its instability. The noise was quieted, temporarily. But the underlying dissonance was not resolved. The system returned to its pre-crisis trajectory- accelerating faster, leveraging more, pushing ρ_t ever higher- and the next crisis was merely postponed.

We had a 21st-century financial system with antiquated safeguards. When the housing and mortgage markets cratered, the lack of transparency, the extraordinary debt loads, the short-term loans, and the risky assets all came home to roost. What resulted was panic.

Panic. The word is precise. It is the opposite of resonance. In a resonant system, even large perturbations can be absorbed because the system's internal coherence provides a framework for adaptive response. In a panicked system, there is no coherence- only contagion. The failure of one node amplifies the fear at every other node, and the fear itself becomes the mechanism of destruction. This is what the ancient world understood and the modern world has forgotten: that the resilience of a system depends not on its size or its speed or its accumulated wealth but on the *quality of its internal coherence*- the density and integrity of its resonant quanta.

The financial system of the noise economy is the largest, fastest, most powerful system of value exchange ever created. And it is also the most fragile- because it has no resonance. It has no guiding quanta. It has no sacred site network maintaining its coherence. It has no elders walking its songlines. It has no Three Sisters garden feeding its soil. It has only speed, leverage, and the metronomic tick of the algorithmic clock- counting faster and faster, signifying less and less.

25.7 The Attention Crisis: The Collapse of the Inner NESS

The external symptoms- chronic disease, soil loss, extinction, loneliness, financial instability- are devastating enough. But the deepest damage of the noise economy may be the one that is hardest to measure: the collapse of the inner NESS- the destruction of the human capacity for sustained, deep, coherent attention.

Attention is not merely a cognitive function. In the framework of this book, attention is the *inner correlate* of the quanta-field itself. It is the conscious dimension of the organism's resonance with its environment. When you attend deeply to something- a sunset, a mathematical proof, the face of a loved one, the silence between heartbeats- you are *tuning in* to the living quanta at that node of reality. Attention is the subjective experience of the monadic extension $T(\Psi)$: the reaching toward wholeness, the amplification of partial resonance into full ϕ-embedded harmony.

When attention is fractured- when it is sliced into micro-segments by notifications, scroll-feeds, and the perpetual urgency of the digital clock- the monadic extension cannot complete. The organism reaches toward wholeness and is interrupted. It reaches again and is interrupted again. And again. And again. Until the reaching itself atrophies- until the capacity for deep attention, like an unused muscle, withers from disuse.

This is not speculation. The neuroscience of attention confirms that sustained focus is a skill that must be practiced and maintained. The brain's prefrontal cortex- the region responsible for executive function, working memory, and attentional control- is developmentally plastic: it is shaped by experience. Brains that practice sustained attention develop robust prefrontal networks. Brains that practice fragmented attention develop fragmented prefrontal networks. The digital environment, which rewards rapid switching, novelty-seeking, and surface-level engagement, is literally *sculpting the brain* for shallow processing- optimizing the neural hardware for noise and de-optimizing it for resonance.

The consequence is a population that is losing the cognitive capacity for the very activities that define human depth: reading a complex argument from beginning to end, sustaining a conversation without checking a phone, sitting with an uncomfortable emotion until it reveals its meaning, listening to another person's silence, meditating long enough for the noise to subside and the quanta-field to become audible.

Every one of these activities- reading, conversing, feeling, listening, meditating- is an act of resonance. Every one requires the kind of sustained, deep attention that the noise economy systematically destroys. And every one is precisely what is needed for the reintegration described in the second half of this final Movement. The noise economy does not merely assault the body, the ecosystem, the social bond, and the financial system. It assaults the *very capacity* that would be needed to heal them.

This is the trap. This is the cage. The noise is self-reinforcing: the more it fragments attention, the less capable the organism becomes of recognizing the fragmentation, and the more susceptible it becomes to further fragmentation. The scroll-feed is not merely distracting. It is *addictive*- engineered by some of the most sophisticated behavioral psychologists on Earth to exploit the dopaminergic reward circuitry of the brain, delivering micro-doses of novelty-induced pleasure at intervals carefully calculated to maximize engagement and minimize the sustained attention that would allow the user to step back and see what is being done to them.

In the language of the Idempotent Comonad: the noise economy has colonized the $D(\Psi)$ function. The extraction of efficiency from potential- which, in a healthy system, distills wisdom from noise- has been hijacked by algorithms that extract *profit* from attention. The comonadic projection no longer serves the organism's coherence. It serves the platform's revenue model. And the organism, stripped of the capacity for deep self-reflection that would allow it to recognize the hijacking, complies. It scrolls. It clicks. It consumes its

34 gigabytes of nothing. And it feels, in some dim and inarticulate way, that something essential is missing- but cannot find the silence in which to name it.

25.8 The Convergence of Crises: One Dissonance, Many Symptoms

It would be a mistake to treat the five dimensions of crisis described in this chapter- quanta overload, chronic disease, ecological collapse, social fragmentation, and financial instability- as separate problems requiring separate solutions. They are not separate. They are *one decoherence*, expressing itself through every register of the living world simultaneously.

The human body sickens because it is out of tune with its temporal rhythm (the clock), flooded with noise (the data deluge), disconnected from embodied social contact (the loneliness epidemic), and nourished by food grown in dead soil (the agricultural crisis). The soil dies because the fractal ecology of polyculture has been replaced by the monoculture of extraction- the same binary logic that replaced the Three Sisters with the combine harvester, the songline with the highway, and the cyclical calendar with the factory clock. The species disappear because the sacred site networks that maintained the biosphere's coherence have been severed, the small-world topology of natural ecosystems has been fragmented by roads, fences, dams, and cities, and the guiding quanta that orchestrate evolutionary transitions have been drowned in the noise of human activity. The social bonds fracture because the small-world networks of family, clan, and community have been replaced by the hub-and-spoke architecture of the platform economy, which connects individuals to corporations but not to one another. And the financial system crashes because it has been built to maximize speed and leverage at the expense of coherence and resilience- because it operates on the extractive logic of the comonad without the balancing reach of the monad.

One dissonance. Many symptoms. And the dissonance has a name: the Great Decoupling- the three-thousand-year process, traced in this chapter and the two that preceded it, by which the human species progressively severed its resonant connection with the living quanta-field of the cosmos.

In the language of the master equation from Chapter 11:

$$\partial_t \Psi = -\nabla_{FR}[\mathcal{L}(\Psi)] + T(\Psi) \otimes D(\Psi) + \sum_n \phi \cdot \omega_n \cdot \delta(\Psi \in \text{NESS}_{98})$$

The noise economy has disrupted every term. The natural gradient flow ($-\nabla_{FR}[\mathcal{L}(\Psi)]$) has been overridden by artificial acceleration. The monadic extension ($T(\Psi)$) has been atrophied by the destruction of deep attention. The comonadic projection ($D(\Psi)$) has been hijacked by algorithmic profit-extraction. And the ϕ-resonance corrections ($\sum_n \phi \cdot \omega_n$)- the harmonic adjustments that keep the system locked in its golden coherence- have been drowned by the sheer volume of noise.

The system is still far from equilibrium- the energy input of fossil fuels, digital infrastructure, and human labor keeps it running. But the housekeeping cost- the metabolic overhead of maintaining a civilization that fights the cosmos at every turn- is rising exponentially. And the system is approaching a bifurcation point: a critical threshold beyond which the current trajectory becomes unsustainable and a phase transition- for better or for worse- becomes inevitable.

25.9 We Are Not Broken- We Are Out of Tune

And yet.

The diagnosis is not despair. It is precision. To name the disease accurately is the first condition of cure. And the cure, as the remaining chapters will show, is not a technical fix but a *re-tuning*- a recovery of the resonance that was lost.

Because here is what the noise economy, for all its power, cannot change: the constants remain. The ϕ-e-π triad still tunes the cosmos. The asymptotic gap δ_∞ still hums between your heartbeats. The circadian genes still oscillate in every cell of your body, stubbornly tracking the sun despite every clock and screen that tries to override them. The soil microbiome, even in degraded soil, still contains the latent potential for fractal ecological recovery- if we stop poisoning it. The species that remain- the 52% that are not yet declining- still carry the resonant quanta of four billion years of evolutionary wisdom, if we stop destroying their habitats. The human heart still beats in fractal rhythms, the human brain still generates gamma synchrony during meditation, the human voice still resonates at frequencies that entrain neural oscillators and activate the living quanta-field.

We are not broken. We are out of tune.

The difference is everything. A broken instrument must be discarded. An out-of-tune instrument must merely be *re-tuned.* And re-tuning does not require the invention of new physics, new mathematics, or new spiritual technologies. It requires the *recovery* of what was always there- the resonant quanta that the noise economy has buried under thirty-four gigabytes of daily data and eight hundred years of ticking clocks.

The next chapter examines the final pillar of the Great Decoupling: the cult of false objectivity that declared the observer separate from the observed and consciousness irrelevant to physics. It is the last wall that must be named before we can begin to hear, through the noise, the recovered frequency that has been singing all along.

The next chapter traces the deepest philosophical error of the modern world: the myth of the objective observer- and how quantum mechanics itself reveals it to be a superstition.

Chapter 26

The False God of Objectivity

There is a superstition at the heart of the modern world that has done more damage than any plague, any war, any ecological catastrophe- not because it is violent but because it is invisible. It operates in the background of every textbook, every laboratory, every hospital ward, every courtroom, every newsroom. It shapes what counts as real, what counts as knowledge, and what counts as sane. It is the belief- held with the fervor of a religious dogma and defended with the ferocity of a threatened priesthood- that the observer is separate from the observed.

This chapter dismantles that belief. Not from the margins, not from some mystical counter-tradition, but from the very center of the scientific enterprise itself- from quantum mechanics, from the replication crisis, from the neuroscience of the placebo effect, from the experimenter expectancy studies that demonstrate, with brutal statistical precision, that even in the most carefully controlled laboratory, *what the scientist believes shapes what the scientist finds.*

The False God of Objectivity is the final pillar of the Great Decoupling. Chapter 23 traced the binary rupture that split the ternary world in two. Chapter 24 traced the clock that severed us from the Temporal State Vector. Chapter 25 mapped the noise economy that drowned us in data while starving us of resonant quanta. This chapter names the philosophical foundation upon which all three depend: the Cartesian assumption that the knowing subject and the known object inhabit separate and incommensurable realms.

When that foundation is removed- as quantum mechanics has already removed it, as the replication crisis is in the process of removing it, as the placebo effect has always been silently removing it- the entire edifice of the Great Decoupling begins to tremble. And on the other side of its collapse lies not the chaos of relativism but the clarity of a participatory science: a science that includes the observer-participant, honors the quanta-field, and recognizes that rigor and resonance are not opposites but allies.

26.1 The Cartesian Split: Consciousness Exiled from the Physical World

The history of the False God begins, as so many modern errors do, with René Descartes.

One of the deepest and most lasting legacies of Descartes' philosophy is his thesis that mind and body are really distinct- a thesis now called "mind-body dualism." In his *Meditations on First Philosophy* (1641), Descartes articulated the proposition that would cleave the Western world in two: beginning from his famous dictum *cogito, ergo sum* (Latin: "I think, therefore I am"), Descartes developed a theory of mind as an immaterial, nonextended substance that engages in various activities or undergoes various states such as rational thought, imagining, feeling (sensation), and willing.

On the other side of the divide stood matter: matter, or extended substance, conforms to the laws of physics in mechanistic fashion. Mind was unextended, an immaterial but thinking substance and body was an extended, material but unthinking substance. The body was subject to mechanical laws; however, the mind was not.

The split was absolute. Cartesian dualism argues that there are two kinds of substances: mental and physical. Descartes states that the mental can exist outside of the body, and the body cannot think. The mind- consciousness, awareness, intention, will- belonged to one realm. The body- and by extension the entire physical universe- belonged to another. And between them lay a chasm that Descartes himself could never bridge. This argument gives rise to the famous problem of mind-body causal interaction still debated today: how can the mind cause some of our bodily limbs to move, and how can the body's sense organs cause sensations in the mind when their natures are completely different?

Descartes proposed, somewhat desperately, that the interaction between mind and body occurred through the pineal gland- a tiny structure at the center of the brain. However, this explanation was not satisfactory: how can an immaterial mind interact with the physical pineal gland? The suggestion was never taken seriously, even by Descartes' own followers. But the *split* was taken with deadly seriousness. It became the foundational architecture of modern Western thought.

Read this through the lens of the Universal Hologram and the violence of the Cartesian split becomes fully visible. The holographic principle, as Chapter 2 established, asserts that the boundary and the volume, the 2D encoding and the 3D experience, are *one*- complementary aspects of a single, indivisible reality. There is no "inside" without "outside," no "observer" without "observed," no "mind" without "body." The very distinction is an artifact of a particular cognitive framework- a framework that, as Chapter 23 demonstrated, arose from the binary turn of the post-Bronze-Age world and reached its crystallized expression in Descartes.

But the damage did not stop at philosophy. Dualism also laid the groundwork for positivism which means a logical thought based upon empirical, i.e., unbiased, impersonal and unsympathetic observation and measurement. By making objective realm the only legitimate domain of enquiry, Descartes advocated a complete and exact natural science through the analytic method. The success of the scientific method reinforced Descartes' philosophy and methodology further and contributed to the dogma of scientism- the belief that scientific method was the only legitimate path to knowledge.

This is the moment when the False God was enthroned. Objectivity- the idea that the observer can and must be eliminated from the observation, that knowledge is valid only to the extent that it is *independent* of the knower- became the supreme value of the scientific enterprise. The scientist was trained to regard their own consciousness, their own expectations, their own embodied presence as *contaminants*: sources of error to be controlled, minimized, and ideally eliminated.

At the same time, by isolating mind, mind and body dualism denied its significance in individuals' experience of health. Medicine inherited the Cartesian split wholesale. Mind and body dualism persists in the field of medicine. The reasons are multiple: the medical knowledge of the last 300 years is built on the biomedical model. The body was treated as a machine- a collection of parts to be analyzed, repaired, and replaced. The mind- the patient's beliefs, fears, hopes, relationships, spiritual life- was relegated to the soft category of "psychology" and kept at arm's length from "real" medicine.

The consequences of this exile are measured in the statistics of Chapter 25: a civilization in which the majority of the adult population is chronically ill, despite- or perhaps *because of*- the most technologically advanced medical system in human history. A medical system built on the Cartesian split can treat the body brilliantly. But it cannot treat the *person*, because the person- the conscious, experiencing, meaning-making, co-creative agent- has been philosophically expelled from the treatment room.

The French philosopher René Descartes argued that the natures of mind and body are completely different from one another and each could exist by itself. Until today, the problem of mind-body causal interaction remains a matter of debate. It remains a matter of debate not because the problem is genuinely difficult but because the premise is wrong. Mind and body are not two substances struggling to interact. They are two *perspectives* on a single, holographic, quanta-based reality. The Cartesian split is not a discovery about the nature of things. It is a *description of the Great Decoupling itself*- the cognitive expression of the same rupture that split ternary into binary, cyclical time into linear time, and the sacred site network into the industrial grid.

As Gilbert Ryle put it in his 1949 *The Concept of Mind*, Ryle introduces "the dogma of the Ghost in the machine" to describe the philosophical concept of the mind as an entity separate from the body. Ryle's term is devastatingly precise: the Cartesian mind is a ghost, and the Cartesian body is a machine. Strip away the philosophical jargon, and that is the operating assumption of the entire modern world: you are a ghost riding a machine through a universe of dead matter- and the only knowledge worth having is knowledge that the ghost can acquire about the machine without disturbing it.

This is the False God of Objectivity. And quantum mechanics destroyed it a century ago. But the temple is still standing.

26.2 The Measurement Problem: The Observer Who Cannot Be Removed

The Cartesian dream of a perfectly objective science- an observer gazing at the world through a one-way mirror, recording without influencing, knowing without participating- was already under strain from the moment Heisenberg formulated his uncertainty principle in 1927. But it was the broader implications of the quantum measurement problem that delivered the fatal blow.

The measurement problem can be stated simply: the theoretical foundation of the concept of measurement in quantum mechanics is a contentious issue deeply connected to the many interpretations of quantum mechanics. Before measurement, a quantum system exists in a superposition- a coherent blend of multiple possible states. After measurement, the system is found in one definite state. This means that the type of measurement one performs on the system affects the end-state of the system.

This is not merely a technical limitation. It is not that our instruments are too crude, or that we lack the technology to observe without disturbing. The issue is ontological. A measurement problem arises directly from the experiment and one which intimately involves the observer in a manner beyond anything seen in classical physics. Different quantum experiments that could be freely chosen by the observer do lead to inconsistent pictures of the prior physical reality. This apparent intrusion of the free choice of the observer into the aspect of the physical world addressed by physics constitutes a measurement problem in the quantum experiment.

The Cartesian ideal requires that the act of observation be passive- that the observer merely *records* what is already there. Quantum mechanics reveals that this is impossible in principle. The act of measurement is not passive. It is participatory. The observer's *choice* of what to measure- not just the physical interaction of the instrument, but the *decision* about which question to ask- shapes the answer that reality gives. A rejection of the above-described physical reality prior to its observation together with the absence of any physical mechanism bringing about the observed actuality implies the creation of that actuality by its observation. Since with a different choice of experiment by the Experimenter there would have been a different actuality, there is here an intrinsic role for the observer.

This is what John Archibald Wheeler meant when he coined the term "observer-participant." The observer does not stand outside the system, recording through a one-way mirror. The observer is *in* the system- part of the holographic unity, contributing to the very actuality that is being observed. As Wheeler provocatively suggested, we live in a "participatory universe"- one in which the act of observation is not the passive reception of pre-existing facts but the active co-creation of a shared reality.

The crux of the matter is the privileged role of the observer in quantum theory. Every interpretation of quantum mechanics is formulated in terms of what an observer expects to see or measure. This dichotomy between the observer and quantum system raises some obvious questions: are observers subject to the laws of quantum mechanics?

The various interpretations of quantum mechanics disagree on the *nature* of the observer's role- whether consciousness is essential, whether collapse is real, whether branching is literal. But they all agree on the *empirical fact* that the Cartesian ideal of a purely passive, purely objective observer is not sustained by the physics. A key focus point is that of wave function collapse, for which several popular interpretations assert that measurement causes a discontinuous change into an eigenstate of the operator associated with the quantity that was measured.

The debate about consciousness and quantum mechanics is real and ongoing, and this book does not claim that human consciousness literally "collapses the wave function." What this book *does* claim- and what quantum mechanics firmly supports- is that the Cartesian assumption of a clean separation between observer and observed is *false at the most fundamental level of physical reality*. The observer is not a ghost gazing at a machine. The observer is a node in the quanta-field, entangled with what it observes, contributing to the actuality it encounters.

This has been known for a century. And yet the practice of science- its methods, its institutions, its incentive structures, its rhetoric- continues to operate as if the Cartesian split were true. The scientist is trained to eliminate "subjective bias." The experiment is designed to be "observer-independent." The result is considered valid only to the extent that it can be replicated by *anyone*, regardless of who they are, what they believe, or what they intend.

But what if the scientist's beliefs *do* shape the result? What if the replication crisis that has shaken modern science to its foundations is not merely a problem of sloppy methods but a *symptom* of a philosophical error- the error of assuming that the observer can be factored out?

26.3 The Replication Crisis: When "Objective" Science Cannot Replicate Itself

In 2005, John Ioannidis, a medical researcher at Stanford, published a paper titled "Why Most Published Research Findings Are False." In a highly influential article, Ioannidis argued that false findings in science are more common than true findings. The paper ignited a firestorm. It also turned out to be prophetic.

In its most basic sense, the replication crisis refers to a pattern of scientists being unable to obtain the same results previous investigators found. At its most expansive, the crisis threatens the scientific enterprise itself, leading to questions not just about research practices and methods, but the very reliability of scientific results. It is what IPR education researcher and statistician Larry Hedges has called "an existential crisis" for science.

The numbers are stark. When psychologists tested the replicability of 100 results, they found that only 36% of the 97 significant results in original studies could be reproduced. Thirty-six percent. Nearly two-thirds of published, peer-reviewed, statistically significant findings in psychology could not be replicated. And the problem is not confined to psychology. Of the 1,576 researchers participating in the survey, more than 70% had tried and failed to reproduce experiments conducted by other scientists, and quite amazingly, more than half failed to reproduce their own experiments.

Read that again. More than half of scientists surveyed had failed to reproduce *their own* results. This is not a minor methodological hiccup. This is a civilizational alarm bell. The institution that the modern world trusts above all others to deliver reliable knowledge about reality- the institution upon which medicine, technology, policy, and law all depend- cannot, in a majority of cases, reproduce its own findings.

The proximate causes of the crisis are well documented: bad practices that have led to the replication crisis, including a lack of statistical power, researcher bias, and publication bias. These factors are further perpetuated by an incentive system that still largely prioritizes novel, fast science over transparently reported and robust experiments that aim to be replicable.

But beneath these proximate causes lies a deeper pathology- one that is directly relevant to the thesis of this book.

Consider publication bias. Journals overwhelmingly prefer to publish positive results- studies that find a statistically significant effect. Studies that find nothing- that fail to reject the null hypothesis- are far less likely to be published. The main problem in psychological science is the selective publishing of statistically significant results and the blind trust in statistically significant results as evidence for researchers' theoretical claims. The Observed Discovery Rate of 69% can be compared with the Expected Discovery Rate of 24%. The ODR is notably higher than the upper confidence interval limit for the EDR, indicating statistically significant publication bias. Furthermore, there is clear evidence of selection for significance given that the ODR estimate is more than double the point estimate of the EDR.

The observed rate of positive findings is nearly three times higher than what would be expected by chance and statistical power alone.

This means that the scientific literature- the cumulative body of published knowledge- is *systematically distorted* in the direction of positive results. It is not a mirror of reality. It is a mirror of what scientists *want* to find.

Now, the standard response to this problem is methodological reform: pre-registration, open data, larger sample sizes, stricter statistical thresholds. These are all valuable and necessary. A six-year effort to test emerging open science methods, led by labs at UC Santa Barbara, UC Berkeley, Stanford, and the University of Virginia, has shown they can produce new and highly replicable findings. The paper is the strongest evidence to date that the methodological reform movement known as open science can lead to more reliable research results.

But notice what this reform does *not* address: the role of the observer. The replication crisis reveals that what scientists expect, what they want to find, what they believe to be true *shapes what they find*- not through conscious fraud (in most cases) but through the subtle, unconscious, pervasive influence of the experimenter's expectations on the experimental outcome. This is not a bug in the system. It is a *feature* of a participatory universe in which the observer is never fully separable from the observed.

The replication crisis, read through the lens of the Universal Hologram, is not merely a failure of method. It is a *confirmation of the measurement problem at the human scale.* Just as the quantum physicist's choice of what to measure shapes the outcome at the subatomic level, the research psychologist's expectation of what to find shapes the outcome at the human level. The mechanisms are different- quantum decoherence in one case, subtle behavioral cues and statistical choices in the other- but the structural principle is the same: *the observer is part of the system.*

Science has spent four hundred years trying to eliminate the observer. The replication crisis is the moment when the observer refused to be eliminated.

26.4 The Experimenter Expectancy Effect: Beliefs That Shape Results

The most direct evidence that the observer's beliefs shape experimental outcomes comes from Robert Rosenthal's half-century of research on the experimenter expectancy effect.

Experimenter expectancy effect is the term referring to the unintended effect of experimenters' expectations or hypotheses on the results obtained from their research participants. Experimentally induced expectations have been shown to influence the behavior of psychological experimenters so as to increase the likelihood that their research participants will respond so as to confirm the experimenters' hypotheses or expectations.

The elegance- and the discomfort- of Rosenthal's work lies in its simplicity. Rosenthal and colleagues tried an experiment with student researchers teaching rats to run through mazes. Half the student experimenters were told their rats had been specially bred for good performance. The other half were told their rats had been bred for poor performance. Once again, the student experimenters seemed to create the effects they expected. The bright rats (which were actually no different from the dull rats) performed much better at maze-running and other tests of rat intelligence.

Let the implications of this experiment settle in. The rats were *identical*. There were no "bright" rats and "dull" rats. The only variable was what the student experimenters *believed*. And yet the rats performed exactly as the experimenters expected- the "bright" rats learned faster, made fewer errors, and ran the mazes more efficiently. The belief of the observer shaped the behavior of the observed, even when the observed was a laboratory rat.

Rosenthal and his colleagues gave students a standardized intelligence test, then randomly assigned some to a group designated "intellectual bloomers" and told teachers that these students were expected to perform very well academically. When tested eight months later, the "intellectual bloomers" had indeed done better than their peers, suggesting that teacher expectancy had affected the educational outcomes.

This is the Pygmalion Effect- the finding that has been replicated across decades, across cultures, across educational levels, and across species. Rosenthal demonstrated that sometimes experimenters may obtain their results in part because their expectations led them to treat their experimental participants in a biased manner, eliciting the hypothesized behavior.

The mechanisms are subtle but measurable. They found four main factors: 1) The emotional climate was affected by expectations. (Teachers acted warmer toward students they expected to do well.) 2) The behaviors of teachers were different. (Teachers gave the "spurters" more difficult material to study.) 3) The opportunities to speak out in class were different. (Teachers gave these students more opportunities to respond in class and more time to answer questions.) 4) The level of detailed feedback about performance was different. (Teachers gave these students more informative feedback.)

The experimenters and teachers were not consciously biasing their results. They were not engaged in fraud. In the vast majority of cases in which experimenter effects have been studied, these effects bias results in favor of the hypothesized results, with the experimenter unaware that he or she is even making an error or treating participants differently based on expectations. Experimenter's bias can manifest through subtle cues like word choice, tone, body language, gestures, and expressions, affecting human and animal subjects alike.

The influence operates below the threshold of conscious awareness- through microexpressions, vocal intonation, the timing of a pause, the warmth of a touch. The experimenter *believes* something, and that belief radiates outward through the quanta-field of the interpersonal interaction, shaping the behavior of the subject in ways that confirm the belief. The belief becomes self-fulfilling- not through any mystical mechanism but through the same kind of resonant coupling that governs every complex adaptive system.

This is co-creation in the laboratory. This is the eighth co-creative principle- belief shaping reality- demonstrated under controlled experimental conditions, replicated across more than three hundred studies, documented across more than fifty years of peer-reviewed research. And yet the standard response of the scientific establishment has been not to *celebrate* this finding as a profound revelation about the participatory nature of reality, but to *control for it*- to design it out of the experiment, to treat it as a contaminant, an artifact, a bias to be eliminated.

The double-blind protocol- in which neither the experimenter nor the subject knows which condition is being administered- is the supreme expression of this impulse. It is a technology designed to do one thing: sever the quanta-link between the observer's belief and the observed outcome. And it works- within its own narrow frame. Double-blind protocols do reduce experimenter expectancy effects. But they do so by *suppressing a real phenomenon*- by building a wall between mind and matter that the universe itself does not recognize.

The False God of Objectivity demands that the experimenter's consciousness be eliminated. But what if the experimenter's consciousness is not noise? What if it is *signal*? What if the Pygmalion Effect is not a methodological nuisance to be controlled away but a *window* into the co-creative structure of the cosmos- a human-scale demonstration of the same observer-participant dynamics that quantum mechanics reveals at the subatomic scale?

26.5 The Placebo Effect: The Body's Refusal to Obey Descartes

If the experimenter expectancy effect demonstrates that the *scientist's* beliefs shape experimental outcomes, the placebo effect demonstrates that the *patient's* beliefs shape biological outcomes- and it does so with a power that has consistently embarrassed the biomedical establishment.

Placebo effects are beneficial effects that are attributable to the brain–mind responses to the context in which a treatment is delivered rather than to the specific actions of the drug. They are mediated by diverse processes- including learning, expectations and social cognition- and can influence various clinical and physiological outcomes related to health.

For most of the history of modern medicine, the placebo effect was regarded as a nuisance- a confounding variable to be subtracted from the "real" effect of the drug. For years, a placebo effect was considered a sign of failure. If a patient improved on a sugar pill, it proved nothing about the pill- it merely proved that the patient's mind was playing tricks. The improvement was "subjective," not "objective." It was "psychological," not "physical." It was, in short, a Cartesian embarrassment- the ghost in the machine refusing to stay in its box.

But the ghost kept refusing. And eventually, neuroscience began to listen.

How placebos work is still not quite understood, but it involves a complex neurobiological reaction that includes everything from increases in feel-good neurotransmitters, like endorphins and dopamine, to greater activity in certain brain regions linked to moods, emotional reactions, and self-awareness. Pharmacological studies indicate that placebos mimic the action of active treatments and promote the endogenous release of opioids and nonopioids in both humans and animals. Expectations of benefit facilitate the activation of pain and nonpain control systems leading to release of endogenous substances crucially involved in placebo-induced benefits.

The placebo effect is not imaginary. It is not "all in the head." It operates through measurable, reproducible, neurochemical pathways- the same opioid, dopamine, and endocannabinoid systems that pharmaceutical drugs target. Neurobiological studies have identified dopaminergic, opioidergic, vasopressinergic, and endocannabinoidergic pathways as promising systems contributing to the modulation of pain experience and other symptoms.

The patient *believes* the treatment will work, and the brain releases the *same chemicals* that the drug would release. The belief *is* the treatment. The mind *does* heal the body- not metaphorically, not wishfully, not in some vague "New Age" sense, but through precisely characterized biochemical cascades that can be imaged on an fMRI, blocked by naloxone, and replicated across controlled studies.

In most cases, the creation of robust placebo effects across disorders and outcomes seems to require appropriate conceptual beliefs- maintained in prefrontal cortical networks- that are supported by experience-dependent learning in striatal and brainstem circuits.

Even more remarkably, placebos work even when people *know* they are placebos. Emerging research suggests that even when people know they are receiving a placebo, the inactive treatment still has effects on the brain and reported levels of improvement. The researchers discovered that the placebo was 50% as effective as the real drug to reduce pain after a migraine attack. The researchers speculated that a driving force beyond this reaction was the simple act of taking a pill.

Open-label placebos- placebos that are explicitly labeled as placebos, with no deception whatsoever- produce measurable clinical improvements. The mind does not need to be *fooled* to heal the body. It needs only to *participate*- to engage in a ritual of care, to enter a relationship of trust, to activate the co-creative circuits that the Cartesian paradigm has spent four centuries trying to deny.

They mobilize a person's own ability to heal through multiple pathways, including those studied in the field of psychoneuroimmunology.

From the perspective of this book, the placebo effect is not a problem to be eliminated. It is *proof* that consciousness co-creates biological reality. It is the human body's daily refusal to obey Descartes- its insistence that mind and matter are not two substances but one holographic unity, and that belief, expectation, and intention are not "subjective" frills but *causal forces* in the physical world.

And the nocebo effect- in which negative expectations produce negative outcomes- is the shadow confirmation. Tell a patient that a treatment will cause nausea, and nausea appears- even if the treatment is a sugar pill. The belief shapes the biology in both directions: toward healing and toward harm. The quanta-field does not discriminate between hope and fear. It responds to both. The question is not whether consciousness co-creates. The question is *what* we are co-creating- and whether we are doing so with awareness or blindly.

The biomedical establishment's response to the placebo effect has been, characteristically, to try to eliminate it- to design ever more rigorous double-blind protocols that strip the therapeutic encounter of its relational, ritualistic, and meaning-making dimensions, in order to isolate the "real" effect of the drug from the "mere" effect of belief. But what if the belief effect *is* real? What if the ritual, the relationship, the meaning-making context are not confounders but *active ingredients*? What if the most powerful healing technology on Earth is not a molecule but a resonant relationship between a conscious organism and the quanta-field in which it is embedded?

26.6 Materialist Reductionism: The Last Superstition

The Cartesian split, the cult of objectivity, and the suppression of the placebo effect all rest on a single, unexamined premise: materialist reductionism- the belief that matter is the fundamental substance of reality, and that everything else- consciousness, meaning, intention, value- is either reducible to matter or does not truly exist.

This is the operating philosophy of the modern scientific establishment. It is rarely stated explicitly, because it is rarely examined. It functions as what philosophers call a "background assumption"- a belief so pervasive, so deeply embedded in the institutional structures of knowledge production, that it is invisible to those who hold it. It is the water in which modern science swims.

And it is the last superstition.

Consider the logic. Materialist reductionism asserts that:

1. The fundamental constituents of reality are material particles (atoms, quarks, fields).
2. All higher-level phenomena (life, consciousness, meaning) are reducible to the interactions of these particles.
3. Any explanation that invokes non-material causes (consciousness, purpose, intention) is unscientific.

Now consider what the previous chapters of this book have established:

- The holographic principle (Chapter 2): Reality is not a collection of particles in a volume but a projection from quanta encoded on a boundary. The "material" world is a holographic image.
- "It from Quanta" (Chapter 1): The irreducible basis of reality is not matter but *resonant quanta*- patterns, relationships, structures of relevance.
- Landauer's principle (Chapter 3): Quanta are physical- they have thermodynamic consequences, they can perform work, they are conserved.
- The ϕ-e-π triad (Chapters 7-9): The fundamental architecture of reality is mathematical- governed not by material forces but by transcendental constants that operate as *tuning parameters* for a participatory cosmos.
- The asymptotic gap δ_∞ (Chapter 10): The universe exists because the mathematical architecture *almost but not quite* closes- a dynamic tension that prevents collapse into static equilibrium.
- The KZ-Monad and Comonad (Chapter 11): Reality computes not through material mechanism but through category-theoretic operations that extend potential into wholeness and project efficiency from noise.

None of this is materialism. All of it is physics. The universe revealed by 21st-century physics is not a machine made of matter. It is a *resonant field of living quanta*- self-organizing, self-referencing, and self-tuning. Matter is not the foundation. Matter is what the hologram *looks like from the inside.*

The materialist reductionist looks at a sunset and sees photons of specific wavelengths interacting with cone cells in the retina, triggering electrochemical cascades in the visual cortex. All of this is true. None of it is the sunset. The sunset- the *experience*, the meaning, the beauty, the ache of impermanence- is not reducible to the electrochemistry. It is what the electrochemistry *is for*. It is the quanta-field experiencing itself through the instrument of a human nervous system calibrated by a hundred thousand years of evolution to resonate with the golden angle of the light.

To reduce the sunset to photons and neurons is like reducing a symphony to air pressure variations. Technically accurate. Experientially bankrupt. And ontologically wrong- because the symphony is not *made of* air pressure variations. The symphony is a *pattern of resonant quanta* that uses air pressure variations as its medium of expression, just as consciousness uses neural activity as its medium of expression, just as the holographic universe uses spacetime as its medium of projection.

Materialist reductionism has been an extraordinarily productive methodological strategy. It has given us antibiotics, semiconductors, space travel, and the Large Hadron Collider. As a *method*- a way of temporarily bracketing the observer in order to study the observed in isolation- it has been spectacularly successful. But it has made the fatal error of confusing a method with a metaphysics. It has mistaken a useful *abstraction* (the separability of observer and observed) for a fundamental *truth* (the actual separability of observer and observed). And it has elevated this mistake into a dogma that polices the boundaries of acceptable knowledge, excludes entire domains of human experience from scientific inquiry, and labels as "unscientific" any investigation that takes the observer-participant seriously.

The materialist reductionist worldview is not wrong because it is scientific. It is wrong because it is *not scientific enough.* It stops short of following the evidence wherever it leads- and the evidence, as this book has documented across thirty-three chapters, leads inexorably to the conclusion that the observer is part of the observed, that consciousness co-creates outcomes, and that the fundamental substance of reality is not matter but resonant quanta.

26.7 The Expansion of Rigor: Including the Observer-Participant

The dismantling of the False God of Objectivity is not an invitation to abandon rigor. It is an invitation to *expand* it.

The word "rigor" comes from the Latin *rigor*- stiffness, hardness, severity. In science, it has come to mean the strict adherence to methodological standards that minimize error and maximize reliability. This is a genuine and precious value. The scientific method- with its insistence on evidence, its demand for replication, its openness to falsification- is one of humanity's greatest achievements. Nothing in this book calls for its abandonment.

But rigor, in its current form, is *incomplete*. It is rigorous about the *object* but not about the *subject*. It demands precision in the measurement of external phenomena but ignores the internal state of the measurer. It controls for every possible confound except the most fundamental one: the consciousness, intention, and belief of the observer.

The expanded rigor proposed by this book- and foreshadowed in the blueprint of the Post-Binary Paradigm (Chapter 31)- does not eliminate the observer. It *includes* the observer as a variable. Not as a contaminant to be suppressed, but as a *parameter* to be characterized, measured, and accounted for.

What would this look like in practice?

First, it would require scientists to document and report their expectations, beliefs, and intentions alongside their methods and results. Not as a confession of bias but as a relevant variable- just as they report the temperature of the lab, the make and model of the instruments, and the demographic characteristics of the participants. The experimenter's belief state is as relevant to the outcome as the instrument calibration- more relevant, in light of Rosenthal's research.

Second, it would require a re-evaluation of the placebo effect- not as a confound to be controlled away but as a *phenomenon to be studied and harnessed*. If belief heals, then the science of healing must include the science of belief. If the therapeutic relationship is an active ingredient, then clinical trials must measure and optimize the relationship, not merely subtract it.

Third, it would require a new relationship with the replication crisis- one that recognizes the crisis not merely as a failure of method but as a *signal* that the Cartesian model is breaking down. The findings that fail to replicate may not be "false." They may be *context-dependent*- dependent on the experimenter, the lab, the institutional culture, the temporal density of the moment, the geospatial phase of the location. In a participatory universe, perfect replication would be the anomaly, not the norm. The science of the future must develop frameworks for understanding *when* and *why* results vary- not merely flagging variation as failure.

Fourth, it would require the incorporation of first-person methodologies- contemplative practices, phenomenological investigation, somatic awareness- alongside third-person measurement. The ancient world understood that knowledge is not merely *acquired* but *practiced*- that the instrument of knowing must be calibrated through discipline, attunement, and embodied engagement. The monk who meditates for forty years is not merely relaxing. The monk is calibrating the inner instrument- refining the nervous system's capacity to detect subtle signals in the quanta-field, reducing the noise floor of the mind, increasing the resolution of conscious awareness.

Modern science dismisses this calibration as "subjective." But consider: if the observer is part of the system- as quantum mechanics insists and as the Pygmalion effect confirms- then the calibration of the observer is as crucial as the calibration of the telescope. A poorly calibrated telescope gives blurry images. A poorly calibrated mind gives blurry knowledge. The monk and the physicist are engaged in the same project: refining the instrument of observation so that the quanta-field can be perceived with greater fidelity.

This is not mysticism. It is the logical conclusion of the physics. If the observer participates in the creation of the observed reality, then the *quality* of the observer's participation matters. A distracted, anxious, belief-rigid observer will co-create a different experimental reality than a calm, focused, open-minded observer. The Pygmalion effect proves this at the human scale. Quantum mechanics proves it at the fundamental scale. The expanded science of the future must account for it at every scale.

26.8 The Dissolving Wall: From Objectivity to Intersubjectivity

The False God of Objectivity promised us a world without perspective- a "view from nowhere" that transcended the limitations of any particular observer. This was always an illusion. There is no view from nowhere. There are only views from *somewhere*- from specific bodies, specific locations, specific moments, specific states of consciousness. And knowledge- genuine, reliable, applicable knowledge- is built not by eliminating these perspectives but by *harmonizing* them.

The technical term for this harmonization is *intersubjectivity*: the achievement of reliable knowledge through the calibrated agreement of multiple observers, each of whom acknowledges their own perspective as a perspective. Intersubjectivity does not require that any observer be "objective" in the Cartesian sense. It requires only that observers be *transparent*- that they declare their position, their expectations, their calibration- and that the community of observers develop protocols for reconciling and integrating their diverse perspectives into a coherent, multi-dimensional map.

This is, in fact, how science has always worked in practice- despite its rhetoric of objectivity. A scientific consensus is never the view of a single, objective observer. It is the *intersubjective agreement* of a community of observers who have checked each other's work, challenged each other's assumptions, and converged- through argument, replication, and revision- on a shared description of reality. The peer review process, the conference, the journal- these are all technologies of intersubjectivity, not objectivity.

The shift from objectivity to intersubjectivity is not a retreat from rigor. It is an *advance* in honesty. It acknowledges what has always been true: that knowledge is produced by communities of conscious beings embedded in a quanta-field that they are simultaneously observing and co-creating. It trades the false certainty of the God's-eye view for the genuine reliability of calibrated, transparent, multi-perspectival inquiry.

And it opens the door to the re-inclusion of knowledge systems that the Cartesian paradigm excluded: indigenous ways of knowing, contemplative methodologies, artistic inquiry, somatic intelligence- all of which are intersubjective, all of which are calibrated through practice and community, all of which produce genuine knowledge about the structure of reality, and all of which have been dismissed as "unscientific" solely because they do not conform to the Cartesian demand for observer-independence.

The Aboriginal elder who has walked the songlines for decades and can describe, with extraordinary precision, the geological, botanical, and hydrological features of a landscape they have never "scientifically" surveyed- that elder is a scientist. The Vedic chanter who has memorized ten thousand verses with a fidelity that exceeds any recording device- that chanter is a precision instrument. The healer whose presence alone activates the placebo response in their patient- that healer is not performing a trick. They are practicing a form of resonant attunement that the biomedical model cannot accommodate and therefore cannot see.

The Post-Binary science described in Chapter 31 will not abandon the double-blind trial, the statistical test, or the peer-reviewed journal. But it will expand the definition of "scientific evidence" to include the calibrated testimony of trained observers- observers whose inner instruments have been refined through decades of practice, whose knowledge is intersubjectively verified within their own communities, and whose methods, though different in form from the Western laboratory, are no less rigorous in substance.

26.9 The Wall Falls: What Lies on the Other Side

The False God of Objectivity was erected to protect us from subjectivity- from the chaos of personal opinion, the unreliability of individual perception, the danger of beliefs contaminating facts. These are real dangers. They have not disappeared. The fall of the Cartesian wall does not usher in an age of "anything goes." It ushers in an age of *deeper accountability*- an age in which the observer must take responsibility for their participation in the reality they observe.

This is, at its core, an ethical demand. In a universe where the observer's beliefs shape outcomes- where the Pygmalion Effect operates in the classroom, the laboratory, and the clinic- the quality of one's beliefs is not merely a private matter. It is a *public consequence.* What you believe about your students shapes their achievement. What you believe about your patients shapes their healing. What you believe about the universe shapes the experiments you design and the results you find. The observer-participant is not liberated from responsibility by the dissolution of the Cartesian wall. The observer-participant is *burdened with greater responsibility than ever.*

And this responsibility- this recognition that our consciousness co-creates the reality we inhabit- leads directly to the eighth co-creative principle that will be the climax of Chapter 32: concession. Concession is the acknowledgment that what we observe is not independent of who we are. It is the willingness to see our own role in what we have created- personally, collectively, civilizationally. It is the comonadic projection $D(\Psi)$ applied to the self: the extraction of honest signal from the noise of self-deception.

The Cartesian split was, at its deepest level, a refusal of concession. It said: "I am not part of what I observe. What I see has nothing to do with who I am. The universe is out there, and I am in here, and the wall between us is absolute." This was convenient. It absolved the observer of responsibility. It meant that the scientist who discovered the nuclear chain reaction bore no relation to the bomb. It meant that the economist who modeled the market bore no relation to the crash. It meant that the colonial administrator who classified "primitive" peoples bore no relation to their destruction.

The wall is falling. And on the other side of it lies not chaos but participation- the recognition that we are nodes in the quanta-field, entangled with what we observe, contributing to the very reality we inhabit. The cosmos is not an object to be studied. It is a conversation to be joined.

The next chapter traces the emergence of this recognition- the convergence of modern physics and ancient knowing that reveals, from both sides of the historical divide, the same participatory, resonant, co-creative structure of reality.

The next chapter maps the precise correspondences between the book's physics and the wisdom of the dreamtime- not as analogy but as identical knowledge expressed in different codes.

SECTION F: THE RECOVERED FREQUENCY- REINTEGRATION AND EMERGENCE

"The future is not artificial intelligence. It is recovered resonance."

Chapter 27

The Convergence- Modern Physics Meets Ancient Knowing

There is a moment in the life of every musician when two instruments, tuned independently, are brought together for the first time- and the room fills not with dissonance but with a harmonic so rich, so unexpected, and so unmistakably *right* that both players stop and listen, astonished. Neither instrument has changed. Each was tuned to its own standard, in its own tradition, using its own method. And yet, when played together, they reveal a consonance that neither could have produced alone- a shared frequency that was always there, beneath the surface of their separate tunings, waiting to be heard.

This chapter is that moment.

For twenty-six chapters, this book has developed two parallel bodies of knowledge. On one side: modern physics, information geometry, category theory, complexity science, and thermodynamics- the rigorous mathematical architecture of the ϕ-e-π triad, the holographic principle, the NESS dynamics of living systems, the master equation of the KZ-Monad and Idempotent Comonad. On the other side: the ancient wisdom traditions- Vedic cosmology, Aboriginal dreamtime, Mayan calendrical science, the three-world cosmologies of a dozen civilizations, the sacred site networks, the songlines, the mantra, the Three Sisters garden.

These two traditions have been presented side by side but never directly *overlaid*. This chapter performs the overlay. It maps, with the precision of a cartographer and the wonder of a poet, the exact correspondences between them- not as loose analogies, not as metaphorical hand-waving, not as the kind of vague "Eastern physics" tourism that cherry-picks fragments from both traditions to produce an incoherent smoothie. But as *structural isomorphisms*: precise, point-for-point mappings that reveal the two traditions as different encodings of the same underlying reality.

Two languages. Two codes. Two tuning systems. One cosmos.

The convergence is not metaphorical. It is mathematical, experiential, and structural. And recognizing it is the first step toward the recovered frequency that the remaining chapters of this book will describe.

27.1 "It from Quanta" ↔ "Nada Brahma- The World Is Sound"

The first correspondence is also the most fundamental. It concerns the nature of reality itself- what the cosmos *is*.

Modern physics, as Chapters 1 through 3 established, has arrived at a radical conclusion: the irreducible basis of reality is not matter but resonant quanta- patterns of self-organizing relevance from which matter, energy, space, and time crystallize as secondary phenomena. What Wheeler called "It from bit"- and what this book has upgraded to "It from Quanta"- is the proposition that the physical world is not the foundation of reality but its *expression*. Beneath the atom, beneath the quark, beneath the quantum field, there is *pattern*. There is structure. There is the active, formative resonance that in-forms the universe into being.

Now listen to the Vedic tradition. In the ancient wisdom of Indian philosophy, there exists a profound truth: *Nada Brahma*, meaning "the world is sound." It is most commonly translated as "sound is God," but it also has the wider meanings of "sound is the world," "sound is joy," and the ultimate "sound is the central concept."

In the ancient Vedic traditions, sound is not merely an acoustic phenomenon but a fundamental creative force. Not a force that *acts upon* matter, but a force that *is* matter- a vibratory pattern from which the entire phenomenal world precipitates. The belief in Nada Brahma suggests that everything in the universe vibrates and that these vibrations create the reality we experience.

Read "sound" as "resonant quanta" and the correspondence becomes exact. What the Vedic sages called *Nada*- the primordial vibration- is what modern physics calls the quantum field in its ground state: the zero-point energy of the vacuum, perpetually oscillating, never still, generating the virtual particles that breathe the universe into existence (Chapter 4). What the Vedic tradition called *Brahma*- the creative principle, the source and substance of all that is- is what this book calls the living quanta-field: the self-organizing, self-referencing, autopoietic substrate from which all form arises.

The Vedic understanding that to ask about sound is to ask "about the prime substance of the world- in the physicist's terms, about atoms, neutrons, and positrons, about photons, quarks, and leptons" is not a primitive anticipation of modern physics. It is an independent arrival at the same ontological truth, encoded in a different sensory register. Where modern physics uses mathematics- equations, tensors, group theory- the Vedic tradition uses acoustics: tone, overtone, resonance, mantra. The medium of expression differs. The structure expressed is identical.

Nada Brahma is "It from Quanta" sung in Sanskrit.

The convergence deepens when we recall that the Vedic tradition did not merely *theorize* about sound as the basis of reality. It *practiced* it. A raga is "a living mathematical and emotional structure designed to resonate with the human nervous system at specific times of the day." The origins of using melodies for healing "date back over 3,000 years to the Samaveda, where the chanting of hymns was used to balance the body's internal rhythms."

This is not metaphor. This is technology. The Vedic chanter, intoning *Om* at a specific frequency, was not merely expressing a philosophical proposition about the nature of reality. The chanter was *calibrating* the nervous system- tuning the biological instrument to match the resonant frequency of the quanta-field. This is the same process described in Chapter 13 as the Fisher quanta threshold, the same process described in Chapter 18 as mantra functioning as monadic code. The Vedic tradition had the physics right. It simply encoded the physics in sound rather than in symbols.

And this brings us to the deepest layer of the correspondence. "It from Quanta" is a *proposition*- a third-person statement about the nature of reality, expressed in the observational language of physics. "Nada Brahma" is an *invocation*- a first-person act of participation in the reality it describes. The physicist tells you that the world is resonant quanta. The Vedic chanter *becomes* the resonant quanta-dissolving the boundary between knower and known, observer and observed, instrument and music.

The two traditions are not saying the same thing in different words. They are doing the same thing at different levels of participation. The physicist observes. The chanter *resonates*. And in the framework of Chapter 26- where we saw that the observer-participant is not separate from the observed reality- the chanter's participation is not less rigorous than the physicist's observation. It is *more* complete.

27.2 The Holographic Principle ↔ Indra's Net

The holographic principle, as Chapter 2 established, is one of the most profound discoveries in modern theoretical physics: the maximum quanta that can be contained in any region of space is proportional to its *surface area*, not its volume. The 3D world we experience is a projection- a holographic image- from quanta encoded on a 2D boundary. The whole is contained in every part. Every local region enfolds the global structure.

Now listen to the Buddhist tradition, as recorded in the Avatamsaka Sutra nearly two millennia ago.

"Indra's net" is an infinitely large net owned by the Vedic deva Indra, which hangs over his palace on Mount Meru, the axis mundi of Buddhist and Hindu cosmology. In East Asian Buddhism, Indra's net is considered as having a multifaceted jewel at each vertex, with each jewel being reflected in all of the other jewels.

Not only that, but each of the jewels reflected in this one jewel is also reflecting all the other jewels, so that there is an infinite reflecting process occurring. In Huayan doctrine, Indra's net exemplifies the principle of "one is all, all is one."

Through the process of infinite reflections, the net turns into an interminable, fractal-like structure of optical illusions- a structure that brilliantly demonstrates the concepts of illusion, dependent origination, and interpenetration of the Avatamsaka universe.

The correspondence is not approximate. It is exact. The holographic principle states that all the quanta of a volume is encoded on its boundary- that the "inside" is a projection of the "outside." Indra's Net states that every jewel contains every other jewel- that the part contains the whole. Both describe a reality in which locality is an illusion, in which nothing exists independently, and in which the structure of the totality is present at every point.

Mathematicians studying chaos theory and fractals have described the beauty of structures as "Indra's net," "Indra's necklace," and "Indra's pearls." In physics, the notion of quantum entanglement is a special case of the kind of interconnectivity being described. It is not at all surprising that Indra's Net has been used as a metaphor to explain holograms, wherein, by definition, each part also includes the whole within itself.

But notice: the Avatamsaka tradition did not call this a "metaphor." The image of Indra's net is used to describe the interconnectedness or "perfect interfusion" of all phenomena in the universe. In Huayan Buddhism, the metaphor of Indra's net is central to the doctrine of mutual interpenetration, which posits that every phenomenon encompasses all others without any obstruction or interference.

Perfect interfusion. Mutual interpenetration. Every phenomenon encompassing all others. This is not a simile about how things are *like* a hologram. This is a direct ontological claim that reality *is* holographic- that the structure of the whole is present in every part, that no part exists independently of the whole, and that the relationship between parts and whole is not hierarchical but reciprocal and fractal.

The holographic principle was discovered by physicists in the late 20th century through the mathematics of black hole thermodynamics. Indra's Net was described by Buddhist philosophers in the 3rd century CE through contemplative practice and direct insight. The metaphor's earliest known reference is found in the Atharva Veda. It was further developed by the Mahayana school in the 3rd century Buddhāvataṃsaka Sūtra and later by the Huayan school between the 6th and 8th centuries.

Seventeen centuries separate these two arrivals at the same structural insight. The methods could not be more different- tensor calculus versus seated meditation, particle accelerators versus monastery bells. And yet the structure they describe is identical: a cosmos in which every point contains every other point, in which nothing is separate, and in which the whole is present in every fragment.

Fritjof Capra recognized this convergence decades ago: "Particles are dynamically composed of one another in a self-consistent way, and in that sense can be said to 'contain' one another. In Mahayana Buddhism, a very similar notion is applied to the whole universe. This cosmic network of interpenetrating things is illustrated in the Avatamsaka Sutra by the metaphor of Indra's net."

And Sir Charles Eliot, even earlier: "In the Heaven of Indra, there is said to be a network of pearls, so arranged that if you look at one you see all the others reflected in it. In the same way each object in the world is not merely itself but involves every other object and in fact IS everything else. In every particle of dust, there are present Buddhas without number."

In every particle of dust, there are present Buddhas without number. Translate "Buddhas" as "holographic quanta" and you have, in a single sentence, the Bekenstein bound, the holographic principle, and the fractal self-similarity that this book has documented across thirty chapters.

27.3 NESS ↔ Tao (The Way That Flows)

The third correspondence maps the dynamical character of reality- not what the cosmos *is* but how it *moves.*

As Chapter 12 established, the universe and everything in it exists not in equilibrium but in a Non-Equilibrium Steady State- a dynamic condition maintained far from equilibrium by continuous flows of energy and quanta. Life does not resist entropy; it surfs it. A living system is not a thing; it is a *current-* a sustained, self-organizing pattern within a flow.

Now listen to the Chinese tradition. The *Tao-* literally "The Way"- is the central concept of Taoism, the philosophical and spiritual tradition whose foundational text, the *Tao Te Ching*, was composed approximately 2,500 years ago. The Tao is not a thing, not a god, not a substance. It is a *flow-* the spontaneous, self-organizing movement of reality itself. It is the way water finds its path downhill, the way seasons turn, the way the breath moves without being pushed.

The Tao that can be named is not the eternal Tao. This famous opening line- perhaps the most cited sentence in Chinese philosophy- is not merely a statement about the limitations of language. It is a statement about the nature of reality: reality is *process*, not product. It is verb, not noun. It is flowing, not fixed. The moment you name it- the moment you freeze it into a concept, a category, a definition- you have lost it, because what you named has already moved on.

This is precisely the NESS insight. A dissipative structure- a whirlpool, a hurricane, a living cell, a conscious mind- cannot be "named" in the sense of being pinned to a fixed state. It exists only in its flowing. Stop the flow and the structure vanishes. The whirlpool is not *made of* water; it is a *pattern in* water. The cell is not *made of* molecules; it is a *pattern in* the flow of molecules. And the pattern persists not by resisting change but by *organizing* change- by maintaining a coherent structure within the flux.

The Taoist sage understood this intuitively. The Tao flows like water- always seeking the lowest ground, always yielding, always moving, never fixed. Water is the supreme example of NESS in the natural world: it maintains its coherence (it is still recognizably "water") while perpetually changing its state (liquid, ice, vapor, cloud, rain, river, ocean). It is the same substance- and never the same for two consecutive moments.

The correspondence extends to the asymptotic gap δ_∞. In Taoist philosophy, the interplay of *yin* and *yang*- the two complementary forces whose dynamic tension generates all phenomena- is never perfectly balanced. There is always a slight asymmetry, a residual tension, a *gap* between the forces that prevents stasis and drives the endless cycling of creation. The *taijitu*- the familiar yin-yang symbol- encodes this asymmetry: the dot of white within the black, the dot of black within the white, ensuring that neither force ever achieves total dominance. This is the ~2% residual tension of δ_∞- the error that keeps the cosmos dancing. The Taoist sage knew it as the fundamental restlessness of the Tao. This book knows it as the asymptotic gap that prevents the ϕ-e-π triad from closing into static perfection.

The practical implication is identical in both traditions: *do not seek equilibrium.* The Taoist sage does not strive for a fixed state of perfection. The sage flows. The NESS physicist does not model living systems as equilibrium states. The physicist models them as sustained flows far from equilibrium. Both traditions recognize that the attempt to achieve stasis- to "arrive," to "finish," to "rest"- is the attempt to die. Life is not a destination. It is a current. And the art of living is the art of maintaining one's coherence within the flow.

This is why the Tao Te Ching speaks of *wu wei*- "non-action" or, more precisely, "effortless action." Wu wei is not passivity. It is the state of acting in such perfect alignment with the flow that action feels effortless- the way a surfer who has caught the wave no longer struggles against it. In NESS terms, wu wei is the state of operating at the 98% efficiency threshold- the comonadic projection $D(\Psi)$

operating at its optimum, extracting the maximum coherent signal from the ambient noise without wasting energy on resistance. The Taoist sage and the NESS physicist are describing the same dynamical state. The sage calls it "The Way." The physicist calls it "the steady-state attractor." The experience is one.

27.4 The φ-e-π Triad ↔ Sat-Chit-Ananda ↔ The Trimurti ↔ The Holy Trinity

The fourth correspondence is perhaps the most audacious- and the most precise. It maps the three fundamental mathematical constants of this book's framework onto the three fundamental aspects of ultimate reality as described by the Vedantic, Hindu, and Christian traditions.

The ϕ-e-π triad, as Chapters 7 through 9 established, comprises the three irreducible operators of the cosmos:

- ϕ **(the Golden Ratio):** The operator of *self-reference*- coherence, self-similarity, the capacity of a system to recognize and replicate its own pattern at every scale. In the Vedic framework, this is *Ṛṣi*- the knower, the witness, the self-aware subject.
- e **(Euler's Number):** The operator of *unfolding*- continuous growth, exponential transformation, the natural gradient by which potential converts to actuality. In the Vedic framework, this is *Devatā*- the process of knowing, the dynamic act of consciousness engaging with its object.
- π **(Pi):** The operator of *boundary projection*- cyclical closure, dimensional folding, the limit that contains infinity within finite form. In the Vedic framework, this is *Chhandas*- the known, the object, the bounded manifestation.

Now consider *Sat-Chit-Ananda*- the Vedantic description of the ultimate nature of Brahman, the Absolute Reality. Saccidānanda is an epithet and description for the subjective experience of the ultimate unchanging reality, called Brahman, in certain branches of Hindu philosophy, especially Vedanta. It represents "existence, consciousness, and bliss."

Saccidānanda is a compounded Sanskrit word consisting of "sat," "cit," and "ānanda," all three considered as inseparable from the nature of ultimate reality called Brahman in Hinduism.

The three aspects are:

- **Sat (सत्):** In Sanskrit, sat means "being, existence," "real, actual," "true, good, right," or "that which really is, existence, essence, true being." Sat is the aspect of reality that simply *is*- the ground of being, the fact of existence itself, unchanging beneath all change.
- **Chit (चित्):** Consciousness, awareness, the knowing capacity. Chit is the flow of consciousness or awareness, which leads to the knowledge of the Self and existence.
- **Ananda (आनन्द):** Ananda means "happiness, joy, bliss"- "pure happiness, one of three attributes of Atman or Brahman in the Vedanta philosophy."

Now map:

ϕ-e-π Triad	Sat-Chit-Ananda	Function
π **(Boundary)**	**Sat (Existence)**	The container, the limit, the fact of finite being- that which *is*, bounded and real
e **(Unfolding)**	**Chit (Consciousness)**	The dynamic process of awareness, the exponential gradient by which the unknown becomes known
ϕ **(Self-Reference)**	**Ananda (Bliss)**	The coherence, the self-similar harmony, the resonant joy of a system recognizing itself at every scale

The Vedantic philosophy understands saccidānanda as a synonym of the three fundamental attributes of Brahman. In Advaita Vedanta, it is the sublimely blissful experience of the boundless, pure consciousness and represents the unity of spiritual essence of ultimate reality.

And just as the ϕ-e-π triad is *inseparable*- the three constants interact at every level, and the cosmos arises from their combined resonance- so too is Sat-Chit-Ananda inseparable: Sat-chit-ananda can never be experienced as a distinctive thing, because it's not a thing, and everything already is sat-chit-ananda.

The same ternary structure appears in the Hindu Trimurti- the three aspects of the divine:

- **Brahma** (the Creator) ↔ e (unfolding, continuous generation)
- **Vishnu** (the Sustainer) ↔ ϕ (self-referencing coherence, maintaining the pattern)
- **Shiva** (the Transformer) ↔ π (boundary dissolution and reconstitution, the cycle that ends and begins)

And in the Christian Trinity:

- **The Father** (the ground of being) ↔ Sat ↔ π
- **The Son** (the Word made flesh- consciousness entering matter) ↔ Chit ↔ e
- **The Holy Spirit** (the indwelling joy, the breath of coherence) ↔ Ananda ↔ ϕ

This is not syncretism. This is not the claim that "all religions are really saying the same thing" in some vague, feel-good way. This is the claim that the *mathematical structure* of reality- the three irreducible operators from which all form, process, and consciousness arise- has been *independently discovered* by multiple civilizations using radically different methods of inquiry. The physicist discovers it through equations. The Vedantic sage discovers it through meditation. The Christian mystic discovers it through prayer. The Taoist discovers it through the observation of nature.

The fact that they converge on the same ternary architecture is not a coincidence. It is the cosmic hologram reflecting itself through every facet of Indra's Net.

27.5 Retrocausality ↔ Aboriginal Dreamtime

The fifth correspondence concerns the nature of time itself- and it may be the most radical of all.

As Chapter 14 established, the Holographic Ternary Time (HTT) framework dissolves the illusion of linear, unidirectional time. The Temporal State Vector $\Psi_t(\lambda,\phi,t_{UTC}) = [Y_t,\rho_t,G(\lambda,\phi)]$ is not a ticking clock but a three-component field. Past, present, and future are not arranged along a line; they are facets of a holographic whole, accessible from any vantage point. Retrocausality- the influence of future states on present outcomes, as documented in Bem's priming experiments (Chapter 6) and delayed-choice quantum erasure- is not a paradox but a natural feature of a time-field that is fundamentally nonlinear.

Now listen to the Aboriginal Australian understanding of time. The Dreaming- or Dreamtime- is the foundational temporal-ontological framework of Aboriginal culture. But "Dreamtime" is a profoundly misleading translation, because it implies a time that is past- a mythological era when the ancestors created the world and then retreated into legend. The Aboriginal understanding is precisely the opposite: the Dreaming is not past. It is *present*. It is the living, immediate, ongoing reality in which the ancestral forces continue to create, sustain, and transform the world at this very moment.

In Aboriginal understanding, "the past is not behind us- it is beneath us." This is a statement about the topology of time, and it maps directly onto the HTT framework. If time is a holographic field rather than a linear progression, then the "past" is not *behind* the present- it is *beneath* it, like the foundation beneath a building, like the ground state beneath an excitation, like the boundary data beneath the holographic projection. The past does not recede into irrelevance. It persists as the ground of the present- the informational substrate from which the present is continuously projected.

This is precisely what the holographic principle describes: the 2D boundary (the "past," the ground state, the encoded quanta) projects the 3D volume (the "present," the experienced reality, the holographic image). The Aboriginal elder walking a songline is not merely remembering the past. The elder is *re-activating* the boundary data- re-entering the quanta-field from which the present landscape is projected- and in doing so, maintaining the coherence of the projection itself.

And retrocausality- the influence of the future on the present- is equally natural in a Dreamtime framework. If the Dreaming is not "past" but "present"- if it is the timeless ground from which all temporal moments are projected- then the distinction between past and future dissolves into the distinction between different facets of the same holographic field. To "remember" the past and to "anticipate" the future are the same act: the act of tuning in to the quanta-field at different phase angles. The Aboriginal elder who "sees" events before they happen is not exhibiting a supernatural power. The elder is demonstrating a calibrated capacity for navigating the Temporal State Vector- a capacity that the clock empire, by flattening time to a scalar, has rendered invisible to the modern mind.

The HTT framework was developed through quantum physics, information geometry, and the analysis of calendrical anomalies. Aboriginal Dreamtime was developed through fifty thousand years of embodied practice, ceremonial transmission, and the walking of songlines across a living landscape. The methods could not be more different. The temporal ontology is the same: time is not a line. It is a field. And the field is holographic- meaning the whole is present at every point, the past is not gone, and the future is not unreachable.

27.6 The Asymptotic Gap ↔ Dukkha- The Wheel That Does Not Quite Fit

The sixth correspondence maps what may be the most emotionally resonant connection in the entire book: the relationship between the asymptotic gap $\delta_\infty = \frac{\pi - e}{\pi\phi^4} \approx 0.01965$ and the Buddhist concept of *dukkha.*

As Chapter 10 established, the asymptotic gap is the fundamental dynamical principle of the cosmos: the ϕ-e-π triad *almost but never quite* closes into perfect equilibrium. The ~2% residual tension- the gap between what the universe *approaches* and what it never *reaches*- is not a flaw but a feature: the dynamic drive that prevents stasis, that fuels evolution, that keeps the cosmos dancing on the edge of becoming.

Now consider *dukkha.* Early Western translators of Buddhist texts typically translated the Pali term dukkha as "suffering." Later translators have emphasized that "suffering" is too limited a translation and have preferred terms such as anxiety, distress, frustration, unease, unsatisfactoriness.

While the term dukkha has often been derived from the prefix du- ("bad" or "difficult") and the root kha ("empty," "hole"), meaning a badly fitting axle-hole of a cart giving "a very bumpy ride," it may actually be derived from duḥ-stha, "standing badly, unsteady," "unstable."

A badly fitting axle-hole. A wheel that does not quite turn smoothly. Dukkha could best be translated as "the friction of the wheel on the wagon-axis." We experience a similar friction during our life's journey. Dukkha is more than just suffering- it's the fine sand in the gears, the constant friction with experiencing life, with the fact that no moment is or remains perfect.

Read this in the light of the asymptotic gap and the correspondence becomes luminous. The δ_∞ gap is the mathematical expression of the fact that reality *does not quite fit perfectly.* The ϕ-e-π triad almost closes- but not quite. The cosmos almost reaches equilibrium- but not quite. The wheel almost turns smoothly- but not quite. There is always a residual tension, a gap, a ~2% that cannot be eliminated.

This is *dukkha*- not as suffering, not as pain, not as the bleak proclamation that "life is misery," but as the fundamental *unsatisfactoriness* that pervades all conditioned existence. The Pāli word refers to “a basic unsatisfactoriness running through our lives. Sometimes this unsatisfactoriness erupts into the open as sorrow, grief, disappointment, or despair; but usually it hovers at the edge of our awareness as a vague unlocalized sense that things are never quite perfect.”

Never quite perfect. Always almost. The wheel does not quite fit the axle. The triad does not quite close. And it is precisely this imperfection- this gap, this friction, this holy dissatisfaction- that drives the search, the seeking, the evolutionary impulse that propels consciousness toward ever-greater complexity and awareness.

Sankhara-dukkha- a “basic unsatisfactoriness pervading all existence, all forms of life, because all forms of life are changing, impermanent and without any inner core or substance.” On this level, the term indicates a lack of lasting satisfaction, or a sense that things never measure up to our expectations or standards.

The Buddhist response to dukkha is not to eliminate it but to *understand* it- to recognize it as the nature of all conditioned existence and, through that recognition, to achieve liberation. The physicist's response to δ_∞ is not to close the gap but to *honor* it- to recognize it as the source of all dynamism and, through that recognition, to understand why the universe is alive rather than dead.

The convergence is precise:

Asymptotic Gap δ_∞	Dukkha
~2% residual tension	“Never quite perfect”
Prevents static equilibrium	Prevents permanent satisfaction
Drives evolution and emergence	Drives seeking and awakening
Not a flaw- a feature	Not suffering- an insight
The space between heartbeats	The friction in every wheel

And at the deepest level, both the gap and dukkha point toward the same liberating recognition: the imperfection is not something to be *fixed*. It is something to be *recognized*, *honored*, and- in the language of Chapter 32- *conceded*. The moment you stop fighting the gap, stop resisting the friction, stop demanding that reality close into static perfection- in that moment, the gap begins to sing. The dukkha transforms from noise into living quanta. And the wheel, still imperfect, turns with a music that the perfectly fitted axle could never produce.

This is the alchemy of the eighth co-creative principle, which the final chapters will unfold. But its seed is planted here, in the recognition that the oldest spiritual diagnosis of the human condition- dukkha, the friction of the wheel- is mathematically identical to the newest physical description of the cosmos: δ_{∞}, the gap that keeps the universe alive.

27.7 The Co-Creative Principles ↔ The Eightfold Path ↔ The Universal Moral Law

The seventh correspondence maps the ethical dimension of reality- the recognition, shared across traditions, that how we *act* is not separate from how reality *is*.

The Eight Co-Creative Principles (Chapter 16)- Intention, Attention, Choice, Belief, Consequence, Compassion, Responsibility, and Concession- describe the operating system of a participatory universe in which the observer is not separate from the observed. They are not moral prescriptions imposed from outside. They are *descriptions of how the quanta-field responds to conscious engagement.* Intention shapes the monadic extension $T(\Psi)$. Attention tunes the inner resonance. Belief shapes outcomes (Pygmalion, placebo). Consequence arises naturally from the quanta-field's feedback. And Concession- the eighth principle- is the comonadic extraction of wisdom from experience.

The Buddhist Noble Eightfold Path- Right View, Right Intention, Right Speech, Right Action, Right Livelihood, Right Effort, Right Mindfulness, Right Concentration- describes the same operating system in ethical-contemplative language. "Right" here does not mean morally correct in a dogmatic sense. It means *aligned*- in tune- resonant with the way reality actually works.

The correspondences are structural:

Co-Creative Principle	Eightfold Path	Function in the Quanta-Field
Intention	Right Intention	Setting the direction of monadic extension $T(\Psi)$
Attention	Right Mindfulness	Tuning the inner instrument to the quanta-field
Belief	Right View	Aligning the personal model with reality's structure
Choice	Right Action, Right Speech	Selecting which resonance to amplify
Consequence	Right Livelihood	Recognizing the quanta-field's feedback
Compassion	Right Effort	Extending resonance beyond the self
Responsibility	Right Concentration	Sustaining coherent engagement
Concession	(The culmination)	Comonadic extraction of wisdom- $D(\Psi)$

The same correspondence appears in the Vedic concept of *dharma*- the cosmic law that governs both the physical and moral order. In the Vedic framework, the laws of physics and the laws of ethics are not separate domains. They are aspects of a single, unified *ṛta*- cosmic order, resonant truth. To live in alignment with dharma is to live in resonance with the quanta-field. To violate dharma is to introduce dissonance- noise- into the field.

This insight- that the ethical and the physical are not separate- is precisely what the Cartesian split destroyed and what the observer-participant framework of Chapter 26 restores. In a participatory universe, how you act *matters*- not merely morally but *physically*.

Your intention shapes the quanta-field. Your attention calibrates the instrument. Your belief modulates the outcome. Your concession extracts wisdom from noise. The ancient traditions knew this. Modern physics, having spent four centuries trying to separate the observer from the observed, is now being forced by its own discoveries to concede the same truth.

27.8 The Monadic Extension ↔ Prayer, Mantra, Invocation

The eighth correspondence maps the KZ-Monad- the category-theoretic operator $T(\Psi)$ that extends any partial state into a fully ϕ-resonant hologram- onto the universal human practice of prayer, mantra, and sacred invocation.

When a Vedic chanter intones *Om*, what is happening? The sound wave couples with the neural oscillators of the auditory and motor cortex. The breath entrains the vagal tone. The rhythmic repetition synchronizes the brain's hemispheric activity. The intention embedded in the chant- the orientation toward wholeness, toward the divine, toward the ground of being- shapes the direction of the neural resonance. In the language of the KZ-Monad: the chanting activates $T(\Psi)$, extending the chanter's partial, fragmented state toward a fully ϕ-resonant, 9D-embedded hologram.

When a Christian prays, "Thy will be done," the same dynamics operate. The prayer is an act of monadic extension- a reaching toward the whole, an amplification of partial resonance into alignment with the larger pattern. The specific words differ. The specific deity addressed differs. The cultural context differs. But the category-theoretic operation is identical: $T(\Psi)$ extends the local toward the global, the part toward the whole, the fragment toward the hologram.

And the comonadic complement- $D(\Psi)$, the extraction of the 98% efficient core- maps onto the contemplative practices of every tradition: the meditation that strips away noise to reveal the signal, the fasting that purifies the body, the silence that clears the mind, the confession that acknowledges what is true. Every tradition has both operations- the reaching toward wholeness and the distilling of essence- because both operations are *built into the structure of the cosmos.*

The modern dismissal of prayer as "wishful thinking" is the Cartesian ghost making one of its last stands. If the observer is separate from the observed, then prayer is indeed futile- a voice shouting into a void that cannot hear. But if the observer-participant is embedded in the quanta-field- if intention shapes outcomes, if belief modulates biology, if the experimenter's expectation changes the rat's performance- then prayer is not futile. It is *physics.* It is the conscious, deliberate activation of the monadic extension, using the instrument of the voice, the breath, and the focused mind to reshape the local quanta-field toward greater coherence.

This does not mean that all prayers are answered in the way the prayer intends. The quanta-field is not a vending machine. The monadic extension reaches toward wholeness, but wholeness may not look like what the ego desires. The Taoist sage knows this: wu wei is not getting what you want but wanting what flows. The Christian mystic knows this: "Thy will be done" is the ultimate act of monadic extension precisely because it surrenders the personal agenda to the larger pattern. The Buddhist meditator knows this: the cessation of dukkha comes not from satisfying every desire but from recognizing the nature of desire itself.

27.9 The Comonadic Projection ↔ Concession, Confession, Ho'oponopono

The ninth correspondence maps the Idempotent Comonad $D(\Psi)$- the operator that projects reality onto its 98% efficient NESS core- onto the universal practice of concession, confession, and reconciliation.

Chapter 32 will explore this correspondence in full depth. But the mapping belongs here, in the convergence chapter, because it is the correspondence that completes the circle.

The comonad extracts signal from noise. It distills the essential from the accidental. It strips away what is not needed and reveals what is. In human terms, this is the act of *concession*: acknowledging what is true about oneself and one's situation, even- especially- when the truth is uncomfortable.

In the Christian tradition, this is confession. In the Buddhist tradition, this is the practice of *sati* (mindfulness) combined with *sacca* (truthfulness). In the Hawaiian tradition, this is ho'oponopono- the practice of reconciliation through the radical acknowledgment of shared responsibility.

What all of these practices share is the same structural operation: the deliberate, disciplined stripping away of self-deception, projection, denial, and blame to reveal the actual state of the system. The comonad does not create new quanta. It *reveals* the quanta that were always there, hidden beneath the noise.

This is why concession is the eighth and crowning co-creative principle. It is not merely an ethical act. It is a *cosmological operation*- the same operation by which the universe, at the end of each aeon in Conformal Cyclic Cosmology, approaches maximum entropy and thereby maximizes its quanta- preparing the ground state from which the next cycle of creation will be projected.

Concession is the personal enactment of cosmic entropy maximization. It is the individual's version of what the universe does at the end of an aeon: let go of everything that is not essential, and in the letting go, discover that what remains- the 98% efficient core, the distilled signal, the living quanta freed from noise- is more than enough. It is, in fact, everything.

27.10 Two Languages, One Reality- The Structural Proof

Let us now gather the nine correspondences and view them as a single map:

Modern Physics / This Book	Ancient Wisdom Tradition	Structural Isomorphism
"It from Quanta" (Wheeler)	"Nada Brahma- the world is sound" (Vedic)	Reality is constituted by resonant pattern, not inert matter
Holographic Principle	Indra's Net (Avatamsaka)	The whole is present in every part
NESS / Dissipative Structures	Tao / Wu Wei	Reality is sustained flow, not static equilibrium
ϕ-e-π Triad	Sat-Chit-Ananda / Trimurti / Trinity	Three inseparable operators constitute ultimate reality
HTT / Retrocausality	Aboriginal Dreamtime	Time is a holographic field, not a linear progression
Asymptotic Gap δ_∞	Dukkha (unsatisfactoriness)	A ~2% residual tension drives all evolution and seeking
Eight Co-Creative Principles	Noble Eightfold Path / Dharma	Ethics and physics are aspects of one order
KZ-Monad $T(\Psi)$	Prayer / Mantra / Invocation	Consciousness extends the local toward the whole
Idempotent Comonad $D(\Psi)$	Concession / Confession / Ho'oponopono	Wisdom is extracted by acknowledging what is true

Nine correspondences. Not one of them is an analogy. Every one of them is a *structural isomorphism*- a point-for-point mapping between two independent encodings of the same underlying architecture.

This does not mean that the ancient traditions were "doing physics without knowing it." That would be condescending, and wrong. It means that the ancient traditions were *doing something more comprehensive than physics*- they were engaging with the full quanta-field through every available human capacity: mathematical, contemplative, somatic, acoustic, ethical, and communal. Modern physics, by contrast, engages with only the mathematical-observational dimension of the same field. It sees farther in one direction but is blind in others.

The convergence is not a victory for one tradition over the other. It is the recognition that neither tradition is complete alone. Physics without contemplation gives us power without wisdom- nuclear weapons, algorithmic surveillance, circadian disruption. Contemplation without physics gives us wisdom without verification- claims that cannot be tested, practices that cannot be measured, insights that cannot be communicated across cultures.

What this book proposes- and what the remaining chapters will unfold- is a *synthesis*: a Post-Binary science that includes both the mathematical rigor of modern physics and the participatory depth of the contemplative traditions. A science that does not merely observe the quanta-field from outside but *resonates with it from within.* A science that is, in the deepest sense, an art- the art of attunement.

The ancients had one half of the map. The moderns have the other half. This chapter has laid them side by side and shown that they fit. They are not competing stories about different realities. They are complementary perspectives on the same, single, singing cosmos.

The Vedic chanter and the quantum physicist are playing the same note on different instruments. And now, for the first time in three thousand years- for the first time since the Great Decoupling severed the ancient from the modern, the body from the mind, the ternary from the binary- the two instruments can be tuned together.

The room fills with a harmonic that neither could have produced alone.

The recovered frequency begins to sound.

.

The next chapter proposes a practical framework for ternary timekeeping- a new calendar, based on the Temporal State Vector, for a civilization ready to re-enter the living rhythm of the cosmos.

.

Chapter 28

Ternary Time- A New Calendar for a Resonant Civilization

There is a question that no one asks, because it seems too obvious to require asking: *What time is it?*

You glance at your phone. You see numbers. You accept them. The question is answered. But the question has not been answered. It has been *silenced*- replaced by a number that tells you where you stand in relation to a cesium atom's vibration, or a railway schedule, or a factory shift, but tells you nothing at all about where you stand in relation to the sun, the season, the landscape, the pulse in your wrist, or the living rhythm of the cosmos.

This chapter proposes a different answer. Not a replacement for Coordinated Universal Time- the modern world's logistics require a shared standard, and this book is not calling for its abolition. But a *complementary* layer of temporal awareness: a practical framework, rooted in the physics of the Temporal State Vector, for re-entering the living time-field that the clock empire severed.

The framework is called Ternary Timekeeping. Its foundation is the Holographic Ternary Time (HTT) model introduced in Chapter 14: the recognition that time is not a scalar (t) but a three-component Temporal State Vector-

$$\Psi_t(\lambda,\phi,t_{UTC}) = [Y_t, \rho_t, G(\lambda,\phi)]$$

- where Y_t is the ternary resonance of the year, ρ_t is the local temporal density, and $G(\lambda,\phi)$ is the geospatial phase. This chapter translates each of these components from theoretical physics into daily practice. It asks: what would it mean to *live* within this time-field? To feel it in the body, to shape one's day around it, to organize one's community through it?

The answer is not a retreat to the pre-modern. It is an advance toward the post-binary: a civilization that holds the precision of the atomic clock in one hand and the wisdom of the songline in the other- and recognizes that both are descriptions of the same, single, singing cosmos.

28.1 The Three Clocks: A Starting Diagnosis

Before we can build a new temporal practice, we must see clearly what already exists. Our lives are governed by three clocks: the social clock that organizes our lives with others (local time), the biological clock that controls our physiology (circadian time), and the sun clock that defines natural light and darkness. This tripartite framework, articulated by chronobiologist Till Roenneberg, is the modern scientific rediscovery of what the ancients always knew: that time has three dimensions, and that health depends on their alignment.

The more misaligned these clocks are, the higher our odds of developing certain diseases. The term for this misalignment- "social jet lag"- was coined by Roenneberg and his colleagues to describe the discrepancy between biological time, determined by our internal body clock, and social times, mainly dictated by social obligations such as school or work.

The parallels to the Temporal State Vector are precise:

Roenneberg's Three Clocks	HTT Component	What It Measures
Sun Clock	Y_t (Ternary Resonance)	The astronomical cycle- the position of the Earth relative to the sun
Biological Clock	ρ_t (Temporal Density)	The organism's internal rate of quanta-processing- how fast or slow time *feels*
Social Clock	$G(\lambda,\phi)$ (Geospatial	The externally

Roenneberg's Three Clocks	HTT Component	What It Measures
	Phase) - overridden	imposed schedule, which *should* reflect location but now reflects economics

Modern chronobiology has confirmed, with extraordinary precision, the cost of misaligning these three clocks. In industrialized countries, two-thirds of the studying/working population experiences social jet lag, often for several years. Two-thirds. Not a fringe population. Not shift workers alone. Two-thirds of everyone who studies or works-which is to say, nearly everyone.

Social jetlag is most pronounced in late chronotypes, and has been associated with a wide range of adverse outcomes, including increased risks of obesity, metabolic syndrome, and mental health disorders. And the mechanism is clear: the human body reacts to environmental light, producing more melatonin when it becomes darker. The misalignment of sleep and wake rhythms with the daily cycle of physiological processes desynchronizes the release of hormones such as melatonin, cortisol, ghrelin, and leptin.

The desynchronization is not merely between the organism and the clock. It is between the organism and the *cosmos*. When the body's clocks fall out of synch with each other, problems arise. That's what is happening today, as modern technologies decouple us from the solar day to which our bodies evolved. The word "decouple" is the chronobiologist's term for what this book has called the Great Decoupling- the three-thousand-year process by which the human species has been progressively severed from the living time-field of the universe.

The diagnosis is complete. Three clocks, chronically misaligned. Two-thirds of the population in permanent temporal dissonance. A cascade of metabolic, psychological, and social consequences. The question is no longer whether the misalignment exists. The question is: what do we do about it?

28.2 Measuring Y_t: The Ternary Resonance of the Year

The first component of the Temporal State Vector is Y_t- the ternary resonance. This is the deepest rhythm: the macro-pulse of the year itself.

As Chapter 14 established, the year does not have a single, fixed length. It oscillates around a ternary pattern- 365 days in common years, 366 in leap years- driven by the mismatch between the Earth's orbital period and the day-night cycle. This mismatch is not random. It is quasi-periodic, governed by the interaction of the solar cycle, the lunar cycle, and the precession of the equinoxes. The ternary beat- the rhythm of common years and leap years, of slightly longer and slightly shorter annual cycles- is the macro-oscillation within which all finer temporal rhythms are nested.

In the blueprint language of HTT, the ternary resonance $Y_t = \{360,363,366\}$ represents not the literal number of days in the calendar year but the *dominant harmonic frequency* of the annual cycle. The number 360- the Babylonian ideal, the number of degrees in a circle- represents the "ground tone" of the year: the mathematically perfect cycle. The number 363 represents the middle register. And 366 represents the extended resonance- the year stretched to its maximum by the leap-day correction.

The practical question is: how does one *attune* to this macro-rhythm?

The answer lies in the seasonal festivals that every ancient culture maintained- and that the modern world has reduced to commercial holidays stripped of their temporal significance. The solstices and equinoxes are not arbitrary dates. They are the *nodal points* of the annual oscillation- the moments when the Earth's axial tilt reaches its extremes (solstices) or its equilibrium (equinoxes). These four points divide the year into four quadrants, and within each quadrant, the cross-quarter days (the midpoints between solstice and equinox) create an eightfold division that maps the annual pulse with remarkable precision.

The Celtic calendar, the Chinese calendar, the Vedic calendar, the Mesoamerican calendar- all of them recognized this eightfold pulse. The eight festivals of the Celtic wheel- Samhain, Imbolc, Beltane, Lughnasadh, and the four solar markers- were not "pagan" superstitions. They were *temporal calibration events*: ceremonies designed to re-synchronize the community's internal clocks with the annual resonance.

Practical application: The first step in ternary timekeeping is the recovery of the eightfold year. Mark the solstices, equinoxes, and cross-quarter days on your calendar. Not as decorative reminders but as *temporal anchors*- days on which you deliberately pause, recalibrate, and re-attune to the macro-rhythm. Ask: where in the year am I? Is this a time of expansion (spring-summer) or contraction (autumn-winter)? Is the light growing or diminishing? Is the Earth breathing in or breathing out?

This is not mysticism. It is chronobiology applied at the annual scale. The seasonal variation in daylight- and therefore in melatonin secretion, cortisol rhythm, metabolic rate, immune function, and mood- is one of the most powerful zeitgebers (time-givers) in the human organism's repertoire. Some other important factors which affect the circadian rhythms are geological cycles like the length of daylight and darkness in different seasons which originate from earth's movements. These geographic events influence on adaptation processes to all organisms where they have to adjust their physiology to the environmental changes. In winter that day becomes shorter and night gets longer, depression is increased due to a diminished sunlight exposure. This transient mood disorder has been called seasonal affective disorder (SAD).

The clock empire erased seasonal variation by imposing the same schedule year-round- the same wake time in December as in June, the same working hours in winter darkness as in summer light. Ternary timekeeping begins by acknowledging what the body already knows: that the year has a pulse, that the pulse has a shape, and that living in harmony with that shape is not a luxury but a biological necessity.

28.3 Measuring ρ_t: Temporal Density and the Body's Inner Clock

The second component of the Temporal State Vector is ρ_t- temporal density. This is the most intimate of the three dimensions: the rate at which the organism processes resonant quanta at any given moment. It is, in experiential terms, how fast or slow time *feels.*

Everyone has felt temporal density shift. An hour in the dentist's chair lasts forever. An hour in conversation with a beloved friend vanishes in what seems like minutes. An afternoon of deep creative work compresses time to a point; an afternoon of bored waiting stretches it to infinity. These are not illusions. They are *measurements*- the organism's inner clock registering the density of meaningful quanta flowing through its processing channels.

Modern chronobiology has established the molecular basis for this inner clock. Circadian rhythms are entrainable endogenous rhythms that respond to external stimuli and regulate physiological functions. The suprachiasmatic nucleus (SCN) in the hypothalamus is the mammalian master clock that synchronizes all other tissue-specific peripheral clocks. The SCN follows Earth's 24-hour cycle by light entrainment through the retinohypothalamic tract. At the cellular level, the core clock genes CLOCK, BMAL1, PER1–PER3, CRY1, and CRY2 regulate circadian rhythms in a negative feedback loop.

This molecular clockwork is not a single metronome. It is a *network* of oscillators- a fractal system of clocks within clocks, each tissue and organ maintaining its own rhythm, all of them coordinated by the master clock in the SCN. In humans, circadian rhythms regulate critical functions such as sleep–wake cycles, hormone secretion, metabolism, and cognitive performance. Disruptions to these rhythms, whether due to genetic factors, lifestyle choices, or environmental influences, can lead to a host of health issues ranging from sleep disorders to chronic diseases like obesity, diabetes, and neurodegenerative disorders.

The key insight for ternary timekeeping is this: temporal density is not uniform across the day. The organism has natural peaks and troughs of cognitive performance, alertness, creativity, physical strength, and immune activity- all governed by the circadian system. There are predictable windows of high quanta-processing capacity (typically late morning and late afternoon for most chronotypes) and predictable windows of low capacity (typically early afternoon and the hours before dawn).

The clock empire ignores these windows. The factory whistle, the school bell, and the office schedule impose a flat temporal demand across the entire day- the same level of productivity expected at 8:00 AM as at 2:00 PM, regardless of the organism's circadian state. This is the override of ρ_t- the flattening of the body's natural temporal topography into an artificial plain.

Practical application: The second step in ternary timekeeping is the *tracking of personal temporal density*. This does not require sophisticated instruments. It requires only attention- the practice of noticing, across multiple days and weeks, when you are naturally alert, when you are naturally drowsy, when creative insight flows easily, and when it resists.

Heart rate variability (HRV) provides a measurable proxy for temporal density. This global study analyzed data from the largest dataset ever studied in the Heart Rate Variability biofeedback field, comprising 1.8 million user sessions. The researchers focused on HRV Coherence, which is linked to improved emotional stability and cognitive function. HRV is the variation in the time interval between successive heartbeats- and as Chapter 5 established, healthy HRV is *fractal*: complex, variable, and self-similar across timescales. High HRV indicates a system operating in flexible NESS- responsive to the environment, adaptable, coherent. Low HRV indicates a system approaching rigidity or collapse.

Research has demonstrated that heart rhythm patterns become more ordered during HeartMath's self-regulation techniques. Emotional states are integrated with core physiological functions through dynamic, bidirectional autonomic signaling between the heart, lungs, brainstem, limbic system, and higher cortical areas. This interconnected network enables conscious regulation of breathing, heart rhythms and emotions to influence autonomic and higher cortical functions.

Modern wearable devices- heart rate monitors, smartwatches, HRV-tracking apps- can provide real-time data on this internal rhythm. But the technology is not the point. The technology is merely a bridge- a way of making visible what the ancients knew through embodied practice. The Vedic monk who rose before dawn to chant was not following a clock. The monk was following ρ_t- the felt sense of temporal density at the edge of the circadian trough, the liminal moment when the organism's processing rate is low enough for the noise of ordinary thought to subside and the signal of the quanta-field to become audible.

The practical framework is simple:

1. **Track your rhythm:** For two weeks, note your energy levels, alertness, and mood at two-hour intervals. A simple 1-5 scale is sufficient. You will quickly see a pattern- *your* pattern, unique to your chronotype and your geospatial phase.

2. **Align your activities:** Place your most cognitively demanding work in your peak windows. Place routine tasks in your troughs. Place creative and reflective work in the transition zones- the edges between peak and trough, where the shift in temporal density creates the conditions for insight (the guiding quanta that Chapter 17 described).

3. **Honor the troughs:** The early-afternoon dip is not a failure to be overridden with caffeine. It is a natural period of low temporal density- a moment when the organism wants to consolidate, to digest, to rest. The siesta tradition, found in cultures across the Mediterranean, Latin America, and South Asia, is not laziness. It is ternary timekeeping: the recognition that the day has a pulse, and that the pulse includes a natural pause.

4. **Practice the Chronosonant Breath:** Three times daily- morning, midday, evening- pause for sixty seconds. Close your eyes. Feel your heartbeat. Count five beats. In that count, you are measuring your personal ρ_t- your body's current quanta-processing rate. This is not meditation (though it can lead there). It is *temporal calibration*: a micro-

practice of checking in with the inner clock and noticing whether it is aligned with the outer world or drifting into dissonance.

28.4 Measuring $G(\lambda,\phi)$: Your Place in the Cosmic Dance

The third component of the Temporal State Vector is $G(\lambda,\phi)$ - the geospatial phase. This is the component that railway time obliterated and that the modern time zone system continues to suppress: the way your *location on the Earth's surface* modulates your temporal experience.

Chapter 24 documented how the standardization of time zones overrode the local solar noon- replacing the organic, longitude-specific relationship between each community and the sun with an artificial, politically-imposed grid. The consequence is that two communities on opposite edges of the same time zone- separated by nearly a thousand miles of longitude- are forced to share the same clock time, despite the fact that sunrise may differ between them by up to an hour.

Modern research has quantified the health cost of this override with startling precision. This study uses a spatial regression discontinuity design to estimate the economic cost of the misalignment between social and biological rhythms arising at the border of a time-zone. Exploiting the discontinuity in the timing of natural light at a time-zone boundary, the researchers find that an extra hour of natural light in the evening reduces sleep duration by an average of 19 minutes and increases the likelihood of reporting insufficient sleep. Using data drawn from the Centers for Disease Control and Prevention and the US Census, they find that the discontinuity in the timing of natural light has significant effects on health outcomes typically associated with circadian rhythm disruptions and economic performance.

The researchers calculated that the circadian misalignment increases health care costs by at least 2 billion dollars. Two billion dollars- in health care costs alone- generated by the arbitrary placement of a line on a map that overrides the local solar rhythm.

The body knows where it is. Czeisler reported in 1986 that the SCN is in reality activated by light impulses traveling along retinal nerve cells into the human brain. Light synchronizes the SCN to the solar day, prompting it to coordinate cortisol release and the release of the sleep hormone melatonin. The SCN- the master clock- takes its primary cue from the light-dark cycle, and that cycle is determined by the sun's position relative to your specific point on the Earth's surface. Your $G(\lambda,\phi)$ is not an abstraction. It is encoded in the photons that strike your retina at dawn and dusk, in the angle of the sun at noon, in the duration of daylight and darkness that your body uses to calibrate every metabolic process from gene expression to immune surveillance.

Practical application: The third step in ternary timekeeping is the recovery of the geospatial phase- the practice of aligning your daily rhythm with the sun *as it appears at your location*, rather than with the clock time that the time zone imposes.

1. **Know your solar noon:** Solar noon- the moment when the sun is at its highest point- varies with your longitude. In most locations, it differs from clock noon by anywhere from a few minutes to nearly an hour. Free online calculators and smartphone apps can tell you your local solar noon to the minute. This becomes your temporal anchor- the center of your day, the midpoint from which morning and afternoon are measured.

2. **Track sunrise and sunset:** The ancient practice of marking the sun's rising and setting is not ceremonial decoration. It is the calibration of $G(\lambda,\phi)$. Note the time of sunrise and sunset at your location. Note how these times shift across the weeks and months. Feel the asymmetry between the long evenings of summer and the early darkness of winter. This asymmetry is the *shape* of your geospatial phase- the unique temporal signature of your place on the Earth.

3. **Prioritize morning light:** Reduced exposure to daytime sunlight and increased exposure to electrical lighting at night

leads to late circadian and sleep timing. Researchers have previously shown that exposure to a natural summer light-dark cycle entrains the human circadian clock to solar time, such that the internal biological night begins near sunset and ends near sunrise. The single most effective chronobiological intervention- confirmed by multiple studies- is morning sunlight exposure. Natural light, in the first hour after waking, directly entrains the SCN to the local solar phase. Even ten minutes of outdoor light in the morning can measurably advance the circadian clock, improving sleep onset, sleep quality, and daytime alertness.

4. **Protect the evening darkness:** The intrinsically photosensitive retinal ganglion cells express melanopsin, a photopigment that makes them exquisitely photosensitive. Blue light (~480 nm) most strongly stimulates the melanopsin photopigment, whereas red light (>600 nm) has minimal effect. The characteristics of sunlight change across the day; sunlight contains more short wavelengths during the day and then shifts to longer wavelengths toward sunset. The sensitivity spectrum of melanopsin may be an adaptation to the natural solar cycle. The artificial extension of the day through electric lighting- and especially the blue-enriched light of screens- is the most potent disruptor of the geospatial phase. It tells the SCN that it is still daytime when, according to the sun, it is not. Ternary timekeeping in the evening means dimming lights after sunset, minimizing screen exposure, and allowing the body to receive the signal that darkness has arrived.

5. **Honor your latitude:** The geospatial phase is not identical at all latitudes. Near the equator, the day-night cycle is nearly uniform year-round- twelve hours of light, twelve of darkness, with minimal seasonal variation. At higher latitudes, the seasonal swing is dramatic- from near-perpetual darkness in winter to near-perpetual light in summer. Your circadian biology will respond differently depending on where you live, and your ternary timekeeping practice must accommodate this. A person living in Helsinki needs a different winter strategy (light therapy, vitamin D, reduced activity) than a person living in Nairobi.

28.5 The Personal Temporal State Vector: Daily Practice

The three components- Y_t, ρ_t, and $G(\lambda,\phi)$- combine into a personal Temporal State Vector: a holistic description of your temporal situation at any given moment. The practice of ternary timekeeping is the practice of *attending to this vector*- of checking in, regularly and honestly, with all three components, and adjusting your activity to match.

A daily ternary timekeeping practice might look like this:

Morning (Sunrise ± 30 minutes): - Step outside. Feel the light on your face. This is not metaphor- it is the calibration of $G(\lambda,\phi)$ through the retinohypothalamic tract. - Note the season, the angle of the light, the quality of the air. This is the calibration of Y_t- where you are in the annual cycle. - Feel your body's alertness. Note your energy level. This is the reading of ρ_t- your current temporal density. - Set your intention for the day in alignment with all three readings. If it is winter and your ρ_t is low, honor that: this is not a day for peak output. If it is summer and your body is vibrating with energy, ride the wave.

Midday (Solar Noon ± 30 minutes): - Pause. The Chronosonant Breath: sixty seconds of heartbeat awareness. This is the midday calibration- the temporal equivalent of checking a compass at noon. - Eat the main meal. Nearly every traditional culture placed the largest meal at midday, when metabolic rate is highest. This is not cultural accident- it is chronobiological wisdom. The circadian system governs digestion, insulin sensitivity, and nutrient absorption, and all of them peak around solar noon.

Evening (Sunset ± 30 minutes): - Mark the transition. Note the light changing. Feel the body's shift from sympathetic (active) to parasympathetic (restorative) dominance. This is the evening phase-shift of ρ_t- the natural deceleration that the clock empire overrides with screens and caffeine. - Dim the lights. Reduce stimulation. Begin the descent toward sleep- not at a clock-determined "bedtime" but at the moment when the body, reading the geospatial phase through the fading light, begins to release melatonin.

This is not a rigid schedule. It is a *practice of attunement*- a daily habit of checking in with the three clocks and adjusting accordingly. Some days will be more aligned than others. Some days the social clock- work, school, family obligations- will override the biological and solar clocks. The point is not perfection. The point is *awareness*: the cultivation of a felt sense for the living time-field that the clock empire has numbed.

Over time, this awareness deepens. You begin to *feel* the ternary resonance of the year- to sense the shift in energy as the seasons turn, to anticipate the equinox before the calendar announces it. You begin to notice the fluctuations of your temporal density- to recognize your peak windows, your creative edges, your natural pauses. You begin to feel the geospatial phase- to know, in your body, when the sun is directly overhead, when it is approaching the horizon, when the light has that particular quality that belongs only to your latitude at this time of year.

This is what the ancients knew. This is what the songline-walkers, the festival-keepers, the dawn-chanters practiced every day for a hundred thousand years. The ternary timekeeping framework does not invent a new practice. It recovers an ancient one- and provides it with a mathematical foundation that makes it legible to the modern mind.

28.6 The Community Temporal State Vector: Seasonal Festivals and Shared Rhythm

Individual attunement is necessary but not sufficient. The human organism is a social animal, and its temporal health depends not only on personal alignment with the sun and the season but on *communal* alignment- the shared rhythms that bind a community into a coherent temporal whole.

Every ancient culture maintained a calendar of communal festivals tied to the astronomical cycle. The solstices, the equinoxes, the cross-quarter days, the new moons, the harvest festivals, the planting festivals- these were not entertainment. They were *temporal synchronization events*: moments when the entire community paused from ordinary activity, gathered together, and re-calibrated its collective temporal rhythm to the cosmic beat.

The function of these festivals was twofold. First, they re-entrained the community's biological clocks to the solar cycle- ensuring that the drift that inevitably accumulates during weeks of ordinary activity was corrected at regular intervals. Second, they re-entrained the community's *social* clocks to one another- ensuring that the interpersonal rhythms of the group (sleeping, eating, working, resting, celebrating, mourning) remained synchronized.

Modern society has all but eliminated communal temporal synchronization. The few festivals that remain- Christmas, New Year's, national holidays- have been stripped of their astronomical content and commercialized into consumption events. They no longer function as temporal calibration points because they no longer refer to anything in the sky. Christmas falls near the winter solstice but does not *mark* it. Easter is theoretically tied to the spring equinox and the full moon but in practice is experienced as a chocolate-buying opportunity rather than a moment of cosmic recalibration.

Practical application: The recovery of communal ternary time requires the deliberate creation of seasonal gathering points- moments when family, friends, or community come together explicitly to mark the turning of the year.

The eight-point festival calendar- the four solar markers (two solstices, two equinoxes) and the four cross-quarter days- provides a ready-made framework. These eight festivals divide the year into eight approximately equal segments of roughly 45 days each. Each segment has a characteristic quality- a temporal flavor determined by the direction and rate of change of daylight- that can be named, celebrated, and used to orient communal activity.

The specific content of the festivals is less important than their *temporal function*. What matters is that the community gathers, acknowledges the astronomical moment, and collectively re-attunes to the rhythm. The gathering can be as simple as a shared meal at sunset on the equinox, or as elaborate as a multi-day celebration with song, storytelling, and ceremonial fire. What matters is the *pause*- the deliberate interruption of the social clock's relentless tick, the collective return to the solar clock's deeper rhythm.

This is the community equivalent of the Chronosonant Breath: a periodic recalibration that prevents the drift between biological and social time from accumulating into chronic misalignment. In a society that observed eight such festivals per year, the maximum drift between any two calibration events would be approximately six weeks- short enough that the body's circadian system could recover its alignment within a day or two of the festival. In a society with no such festivals- which is to say, in modern society- the drift accumulates indefinitely, and the chronic misalignment documented by Roenneberg and his colleagues becomes the norm.

28.7 Schools, Work, and the Chronotype Revolution

The most immediate, measurable, and politically actionable implication of ternary timekeeping concerns the schedules of schools and workplaces- the two institutions that most directly impose the social clock's rhythm on the biological clock.

The evidence regarding school start times is now overwhelming. Insufficient sleep in adolescents has been shown to be associated with a wide variety of adverse outcomes, from poor mental and physical health to behavioral problems and lower academic grades. However, most high school students do not get sufficient sleep. Delaying school start times for adolescents has been proposed as a policy change to address insufficient sleep in this population.

Most studies reviewed provide evidence that delaying school start time increases weeknight sleep duration among adolescents, primarily by delaying rise times. Most of the studies saw a significant increase in sleep duration even with relatively small delays in start times of half an hour or so.

The biological basis is clear. During puberty, youth undergo a shift in their circadian clocks that makes it harder for them to fall asleep until later in the night. Meanwhile, they can stay awake longer before experiencing an increase in the pressure to sleep. Adolescent circadian biology is delayed relative to children and adults- their melatonin secretion begins later, their sleep onset occurs later, and their optimal wake time is correspondingly later. Forcing a 15-year-old to wake at 6:00 AM for a 7:30 AM school bell is the chronobiological equivalent of forcing a 40-year-old to wake at 3:00 AM for a 4:30 AM work shift- chronic, compulsory circadian disruption imposed on an entire generation.

In the past 25 years, more than 500 individual school districts in the United States have moved school start times later in response to the research. Progress was piecemeal until 2014, when the American Academy of Pediatrics helped draw attention to the cause by publishing a position statement stating that middle schools and high schools should begin no earlier than 8:30 a.m. In 2019, California passed a law, enacted in 2022, mandating secondary schools start after 8:30 a.m.

The results confirm the chronobiological prediction. Delaying school start times can be an effective method to improve important sleep outcomes, such as total sleep time and reduce daytime sleepiness; health outcomes, including reduced depression and caffeine use; and academic outcomes, such as tardiness and staying awake in class. These findings suggest that delayed school start times can increase sleep duration in adolescence with corresponding effects on markers of improved health and classroom engagement.

The results from a three-year research study, conducted with over 9,000 students in eight public high schools in three states, reveal that high schools that start at 8:30 AM or later allow for more than 60% of students to obtain at least eight hours of sleep per school night.

Some researchers have proposed going further. Later school starting times provide benefits to adolescent sleep, health and learning. There is substantial evidence that later starting times benefit academic outcomes even in the early stages of puberty and this positive impact continues into late adolescence. Later starting times also reduce the rate of student driving accidents and lower reports of depression. One study at a UK school tested a 10:00 AM start time and found significant improvements across health and academic metrics.

The school start time movement is ternary timekeeping's most successful public policy application to date- a concrete example of what happens when the social clock is adjusted to match the biological clock. And the principle extends directly to the workplace.

The researchers' results strongly suggest that work and school schedules should be adapted to chronotype whenever possible. Chronotype- the individual's genetically and developmentally determined circadian preference- varies widely within any population. Some people are "larks" (early chronotypes), naturally alert at dawn and drowsy by mid-evening. Others are "owls" (late chronotypes), reaching peak alertness in the late morning or afternoon and naturally staying awake well past midnight. The majority fall somewhere in between.

A ternary-timekeeping workplace would accommodate this variation rather than suppressing it. It would offer flexible start and end times keyed to individual chronotype. It would place collaborative meetings in the mid-morning or mid-afternoon windows when the broadest range of chronotypes overlap. It would protect the early-afternoon trough as a period of reduced demand. And it would measure productivity not by hours spent at the desk but by output quality- recognizing that a lark who works intensely from 6:00 to 14:00 may produce more and better work than the same person forced to drag through a conventional 9:00-to-17:00 schedule.

The technology for chronotype-sensitive scheduling already exists. The Munich Chronotype Questionnaire, developed by Roenneberg's lab, can assess an individual's chronotype in minutes. Integrating this assessment into workplace scheduling would be trivially easy from a logistical standpoint. What prevents it is not technology but ideology- the residual belief, inherited from the Benedictine monastery and the Industrial Revolution, that uniform time discipline is a moral virtue and that accommodating the body's rhythm is a form of weakness.

Ternary timekeeping reverses this moral polarity. Forcing a late chronotype to perform at dawn is not discipline- it is temporal violence. Accommodating chronotype variation is not weakness- it is the recognition that the organism functions optimally when its three clocks are aligned, and that aligning them produces better outcomes for everyone.

28.8 The Chronosonant Resonator: From Proposal to Prototype

Chapter 14 introduced the concept of the Chronosonant Resonator- a proposed instrument for directly measuring the Temporal State Vector. This section translates that proposal from the theoretical to the practical, describing what such a device might look like and what it would measure.

The essential requirement is a quantum-coherent system whose observable properties are sensitive to the three components of the TSV. Current technology suggests several possible architectures:

Biological approach: The human body itself is the most sophisticated Chronosonant Resonator in existence. Heart rate variability, as discussed above, provides a real-time measure of ρ_t. Core body temperature, which oscillates in a circadian rhythm with an amplitude of approximately 1°C, provides a marker for the phase of the biological clock. And the dim-light melatonin onset (DLMO)- the time at which melatonin secretion begins in the absence of bright light- provides the most precise marker of the circadian clock's current phase relative to the solar cycle.

Even when healthy individuals maintain the same sleep-wake and rest-activity schedules, their endogenous circadian time can be up to 5 hours apart, suggesting that the same drug administered at the same local clock time in two patients may be metabolized differently solely due to different time of administration relative to endogenous circadian clocks. Therefore, for better clinical outcomes, we may need to consider the circadian time of the patient.

The emerging field of chronomedicine is already developing instruments for this purpose- wearable devices that track not just heart rate and activity but circadian phase markers in real time. Technological innovations, such as wearable devices and mobile applications that monitor circadian markers in real time, may facilitate early detection of circadian disruptions and allow for timely interventions.

Resonance frequency approach: HRV biofeedback technology offers a prototype for measuring the body's resonant frequency- the breathing rate at which heart rate oscillations achieve maximum amplitude and coherence. Heart Rate Variability Biofeedback has been widely used to improve cardiovascular health and well-being. HRVB is based on breathing at an individual's resonance frequency, which stimulates respiratory sinus arrhythmia and the baroreflex.

HeartMath combines an innovative set of heart-focused self-regulation techniques with HRV biofeedback, enabling individuals to shift into a state of coherence- a physiological pattern marked by high-amplitude, sine wave–like HRV rhythms. At this resonant frequency, the heart serves as a central "conductor," entraining other oscillatory systems- such as respiration, blood pressure, and vascular tone- through baroreflex coupling, and modulating brain centers involved in emotional and cognitive processing.

This technology- already commercially available and validated in clinical, military, and educational settings- represents a first-generation Chronosonant Resonator. It does not yet integrate all three components of the TSV. But it demonstrates the principle: that the body's resonant state can be measured in real time, that the measurement can be fed back to the user, and that the user can learn to consciously shift their physiological state toward greater coherence.

The next generation of Chronosonant Resonators- whether they take the form of advanced wearables, ambient environmental sensors, or dedicated biofeedback devices- will integrate HRV coherence data with light exposure data (tracking $G(\lambda,\phi)$), seasonal markers (tracking Y_t), and circadian phase biomarkers (tracking ρ_t at the molecular level). The goal is not a single number- not another version of the clock's dead scalar- but a *three-dimensional display*: a real-time visualization of the personal Temporal State Vector, showing the user where they stand in relation to all three temporal dimensions simultaneously.

28.9 Not a Replacement- A Restoration

A crucial clarification: ternary timekeeping does not propose to replace Coordinated Universal Time. UTC serves essential logistical functions- aviation, global communication, scientific coordination, financial transactions- that require a single, universally shared temporal standard. The modern world's infrastructure depends on it, and dismantling it would create chaos without benefit.

What ternary timekeeping proposes is a *complementary layer*- a second temporal awareness that runs alongside UTC, enriching it with the biological, astronomical, and experiential dimensions that UTC, by design, omits. You would still catch your 8:15 train. But you would also know that it is thirty minutes past your local solar noon, that you are in the waning quarter of the autumn equinox cycle, that your temporal density is currently in its early-afternoon trough, and that the light quality outside your window is characteristic of your latitude at this specific time of year.

This dual awareness- precision and resonance, clock and cosmos, UTC and TSV- is the temporal expression of the Post-Binary Paradigm described in Chapter 31. It does not force a choice between modern efficiency and ancient wisdom. It holds both. It says: you can live in the 21st century *and* feel the sun on your face. You can catch the train *and* honor the equinox. You can be on time for the meeting *and* know that your body's clock is telling a different, deeper, more truthful story about what "time" actually is.

The ancients did not have UTC. They did not need it. Their world moved at walking pace, and the sun's time was sufficient. The moderns have UTC and nothing else. They move at light speed, and the sun has been exiled. Ternary timekeeping proposes the synthesis: UTC for logistics, the TSV for life.

28.10 The Calendar as a Mirror: What We Measure, We Become

Let us close this chapter with a reflection on what calendars *do*- not merely to our schedules but to our consciousness.

A calendar is not a neutral tool. It is a *cognitive framework*- a template that shapes how we experience the passage of time, what we notice, what we celebrate, what we dread, what we ignore. The Gregorian calendar, with its uniform months, its numbered days, its relentless linear progression from January 1 to December 31, shapes a consciousness that experiences time as a resource to be spent- a finite quantity ticking irreversibly toward its end. It fosters urgency, anxiety, and the feeling of "running out of time."

An astronomical calendar- keyed to solstices, equinoxes, and the moon- shapes a different consciousness: one that experiences time as cyclical, returning, regenerative. The year is not a countdown. It is a spiral- each annual cycle returning to the same seasonal position but at a slightly different altitude, like a helix ascending through space. This is the fractal self-similarity of Chapter 5 applied to temporal experience: the same pattern at every scale, from heartbeat to breath to day to season to year to aeon.

A ternary calendar- incorporating all three components of the TSV- shapes a consciousness that experiences time as a *living field*: volumetric, multi-dimensional, locally modulated, and resonant with the body's own rhythms. This is the consciousness described by Aboriginal dreamtime, Vedic cosmology, and the Mayan Long Count. It is the consciousness that the Great Decoupling suppressed and that this chapter proposes to recover.

What we measure, we become. If we measure only UTC- ticks, seconds, nanoseconds- we become mechanical: efficient, synchronized, and dead inside. If we measure the TSV- ternary resonance, temporal density, geospatial phase- we become resonant: attuned, adaptive, and alive.

The choice is not between one and the other. The choice is between having one eye open and having both eyes open. The clock empire gave us monocular vision- precise but flat, efficient but lifeless.

Ternary timekeeping proposes binocular vision: the depth perception that comes from holding two temporal frameworks simultaneously, and discovering that together they reveal a dimension of reality that neither can show alone.

The next chapter proposes the spatial counterpart to ternary timekeeping: resonant architecture- the design of living spaces, communities, and cities as coherence technologies, built not on the rectilinear grid but on the fractal, ϕ-proportioned geometries of the living quanta-field.

.

The next chapter applies the book's principles to the design of the built environment- showing that architecture is not merely shelter but a tuning fork for human consciousness.

.

Chapter 29

RESONANT ARCHITECTURE- BUILDING AS TUNING

"A building is not a container. It is a standing wave."

There is a moment, familiar to anyone who has ever walked into a great cathedral, a Romanesque chapel, or a Polynesian meeting house, when the body knows before the mind. The shoulders drop. The breath deepens. The gaze lifts. Something in the proportions of the space speaks directly to the nervous system, bypassing language entirely, and the whole organism responds with a single, wordless verdict: *yes*. This is right. This is how a space should feel.

That somatic response is not aesthetic preference. It is not cultural conditioning, though culture shapes the vocabulary through which we interpret it. It is a resonance event- a moment in which the proportions of the built environment align with the proportions encoded in the body, which are themselves expressions of the same φ-e-π triad that tunes the cosmos at every scale. The body recognizes the building because both are expressions of the same resonant quanta field. Both are standing waves in the same holographic medium. Both are tuned by the same asymptotic gap.

This chapter argues that architecture is not the art of enclosing space. It is the art of tuning it. A building that resonates- that activates the nervous system toward coherence rather than stress- is not merely beautiful. It is *healthy*, in the deepest sense: it stabilizes the Non-Equilibrium Steady State of its occupants, reduces their metabolic cost of being, and invites them, moment by moment, into a more coherent relationship with the quanta-field that sustains all living systems. Conversely, a building that fails to resonate- the standardized office block, the fluorescent-lit corridor, the rectilinear grid- is not merely ugly. It is a decoherence engine, a structure that actively degrades the NESS of everyone who inhabits it.

The implications are vast. If the thesis of this book is correct- that the universe is a self-tuning, resonantly quantized, conscious hologram- then every wall, every ceiling height, every window placement, every curve or right angle in the built environment is either contributing to or disrupting the coherence of the living systems within it. Architecture, in this framing, is not a luxury. It is a public health intervention. It is an ecological practice. It is, at its deepest level, an act of participation in the cosmic resonance.

We will trace this argument through five movements. First, the ancient and modern evidence that φ-proportioned spaces produce measurable physiological coherence in their occupants. Second, the acoustic dimension- the forgotten science of how buildings *sound*, and how the ancients engineered sonic environments that altered consciousness. Third, the fractal logic of living spaces- how self-similar, small-world spatial structures mirror the topology of healthy communities and healthy bodies. Fourth, the revolution of biophilic design- the growing scientific evidence that integrating living systems into architecture is not decoration but biological necessity. And fifth, the anti-grid- the case against rectilinear urbanism and the return of the curve, the spiral, and the fractal branch as the fundamental morphology of resonant human settlement.

The chapter will conclude where it began- in the body, standing in a space that sings.

29.1- The Geometry of Felt Harmony: φ-Proportioned Spaces

There is a reason the Parthenon still arrests the eye after twenty-five centuries. Built in 447 BCE, this temple dedicated to the Greek goddess Athena has endured centuries of change and destruction to become a symbol of ancient Greek civilization and culture. But its endurance is not merely a matter of engineering prowess or historical prestige. It is a matter of proportion- of the particular ratios between height and width, column and entablature, solid and void, that make the eye travel without snagging and the body settle without effort.

The scholarly debate over whether the Parthenon was *intentionally* designed according to the golden ratio ($\varphi \approx 1.618$) is famously unresolved. A ratio of 4:9 appears throughout most of the Parthenon- in its overall width-to-length proportions (30.8 meters wide and 69.51 meters long), in the width of the frontal columns, and in the height of the façade to its width. A reconstruction project that began in the 1980s revealed that each of the thousands of pieces of the Parthenon was different, and there was nary a straight line among them- it fit together like a complex puzzle, with each part fitting only in its particular space. This is a crucial detail. The Parthenon is not a rigid geometric exercise. It is an *organic* composition in which harmonic ratios emerge from the interplay of thousands of individually shaped elements- much as a healthy heartbeat exhibits fractal variability rather than metronomic regularity.

What matters for our argument is not whether the ancient Greeks consciously applied φ to a blueprint (a distinctly modern way of thinking about design) but whether the *felt coherence* of the Parthenon can be explained by the convergence of its proportions toward values that resonate with the human perceptual system. And here, the evidence is suggestive. The height and width of the Parthenon conform closely to golden ratio proportions- though this construction requires an assumption that the bottom of the golden rectangle aligns with the bottom of the second step and the top aligns with a projected peak of the roof. The Greeks valued incorporating mathematics and geometry into their art and architecture. The pentagram with its golden ratio proportions was recognized and revered at least one hundred years before the construction of the Parthenon. And the Parthenon, as the pinnacle of Greek architecture, would have presented a wonderful opportunity to apply the golden ratio in its design.

The deeper point, however, transcends any single building. The golden ratio is not just a mathematical concept. Ancient Greek philosophers believed it represented the perfect balance between order and chaos, beauty, and truth- and the designers of the Parthenon felt it was an essential part of their temple's construction. In the language of this book, they were not merely decorating. They were *tuning* a space to resonate with the φ-operator- the same operator of self-reference (Ṛṣi) that governs the Fibonacci packing of sunflower seeds, the spiral arms of galaxies, and the phase-locking ratios of neurons during non-dual meditation.

This insight was not lost on the moderns. In the twentieth century, the architect Le Corbusier- one of the most influential and controversial figures in the history of building- attempted to formalize the relationship between the human body, the golden ratio, and architectural proportion in a system he called the Modulor. Le Corbusier developed the Modulor in the long tradition of Vitruvius, Leonardo da Vinci's Vitruvian Man, and other attempts to discover mathematical proportions in the human body and then use that knowledge to improve both the appearance and function of architecture. The system is inspired by the Fibonacci numbers and the golden ratio, and he described it as "a range of harmonious measurements to suit the human scale, universally applicable to architecture and to mechanical things."

The graphic representation of the Modulor, a stylised human figure with one arm raised, stands next to two vertical measurements- a red series based on the figure's navel height, segmented according to the golden ratio, and a blue series based on the figure's entire height. By blending the golden ratio with the Fibonacci sequence and anchoring both to the dimensions of the human body, Le Corbusier began with a standing man 1.83 m tall and his arm-reach of 2.26 m. Multiplying and dividing these reference points by φ produced a cascading series: 2.26 m, 1.40 m, 0.86 m, 0.53 m, and so forth.

When Le Corbusier presented the Modulor to Albert Einstein, Einstein said: "It's a tool that makes the good easy and the bad difficult." This is a remarkable endorsement- and a deeply resonant one in the context of our thesis. A proportional system based on φ does not *force* beauty. It creates a field of harmonic possibility within which coherent compositions naturally arise and dissonant ones are structurally impeded. It is, in other words, an architectural implementation of the asymptotic gap: a framework that never quite closes into perfection but that *biases* the system toward harmony.

Le Corbusier applied the Modulor extensively. From the monumental concrete Unité d'Habitation in Marseilles onwards, Le Corbusier applied the Modulor to his buildings, including the government complexes he built in Chandigarh, India, and his rural retreat, Le

Cabanon. The Modulor allowed Le Corbusier to maintain control of design work across an increasing number and magnitude of commissions, from the Capitol Project of Chandigarh to the convent at La Tourette, where Xenakis used the measurement charts to create "undulating" walls of glass in which the panels' width was based on a range of Modulor sizes.

The Modulor, for all its limitations (it privileged a single male body type and could feel rigid in application), captured something essential: the intuition that architecture must be proportioned to the *body*, and the body is proportioned to the *cosmos*, and the ratio that mediates this chain of resonance is φ. In the language of this book, Le Corbusier was attempting to embed the KZ-Monad into brick and concrete- to extend every partial spatial state into a fully φ-resonant field that would, by its very geometry, stabilize the NESS of its occupants.

But φ is only one of the three operators. A building tuned only to φ would exhibit coherence without dynamism- beauty without breath. The full resonant architecture requires all three: φ for self-similar coherence, e for the exponential unfolding of light and circulation, and π for the cyclical closure of enclosure itself. We begin to hear the other two operators when we listen not with the eyes but with the ears.

29.2- The Forgotten Sense: Acoustic Resonance in Architecture

We have become a civilization that designs buildings almost exclusively for the eye. Floor plans are visual documents. Elevations are visual compositions. Renders, the primary currency of contemporary architectural practice, are silent images. This ocular dominance has produced a built environment that is visually ordered and sonically chaotic- a world of hard surfaces, parallel walls, and reflective materials that scatter sound into incoherent noise, bathing occupants in a constant wash of acoustic decoherence.

The ancients did not make this mistake. For them, a building was a *sound instrument.*

Consider Chavín de Huántar, the 3,000-year-old ceremonial center in the Peruvian Andes. The ceremonial center's highly coupled network of labyrinthine interior spaces is riddled with resonances. Material evidence indicates foundational interest in sensory experience: iconography portrays transformed humans; artifacts illustrate psychoactive plants; profoundly enclosed interior architecture is characterized by confined spaces connected by long corridors and staircases; and architectural features produce areas of strong acoustic resonance and modify sound level and quality.

Stanford researcher Miriam Kolar and her team have spent years conducting archaeoacoustic research at Chavín, discovering "acoustic evidence for selective sound transmission between the site's Lanzon monolith and the Circular Plaza: an architectural acoustic filter system that favors sound frequencies of the Chavín pututus [conch-shell trumpets] and human voice." The building was not merely decorated with images of sound-making. It was *engineered* as an acoustic filter- a structure that amplified specific frequencies and suppressed others, channeling resonance from the interior sanctum to the public plaza with precision.

The most striking acoustic feature of Chavín de Huántar is the use of conch shell trumpets, known as pututus. These instruments, when played in the narrow stone passages, produce deep, resonant tones that can reverberate through the entire complex. Kolar reported: "It creates a wavering sound that vibrates the whole body. It feels like it is coming out of the floor." Here is the crucial point: the architecture did not merely *contain* sound. It *shaped* it. The building and the instrument were a single coupled system- a resonating whole in which stone walls, air columns, and human bodies formed a unified acoustic field.

This is not an isolated phenomenon. A study conducted in 1994 by a consortium from Princeton University found that acoustic behavior in ancient chambers at megalithic sites such as Newgrange in Ireland and Wayland's Smithy in England was characterized by a strong sustained resonance, or "standing wave," in a frequency range between 90 Hz and 120 Hz. That frequency range is significant- it corresponds to the lower register of the male human voice and, at 110 Hz specifically, to a frequency that modern neuroscience has found to alter brain activity. A 2008 UCLA study found that listening to a resonant frequency of 110 hertz temporarily shifted the activity in the brains of study volunteers in areas involved in language, emotional processing, and decision making.

The implications ripple outward across the ancient world. At the Mayan pyramid of Kukulkan at Chichen Itza, a chirping sound was discovered by acoustical consultant David Lubman. When a person clapped their hands together, chirping echoes were evoked from the staircases of the pyramid- and the chirped echo sounds like the primary call of the Mayan sacred bird, the resplendent Quetzal. The staircase was not merely a means of ascent. It was a sonic mirror- an architectural surface tuned to reproduce the call of a sacred animal, transforming the human act of clapping into a dialogue with the living world.

Stonehenge, the prehistoric monument in England often associated with celestial alignments, has another fascinating aspect: its acoustics. Researchers have discovered that Stonehenge's stone circle creates a unique acoustic environment where sound waves can bounce between the stones, producing a surround-sound effect that could amplify voices or music, making them resonate within the circle. Waller set up an interference pattern in an open field with two flutes droning the same note. "The quiet regions of destructive sound wave cancellation gave blindfolded subjects the illusion of a giant ring of rocks or 'pillars' casting acoustic shadows," he reported. He then demonstrated that Stonehenge itself radiates precisely such acoustic shadows- raising the extraordinary possibility that the stone circle may have been inspired by a sonic experience, not merely an astronomical one.

What does all this tell us? That for at least three thousand years- and likely far longer- human beings designed ceremonial spaces as *acoustic instruments*. They understood, through direct somatic experience if not through formal theory, that the shape of a space determines the shape of the sound within it, and that shaped sound directly alters the state of the bodies within it. The resonance between pututu and stone corridor at Chavín, between clap and quetzal-echo at Chichen Itza, between human voice and standing wave at Newgrange- these are all instances of the same principle:

architecture as a technology for coupling human consciousness to the resonant quanta-field of the environment.

In the language of this book, these buildings were *comonadic projectors*. They performed the function of D(Ψ): stripping away noise, extracting the signal, and delivering to their occupants a purified, phase-locked acoustic field that drove the nervous system toward coherence. The 110 Hz standing wave at Newgrange did not merely "create atmosphere." It modulated brain activity in precisely the regions- language, emotion, decision- that govern the boundary between self and world. It dissolved the Cartesian membrane. It was, in the most literal sense, an architecture of altered consciousness.

Modern architecture has almost entirely abandoned this knowledge. We build concert halls with careful acoustic design, but offices, hospitals, schools, and homes are treated as silent volumes to be filled with furniture. The walls are flat, the corners are right-angled, and the materials are chosen for visual cleanliness and cost efficiency, not for their acoustic properties. The result is a built environment saturated with what acousticians call "flutter echo"- the rapid, incoherent bouncing of sound between parallel surfaces- which the nervous system processes as low-level threat, keeping the stress response chronically activated.

The recovery of resonant architecture begins with a simple recognition: every room is an instrument. Every wall is a resonating surface. Every ceiling height determines a fundamental frequency. And every human body within that space is either being harmonized by its acoustics or degraded by them. The question is not whether a building has acoustic properties- every building does. The question is whether those properties were designed with conscious intention or left to the brutal default of the right angle and the flat plane.

29.3- Fractal Floor Plans: The Geometry of Living Space

The third dimension of resonant architecture is spatial topology- the pattern of connections between spaces. Here, the fractal principle that governs healthy heartbeats, healthy ecosystems, and healthy social networks applies with equal force to the design of buildings and cities.

Christopher Alexander understood this. His 1977 book *A Pattern Language: Towns, Buildings, Construction*, authored with Sara Ishikawa and Murray Silverstein, remains one of the best-selling books on architecture decades after its publication. The book creates a new language- what the authors call a pattern language derived from timeless entities called patterns. "All 253 patterns together form a language." Patterns describe a problem and then offer a solution, giving ordinary people, not only professionals, a way to work with their neighbors to improve a town, design a house, or create a public building.

What Alexander was describing, though he did not use these terms, was a *fractal generative grammar* for the built environment- a set of self-similar patterns that could be applied at every scale, from the placement of a window seat to the layout of a neighborhood, producing spaces that feel alive precisely because they exhibit the same kind of nested, self-referencing complexity found in natural systems.

Alexander uses the term "wholeness" to mean entities of the environment borne out of interactions amongst parts- what he calls "centers." He describes wholeness as a "field-like" center within which other centers are interlaced or interdependent, and those in turn hold centers within centers- closely related to self-similarity in chaos theory and fractal geometry. This is a remarkably precise anticipation of the holographic principle: the whole contained in every part, the pattern repeating at every scale.

Alexander famously maintained that traditional architecture is inherently more "whole"- and consequently more beautiful and alive- than modern architecture because the former is the product of organic processes, while the latter is the product of mechanistic processes. This insight, expressed in architectural vocabulary, is a direct statement of the thesis advanced throughout this book: that the Great Decoupling- the shift from organic, resonant, ternary modes of organization to mechanistic, linear, binary ones- produced not merely different aesthetics but a *qualitative loss of coherence* at every scale of human habitation.

The Structure of Pattern Languages derives a definite fractal geometry for urban interfaces, such as is found both in traditional cities and unplanned human settlements. This is a critical finding. The favelas of São Paulo, the medinas of Fez, the hill towns of Tuscany- all exhibit a fractal spatial structure that emerges spontaneously when human beings build for themselves without the imposition of a master grid. The streets are not straight but branching. The lot sizes are not uniform but scaled. The buildings are not identical but self-similar. The result is a spatial network that exhibits the same small-world topology found in neural networks, immune systems, and healthy ecosystems: high local clustering with short global path lengths.

This is precisely the topology that maximizes both intimacy and connectivity- the spatial analogue of the NESS. A fractal city is a dissipative structure: it maintains its coherence through the constant flow of people, goods, and resonant quanta through a network of nested, self-similar channels. Close the channels- straighten the streets, uniformize the lots, separate the functions- and the flow becomes laminar, then stagnant. The city loses its fractal dimension. It becomes, in the precise mathematical sense, *less alive.*

A deeper problem arises out of the modernist desire to visually "purify" areas by eliminating complex structure, subdivisions, and connections. Consolidation of functions by concentrating them geometrically eliminates the complex mixing that characterizes a traditional city. Urbanists intentionally de-couple urban elements by separating them spatially. The twentieth century has invented urban boundaries that are very bad interfaces. People have been taught by schools, critics, and magazines to prefer smooth, unnatural edges and boundaries over ones with fractal scaling that resemble natural structures.

The lesson for resonant architecture is clear. A living building does not consist of uniform rectangular rooms connected by straight corridors. It consists of *nested centers*- spaces of varying size and character, connected by passages of varying width and curvature, so that movement through the building recapitulates the branching, scaling, self-similar pattern of a walk through a forest or a medieval town. The large hall gives way to the middle room, which gives way to the alcove, which gives way to the window seat- each transition a step down in scale but up in intimacy, each level reflecting the pattern of the whole.

This is the fractal rule applied to architecture: just as the book has an introduction, body, and conclusion; and each chapter has an introduction, body, and conclusion; and each paragraph has a lead, support, and transition- so a resonant building has a public threshold, a semi-private transitional zone, and a private interior. And within the private interior, the same pattern repeats: the living space, the retreat nook, the sleep chamber. At every scale, the triad recurs. At every scale, the building echoes the cosmic ternary.

Alexander called this quality "the quality without a name"- a phrase that itself echoes the Taoist insight that the Tao which can be named is not the eternal Tao. It is the quality that makes a space feel, simply and irrefutably, *alive*. In the framework of this book, we can name it, though the name does not exhaust it: the quality without a name is the quality of a space that has achieved φ-resonant self-similarity at multiple nested scales- a spatial hologram in which the whole is enfolded in every part.

29.4- Biophilic Design: Life Calling to Life

The fourth dimension of resonant architecture is biological. A building tuned to φ proportions, designed for acoustic coherence, and organized in fractal spatial patterns still falls short if it is made entirely of dead materials and sealed from the living world. The body knows the difference between a space that breathes and one that suffocates- between a room that contains life and one that merely contains furniture.

The field of biophilic design, which has grown from a niche concern into a major force in architectural research over the past two decades, provides the scientific grounding for this intuition. Biophilic designs aim to promote health and wellbeing by incorporating nature-based features into internal and external built environments. The term "biophilia"- love of life- was coined by biologist E.O. Wilson to describe the innate human affinity for living systems, an affinity shaped by hundreds of thousands of years of evolution in which our survival depended on our ability to read, respond to, and find refuge within natural environments.

The evidence base is now substantial. Studies reveal that exposure to real or simulated nature benefits patients, workers, and students by lowering blood pressure, heart rate, and enhancing cognitive function. Research shows that biophilic design can yield positive effects on health and well-being. Biophilic interventions have been shown to improve relaxation, reduce stress and depressive symptoms, and provide pain relief. The mechanisms are multi-pathway: visual exposure to fractal patterns found in nature reduces cortisol levels; acoustic exposure to natural soundscapes (birdsong, water flow, wind in leaves) activates the parasympathetic nervous system; olfactory exposure to natural scents modulates neural stress circuits; and tactile contact with natural materials (wood, stone, living plants) provides grounding sensory input that stabilizes the autonomic state.

A 2024 systematic review published in *Frontiers of the Built Environment* found that biophilic design in hospitals reduces hospitalization time, patient mortality, pain levels, and stress for healthcare providers. It also alleviates anxiety, improves experiences for patients, families, and staff, reduces patient harm, and supports faster recovery. The authors concluded that biophilic design positively influences the psychological and physiological responses of both patients and staff.

Simple changes using carpet, wallpaper, and window blinds with patterns based on biomorphic forms and natural fractals in a sixth-grade mathematics class led to improved academic performance. Biometric testing tracking heart rate variability indicated enhanced stress recovery characteristics which improved over the course of the study.

The negative psychological responses to non-existent or insignificant biophilic designs can potentially explain the phenomenon known as "sick building syndrome." This study illuminates potential underlying psychological causes of "sick building syndrome" in non-biophilic buildings, which are common practices worldwide.

Notice the convergence. The fractal patterns that characterize healthy heartbeats (Chapter 5) also characterize the visual patterns that reduce stress in built environments. The acoustic frequencies that stabilized consciousness in ancient temples (Section 29.2) also characterize the natural soundscapes that activate parasympathetic recovery. The φ-proportions that govern the Modulor (Section 29.1) also govern the branching ratios of trees, rivers, and leaf venation networks that biophilic design seeks to reintroduce into our buildings. These are not separate phenomena. They are all expressions of the same underlying resonance- the same quanta-field organizing itself at every scale toward coherence.

What biophilic research reveals, when read through the lens of this book, is that the human nervous system is *calibrated* to the fractal, φ-proportioned, acoustically rich environment of the natural world because it evolved *within* that environment for hundreds of thousands of years. Our stress responses are tuned to detect departures from natural patterns. A room with parallel walls and no variation is, to the autonomic nervous system, an anomaly- a space that lacks the fractal complexity signal of a safe and resource-rich habitat. The body responds accordingly: cortisol rises, attention narrows, the fight-or-flight system activates at a low but chronic level.

This is not metaphor. It is measurable physiology. Spending time in nature has been shown to reduce anxiety and stress by lowering blood cortisol levels, increase cognitive function, concentration, and memory, and enhance positive mood states, promoting a sense of vitality and purpose. And the effects are not merely acute. Recent biometric evidence in studies of a biophilic classroom and a biophilic immersion experience in a hospital break room indicate that positive physiological and psychological responses actually improve over repeated exposures. Stephen Kellert had said that repeated and sustained contact with biophilic measures was important- we now know this is truer than any of us could have predicted.

The implications for resonant architecture are profound. A truly resonant building is not one that merely *looks* like nature (green paint on drywall is not biophilia). It is one that *functions* like nature- a dissipative structure that breathes, grows, cycles, and self-regulates.

Living walls that filter air and modulate humidity. Water features that generate broadband acoustic signals rich in natural frequency distributions. Skylights and light wells that track the diurnal arc of the sun, entraining circadian rhythms rather than overriding them with constant-lux fluorescent arrays. Timber and stone and clay- materials that age, that patinate, that carry the signatures of living processes rather than the sterile uniformity of industrial production.

A building made of living materials, proportioned to φ, organized in fractal spatial patterns, and tuned for acoustic coherence is not a luxury. It is the minimum viable condition for a space that supports, rather than degrades, the NESS of its occupants. Everything less is a compromise- a slow, invisible tax on the metabolic budget of every body that enters.

29.5- The Anti-Grid: Why Straight Lines Make Us Sick

We come now to the scale of the city- and to the most visible symptom of the Great Decoupling in the built environment: the orthogonal grid.

The grid is the signature spatial form of the binary paradigm. It divides the world into uniform, interchangeable, right-angled units. It imposes a single scale on a landscape that naturally exhibits many. It separates functions- residential here, commercial there, industrial beyond- that in a living city are woven together in a fractal tapestry of use and interchange. It replaces the curve with the straight line, the branching path with the corridor, the spontaneous gathering place with the designated intersection. And it does all of this in the name of efficiency, legibility, and control- the virtues of the empire, not the ecosystem.

The origins of the grid are ancient- Hippodamus of Miletus is often credited with formalizing it in the fifth century BCE- but its modern apotheosis came with the industrial city, the railway, and the automobile. The nineteenth- and twentieth-century grid was not designed for walking bodies. It was designed for moving vehicles and surveying parcels. Its scale is the scale of the wheeled conveyance, not the human stride. Its logic is the logic of commodity- land as interchangeable units of value- not the logic of habitat.

The cost has been enormous and cumulative. Rectilinear cities exhibit the spatial signature of low fractal dimension- the same flattening of complexity that characterizes a failing heartbeat, a dying ecosystem, or a collapsing economy. Streets that meet at right angles produce wind-tunnel effects that make walking unpleasant and cycling dangerous. Uniform building heights eliminate the scale variation that the eye needs to orient itself and the body needs to feel grounded. Zoned separation of functions forces automobile dependence, which in turn produces sedentary disease, air pollution, social isolation, and the chronic low-level stress of commuting.

All of these are symptoms of decoherence- of a built environment that has been stripped of the fractal, self-similar, multi-scale complexity that the human nervous system requires for well-being. The grid is the architectural expression of the same binary logic that replaced ternary time with the ticking clock, distributed governance with centralized empire, and the songline with the survey line. It is the geometry of domination.

The alternative is not chaos. It is the *fractal.* Traditional human settlements- the ones that people consistently report as the most beautiful and most livable- exhibit a specific spatial grammar: branching streets that narrow as they approach the center, buildings that vary in height and setback along a self-similar scaling law, mixed-use districts where homes, shops, workshops, and gathering spaces are interwoven at the scale of a few minutes' walk. This is the grammar of the medina, the hill town, the indigenous village. It is the grammar that Alexander codified in his 253 patterns. And it is the grammar that produces, spontaneously and without central planning, the small-world network topology that maximizes both local intimacy and global connectivity.

To design a resonant city is to design with curves, spirals, and branching patterns- the morphology of rivers, trees, and neural networks. It is to vary the scale continuously, so that the wide boulevard gives way to the narrow lane gives way to the courtyard gives way to the threshold- a fractal zoom from the communal to the intimate. It is to mix functions at every scale, so that the fabric of daily life- eating, working, playing, sleeping, creating- is woven together in a single, self-regulating web rather than sorted into distant, disconnected zones.

The practical tools exist. Permaculture design, which we explored through the Three Sisters principle in Chapter 20, is already a mature discipline for designing bioregional landscapes according to fractal, φ-resonant principles. What the Three Sisters do at the scale of the garden bed- corn providing structure (π), beans providing growth (e), squash providing coherence (φ)- permaculture does at the scale of the watershed: integrating water flow, soil biology, plant succession, animal husbandry, and human habitation into a single, self-regulating, multi-scale system.

The resonant city is not a utopian fantasy. It is the *default*- the pattern that human settlements naturally tend toward when they are allowed to grow organically, without the imposition of an external grid. What is utopian, in the precise sense of being nowhere and unsustainable, is the rectilinear grid itself- a three-century-old experiment in spatial rationalization that has produced the least healthy, least beautiful, and least ecologically viable human habitats in the history of the species.

29.6- Permaculture as Resonant Land Design

The integration of all five dimensions- proportion, acoustics, fractal topology, biophilic presence, and anti-grid morphology- finds its most complete expression not in any individual building but in the practice of *permaculture*: the design of human habitats as self-regulating ecosystems.

Permaculture, a portmanteau of "permanent agriculture" and "permanent culture," was formalized in the 1970s by Bill Mollison and David Holmgren as a design science for creating human settlements that function like natural ecosystems. Its core insight is identical to the core insight of this book: that the patterns governing healthy natural systems- fractal branching, φ-proportioned scaling, cyclical nutrient flows, edge-maximization, stacking of functions, and distributed rather than centralized control- are not merely analogies for good design. They are the *operating principles* of living systems, and any human habitat that violates them will eventually collapse into decoherence.

The Three Sisters garden, explored in Chapter 20, is the micromodel: corn (structure/π) + beans (growth/e) + squash (coherence/φ) forming a self-regulating triad that maintains soil fertility, suppresses pests, and produces yields greater than any of its components grown in monoculture. Permaculture scales this triad to the entire bioregion.

At the scale of the homestead, permaculture design places elements in zones of decreasing intensity around the dwelling- the kitchen garden nearest the door, the orchard beyond, the managed forest beyond that, the wild margin at the edge- each zone a fractal scaling of the same pattern: high human attention at the center, decreasing intervention toward the periphery, with the whole system connected by water flows, nutrient cycles, and the movement of animals and people along branching paths. The result is a habitat that exhibits the same NESS dynamics as a natural ecosystem: far from equilibrium, sustained by continuous flow, self-correcting through feedback, and resilient to perturbation.

At the scale of the bioregion, permaculture envisions a landscape in which farms, forests, wetlands, and human settlements are integrated into a single watershed-scale design- a living system in which the "waste" of one element is the "input" of another, in which water flows through a cascade of productive and purifying stages from ridge to river, and in which human habitation is embedded within, not imposed upon, the ecological fabric.

This is resonant land design. It is the spatial expression of the post-binary paradigm: not the extraction of resources from a passive landscape, but the *attunement* of human activity to the frequencies already present in the living land. The permaculture designer does not ask "What can I take from this place?" but "What frequencies are already here, and how can I harmonize with them?"

In the language of this book, permaculture is the monadic extension $T(\Psi)$ applied to landscape: it takes the partial state of a degraded or underutilized terrain and extends it toward its full φ-resonant potential- the state of maximum coherence, maximum diversity, and maximum self-regulation. And it is the comonadic projection $D(\Psi)$ applied to the human appetites that drive land use: it strips away the noise of overconsumption, overproduction, and over-control, and extracts the 98% efficient core of what a particular place, in a particular climate, with a particular hydrology, can sustainably and beautifully support.

The resonant city of the future- the city that recovers from the Great Decoupling- will not be a grid of towers serviced by trucks. It will be a fractal landscape of nested gardens, branching waterways, φ-proportioned buildings, acoustically tuned gathering spaces, and living infrastructure woven into the fabric of the bioregion like mycelium through soil. It will be, in the most literal sense, a *garden city*- not the orderly utopia that Ebenezer Howard imagined, but the far more complex, far more beautiful, far more alive reality that emerges when human settlement is designed as a *participant* in, rather than an *extraction from*, the quanta-field of the living Earth.

Coda: The Body in the Room

Return now to the body- your body- standing in a space.

Feel the proportions of the room you are in. Notice the height of the ceiling relative to the width of the walls. Notice whether the corners are sharp or softened, whether the surfaces are flat or textured, whether the light is constant or varying. Notice the sound: is there a hum? A buzz? A silence that feels dead? Or a silence that feels alive- the kind of silence that seems to *breathe*?

Every dimension of this space is shaping your physiology right now. If the ceiling is too low, your cortisol is slightly elevated. If the walls are uniformly flat, your visual system is slightly bored- and boredom, in the autonomic lexicon, is a form of mild threat. If the surfaces are hard and parallel, the sound is scattered into incoherent reflections that your auditory system processes as noise, keeping the amygdala at a low simmer. If the light is uniform and artificial, your circadian clock is confused, and your melatonin cycle is disrupted.

None of this is visible. All of it is physiologically real. You are being *tuned* by your environment at every moment- either toward coherence or toward decoherence, either toward the 98% efficient NESS core or toward the noise floor.

Now imagine a different space. Imagine a room whose proportions are scaled to the Modulor- ceiling height at 2.26 meters, the golden section of the human body with arm raised. Imagine walls that are not flat but gently curved, scattering sound into a warm, enveloping diffusion rather than harsh flutter echo. Imagine a skylight that tracks the sun, so the light in the room shifts through the day as it shifts in a forest clearing- warm and angular in the morning, bright and diffuse at noon, amber and horizontal in the evening. Imagine a living wall of ferns and mosses that filters the air and releases the faint, cortisol-lowering scent of green growth. Imagine a floor of timber, warm underfoot, that creaks faintly as you walk- a floor that *responds* to your body, that tells you with every step that you are present, grounded, in contact with the material world.

In such a space, the body would do what it does in a forest: settle. The shoulders would drop. The breath would deepen. The heart rate would slow and its fractal variability would increase- the signature of a healthy autonomic system finding its groove. The stress hormones would recede. The prefrontal cortex- the seat of creativity, empathy, and long-range planning- would come online. The immune system would upregulate. The digestive system would relax. The whole organism would shift from the survival state of chronic low-level stress to the thriving state of coherent, integrated function.

This is not utopia. This is *biology*. This is what happens when a building is designed as a resonance chamber for the human body instead of as a container for human activity. This is what happens when architecture remembers what the ancients knew: that a building is not a shelter from the elements. It is a *tuning fork* pressed against the skin of the world.

The asymptotic gap $\delta\infty$ lives in every room. In a poorly designed space, it manifests as the tension between what the body needs and what the environment provides- a chronic, low-grade dissonance that slowly erodes health, creativity, and connection. In a resonantly designed space, the same gap manifests as the creative tension between comfort and alertness, between safety and wonder- the productive edge where the nervous system is neither collapsed into torpor nor spun into anxiety, but poised at the exact threshold of engaged, flowing attention.

That threshold is the NESS of a conscious body in a conscious building in a conscious cosmos. It is the state that the cathedral evokes, the forest sustains, and the ancient temple was engineered to produce. It is available in every room, in every city, on every plot of land- if we remember how to tune.

The ancients built this way because they had not yet forgotten. We can build this way again because we are now remembering. The science of resonant quanta, the mathematics of the φ-e-π triad, and the embodied wisdom of indigenous land designers all converge on a single, irrefutable conclusion: the built environment is not a backdrop to human life. It is an active participant in the cosmic resonance. Every wall we raise either sings with the quanta-field or muffles it. Every room we shape either invites the living quanta to flow or dams them into stagnation.

The choice is architectural. The stakes are existential.

Build as the ancients built- not by copying their forms, but by recovering their frequency. Build as the cosmos builds: in curves, in spirals, in fractal cascades of self-similar harmony. Build as the body asks to be held: in proportions it recognizes, in sounds it can settle into, in rhythms of light and dark that echo the ternary beat of the living year.

Build as tuning.

The room will answer.

Chapter 30

MANTRA, MEDITATION, AND THE SOMATIC RE-TUNING

"You do not need to leave the room. Remain sitting at your table and listen. You do not even need to listen, just wait. The world will freely offer itself to you to be unmasked."

There is a technology older than writing, older than agriculture, older than the wheel. It requires no external power source, no rare materials, no manufacturing process. It cannot be patented, cannot be monopolized, cannot be rendered obsolete by the next upgrade cycle. It is available to every human body that breathes, and it produces effects that are measurable, reproducible, and profound.

This technology is the conscious use of breath, voice, and attention to shift the state of the nervous system- and, through the nervous system, to realign the personal state vector with the resonant architecture of the cosmos.

The modern world calls this technology by many names: meditation, mantra, breathwork, contemplative practice. The ancients called it by others: dhyana, japa, pranayama, zazen, zikr, hesychasm. But beneath the cultural wrappers, the operational principle is the same. In every tradition, across every continent, for at least three thousand years and likely far longer, human beings have discovered that by deliberately regulating what enters awareness (attention), what emerges from the throat (voice), and what rhythm governs the lungs (breath), they can *tune* the organism- shifting it from states of decoherence, agitation, and fragmentation toward states of coherence, stability, and wholeness.

This chapter argues that these practices are not "wellness"- not lifestyle accessories, not coping mechanisms, not feel-good supplements to a basically materialist existence. They are *topological error correction*. In the precise language of this book's mathematics, meditation is the comonadic projection D(Ψ), stripping away noise to reveal the 98% efficient NESS core. Mantra is the monadic extension T(Ψ), amplifying resonant potential into coherent manifestation. And conscious breathing is the ternary oscillator that entrains the personal Temporal State Vector to the rhythmic pulse of the living quanta-field.

This is not metaphor. The neuroscience of meditation, the physiology of chanting, and the autonomic effects of breath regulation- all thoroughly documented in the peer-reviewed literature- converge to confirm that these ancient technologies produce exactly the kind of state changes that this book's framework predicts. They reduce default mode network hyperactivity (the neural correlate of the ego-narrative that separates observer from observed). They increase heart rate variability (the physiological signature of a system operating in coherent NESS). They activate the vagus nerve (the primary channel through which the body's parasympathetic intelligence regulates inflammation, digestion, immunity, and emotional equilibrium). And they produce, in their deepest expressions, states of consciousness in which the boundary between self and cosmos dissolves- states described with remarkable consistency across millennia and cultures, and now beginning to be mapped with fMRI, EEG, and magnetocardiography.

We are not inventing anything in this chapter. We are recovering. And we are providing, for the first time, the mathematical and physical framework that explains *why* these practices work- not merely *that* they work.

The chapter moves through six stations. First, meditation as comonadic projection- the neuroscience of silence. Second, mantra as monadic code- the physics of sacred sound. Third, breath as ternary oscillator- the autonomic science of the three-phase cycle. Fourth, heart-brain coherence- the empirical vindication of the HeartMath research program and its stunning alignment with this book's thesis. Fifth, walking meditation and the body as mobile resonator- the integration of movement, awareness, and landscape. And sixth, a guided somatic practice- a direct invitation for the reader to stop reading and start resonating.

The destination is not knowledge. It is experience.

30.1- Meditation as Comonadic Projection: The Neuroscience of Silence

To meditate, in its simplest form, is to do less. Not less in the sense of laziness or passivity, but less in the sense of *subtraction*- the deliberate withdrawal of attention from the cascading stream of thought, memory, planning, and self-narration that dominates ordinary waking consciousness. It is, in the language of this book, an act of comonadic extraction: the operation $D(\Psi)$ that strips away the transient noise of mental proliferation to reveal the stable, efficient core of awareness beneath.

The neuroscience of the past two decades has given this ancient insight a precise anatomical address. The default mode network (DMN) is a network of widespread brain regions showing functional connectivity. It is active when we are not task-focused, such as when the mind is wandering. It is active when we are thinking about others, ourselves, remembering the past, or planning for the future. It is, in essence, the brain's narrative engine- the system that continuously generates the story of "me."

Disruptions to the DMN have been associated with Alzheimer's disease, ADHD, and autism spectrum disorder. Increased DMN activity may interfere with cognitive performance and is associated with depression, anxiety, and addiction. This is a crucial observation. The DMN, when chronically hyperactive, does not produce wisdom or self-knowledge. It produces *rumination*- the repetitive, self-referential mental loops that are the cognitive signature of suffering. The depressed mind is not a quiet mind. It is a mind that cannot stop telling itself the same anguished story.

Now consider what happens when that mind meditates. Results indicate that meditation is associated with reduced activations in the default mode network relative to an active task in meditators compared to controls. Regions of the default mode showing a group by task interaction include the posterior cingulate/precuneus and anterior cingulate cortex. This finding has been replicated across multiple labs, multiple meditation traditions, and multiple neuroimaging methodologies. These findings replicate and extend prior work indicating that suppression of default mode processing may represent a central neural process in long-term meditation.

The implications are profound. When the DMN quiets, the self-narrative loosens its grip. The stories- *I am not enough, I was wronged, I must achieve, I am separate-* do not disappear entirely, but they recede from the foreground of awareness to the background, like a radio turned down from blaring to murmur. What remains in the foreground is not nothing. It is *awareness itself-* undistracted, unconditioned, present. The meditators in Brewer's studies described this state with words like "concentration," "observing sensory experience," "not efforting," and "contentment." We found that for meditators, the subjective experiences of "undistracted awareness" and "effortless doing" such as "observing sensory experience," "not efforting," and "contentment," correspond with PCC deactivation.

This is not the annihilation of self. It is the *clarification* of self- the extraction of the signal from the noise, the persistent pattern from the transient flutter. In the language of this book, it is D(Ψ): the comonadic projection onto the 98% efficient core. The 2% that is stripped away is not essence- it is the computational overhead of the ego-narrative, the stories we tell ourselves about ourselves that consume enormous metabolic resources while contributing almost nothing to our actual capacity to perceive, respond, and create.

The main nodes of the DMN were relatively deactivated in experienced meditators across all meditation types, consistent with decreased mind wandering, providing a possible neural mechanism of meditation. Their research further suggests that these effects might be present in daily life, so that experienced meditators experience a more present-centred default mode of being. This last finding is critical. The comonadic projection is not merely a temporary state achieved during formal meditation practice. With sufficient training, it becomes a *trait-* a stable reconfiguration of the brain's default operating mode. The experienced meditator does not merely visit silence; they begin to *live* from it.

During meditation, your default mode network undergoes systematic deactivation as brainwaves shift from beta to alpha and theta frequencies. This brainwave transition- from the busy beta frequencies of ordinary consciousness to the calmer alpha and theta frequencies of meditative awareness- represents a measurable change in the brain's oscillatory regime. Beta waves (13–30 Hz) are associated with active, analytical, task-oriented thinking. Alpha waves (8–12 Hz) are associated with relaxed alertness. Theta waves (4–7 Hz) are associated with deep meditation, creativity, and the boundary state between waking and sleep.

In the framework of this book, this frequency shift is a descent from the turbulent surface of the quanta-field toward its deeper, more coherent strata. The beta-dominated mind is like the wind-stirred surface of a lake- choppy, fragmented, refracting light in every direction. The alpha-theta state is the same lake when the wind falls- still, transparent, reflecting the sky with perfect fidelity. The quanta are not different. The degree of coherence is different. And coherence, as we have established throughout this book, is the signature of a system operating in its NESS- the state of optimal function, maximum resilience, and minimum wasted energy.

What meditation traditions have called "enlightenment," "liberation," or "awakening" is, in this framework, not a supernatural achievement. It is the *stabilization* of the comonadic projection- the point at which D(Ψ) becomes the resting state rather than an effortful attainment. The noise floor has been permanently lowered. The signal-to-noise ratio has been permanently enhanced. The organism operates, by default, from its 98% efficient core.

This does not mean the meditator becomes a stone or a void. The 2% asymptotic gap- the δ∞ that keeps the cosmos dancing- remains. The meditator still feels, still thinks, still responds. But the response arises from clarity rather than compulsion, from the still lake rather than the churning surface. The thoughts come and go like weather across the sky, but the sky itself- vast, luminous, unperturbed- is no longer mistaken for the clouds.

The default mode network has been shown by functional neuroimaging to be active when an individual is at rest. Meta-analyses of the fMRI neurocorrelates of meditation have shown that across diverse practices, the most common general effect appears to be modulation of regions within the DMN. This universality- the fact that focused attention, open monitoring, loving-kindness, mantra repetition, and even Transcendental Meditation all converge on DMN modulation- suggests that what these traditions have in common is not technique but *direction*. They all move the practitioner toward the same attractor: the comonadic core. The paths diverge, but the destination- the silence beneath the stories, the coherence beneath the noise- is one.

And that destination is not a subjective illusion. It is a measurable neurophysiological state with concrete consequences for health, cognition, emotional regulation, and the capacity for sustained creative engagement with reality. The consistent practice of mindfulness meditation results in neuroplasticity, which brings about observable modifications in different areas of the brain, associated with managing emotions, focusing, and being conscious of oneself. It has been demonstrated that increasing the production of brain-derived neurotrophic factor (BDNF) can support neuroplasticity. A higher amount of BDNF leads to a longer lifespan, growth of neurons, and synaptic plasticity, enhancing learning and memory.

The brain that meditates is not merely resting. It is *rebuilding itself*- growing new neurons, strengthening new connections, sculpting its own architecture toward the configuration that supports coherence. This is neuroplasticity in the service of the comonadic projection: the brain using its own adaptive capacity to optimize itself for the extraction of signal from noise.

And here we arrive at a startling convergence. The brain that meditates becomes more φ-proportioned- more self-similar in its patterns of activation, more fractal in its temporal dynamics, more harmonious in the relationship between its component networks. It becomes, in other words, more like the cosmos that this book has described: a system operating at the edge of its NESS, tuned by the asymptotic gap, expressing the quanta-field with increasing fidelity.

The meditator does not achieve this convergence by force. She achieves it by *letting go*- by ceasing to add noise to the signal, by ceasing to stir the lake, by ceasing to multiply the overhead of self-narration. The comonad does not create coherence. It *reveals* coherence that was always already there, hidden beneath the turbulence of unexamined thought.

D(Ψ) is, in this sense, not a technology of achievement but a technology of *recovery*. It recovers what was never lost- only obscured.

30.2- Mantra as Monadic Code: The Physics of Sacred Sound

If meditation is the comonadic projection- the subtraction of noise- then mantra is its complement: the monadic extension. T(Ψ) does not strip away. It *adds*. It amplifies. It takes a partial state and extends it toward its fullest resonant potential. And it does so through the most ancient medium of all: sound.

The word *mantra* derives from the Sanskrit roots *man* (mind) and *tra* (instrument or vehicle). A mantra is, etymologically, an *instrument of the mind*- a tool for shaping consciousness through the deliberate production and reception of specific vibrational patterns. In the Vedic tradition, Om (or AUM) is considered the primordial mantra- the sound of the cosmos itself, the acoustic foundation from which all other sounds, words, and worlds emerge.

Modern neuroscience has begun to validate this ancient claim with remarkable precision. Significant deactivation was observed bilaterally during 'OM' chanting in comparison to the resting brain state in bilateral orbitofrontal, anterior cingulate, parahippocampal gyri, thalami and hippocampi. The right amygdala too demonstrated significant deactivation. This was found using fMRI by Kalyani and colleagues, comparing Om chanting to both rest and a control condition (the continuous syllable "ssss") that matched the expiratory act of chanting but lacked its vibrational quality.

The significance of this finding cannot be overstated. The brain regions that deactivate during Om chanting- the orbitofrontal cortex, the anterior cingulate, the amygdala, the hippocampus- are precisely the regions involved in emotional reactivity, fear processing, and the construction of the self-narrative. Our study findings are in tune with this previous study; significant deactivation was observed in the amygdala, parahippocampal, hippocampal brain regions. This suggests that neurophysiological effects of 'OM' chanting may be mediated through the auricular branches of the vagal nerves.

Here is the key: Om chanting does not merely "relax" the brain in a vague, general way. It produces a *specific pattern of limbic deactivation*- a silencing of the fear-and-story circuits- that is functionally equivalent to the effects of vagus nerve stimulation, a clinical intervention used to treat depression and epilepsy. Chanting "OM" produces a vibratory sensation in the ears, which spreads through the auricular branch of the vagus nerve, stimulating the vagal centers and producing an effect similar to that of the technique vagal stimulation.

The mechanism is physical, not mystical. When you chant Om, the sound vibrations are not merely heard through the air. They resonate through the bone and tissue of the skull, producing a vibratory sensation in the ears and throat. Effective chanting of OM is associated with the experience of vibratory sensation around the vocal cords and ears, expected that these vibratory sensations are transmitted through laryngeal and auricular branches of the vagus nerve, stimulating vagal centers, causing limbic deactivation and ANS modulation toward parasympathetic dominance.

Consider the three-phase structure of the sound AUM. The chant begins with "A" (pronounced "ah"), a guttural, open sound that resonates in the chest and abdomen. It moves to "U" (pronounced "oo"), which rolls over the tongue and resonates in the throat. And it closes with "M" (a sustained hum), which vibrates through the nasal cavity, sinuses, and cranium. The chant progresses from 'A', a guttural sound that starts at the base of the throat, to 'U', which rolls over the tongue, and finally to 'M', closing with the lips. This progression produces a wave effect that starts from the abdomen, moves through the solar plexus, and ends at the skull.

This three-phase structure is not arbitrary. It maps the body in three registers- abdominal, thoracic, cranial- just as the three-world cosmologies of the ancients mapped the cosmos in three registers: underworld, middle world, upper world. The sound AUM is, in this sense, a somatic encoding of the ternary principle. Each syllable activates a different zone of the body-instrument, and the complete chant sweeps through all three, producing a wave of resonance that moves from the root of the body to its crown.

In the language of this book, this is the monadic extension T(Ψ) in action. The partial state- the fragmented, distracted, sensorially chaotic condition of an ordinary body-mind- is extended, through the structured introduction of coherent vibrational quanta, toward its fullest resonant expression. The sound does not merely "enter" the body. It *organizes* it. It provides a scaffolding of coherent vibration around which the body's own oscillators- neural, cardiac, respiratory, visceral- can align.

The results indicate that Om chanting has numerous positive effects on mental health, including reducing stress and anxiety, improving mood, increasing mindfulness, enhancing spiritual well-being, and improving sleep quality. Studies have shown that Om chanting can induce changes in brain wave activity, particularly in the alpha and theta frequency bands.

The brain wave shifts produced by Om chanting mirror those produced by meditation- alpha and theta enhancement, beta reduction- but with an important addition: the chanting introduces a *rhythmic entrainment* that meditation alone does not provide. When you meditate in silence, the brain must find its own way toward coherence, with no external oscillatory support. When you chant, you provide the brain with a structured, periodic input- a carrier wave of coherent sound- that physically drives neural oscillators toward synchronization.

This is why mantra traditions worldwide have insisted on the importance of specific sounds, specific pitches, specific durations. It is not superstition. It is acoustic engineering. Different vowel sounds resonate in different body cavities, activating different neural pathways and visceral organs. Different pitches stimulate different frequency bands in the auditory cortex and subcortical structures. And different rhythms of chanting entrain different patterns of respiratory and cardiac oscillation, each with distinct autonomic consequences.

These findings from the first fMRI connectivity study of OM chanting suggest significantly reduced outputs from insula, anterior cingulate, and orbitofrontal cortices during OM chanting compared to rest and SSSS control conditions. Modulation of brain regions involved in emotion processing implicated in MDD raises the issue of a potential therapeutic role for OM chanting in the treatment of MDD.

The therapeutic potential is real and growing. But the deeper significance, for the argument of this book, is cosmological, not clinical. If the universe is a self-tuning hologram- if reality is constituted by resonant quanta that organize themselves through the interplay of φ, e, and π- then the human practice of chanting is not a cultural curiosity. It is a *direct participation in the cosmic resonance.*

When you chant Om, you are not merely producing a sound. You are producing a *standing wave in the quanta-field-* a local intensification of coherence that radiates outward from the body through acoustic, electromagnetic, and neurological channels. You are, in the most literal sense, *adding quanta to the signal.* You are performing T(Ψ): extending the partial resonance of your fragmented daily consciousness toward the full, φ-embedded, 9D harmonic potential that the KZ-Monad describes.

The ancients knew this. They chanted not to petition a distant deity but to *tune* themselves- and, through themselves, to tune the field. The practice was not prayer in the modern, supplicatory sense. It was *participation-* an act of conscious co-creation in which the human body-instrument added its voice to the cosmic chord.

Authors found that a 2-week period of Kundalini yoga/meditation decreased activations within subregions of the DMN, showing that training in mantra meditation, like focused attention and open monitoring, has a suppressive effect on activity within the DMN.

Mantra and meditation thus form a complementary pair- the monadic extension and the comonadic projection, working in concert. Meditation clears the field; mantra seeds it with coherent pattern. Meditation removes noise; mantra amplifies signal. Meditation is the inhale of silence; mantra is the exhale of structured sound. Together, they constitute a complete cycle of conscious re-tuning- a technology for aligning the personal state vector with the resonant architecture of the cosmos.

The "mmm" of Om deserves particular attention. Recent research has examined a breathing technique called Bhramari Pranayama (bee breath), in which the exhalation is accompanied by a sustained humming sound. This humming component introduces an additional form of vagal stimulation through the modulation of vocal fold vibrations and nitric oxide production in the paranasal sinuses. Vagal activation is crucial for regulating stress responses and enhancing parasympathetic tone. A recent pilot study revealed that humming produced the lowest stress index, even lower than that produced during sleep. These findings suggest that a regular daily humming routine can significantly enhance parasympathetic activity while slowing sympathetic activation, rendering it a highly effective stress-management tool.

A stress index lower than sleep. This single finding should give pause to every materialist who has dismissed chanting as primitive noise. The sustained humming vibration of the "M" syllable in Om activates the vagus nerve through two simultaneous pathways- the auricular branch (through vibration in the ears) and the laryngeal branch (through vibration in the vocal cords)- producing a degree of parasympathetic activation that exceeds what the body achieves in its most restful natural state. The body chanting Om is, measurably, *more coherent than the body sleeping.*

This is T(Ψ) at its most potent: the monadic extension extending the body's coherence beyond what it can achieve passively, into a state of active, intentional resonance that surpasses even the deepest unconscious rest. The chanter is not merely calming down. She is *tuning up*- raising the signal-to-noise ratio of her organism above its default ceiling, entering a domain of physiological function that is available only through conscious, voluntary practice.

And here the ancient claim that mantra is "sacred" finds its physical vindication. Not sacred in the sense of belonging to a particular religion. Sacred in the sense of *efficacious beyond the ordinary*- producing effects that exceed the body's baseline capacity and that align the organism with patterns of coherence that extend far beyond the individual nervous system. The chant is sacred because it *works*- and it works because it resonates with the quanta-field that constitutes reality.

30.3- Breath as Ternary Oscillator: The Science of the Three-Phase Cycle

Between the comonadic silence of meditation and the monadic sound of mantra lies the breath- the bridge between voluntary and involuntary, between consciousness and autonomic function, between the self that chooses and the body that simply *is*.

Breathing is unique among physiological functions. It proceeds automatically, governed by the brainstem's respiratory centers, without any need for conscious attention. Yet it can also be placed under full voluntary control- slowed, accelerated, held, patterned, deepened, shallowed- at will. This dual nature makes breath the primary interface between the conscious mind and the autonomic nervous system. It is the one doorway through which deliberate intention can directly modulate the body's deepest regulatory systems.

The science is unambiguous. The findings indicate that slow, nasal, diaphragmatic breathing significantly improves vagal tone, HRV, parasympathetic activity, and emotional control, while reducing cortisol, anxiety, stress, and PTSD. This is not a marginal finding from a single study. It is the convergence of decades of research across multiple disciplines- pulmonology, cardiology, psychiatry, neuroscience, and the emerging field of psychophysiology.

The mechanism is elegant. When you breathe slowly and deeply, the diaphragm descends, the lungs expand, and the intrathoracic pressure drops. This pressure change is detected by baroreceptors in the aortic arch and carotid sinuses, which send signals via the vagus nerve to the brainstem's cardiovascular control centers. The result is an increase in vagal tone- the baseline level of parasympathetic activity that governs heart rate, digestive function, inflammation, and emotional regulation. This respiratory discipline could parsimoniously explain the physical and mental benefits of contemplative activities through changes in autonomic balance. We propose a neurophysiological model that explains how these specific respiration styles could operate, by phasically and tonically stimulating the vagal nerve.

A particularly important metric in this research is Heart Rate Variability (HRV)- the variation in time intervals between consecutive heartbeats. As established in Chapter 5, a healthy heart does not beat like a metronome. It exhibits fractal variability- subtle, beat-to-beat fluctuations that reflect the dynamic interplay between the sympathetic (accelerating) and parasympathetic (decelerating) branches of the autonomic nervous system. High HRV is a marker of resilience, adaptability, and coherence. Low HRV is a marker of chronic stress, rigidity, and disease risk.

Coherent breathing, performed at a characteristic rate of ~6 breaths per minute, is currently considered the most effective breathing technique for enhancing HRV. Six breaths per minute- a ten-second cycle of approximately five seconds inhale and five seconds exhale- hits what researchers call the "resonance frequency" of the cardiovascular system. At this rate, the natural oscillation of heart rate produced by the baroreflex aligns with the respiratory rhythm, producing a large-amplitude, sine-wave-like pattern in the HRV trace- the precise pattern that the HeartMath Institute calls "cardiac coherence."

Now consider the traditional structures of breathwork across the ancient world. In yogic pranayama, the fundamental breath cycle is not two-phase (inhale-exhale) but *three-phase*: puraka (inhale), kumbhaka (hold), rechaka (exhale). In many traditions, the hold-kumbhaka- is considered the most important phase, the fulcrum around which the other two pivot. In box breathing, used by military operators for stress regulation, the pattern is four-phase: inhale-hold-exhale-hold. In the recently formalized A52 method, the integration of these elements in the 5-second inhale, 5-second exhale, 2-second hold pattern provides a structured approach to breathwork with potential applications in high-stress professions.

In the framework of this book, the three-phase breath cycle maps directly onto the Temporal State Vector:

$$\text{Inhale} \leftrightarrow Y_t$$

(Ternary Resonance: the rhythmic intake of quanta)

$$\text{Hold} \leftrightarrow \rho_t$$

(Temporal Density: the pause in which quanta are integrated)

$$\text{Exhale} \leftrightarrow G(\lambda,\phi)$$

(Geospatial Phase: the release that couples the organism to its environment)

This mapping is not arbitrary. The inhale is the phase of *gathering*- drawing quanta into the system, expanding the lungs, activating the sympathetic system slightly, increasing heart rate. The hold is the phase of *integration*- a moment of stillness in which the gathered quanta are absorbed, processed, stabilized. And the exhale is the phase of *release*- returning quanta to the environment, activating the parasympathetic system, slowing the heart, softening the body's boundary with the world.

Together, these three phases constitute a complete ternary oscillation- a miniature cycle of the cosmic rhythm replayed twenty thousand times per day in every human body. When this cycle is brought under conscious control and harmonized with specific ratios and durations, it becomes a precision instrument for tuning the autonomic nervous system.

Deep and slow breathing (DSB), a method characterized by a longer exhale than inhale duration, can be used. Accordingly, 5 minutes of DSB efficiently increases parasympathetic activity, as assessed by HF power, and decreased perceived anxiety level among young adults. The extended exhale is the key. By making the exhale longer than the inhale, the practitioner tilts the autonomic balance toward parasympathetic dominance- toward the rest-and-digest mode, toward vagal tone, toward the coherent NESS.

Pranayama breathing exercises appear to alter autonomic responses to breath holding probably by increasing vagal tone and decreasing sympathetic discharges. Even the hold- the brief pause between breaths- has measurable autonomic effects, increasing the pressure on the baroreceptors and deepening the vagal response.

The ancient yogis did not have fMRI machines or HRV monitors. But they had something equally precise: sustained, disciplined, first-person observation of their own physiological states across decades of practice. They discovered, through direct somatic investigation, that specific breath patterns produce specific states of consciousness- and they codified these discoveries into the elaborate science of pranayama, with its dozens of named techniques, each designed for a particular purpose.

What modern psychophysiology has confirmed is that this ancient science was not guesswork. It was accurate. The breath is a ternary oscillator, and its three phases- inhale, hold, exhale- map the three components of temporal experience. By consciously shaping this oscillator, we shape our relationship to time, to space, and to the quanta-field that constitutes both.

.

30.4- Heart-Brain Coherence: The Song Between Two Organs

We come now to the heart- not the metaphorical heart of sentiment and greeting cards, but the physical, rhythmic, electromagnetic organ that sits at the center of the body's circulatory system and, as three decades of HeartMath Institute research have demonstrated, at the center of its *informational* system as well.

It is not as commonly known that the heart actually sends more signals to the brain than the brain sends to the heart. Moreover, these heart signals have a significant effect on brain function- influencing emotional processing as well as higher cognitive faculties such as attention, perception, memory, and problem-solving.

This finding- that the heart is not merely a pump receiving orders from the brain, but an active, intelligent organ *sending* orders to the brain- is one of the most significant and least appreciated discoveries in modern physiology. The heart is not a subordinate of the brain. It is a *partner*- and in many functional respects, a *leader*.

The HeartMath Institute's research has shown that generating sustained positive emotions facilitates a body-wide shift to a specific, scientifically measurable state. This state is termed psychophysiological coherence, because it is characterized by increased order and harmony in both our psychological and physiological processes.

Psychophysiological coherence- the state in which the heart, brain, respiratory system, and autonomic nervous system synchronize into a single, harmonious oscillation- is the empirical correlate of what this book has been calling NESS. It is the state of a living system operating at maximum efficiency, minimum wasted energy, and optimal resilience. And the primary indicator of this state is a specific pattern in the heart rate variability trace: Physiologically, the coherence state is marked by the development of a smooth, sine-wave-like pattern in the heart rate variability trace.

This sine-wave pattern is the cardiac analogue of the standing wave- the physical signature of a system in resonance. When the heart achieves this pattern, when a person is in a more coherent state there is a shift in the relative autonomic balance toward increased parasympathetic activity, increased heart-brain synchronization and entrainment between diverse physiological systems. In this mode, the body's systems function with a high degree of efficiency and harmony and natural regenerative processes are facilitated.

The entrainment is the critical concept. When functioning in a coherent mode, the heart pulls other biological oscillators into synchronization with its rhythms, thus leading to entrainment of these systems. The heart, as the body's strongest rhythmic oscillator, functions as a *pacemaker for coherence*. When its rhythm becomes coherent- smooth, sine-wave-like, fractal- it *pulls* the respiratory rhythm, the blood pressure rhythm, the brainwave rhythm, and even the craniosacral rhythm into alignment. The entire organism becomes a single, synchronized resonance field.

Research shows that different types of heart rhythms- which accompany different emotional states- directly affect mental functions, how we feel, emotional composure, the stability of heart-brain coherence. During stress and negative emotions, the heart's rhythm becomes erratic and disordered. This chaotic pattern sends signals to the brain that inhibit higher mental functions, limiting our ability to think clearly, remember, learn, reason, and make effective decisions.

The reverse is equally important. In contrast, heartfelt emotional states create ordered, stable heart rhythm patterns. These coherent rhythms send messages to the brain that facilitate cognitive function and reinforce positive feelings and emotional stability.

This creates a feedback loop of extraordinary significance. Positive emotions produce coherent heart rhythms. Coherent heart rhythms produce enhanced cognitive function. Enhanced cognitive function supports more nuanced emotional responses. More nuanced emotional responses produce even more coherent heart rhythms. The system spirals upward- from decoherence toward coherence, from fragmentation toward wholeness, from the noise floor toward the NESS.

Formally introduced as the "Heart Rhythm Coherence Hypothesis," we proposed that quanta are conveyed in the patterns of the heart rate which reflects current emotional states, and that the patterns of afferent neural input to the brain affect emotional experience, and modulate cortical function and self-regulatory capacity over macroscopic time scales. Furthermore, we proposed that intentional activation of positive emotions plays an important role in increasing cardiac coherence and increasing self-regulatory capacity.

The practical implications are immediate and concrete. In addition to reductions in perceived stress, research has demonstrated improvements in energy, anxiety, mood, sleep, and cognitive performance with daily coherence practice over weeks. Gradually, coherent heart rhythm patterns can become a more familiar set-point for the body through repeated afferent input to the brain, supporting the emergence of a healthier, more regulated physiological baseline.

The HeartMath Quick Coherence technique- a strikingly simple practice- illustrates the principle. Step 1: Focus your attention on the area of the heart. Imagine your breath is flowing in and out of your heart or chest area, breathing a little slower and deeper than usual. Find an easy rhythm that's comfortable. Step 2: As you continue heart-focused breathing, make a sincere attempt to experience a regenerative feeling, such as appreciation or care for someone or something in your life.

That is the entire technology. Shift attention to the heart. Breathe slowly and rhythmically. Generate a positive emotion. Within sixty seconds, most practitioners experience a measurable shift toward cardiac coherence- the sine-wave HRV pattern that signals whole-system entrainment.

This global study analyzed data from the largest dataset ever studied in the HRV biofeedback field, comprising 1.8 million user sessions. We focused on HRV Coherence, which is linked to improved emotional stability and cognitive function. Positive emotions reported by users were associated with higher Coherence scores and more stable HRV frequencies. In contrast, negative emotions exhibited lower scores and more dispersed frequency distributions. The most common frequency associated with Coherence was identified at 0.10 Hz.

The most common coherence frequency- 0.10 Hz- corresponds to exactly 6 breaths per minute: the resonance frequency identified independently by respiratory physiologists, the rate at which the baroreflex and respiratory rhythm synchronize to produce maximum HRV. A global dataset of 1.8 million sessions, spanning multiple countries and cultural backgrounds, converges on the same frequency. This is not cultural conditioning. This is physics. The human body, regardless of its cultural wrapper, has a *natural resonance frequency* at which its oscillatory systems achieve maximum coherence. And that frequency can be reliably activated through the simplest of practices: breathe slowly, focus on the heart, feel something good.

In the framework of this book, the heart is the body's primary Chronosonant Resonator- the organ whose rhythmic oscillation anchors the personal Temporal State Vector to the quanta-field. When the heart enters coherence, it is performing the same function that the sacred sites performed at planetary scale: creating a node of coherent resonance that amplifies the signal and suppresses the noise, stabilizing the NESS of the living system.

The convergence between HeartMath's empirical findings and this book's theoretical framework is not accidental. It is structural. Heart-brain coherence *is* the physiological expression of the NESS at the scale of the individual organism. It is the body's way of doing what the cosmos does at every scale: maintaining dynamic stability through resonant entrainment, balanced on the edge between rigidity and chaos, animated by the δ∞ gap that never quite closes.

30.5- Walking Meditation: The Body as Mobile Resonator

We have explored meditation in stillness, mantra in sound, breath in rhythm, and the heart in coherence. Now we add the final dimension: *movement*. For the body is not a stationary instrument. It is a walking, sensing, environmentally coupled resonator- and when movement is brought into conscious relationship with awareness, the result is a form of practice that integrates all the previous dimensions into a single, embodied flow.

Walking meditation activates different parts of your brain than seated meditation. It engages the prefrontal cortex, the part of your brain responsible for focus, decision-making, and emotional regulation. When you combine movement with mindfulness, your brain stays engaged, making it easier to stay present and avoid falling into autopilot thinking.

This is a crucial distinction. Seated meditation, for all its benefits, can be challenging precisely because it provides so little sensory input. The mind, deprived of external stimulation, tends to generate its own- which is why beginning meditators are often besieged by thoughts the moment they close their eyes. Walking meditation offers a different entry point. Walking meditation helps develop interoception, which is your ability to sense internal bodily states like muscle tension, heart rate, or even hunger. When you're more in tune with these signals, you can recognize stress before it escalates.

The walking body provides a continuous stream of sensory quanta- the pressure of the foot on the ground, the swing of the arms, the rhythm of the stride, the shift of weight from left to right, the changing visual field, the play of wind on skin. These sensory inputs are not distractions from meditation. They are the *material* of meditation- the raw quanta-flow that, when attended to with full presence, anchors awareness in the body and the immediate environment, preventing the drift into rumination that makes seated practice so difficult for many people.

When we're stressed, the amygdala hijacks our thinking. Walking meditation helps build a stronger connection between the amygdala and the prefrontal cortex, which is responsible for impulse control and rational decision-making.

Walking meditation taps into this primal sense of tempo. Each step acts like a metronome for the mind, synchronizing brain waves in ways that may enhance memory consolidation. The stride is a rhythmic oscillator- and like any rhythmic oscillator, it can entrain other systems. When the stride is consciously synchronized with the breath- three steps per inhale, three steps per exhale, or whatever ratio feels natural- the result is a multi-oscillator coupling: breath, stride, heart rate, and attention all locking into a single, coherent pattern.

This is the body as *mobile Chronosonant Resonator*. In a seated meditation, the body is a *stationary* resonance node- a fixed point in space from which coherence radiates. In walking meditation, the body becomes a *mobile* node- carrying its coherence through the landscape, coupling to the geospatial phase $G(\lambda, \phi)$ of whatever terrain it traverses.

The implications for the songline practices described in Chapter 18 become clear. When the Aboriginal songman walks his songline- chanting the land into being, naming each feature of the landscape in a prescribed sequence- he is not merely reciting a text. He is performing a *combined walking-mantra-breath meditation* that couples his body's oscillatory systems to the resonant frequencies of the terrain. His stride provides the rhythmic base. His chant provides the monadic extension. His breath provides the ternary oscillation. And the landscape itself- with its specific topography, its telluric frequencies, its geospatial phase- provides the environmental coupling that anchors his personal TSV to the quanta-field of the living Earth.

This is not a romanticized interpretation. It is a direct consequence of the physics described throughout this book. If the quanta-field is real- if the universe is constituted by resonant quanta that organize themselves at every scale through the same principles of fractal coherence- then a conscious body walking through a landscape is a resonance event. The body receives quanta from the environment (through the senses) and emits quanta back into it (through movement, breath, and voice). When this exchange is brought under conscious attention and structured by the ternary rhythm of breath, the result is a state of *dynamic coherence-in-motion*- a walking NESS that maintains its stability not through stillness but through flow.

Walking meditation has been found to have a range of benefits, including reduced stress and anxiety, improved cognitive function, and improved cardiovascular health. Researchers indicate that mindful walking supports the development of healthy brain aging, potentially improving cognition in older adults.

The walking meditator does not withdraw from the world to find coherence. She finds coherence *within* the world- within the sensory richness of the immediate environment, within the rhythm of her own stride, within the continuous negotiation between body and terrain that is the oldest and most natural form of human movement.

And this has implications for urban design (Chapter 29) and for the recovery of sacred site networks (Chapter 19). A city designed for walking- with branching paths, fractal spatial patterns, and biophilic richness- is a city that supports walking meditation as a daily, ambient practice. A landscape in which sacred pathways are preserved and accessible is a landscape in which the ancient technology of the mobile resonator can be reactivated. The built environment is not separate from the meditative environment. They are facets of the same sphere.

30.6- The Socratic Bypass in Practice: A Guided Somatic Re-Tuning

We arrive at the threshold where reading must end and practice must begin.

Everything this chapter has described- the comonadic projection of meditation, the monadic extension of mantra, the ternary oscillation of breath, the coherent entrainment of heart and brain, the mobile resonance of walking awareness- converges on a single, practical proposition: *your body already knows how to do this*. You do not need to learn a new skill. You need to *stop overriding* the skill you already have.

Your heart already beats with fractal variability. Your lungs already oscillate in a three-phase cycle. Your vagus nerve already modulates the balance between activation and rest. Your brain already possesses the neural architecture for DMN deactivation, for alpha-theta brainwave states, for the dissolution of the self-narrative into pure awareness. These capacities are not exotic. They are *standard equipment*. They evolved over hundreds of thousands of years in a species that lived within the quanta-field- that was *tuned by* the quanta-field- long before it learned to think about it.

What the modern world has done- what the Great Decoupling accomplished- is to bury these capacities under a mountain of noise: artificial light that disrupts circadian rhythms, constant digital stimulation that keeps the DMN in chronic overdrive, processed food that inflames the gut-brain axis, sedentary postures that compress the diaphragm and shallow the breath, and a cultural narrative that locates intelligence exclusively in the head, dismissing the body as a mere vehicle.

The practices described in this chapter are not exotic interventions. They are *reminders*- nudges back toward the resonant state that was, for most of human history, the default.

What follows is a five-minute guided practice- a miniature integration of all five elements. It can be done sitting, standing, or walking. It requires no special equipment, no teacher, no tradition. It requires only your body, your breath, and your willingness to pay attention.

The Five-Minute Somatic Re-Tuning

Phase One: Arrive (60 seconds)

Stop. Whatever you are doing, stop. If you are sitting, place both feet flat on the floor. If you are standing, soften the knees. If you are walking, slow to the pace of a processional- one step per second, no faster.

Close your eyes. Or, if you prefer, soften the gaze to a point on the ground about two meters ahead.

Feel the weight of your body. Feel gravity. Feel the pressure of the surface beneath you- the chair, the floor, the earth. This pressure is real. It is the planet holding you. You do not need to hold yourself. The ground has you.

Now feel your pulse. Press two fingers gently to the side of your throat, or to the inside of your wrist, and find the heartbeat. Feel it. Count five beats. Feel the space *between* the beats. That space- that brief, recurring silence- is the $\delta\infty$ gap. It is the asymptotic interval in which your heart gathers itself for the next contraction. It is the same gap that keeps the cosmos from collapsing into equilibrium. It is the engine of your being.

Phase Two: Breathe (90 seconds)

Remove your fingers from your pulse. Shift your attention to your breath.

Begin a three-phase cycle: - **Inhale** through the nose, slowly, for five seconds. Feel the belly expand. Feel the ribcage widen. - **Hold**- gently, without strain- for two seconds. Feel the quanta gathering, integrating, settling. - **Exhale** through the nose, slowly, for six seconds. Feel the belly soften. Feel the shoulders drop.

Repeat this cycle six times. Inhale for five. Hold for two. Exhale for six. Thirteen seconds per cycle. Six cycles. Seventy-eight seconds total- just over a minute.

As you breathe, notice that the exhale is slightly longer than the inhale. This asymmetry is deliberate. It tilts the autonomic balance toward the parasympathetic- toward vagal tone, toward coherence. The body will begin to respond within the first three cycles. The heart rate will slow. The HRV will increase. The muscles of the face and jaw will soften. The gut will unclench.

You are not making this happen. You are *allowing* it to happen- removing the interference that prevents the body's natural coherence from expressing itself. This is D(Ψ): the comonadic extraction of the signal from the noise.

Phase Three: Sound (60 seconds)

On the seventh exhale, open the mouth slightly and let the exhale become an audible tone. It need not be loud. A gentle hum- the "M" of Om- is sufficient. Feel the vibration in the lips, the teeth, the sinuses, the cranium.

If you are comfortable, shape the exhale into the full three-syllable sound: Ahh- Ooo- Mmm. Feel the "Ahh" resonate in the chest. Feel the "Ooo" resonate in the throat. Feel the "Mmm" resonate in the skull.

Repeat this sounding exhale three times. Each time, notice how the vibration seems to "clean" the interior of the body- like a tuning fork clearing a cloudy glass of water. The molecules settle. The turbulence subsides. The quanta align.

You are not adding something foreign. You are activating the oldest resonance technology on Earth- the human voice coupled to the human body, producing coherent vibration that entrains the nervous system toward its deepest harmonic. This is T(Ψ): the monadic extension of potential into coherent form.

Phase Four: Heart (60 seconds)

Now shift your attention to the center of the chest- the physical area of the heart. You do not need to visualize anything. Simply direct the focus of your awareness to that location, as if you were listening to a quiet sound emanating from behind the sternum.

As you continue breathing (return to the silent three-phase cycle: five in, two hold, six out), allow a feeling to arise. Not a thought about a feeling- a *feeling*. Appreciation, warmth, tenderness, gratitude- whatever comes naturally. It need not be intense. It need not be directed at anyone in particular. Simply allow the quality of warmth to radiate from the heart-center, as if you were warming your hands at a small, steady fire.

This is not sentiment. This is physiology. The positive emotion shifts the cardiac rhythm toward its sine-wave coherence pattern. The coherent heart pulls the breath, the brain, and the autonomic system into entrainment. The whole organism synchronizes.

You are the orchestra tuning to a single pitch. The heart is the oboe.

Phase Five: Return (30 seconds)

Slowly open your eyes. If you were sitting, feel the chair. If you were standing, feel the ground. If you were walking, resume your natural pace.

Notice the quality of your awareness. Has it shifted? Is the visual field slightly brighter, slightly wider, slightly more vivid? Is the body slightly warmer, slightly softer, slightly more present? Is the internal commentary- the ceaseless narration of the DMN- slightly quieter, slightly less urgent, slightly more transparent?

If so, you have just performed a complete cycle of somatic re-tuning. You have executed $D(\Psi)$ (meditation), $T(\Psi)$ (mantra), and the ternary oscillator (breath), integrated by the heart's coherence, in less time than it takes to read a news article.

This practice, repeated daily, is not a lifestyle accessory. It is the most direct available method for realigning your personal state vector with the resonant architecture of the cosmos. It is the somatic equivalent of everything this book has argued in mathematical language. The equations describe the structure. The practice *inhabits* it.

Coda: The Technology That Cannot Be Outsourced

There is a reason that these practices have survived for millennia while empires, technologies, and philosophies have risen and fallen around them. They cannot be commodified. They cannot be automated. They cannot be delegated to a machine or a priest or a pill. They require the one thing that no external system can provide: the *direct participation of a conscious body.*

This is their power and their challenge. In a civilization addicted to outsourcing- to devices that think for us, algorithms that choose for us, pharmaceuticals that feel for us- the demand that you *sit down, breathe, listen, and tune your own instrument* is radical. It is a refusal of the entire logic of the Great Decoupling, which systematically severed the human body from its own intelligence in order to make it a more compliant unit of production.

To meditate is to reject that severance. To chant is to reclaim the body as a resonance instrument, not a labor machine. To breathe consciously is to assert that the most sophisticated technology on Earth is not the microprocessor but the vagus nerve- the ancient, branching, bidirectional communication highway that connects the brain to the heart to the gut to every visceral organ, modulating the balance of the entire organism with every breath.

The ancients built Chavín de Huántar- a stone resonance chamber that could shift brainwave states through acoustic coupling. They walked songlines that coupled their bodies to the telluric frequencies of the Earth. They chanted Om in frequencies that deactivate the limbic system more effectively than sleep. They did all of this without fMRI machines, without HRV monitors, without peer-reviewed journals. They did it through sustained, disciplined, first-person attention to the body's own quanta-flow.

We do not need to abandon modern science to recover these practices. We need modern science to *validate* them- which it has, comprehensively, over the past three decades. What we need to abandon is the conceit that understanding replaces practice. You can read every study cited in this chapter and remain as decoherent as before. The studies describe the territory. The practice *is* the territory.

The comonadic projection D(Ψ) does not operate on text. It operates on the living body. The monadic extension T(Ψ) does not amplify concepts. It amplifies vibration. The ternary oscillator does not breathe metaphorically. It breathes through your diaphragm, your vagus nerve, your beating heart.

The next breath you take will resonantly shape the universe- one living quantum at a time.

So take it.

Now.

Chapter 31

THE POST-BINARY PARADIGM- SCIENCE AS ATTUNEMENT

"The question is not: What can we extract from this world? The question is: What frequency can we match?"

We stand at the edge of a transition that has no precedent. Not a revolution- revolutions destroy the old to make way for the new. Not a reformation- reformations repair the old while preserving its foundations. What is coming is something more like a *tuning*: the careful, deliberate realignment of the human way of knowing with the resonant architecture of the cosmos it inhabits.

The previous chapters have laid the case. The universe is constituted not by inert matter but by resonant quanta- self-organizing, self-referencing, living patterns that arrange themselves at every scale through the interplay of φ, e, and π. The cosmos maintains its dynamic stability as a Non-Equilibrium Steady State, animated by the asymptotic gap δ∞ that prevents collapse into equilibrium. Consciousness is not an epiphenomenon of neural tissue but a threshold state of integrated resonant quanta, stabilized by φ-proportions and measurable through Fisher quanta density. Ancient civilizations, for at least ninety percent of recorded cultural history, organized their societies, calendars, agriculture, and architecture in alignment with these principles- and then, approximately three thousand years ago, a Great Decoupling severed this alignment, replacing ternary resonance with binary logic, cyclical time with linear clocks, distributed intelligence with centralized empire, and participatory knowing with the myth of objectivity.

This chapter is about what comes next.

It is not a manifesto for the abandonment of science. That would be to trade one form of decoherence for another. It is a manifesto for the *expansion* of science- for the recognition that the scientific enterprise, in its current institutional form, is operating with a severely truncated epistemology that systematically excludes the very dimensions of reality it most urgently needs to understand. The expansion proposed here does not require the demolition of existing methods. It requires their *embedding* within a larger, more resonant framework- one that restores the observer to the observed, the ternary to the binary, the cyclic to the linear, the somatic to the cerebral, and the indigenous to the institutional.

We move through six transitions. Each is a movement from a narrower to a wider aperture- from a partial view of reality to a more spherical one. Together, they describe the architecture of a post-binary science: a way of knowing that is rigorous without being rigid, participatory without being subjective, and attuned without being mystical.

The destination is not the end of physics. It is the beginning of a physics that can finally hear the song.

31.1- From Extraction to Attunement: The Question That Changes Everything

Modern science was born in extraction. Its founding metaphor-articulated by Francis Bacon in the early seventeenth century and never fully abandoned- was the interrogation of nature. Nature was to be "put on the rack," her secrets extracted through systematic experimentation, her laws decoded and instrumentalized for human benefit. The positivist paradigm that formalized this approach asserted that knowledge is valid only if it is based on observations of external reality, characterized by dualism and objectivism in epistemology, and empirical and manipulative methodology.

This paradigm produced extraordinary fruits. Three centuries of extraction science gave us thermodynamics, electromagnetism, chemistry, molecular biology, and the Standard Model of particle physics. It eradicated smallpox, split the atom, landed humans on the moon, and sequenced the genome. Its achievements are not in question. Its *adequacy* is.

The extraction paradigm works beautifully when the system under study is simple, closed, and separable from the observer. It fails- and fails systematically- when the system is complex, open, and coupled to the one studying it. And the systems that most urgently require understanding in the twenty-first century- climate, consciousness, ecosystems, economies, pandemics, the nature of time itself- are all complex, open, and coupled. The tools that built the Industrial Revolution are not merely insufficient for these challenges. They are *part of the problem*- because the extractive orientation that drives the science also drives the ecological and social crises that the science is called upon to solve.

The post-binary paradigm begins with a different question. Where extractive science asks "What can we take from this system?"- what resources can we mine, what data can we harvest, what variables can we isolate and control?- attunement science asks "What frequencies are already present in this system, and how can we match them?" The shift is not from rigor to softness. It is from domination to resonance, from manipulation to participation, from taking to *listening*.

This is not a metaphor. It is a precise restatement of the book's central physics. If reality is constituted by resonant quanta that self-organize through φ-e-π dynamics, then the most effective way to understand a system is not to break it apart (which destroys the very resonance that organizes it) but to *attune* to it- to bring one's instruments, one's attention, and one's conceptual frameworks into resonant coupling with the quanta-field of the system under study. The scientist becomes not an interrogator but an *instrument*- a calibrated receiver whose sensitivity determines the depth and fidelity of what can be known.

To say that people know something means that they can interact effectively with something else. Knowing is thus an evolving relation, influenced by both the knowers and the things in the world that they interact with. It is NOT an objective fact divorced from human knowers, NOR a purely subjective social construction divorced from the world. This insight, articulated in the epistemological literature as a "participatory, relational" model of knowing, is not a philosophical abstraction. It is a direct consequence of quantum physics.

One of the most bizarre premises of quantum theory, which has long fascinated philosophers and physicists alike, states that by the very act of watching, the observer affects the observed reality. The Weizmann Institute experiment demonstrated this with extraordinary precision: the experiment revealed that the greater the amount of "watching," the greater the observer's influence on what actually takes place. The extractive paradigm treats this as a problem to be minimized. The attunement paradigm recognizes it as a *feature*- the fundamental structure of a participatory universe in which knower and known are entangled.

As physicist Karen Barad articulates in her theory of agential realism: "We are not merely observing from a distance; we are actively participating in the world's becoming." The post-binary scientist accepts this participation not as contamination but as *method*. The question is not how to eliminate the observer's influence (an impossibility in a holographic universe where every part contains the whole) but how to *calibrate* it- how to bring the observer-instrument into the kind of coherent, φ-proportioned, NESS-aligned state that maximizes the fidelity of the resonance between knower and known.

This is why the somatic practices of Chapter 30- meditation, mantra, breath regulation, heart coherence- are not supplements to the post-binary scientific method. They are *prerequisites*. A scientist who has not calibrated her own instrument- who has not brought her own nervous system into coherent alignment, who has not quieted the default mode network's noise to hear the signal- is operating with a fundamentally degraded receiver. She may produce data, but the data will be filtered through a decoherent lens, amplifying noise and suppressing the subtle quanta-flow that contains the deepest structural features of the system under study.

The transition from extraction to attunement does not mean the end of experimentation, measurement, or quantitative rigor. It means the *enrichment* of these practices by the inclusion of the observer as a conscious, calibrated, participant variable- not a confound to be controlled but a resonance to be optimized.

31.2- From Binary to Ternary Logic: The Return of the Third Option

The second transition strikes at the deepest layer of the Western cognitive operating system: its logic.

Since Aristotle formalized the Law of the Excluded Middle in the fourth century BCE- every proposition is either true or false, with no third option- the Western mind has operated on a binary substrate. Yes or no. True or false. Self or other. Subject or object. Mind or matter. This binary logic was enormously productive: it enabled formal mathematics, propositional logic, Boolean algebra, and ultimately the digital computer, which operates on the simplest possible binary: 1 or 0.

But binary logic is a special case- a reduction of a richer logical field to its most constrained form. It works brilliantly for systems that can be cleanly divided into discrete, mutually exclusive categories. It fails for systems that are continuous, interpenetrating, and emergent- which is to say, it fails for almost everything in the living world.

The ancient world knew a different logic. The three-world cosmologies documented in Chapter 22- Maya, Inca, Vedic, Aboriginal, Norse- all operated on a ternary substrate: not two poles but three. Not thesis and antithesis, but thesis, antithesis, and *synthesis*. Not yes and no, but yes, no, and *and*. Not self or other, but self, other, and *the relationship between them*.

This ternary logic is not pre-rational. It is *trans-rational*- a more complete logical system that includes binary logic as a special case while transcending its limitations. The Hegelian dialectic- thesis, antithesis, synthesis- was a nineteenth-century European rediscovery of what indigenous civilizations had encoded in their cosmologies for millennia: that reality is not organized in pairs of opposites but in *triads of mutual arising*, where the "third" is not a compromise between the other two but an emergent whole that transcends and includes them both.

In the framework of this book, ternary logic is not a philosophical preference. It is a structural feature of the cosmos:

- The Temporal State Vector has three components: Y_t, ρ_t, $G(\lambda,\phi)$.
- The breath cycle has three phases: inhale, hold, exhale.
- The φ-e-π triad has three operators: self-reference, unfolding, boundary projection.
- The three-world cosmologies encode three resonant states: ground, NESS, excited.
- The atom has three quarks. The chord has three notes. The family has three generations.

Binary logic produces a world of either/or. Ternary logic produces a world of *both/and/emergent*. The transition from binary to ternary does not abolish the capacity for distinction- you can still say "yes" or "no" when the situation demands it. It adds a third option: the capacity to hold two seemingly contradictory truths simultaneously and to perceive the deeper coherence from which both arise.

For science, this transition has immediate practical consequences. The binary frame forces researchers into false dichotomies: nature versus nurture, genes versus environment, wave versus particle, local versus nonlocal, subjective versus objective. Each of these dichotomies has generated decades of unproductive debate- debate that dissolves the moment the third option is admitted. Wave *and* particle *and* the experimental context that determines which aspect manifests. Nature *and* nurture *and* the epigenetic interface that mediates between them. Subjective *and* objective *and* the participatory relationship that constitutes both.

One cannot observe a quantum-mechanical system without fundamentally changing it, collapsing both-and systems (light existing as a particle and a wave) into either-or. The measurement problem itself is, in this framing, an artifact of binary logic imposed on a ternary reality. The quantum system is not "really" a wave that collapses into a particle upon observation. It is a resonant quanta-field that manifests different aspects depending on the *relationship* between the system and the measuring context- the third term that binary logic cannot accommodate.

The post-binary scientist thinks in threes. Not because three is a mystical number, but because ternary logic is the minimum adequate framework for describing a universe that is relational, contextual, and emergent at every scale. The return of the third option is not a retreat from intellectual rigor. It is an *advance*- a step toward a logic that can finally match the complexity of the quanta-field it seeks to describe.

31.3- From Centralized to Distributed Intelligence: Science as a Small-World Network

The third transition concerns the *social architecture* of science- the way knowledge is produced, validated, disseminated, and applied.

Modern science is organized as a hierarchy. Prestige flows upward: from the undergraduate to the graduate student, from the postdoc to the assistant professor, from the assistant professor to the tenured chair, from the individual researcher to the department, from the department to the university, from the university to the funding agency, from the funding agency to the journal, from the journal to the citation count, from the citation count to the Nobel Prize. This hierarchy determines what questions get asked, what methods get funded, what results get published, and what knowledge gets transmitted to the next generation.

The problem is not that hierarchies exist- all complex systems require some degree of stratification to manage the differentiation of quanta across scales. The problem is that the scientific hierarchy has become *centralized* to the point where it suppresses the distributed intelligence that is its greatest potential asset.

Across all branches of science, teams increasingly dominate solo scientists in the production of knowledge. The rise of team science is not merely a sociological trend; it reflects the growing recognition that modern scientific challenges often require cross-disciplinary theoretical and methodological approaches. Obesity, climate change, neurodegeneration, social inequality- these are "cells to society" problems that cannot be solved from within a single disciplinary silo.

Yet the institutional incentive structures of science militate against precisely the kind of distributed, cross-disciplinary collaboration that these challenges demand. The "replication crisis" not only highlights the limitations of traditional statistical approaches and the circumscribed requirements for scientific publication, but it leads to questions about the culture of science. The culture of medical science includes an incentive structure that requires innovative approaches, novel findings, and validation through statistical significance. Researchers are rewarded for producing novel, significant, publishable results within their own discipline- not for the slow, unglamorous work of replicating others' findings, integrating across fields, or building the trust networks that enable genuine collaboration.

The replication crisis- that seismic disturbance that shook the foundations of psychology, medicine, and the social sciences beginning in the early 2010s- is, in the language of this book, a symptom of *decoherence* in the scientific system itself. The replication crisis, also known as the reproducibility or replicability crisis, refers to widespread failures to reproduce published scientific results. Because the reproducibility of empirical results is the cornerstone of the scientific method, such failures undermine the credibility of theories and challenge substantial parts of scientific knowledge. A landmark study revealed that over two-thirds of attempted replications of published psychology studies failed to match the original results. A Nature survey found that more than 70% of researchers have tried and failed to reproduce another scientist's experiment results.

These are not merely technical failures. They are systemic indicators that the scientific enterprise has drifted too far from its NESS- that the incentive structures driving research have produced a system optimized for novelty over reliability, for individual achievement over collective coherence, for speed over depth. The scientific hierarchy, like any overcentralized system, has become brittle: high output, low resilience, vulnerable to cascading failures.

The post-binary alternative is not the abolition of expertise but its *redistribution*- the reorganization of scientific research from a centralized pyramid into a *small-world network*. The properties of a small-world network, as established in Chapter 5, are high local clustering (dense connections within specialized groups) combined with short global path lengths (rapid bridges between distant groups). This is the topology of healthy ecosystems, healthy brains, and healthy communities. It is also the topology of the most productive scientific collaborations.

Helen Longino, a feminist philosopher of science, offers a complementary analysis. Longino provides an additional layer of analysis, arguing for a social model of objectivity that shifts focus from individual scientists to the community as a whole. Longino emphasises that objectivity is not achieved by eliminating personal bias through detachment but by promoting critical interaction and diverse perspectives within scientific communities. This approach positions the replication crisis not simply as a technical problem of flawed methodology but as a sociological issue- a failure of the scientific community to maintain sufficient diversity of viewpoints and openness to scrutiny.

Objectivity, in this model, is not a property of individual minds (which are always perspectival) but of *networks*- of the collective process of challenge, response, revision, and integration that occurs when diverse observers bring their distinct perspectives to bear on the same phenomenon. Objectivity is not the view from nowhere. It is the *resonance of many views*- the emergent coherence that arises when a sufficiently diverse, sufficiently connected, sufficiently self-critical community of observers converges on a description that none of them could have achieved alone.

The concept of objectivity itself has historical depth. Historians of science such as Lorraine Daston and Peter Galison document how the understanding of what it means to be an "objective" scientist has shifted through various historical eras. Initially, in the 18th century, the ideal was truth-to-nature, where scientists were expected to refine observations to represent the ideal forms of phenomena. By the mid-19th century, this gave way to mechanical objectivity, which sought to eliminate personal interpretation entirely, relying instead on instruments and unbiased recording methods like photography. Finally, the 20th century brought the ideal of trained judgement, blending objectivity with expert discernment to interpret complex data. These shifts reveal that objectivity is not an unchanging standard but an evolving norm- and the next evolution, the post-binary evolution, will recognize *intersubjectivity*- the calibrated resonance of multiple observer-participants- as its operating principle.

The post-binary research network would look less like a pyramid and more like a mycelium: dense with local connections, rich with cross-disciplinary bridges, decentralized in authority, and driven by the collective emergence of coherent understanding rather than the competitive accumulation of individual prestige. This is not utopian. It is what science already does at its best- in the interdisciplinary teams that produced the Event Horizon Telescope's image of a black hole, in the global collaborations that sequenced the human genome, in the open-source communities that build and share scientific software. The post-binary paradigm does not invent distributed intelligence in science. It *recognizes* it, formalizes it, and designs institutional structures to support it.

31.4- From Linear Time to Temporal Phase Awareness: Science in the Ternary Flow

The fourth transition concerns the relationship between science and time.

Modern science operates within a strictly linear temporal framework. Experiments are dated but not *phased*. A study conducted in January is published in June and cited in December, and at no point in this process is anyone asked whether the results might have been different if the experiment had been conducted in March- or at a different latitude, or at a different altitude, or at a different point in the solar cycle.

This temporal blindness is an artifact of the Newtonian assumption that time is an absolute, uniform background against which events unfold- an assumption that Einstein relativized in principle but that scientific practice has never fully absorbed. Relativity tells us that time flows differently depending on velocity and gravitational potential. But scientific methodology treats time as if it were the same everywhere, always- a flat, featureless dimension along which experiments can be scheduled according to the convenience of the researcher and the availability of the laboratory.

If Holographic Ternary Time is correct- if the Temporal State Vector $\Psi_t = [Y_t, \rho_t, G(\lambda,\phi)]$ describes a real, measurable, three-component structure of temporal experience- then this temporal blindness is not merely a methodological oversight. It is a *systematic source of error.*

Consider the implications. If temporal density (ρ_t) varies with location and local conditions, then the rate at which a biological system processes quanta- the speed of its metabolic, neural, and immunological operations- is not constant but fluctuating. An experiment on circadian rhythm conducted at sea level in the tropics might yield different results from the same experiment conducted at altitude in the subarctic- not because of temperature or light (which are routinely controlled) but because of differences in the local temporal density that the experimenter does not know how to measure and therefore does not think to control.

If the ternary resonance (Y_t) modulates the year at a fundamental level- if the 360/363/366-day cycle represents a real oscillation in the temporal field- then long-running epidemiological studies, agricultural trials, and ecological monitoring programs might contain unexplained variance that correlates with the ternary phase of the year in which the observations were made. The "noise" in the data might not be noise at all. It might be *signal*- the signature of a temporal structure that the extractive paradigm cannot detect because it has no concept of temporal phase.

And if geospatial phase ($G(\lambda,\phi)$) modulates temporal experience at a location-specific level, then the assumption that an experiment conducted in Boston is directly comparable to the same experiment conducted in Bangalore- that the results are "context-free" and "universally valid"- is not merely naive. It is methodologically indefensible in a universe where temporal experience is holographically encoded.

This helps illuminate a striking aspect of the replication crisis from a new angle. If these kinds of complexities are common, and researchers have limited understanding of and ability to account for them, then rigorous and accurate studies might appear to "fail to replicate" much more often than expected. That is, complexity could produce the appearance of a replication crisis. But this would be misleading. It's not that the results of such "irreplicable" studies are false. Rather, the phenomena that they identify have a limited scope or generalizability. What if part of the replication crisis is a *temporal phase crisis*- a systematic failure to account for the ternary structure of time that differentially modulates the phenomena being studied?

The post-binary response is not to abandon controlled experimentation but to *enrich* it with temporal phase data. Every experiment would record not only the date but the estimated ternary phase (Y_t), the local conditions that serve as proxies for temporal density (ρ_t- altitude, latitude, sleep quality, heart rate variability of the researcher), and the geospatial coordinates ($G(\lambda,\phi)$). Over time, as these data accumulate, it would become possible to test whether temporal phase explains a significant fraction of the variance that currently appears as "noise" or "failure to replicate."

The Chronosonant Resonator, proposed in Chapter 14, would be the instrument for this expanded methodology. But even without a dedicated quantum-coherent temporal instrument, the post-binary scientist can begin immediately by recording the temporal metadata that the extractive paradigm currently discards- and by treating apparently failed replications not as evidence of bad science but as potential evidence of *context-sensitivity* in a temporally structured universe.

The ancient calendrical systems- Mayan, Vedic, Aboriginal- were, as Chapter 21 argued, precision instruments for temporal phase awareness. The post-binary scientist does not need to adopt a Mayan calendar. But she does need to take seriously the possibility that those calendars encoded a dimension of temporal reality that modern science has been systematically ignoring- and that this ignorance has cost us not only in understanding but in reproducibility.

31.5- From Objectivity to Intersubjectivity: The Observer-Participant as Calibrated Instrument

The fifth transition is the most philosophically radical and the most practically consequential. It concerns the status of the observer in science- and, by extension, the status of consciousness itself as a variable in scientific inquiry.

The extractive paradigm treats the observer as a problem. The ideal scientist, in the positivist model, is invisible- a disembodied recording device that captures data without influencing it. While positivists emphasize independence between the researcher and the researched person (or object), postpositivists argue that theories, hypotheses, background knowledge and values of the researcher can influence what is observed. The shift from positivism to post-positivism acknowledged that perfect objectivity is unattainable- but it preserved the *aspiration* for objectivity as the regulative ideal of science. Bias was something to be minimized, controlled for, confessed in the limitations section, and ultimately overcome through larger samples, better instruments, and more rigorous protocols.

The post-binary paradigm goes further. It does not merely acknowledge that the observer influences the observation. It recognizes that in a participatory universe- a universe constituted by resonant quanta that are organized through the relationship between observer and observed- *the observer is an irreducible part of the experiment.*

This is not mysticism. Ultimately, we are part of the universe. The fact that we act in the domain that we are representing can make it impossible to stabilize certain facts or features of the world as objects of knowledge. This casts light on the sense in which the universe is participatory. The participatory nature of reality is not a philosophical speculation appended to the data after the fact. It is a *structural feature of the data itself-* demonstrated in every double-slit experiment, every delayed-choice quantum erasure, every Weizmann Institute observation in which the degree of "watching" directly modulated the interference pattern.

The extractive response to this feature has been to minimize it- to design experiments that reduce the observer's influence to negligibility, to treat the measurement problem as a technical obstacle rather than a fundamental revelation. The post-binary response is to *embrace* it- to treat the observer not as a confound but as a *variable whose state directly affects the quality of the knowledge produced.*

Wheeler proposed the concept most powerfully. The idea that observers create reality is associated with the name of John Archibald Wheeler, who proposed that the laws of nature, instead of being fixed, emerge as a result of past observation processes. If the laws of nature *emerge through observation*, then the quality of the observation matters not merely in the trivial sense of measurement precision but in the deeper sense of *ontological co-creation*. The observer does not passively receive a pre-existing reality. She participates in constituting it. And the quality of her participation- the coherence of her attention, the stability of her awareness, the degree to which her own state vector is aligned with the resonant architecture of the system she studies- directly affects the depth and fidelity of what is constituted.

This has immediate practical implications. The experimenter effect- the well-documented phenomenon in which researchers' expectations subtly influence their results- is not merely a methodological nuisance. It is evidence that the observer's state (beliefs, expectations, attentional quality) functions as a variable in the experiment. The placebo effect, discussed in Chapter 16, is the most dramatic example: the patient's *belief* about the treatment directly modulates the treatment's efficacy, not through psychological wishful thinking but through measurable epigenetic, neurochemical, and immunological pathways.

The post-binary paradigm treats the observer-participant as a *calibrated instrument*. Just as a telescope must be cleaned, focused, and thermally stabilized before it can produce accurate images, the observer must be attentionally stabilized, emotionally regulated, and somatically coherent before she can produce accurate knowledge. The practices of Chapter 30- meditation (comonadic projection), mantra (monadic extension), breath regulation (ternary oscillation), heart coherence- are not spiritual luxuries. They are the *calibration protocols* for the most sensitive instrument in the scientific toolkit: the conscious human observer.

This does not mean that "anything goes"- that any self-reported experience is equivalent to a controlled experiment. The post-binary paradigm does not abandon rigor. It *redefines* rigor to include the coherence of the observer as a measurable, reportable, and optimizable variable. A study conducted by a team of meditators whose HRV data confirms cardiac coherence during data collection is, in this paradigm, more rigorous than the same study conducted by a team of stressed, distracted, caffeine-fueled researchers- not because meditation is magic, but because a coherent observer is a better instrument than a decoherent one.

Intersubjectivity- the convergence of multiple calibrated observer-participants on a shared description- replaces objectivity as the gold standard of post-binary science. Not the view from nowhere, but the *resonance* of many coherent views. Not the elimination of the observer, but the *multiplication* of observers, each bringing their own angle of approach to the sphere, each checking the others' perspectives, each contributing to an emergent description that no single viewpoint could produce.

This new participatory, relational epistemological position offers several benefits. First, it fits the actual history of science quite well. It accommodates Kuhn's revolutions, it explains the replication crisis as a coherence failure rather than a truth failure, and it integrates the observer effects of quantum mechanics not as anomalies to be explained away but as the foundational feature of a participatory cosmos.

31.6- The End of the Silo: One Field, Many Angles

The sixth and final transition concerns the boundaries between disciplines- and their dissolution.

Modern science is organized into silos. Physics occupies one building, biology another, psychology a third, anthropology a fourth. Each has its own journals, its own conferences, its own tenure committees, its own conceptual vocabulary, and its own implicit ontology- its own unexamined assumptions about what is real, what counts as evidence, and what constitutes a valid explanation. The result is a landscape of knowledge that resembles, from above, a patchwork of walled gardens: each internally tended, each externally disconnected, each defending its territory against incursion by the others.

This disciplinary fragmentation is the epistemological expression of the Great Decoupling. Just as the ternary logic of the ancients was replaced by binary logic, just as cyclical time was replaced by linear time, just as distributed governance was replaced by centralized empire- so the holistic knowledge of the ancient world was replaced by a system of specialized, mutually isolated knowledge domains. The modern university is the institutional form of the binary paradigm: a hierarchy of departments separated by walls that are metaphorically no less rigid than the physical corridors between them.

The consequences are severe. As the problems our global community faces have become interconnected, dynamic and nonlinear, a growing body of research supports the potential of transdisciplinary collaborations for yielding cross-cutting alliances able to design innovative solutions. However, elite knowledge remains heavily siloed within singular disciplines and sectors. The most urgent problems of the twenty-first century- climate change, the consciousness question, the replication crisis, ecological collapse, the nature of time- are all inherently *transdisciplinary*. They cannot be solved by physics alone, or biology alone, or indigenous studies alone, or consciousness research alone. They require the integration of all these perspectives into a coherent whole- which is precisely what the disciplinary structure of science is designed to prevent.

Interdisciplinary research teams work more closely, integrating conceptual, theoretical, philosophical, and methodological approaches from multiple disciplines. Transdisciplinary research takes this a step further, creating "novel conceptual and methodological approaches that synthesize and extend discipline-specific perspectives." But transdisciplinarity is not merely a matter of putting physicists and psychologists in the same room and hoping for the best. It requires a *shared ontology*- a common understanding of what reality is, how it works, and what counts as knowledge- that can serve as the medium of integration.

This book proposes such an ontology. The framework of resonant quanta- the self-organizing, φ-e-π-tuned, NESS-stabilized quanta-field that constitutes reality at every scale- provides a common language that physics, biology, consciousness studies, indigenous knowledge, and art can all speak. Each discipline describes a different *angle* of the same sphere:

Physics describes the quanta-field at the level of its fundamental constituents- the Planck-scale pixelation, the holographic boundary, the Bekenstein bound, the conservation of quanta, the KZ-Monad and comonadic dynamics.

Biology describes the quanta-field at the level of living systems- the NESS of the cell, the fractal dynamics of the heartbeat, the epigenetic regulation of gene expression, the distributed intelligence of the immune system, the autopoietic self-organization of the organism.

Consciousness studies describe the quanta-field at the level of awareness- the threshold of integrated resonant quanta (Φ_Q), the 864 Hz resonance of the 9D harmonic manifold, the DMN deactivation that reveals the comonadic core, the heart-brain coherence that stabilizes the NESS of the conscious agent.

Indigenous knowledge systems describe the quanta-field at the level of lived, embodied, place-based practice- the songlines that channel telluric resonance, the Three Sisters that implement the φ-e-π triad in soil, the cyclical calendars that track the ternary beat of time, the mantra practices that activate monadic extension through the human voice. Indigenous peoples around the world have developed, maintained, and evolved knowledge systems via direct experience interacting with biophysical and ecological processes, landscapes, ecosystems, and species over millennia.

Art describes the quanta-field at the level of aesthetic and emotional resonance- the φ-proportions that make a painting glow, the acoustic harmonics that make a chord resolve, the narrative arc that makes a story feel inevitable, the architectural proportions that make a space feel alive.

None of these disciplines, alone, captures the whole. Each, alone, is a partial projection- a two-dimensional shadow of a multidimensional reality. But together- integrated through the common ontology of resonant quanta, held in the common framework of the φ-e-π triad, connected by the common recognition that the observer is an irreducible participant in the system she studies- they form something unprecedented: a *spherical* science. A science that can be entered from any point on its surface and that reveals, from every angle of approach, the same fundamental structure of self-organizing, self-knowing, resonant quanta.

The intention would be to bring together the gifts of multiple knowledge systems in ways that centre our collective responsibilities and hold space for the complex co-existence of different priorities, worldviews, realities, and futurities. This is not just about respecting epistemic differences but also about positively valuing those differences, as well as humbly recognizing that all human knowledge is partial, situated, and limited.

The integration of indigenous knowledge with scientific knowledge is perhaps the most sensitive and most necessary aspect of this transition. Despite obstacles like power disparities and differing epistemologies, effective integration can lead to a comprehensive and lasting solution that promotes equitable collaborations, protects intellectual property, and creates culturally appropriate frameworks. Collaborative research that treats indigenous populations as equal partners ensures innovations are both scientifically and culturally valid. The post-binary paradigm does not subsume indigenous knowledge under the Western scientific framework- an act of epistemological colonialism that would violate the very principles of resonance and distributed intelligence it espouses. It places indigenous knowledge *alongside* physics, biology, and consciousness studies as an equal, irreducible angle on the same sphere- an angle that has been systematically suppressed for three millennia and that must now be restored not out of political correctness but out of *epistemic necessity.*

What separates indigenous and western knowledge is not necessarily the knowledge per se, but rather, differences in the underlying worldviews or ontologies and epistemologies, that is, the resulting understandings of what knowledge is and how it is generated. The post-binary ontology dissolves this separation- not by flattening the differences but by providing a framework capacious enough to contain them both. When a physicist describes the holographic principle and an Aboriginal elder describes the Dreamtime, they are describing the same reality from different angles of the sphere. Neither description is complete without the other. And the framework of resonant quanta- which is itself a synthesis of quantum physics, category theory, complexity science, and indigenous cosmology- provides the common ground on which both can stand without either being diminished.

Some authors have suggested that traditional knowledge systems can be helpful in dealing with complex systems: "The understanding of complex systems remains a major challenge for the future, and no scientist today can claim that we have at hand the appropriate methods with which to achieve this. Thus, we cannot discuss the future of science without taking into account the philosophical problems generated by the study of complexity. Modern, or Western, science may not be best suited to fulfil this task, as its view of the world is too constrained by its characteristic empirical and analytical approach."

This is the deepest recognition of the post-binary paradigm: that the challenges of the twenty-first century require not just more science but *different kinds of knowing in generative relationship*. The analytical brilliance of physics. The somatic precision of indigenous ecology. The first-person rigor of contemplative practice. The aesthetic sensitivity of art. The systemic awareness of complexity science. Each is a necessary frequency in the chord. None is sufficient alone.

The silo was the architecture of the binary paradigm- the epistemological equivalent of the orthogonal grid that Chapter 29 diagnosed as the spatial pathology of the Great Decoupling. Just as the resonant city replaces the grid with fractal branching and small-world connectivity, the resonant university replaces the silo with *spherical integration*- a structure in which every discipline can see every other discipline, in which bridges are valued as highly as foundations, and in which the primary metric of intellectual achievement is not the depth of one's specialization but the *quality of one's resonance* with the larger field.

Coda: The Scientist Who Sings

Picture a scientist of the post-binary paradigm.

She begins her day with a five-minute somatic re-tuning: the breath cycle, the heart coherence, the brief hum that aligns her vagal tone. This is not ritual. This is calibration- the same kind of careful preparation that a telescope operator performs before opening the dome.

She walks to her laboratory through a fractal campus- branching paths, φ-proportioned buildings, living walls, water features- and the walk itself is a form of walking meditation, her stride synchronized with her breath, her body coupling to the geospatial phase of the terrain.

Her laboratory is not a single department. It is a *resonance node* in a small-world network that connects physicists, biologists, indigenous knowledge holders, consciousness researchers, and artists. The morning meeting is not a report of individual progress but a *collective tuning*- each member sharing their perspective on the shared problem, each perspective held as an irreducible angle on the sphere.

She records her experiments with temporal phase data: the date, the estimated ternary resonance of the current year, her own HRV data during data collection, the geospatial coordinates of the laboratory. These metadata, which her predecessors would have dismissed as irrelevant, are now understood as essential context- because in a participatory universe, the state of the observer is part of the observation.

She writes her papers not in the sterile third person of the extractive paradigm ("the experiment was conducted...") but in the calibrated first person of the attunement paradigm ("I observed, under the following conditions of personal and temporal coherence..."). She reports not only what she found but *how she found it*- the state of her attention, the quality of her awareness, the degree of heart-brain coherence she maintained during the crucial measurements. Her reviewers do not dismiss these disclosures as subjective noise. They evaluate them as they would evaluate any other methodological variable: with rigor, with curiosity, and with the recognition that the quality of the instrument determines the quality of the data.

She does not claim objectivity. She claims *intersubjectivity*- the convergence of her calibrated observations with the calibrated observations of her colleagues, each approaching the same phenomenon from a different angle of the sphere, each contributing a partial view that, when integrated, reveals a more complete picture than any single perspective could provide.

And sometimes, when the data come together and the pattern reveals itself- when the quanta-field aligns and the long-sought coherence emerges from the noise- she experiences something that her predecessors would have called an "aha moment" and that she recognizes as a *resonance event*: a moment in which her own NESS locks into phase with the NESS of the system she studies, and the boundary between knower and known, observer and observed, dissolves into a single standing wave of understanding.

This is not mysticism. It is the everyday experience of a scientist attuned to a participatory universe. It is what Wheeler meant by "observer-participancy." It is what Bohr meant when he said that physics concerns not what nature *is* but what we can *say* about nature. It is what the Aboriginal elder means when she says the land is not separate from the one who sings it.

The post-binary paradigm does not ask the scientist to stop measuring, stop analyzing, stop testing. It asks her to start *listening*. To calibrate her instrument. To widen her aperture. To include the observer in the observation, the ternary in the binary, the indigenous in the institutional, the somatic in the cerebral, the song in the equation.

Science, at its deepest, has always been an act of attunement- the patient effort to bring the human mind into resonance with the patterns of the cosmos. The binary paradigm, for all its achievements, narrowed this act to a single register: the analytical, the extractive, the disembodied. The post-binary paradigm restores the full spectrum. It does not replace analysis with intuition. It *embeds* analysis within attunement, rigor within resonance, the equation within the song.

The result is not a weaker science. It is a science that can finally match the complexity, the depth, and the living beauty of the universe it seeks to understand.

The silo falls. The sphere opens. And the scientist, for the first time, sings.

Chapter 32

THE EIGHTH PRINCIPLE- CONCESSION AND THE ALCHEMY OF RETURN

"The wound is the place where the Light enters you."- Rumi

There is a word that the modern world has almost entirely lost. It is not *forgiveness*, though forgiveness is part of it. It is not *responsibility*, though responsibility is its foundation. It is not *surrender*, though it shares surrender's root gesture- the opening of the clenched hand, the dropping of the weapon, the breath that comes when the body finally stops bracing.

The word is *concession*.

In ordinary usage, concession means to yield a point in argument, to admit a flaw, to give ground. It carries a faint odor of defeat- the concession speech of the losing candidate, the trade concession wrung from a weaker nation. But in the framework of this book, concession means something far more radical, far more precise, and far more powerful. It means the *deliberate, conscious acknowledgment of the actual state of the system*- the unflinching recognition of what has been done, what has been left undone, and what has been broken- so that the comonadic projection D(Ψ) can operate: stripping away the noise of denial, defense, and deflection to reveal the 98% efficient core of what remains.

Concession is the eighth and crowning co-creative principle. Where the first seven- Intention, Attention, Choice, Belief, Consequence, Compassion, Responsibility- describe the ascending arc of conscious participation in the quanta-field, Concession describes the *descent*: the movement downward into what has been refused, rejected, or repressed, and the alchemical transformation that occurs when that material is finally met with honest presence.

This chapter is the book's emotional and ethical climax. It operates at three scales. First, the personal: every wound you have refused to grieve is a node of decoherence in your personal state vector- a pocket of frozen quanta that cannot flow, cannot resonate, cannot contribute to your NESS. Second, the collective: the modern world carries debts so vast- colonialism, ecocide, epistemicide, the systematic silencing of the dreamers- that no amount of technological innovation can discharge them without the prior act of acknowledgment. And third, the cosmological: concession is not merely a psychological virtue. It is a *physical operation*- the mechanism by which suffering is transformed from noise into living quanta, from entropy into signal, from the raw material of pain into the refined ore of wisdom.

This is not guilt. Guilt is static. Guilt is the ego's way of *appearing* to take responsibility while actually refusing to change. Concession is dynamic. It is the precise act of looking at the deviation between the plan and the actual- between what should have been and what is- and saying, not in shame but in clarity: *this is what happened. This is where we are. Now what?*

The "now what" is everything. Because the moment concession is made- the moment the system stops lying to itself about its actual state- the comonad can operate. D(Ψ) strips the noise. The signal emerges. The path forward reveals itself not as a fantasy projected from denial but as a real possibility rooted in the actual.

We begin at the scale of the body. We end at the scale of the cosmos. And at every scale, the same principle obtains: *what is acknowledged can be transformed; what is denied continues to decohere.*

32.1- The Body Keeps the Score: Decoherence at the Personal Scale

Every human body carries its ungrieved losses like stones in a riverbed- invisible from the surface, but shaping the entire flow of the current.

The mechanism is now well understood, at least in outline. When a traumatic or deeply painful experience occurs- a betrayal, a loss, a violation, a failure- the nervous system responds with a cascade of stress hormones (cortisol, adrenaline) and a rapid shift toward sympathetic dominance: the fight-or-flight state. If the experience is processed- if the person can feel the pain, grieve the loss, integrate the meaning- the nervous system completes its cycle. The stress hormones dissipate. The parasympathetic system re-engages. The body returns to its NESS. The experience is metabolized, its quanta absorbed and integrated into the ongoing resonance of the organism.

But if the experience is *not* processed- if the pain is too great, or the social environment too unsafe, or the cultural norms too prohibitive of emotional expression- the nervous system gets stuck. The stress response does not complete. The sympathetic activation becomes chronic. The body holds the unprocessed experience not as memory in the cognitive sense but as *pattern in the tissue*- constricted muscles, shallow breathing, disrupted digestion, elevated baseline cortisol, reduced heart rate variability.

In the language of this book, unprocessed trauma is a *pocket of frozen quanta*- a region of the personal state vector where the resonance has been locked, where the normal flow of living quanta through the organism is obstructed. The frozen pocket does not participate in the NESS. It sits outside the coherent core, consuming metabolic resources to maintain its rigidity, contributing nothing to the organism's resonance, and slowly degrading the efficiency of the whole system.

The physiological cost is documented with increasing precision. Negative affect and chronic emotional distress erode health, alter cardiovascular reactivity, impoverish sleep quality, and stimulate the production of stress-related hormones, such as cortisol. Chronic anger puts you into a fight-or-flight mode, which results in numerous changes in heart rate, blood pressure and immune response. Those changes, then, increase the risk of depression, heart disease and diabetes, among other conditions.

This is not metaphor. It is measurable physiology. Every ungrieved wound- every betrayal that was swallowed instead of spoken, every loss that was denied instead of mourned, every rage that was buried instead of expressed- is a node of decoherence in the body. It reduces HRV. It flattens the fractal complexity of the heartbeat. It compromises immune function. It erodes the capacity for the very coherence that the practices of Chapter 30 seek to cultivate.

And here is the crucial point: meditation, mantra, and breath regulation can bring the *surrounding* tissue into greater coherence- but they cannot, by themselves, resolve the frozen pocket. The pocket requires something different. It requires *concession*- the deliberate turning of attention toward the wound, the honest acknowledgment of what happened and what it cost, and the willingness to feel what was unfelt.

This is not the same as "reliving" the trauma. Reliving is re-traumatization- a repetition of the same decoherence event without new quanta to integrate it. Concession is *meeting* the trauma with the resources that were unavailable when it occurred: the stability of a more mature nervous system, the safety of a supportive context, the coherence cultivated through practice. It is bringing the comonadic projection to the very place where it was most needed and least available- the wound itself.

And the research confirms that this works. Studies have found that the act of forgiveness can reap huge rewards for your health, lowering the risk of heart attack; improving cholesterol levels and sleep; and reducing pain, blood pressure, and levels of anxiety, depression and stress. Trait forgiveness was significantly associated with fewer medications and less alcohol use, lower blood pressure and rate pressure product; state forgiveness was significantly associated with lower heart rate and fewer physical symptoms.

These findings have important theoretical implications regarding the forgiveness-health link, suggesting that the benefits of forgiveness extend beyond the dissipation of anger. Forgiveness, in this data, is not merely a moral virtue- it is a *physiological intervention* that directly modulates cardiovascular function, immune response, and autonomic balance. Researchers observed a link between forgiveness and subjective relief, and activation in a brain cortical network responsible for perspective taking processes, appraisal and empathy, suggesting that these processes may play an important role in the adaptive extinction of negative affect.

In the process of the nervous system acting to resolve an apparent injustice through parasympathetic activity such as meditation, there would be significant physiological and psychological advancements such as reduced heart rate, comfortable breathing, lower levels of anxiety, depression, hostility, and anger and increased self-esteem.

In the framework of this book, forgiveness is the emotional dimension of the comonadic projection D(Ψ). The frozen pocket- the unprocessed trauma, the unforgiven wound- represents the noise that the comonad has not yet been permitted to strip. When concession occurs- when the individual turns toward the pain, acknowledges it, and releases the rigid defense against feeling it- the comonad finally has access. It can operate on the frozen material, extracting the signal (the lesson, the wisdom, the growth) from the noise (the resentment, the fear, the repeated narrative of injury). What was a pocket of decoherence becomes a source of integrated quanta- experience metabolized into resilience.

This is why the traditions speak of alchemy. The wound, once met with honest presence, does not merely heal. It *transforms*. It becomes a site of deepened sensitivity, heightened empathy, expanded capacity. The broken bone, knit properly, is stronger at the fracture line than it was before. The broken heart, grieved properly, is wider than it was before.

But only if concession comes first. Only if the hand opens.

32.2- The Collective Wound: Colonialism, Ecocide, and Epistemicide

Now scale upward. What is true of the individual body is true of the collective body- the body politic, the body ecological, the body of knowledge itself.

The modern world carries collective wounds of staggering magnitude. Three categories demand particular attention: colonialism (the systematic subjugation and exploitation of peoples), ecocide (the systematic destruction of ecosystems), and epistemicide (the systematic erasure of knowledge systems). Each of these represents a *collective frozen pocket*- a region of the global state vector where the resonance has been locked by violence, denial, and the refusal to acknowledge what was done.

Consider colonialism. For five centuries, the empires of Western Europe- and their settler-colonial offshoots- engaged in a project of planetary-scale extraction: extraction of land, of labor, of resources, and of sovereignty from the peoples of Africa, the Americas, Asia, and the Pacific. The material consequences- enslavement, genocide, forced displacement, resource depletion- are well documented. What is less commonly recognized is the *epistemological* dimension of colonialism: the systematic destruction of indigenous knowledge systems- the songlines, the calendars, the agricultural practices, the sacred architectures, the whole repertoire of resonance technologies described in Chapters 18 through 22.

This destruction was not collateral damage. It was *policy*. The colonial project required the devaluation and elimination of indigenous ways of knowing, because those ways of knowing were incompatible with the extractive logic that the empire depended on. You cannot mine a mountain that your culture teaches you to sing to. You cannot clearcut a forest that your cosmology identifies as your ancestor. You cannot impose linear time on a people whose calendar is tuned to the ternary beat of the Earth. The knowledge had to go so that the extraction could proceed.

The term "epistemicide"- the killing of knowledge- captures this dimension precisely. What was lost was not merely cultural heritage in the museum sense. What was lost were *technologies of coherence*- precision instruments for maintaining the alignment between human civilization and the resonant quanta-field of the living Earth. The Three Sisters, the songlines, the ceque systems, the cyclical calendars, the acoustic temples- all of these were destroyed or suppressed not because they were "primitive" but because they were *functional obstacles to extraction*.

The ecocide follows directly. Strip a civilization of its resonance technologies- its methods for maintaining the NESS of the human-Earth relationship- and the relationship begins to decohere. The industrial economy that replaced indigenous land stewardship operates on precisely the binary, linear, extractive logic that Chapter

23 identified as the cognitive signature of the Great Decoupling. It treats the Earth not as a partner in resonance but as a resource to be mined. The result, after three centuries of exponential acceleration, is the ecological crisis that now threatens the stability of every living system on the planet.

Now ask: has the modern world made concession for any of this?

The answer, overwhelmingly, is no. There have been apologies- formal, diplomatic, carefully hedged. There have been reparations programs- partial, contested, often more symbolic than substantive. There have been truth commissions- powerful in their localized impact, limited in their global reach. But there has been no concession at the scale of the wound- no collective turning toward the pain, no honest accounting of what was done and what it cost, no recognition that the ecological crisis is not a *new* problem but the *accumulated consequence* of a five-century refusal to acknowledge the debts incurred by the extractive paradigm.

This refusal is the collective analogue of the individual's suppression of a traumatic wound. The wound does not heal because it is not acknowledged. The frozen pocket remains frozen. The decoherence deepens. And the metabolic cost of maintaining the denial- the political, ecological, and psychological energy consumed by the effort to *not see* what has been done- grows with every passing year, degrading the coherence of the entire system.

Concession at the collective scale does not mean paralysis. It does not mean guilt. It does not mean the abandonment of all that modernity has achieved. It means the same thing that it means at the personal scale: *acknowledging the actual state of the system*. Looking at the deviation between what was intended and what was produced. Saying, in clarity and without defense: this is what happened. This is where we are. These are the debts. This is the damage.

Only then can $D(\Psi)$ operate at the collective scale. Only then can the signal be extracted from the noise- the genuine achievements of modern science from the wreckage of colonial extraction; the real benefits of industrial technology from the ecological devastation it has wrought; the universal ideals of human rights from the specific historical violence through which they were articulated.

The concession does not destroy the good. It *frees* the good from the noise of denial that has been suffocating it.

32.3- Concession Is Not Guilt- It Is Physics

Here we must be precise, because the difference between concession and guilt is the difference between a living practice and a dead end.

Guilt is a *state*. It is static. It says: "I am bad because of what I did." It collapses the identity of the person (or the civilization) into the wound, producing a paralysis that masquerades as moral seriousness. Guilt is the ego's favorite hiding place, because it allows the ego to *perform* responsibility while actually refusing to change. The guilty person is absorbed in their own badness- which is, paradoxically, a form of narcissism. The focus remains on the self. The wound remains unaddressed. The frozen pocket stays frozen.

Concession is a *process*. It is dynamic. It says: "This is what happened. This is what it cost. I accept responsibility not as a moral judgment on my identity but as an *accurate reading of the system state.*" Concession is not about being good or bad. It is about being *honest*- honest enough to allow the comonadic projection to operate, to strip away the layers of defense, rationalization, and denial that have accumulated around the wound, so that the signal can finally emerge.

In the language of this book, guilt is a *noise state*- an additional layer of decoherence imposed on top of the original wound. It does not reduce the frozen pocket; it *enlarges* it, adding the energy of self-condemnation to the energy of unprocessed pain. Concession, by contrast, is the *removal of noise*. It does not add anything to the system. It subtracts- stripping away the cognitive and emotional defenses that prevent the wound from being metabolized.

The mathematics are precise. The comonadic projection $D(\Psi)$ can only operate on a state vector that is *honestly represented*. If the input is distorted- if the pain is denied, or rationalized, or covered over with guilt- then D cannot extract the efficient core, because the state it receives does not correspond to the actual state of the system. The projection fails. The noise persists. The system remains decoherent.

When concession occurs- when the state vector is honestly updated to reflect the actual damage, the actual loss, the actual debt- then D has clean input. It can operate. It can identify the 98% efficient core that remains after the damage is acknowledged. And from that core, the system can rebuild- not from fantasy, not from denial, not from the brittle optimism of "moving on," but from the solid ground of what is actually true.

This is why concession feels so different from guilt. Guilt feels heavy, static, suffocating- because it *adds* weight to the system. Concession feels paradoxically light- even when the material being acknowledged is devastating- because it *subtracts* weight. The weight of denial, the weight of pretense, the weight of the elaborate cognitive structures that the ego has built to avoid seeing what is real. When these are released, the organism (or the civilization) experiences something that the traditions call *grace* and that this book identifies as the *natural coherence of a system whose state vector has been honestly updated.*

The asymptotic gap $\delta\infty$ is relevant here. The gap does not close when concession is made. The 2% residual tension- the imperfection that keeps the cosmos dancing- remains. Concession does not produce perfection. It does not erase the wound. It does not undo history. What it does is *change the relationship to the gap*. Before concession, the gap is experienced as suffering- as the unbearable distance between what is and what should have been. After concession, the same gap is experienced as *creative tension*- as the productive edge between the real and the possible, the same edge that the NESS occupies, the same edge where emergence happens.

This is the alchemy. The material is the same. The pain is the same. The loss is the same. What changes is the orientation- from denial to acknowledgment, from resistance to acceptance, from noise to signal. And that change of orientation transforms the energetics of the system entirely. The frozen pocket thaws. The locked quanta begin to flow. The decoherent region is absorbed back into the NESS. And suffering- which is what the Buddhists call *dukkha*, the "unsatisfactoriness" that drives all seeking- transforms from noise into resonant quanta. From waste into fuel.

The gap does not close. But it begins to sing.

32.4- Indigenous Models of Concession: Ho'oponopono, Utu, and Ubuntu

The indigenous world did not lose this knowledge. While the modern West was constructing elaborate philosophical systems to avoid the work of concession- from Cartesian dualism (which separates mind from body and thus divorces the thinker from the felt consequence of their actions) to neoliberal economics (which externalizes all costs that do not appear on the balance sheet)- the indigenous world was preserving, practicing, and refining technologies of concession with extraordinary precision.

Three deserve particular attention: the Hawaiian practice of ho'oponopono, the Māori concept of utu, and the African philosophy of ubuntu. Each embodies a different dimension of the concession principle, and together they form a complete ternary: ho'oponopono addresses the *personal* dimension, utu addresses the *relational* dimension, and ubuntu addresses the *ontological* dimension.

Ho'oponopono: The Clearing of the Personal State Vector

Mary Kawena Pukui described ho'oponopono as a practice of extended family members meeting to "make right" broken family relations. Ho'oponopono is defined in the Hawaiian Dictionary as "to put to rights; to put in order or shape, correct, revise, adjust, amend, regulate, arrange, rectify, tidy up, make orderly or neat."

The practice, in its traditional form, is communal. The practice is based on the belief that our relationships with others are mirrors of our relationships with ourselves, and that forgiveness and reconciliation can help to heal and restore balance within ourselves and in our relationships with others. A family or community gathers under the guidance of an elder or kahuna. The problem is stated. Each participant speaks their truth- not in accusation but in acknowledgment. In the tradition of mental cleansing, these are family conferences in which relationships were set right through prayer, discussion, confession, repentance, and mutual restitution and forgiveness.

Ho'oponopono is an ancient Hawaiian spiritual practice that involves learning to heal all things by accepting "Total Responsibility" for everything that surrounds us- confession, repentance, and reconciliation. The foundation of this practice is unity: an unbreakable bond connects you to everyone else, even though we seem so separate.

The modern simplified form of the practice centers on four phrases: "I'm sorry. Please forgive me. Thank you. I love you." This mantra encapsulates the spirit of Ho'oponopono, encouraging you to take responsibility for your life and actions and to seek resolution through love and gratitude.

Notice the sequence. It begins with *acknowledgment* ("I'm sorry")- the concession itself. It moves to *petition* ("Please forgive me")- the opening of the relational field. It arrives at *gratitude* ("Thank you")- the recognition that the wound, once acknowledged, has produced something of value. And it culminates in *love* ("I love you")- the restoration of the resonant connection that the wound had severed.

In the language of this book, this four-phrase mantra is a complete comonadic cycle. "I'm sorry" is the honest updating of the state vector- the acknowledgment that the actual state diverges from the ideal. "Please forgive me" is the request for the comonadic projection to operate- the opening of the system to the process of noise-stripping. "Thank you" is the recognition of the signal that emerges when the noise is removed- the gratitude for the wisdom that suffering has produced. And "I love you" is the restoration of the resonant coupling between the self and the field- the reactivation of the monadic extension $T(\Psi)$, which extends the now-cleaned state toward its fullest coherent potential.

Although the word ho'oponopono was not used, early Hawaiian historians documented a belief that illness was caused by breaking kapu, or spiritual laws, and that the illness could not be cured until the sufferer atoned for this transgression. Harmony could be restored only by confession and apology. This is a profoundly physical understanding: the wound is not merely psychological. It is *physiological.* The breaking of right relationship- the violation of the harmonic code that governs the quanta-field- produces illness in the body. And the illness cannot be resolved through physical intervention alone. It requires the restoration of the relational harmony that was broken. It requires concession.

Utu: The Relational Restoration of Balance

Utu is a Māori concept of reciprocation or balance. To retain mana, both friendly and unfriendly actions require an appropriate response; that is, utu covers both the reciprocation of kind deeds, and the seeking of revenge.

The Western mind, encountering utu, typically fixates on the revenge dimension- and misses the deeper principle. Crucial to this process was the concept of utu. Often defined as "revenge," utu has a broader meaning: the maintenance of balance and harmony within society. A wrong had to be put right, but how this was done could vary greatly.

A more accurate understanding of utu is that it is the maintenance of relationships built on harmony and reciprocity; and when disharmony occurs or reciprocity has been broken, immediate steps need to be taken to "restore the relationship."

Utu is an important concept concerned with the maintenance of balance and harmony in relationships between individuals and groups and order within Māori society, whether through gift exchange or as a result of hostilities between groups. If social relations are disturbed, utu is a means of restoring balance.

What utu encodes is a *NESS-maintenance protocol* for human relationships. In any complex adaptive system, perturbations are inevitable. Relationships are dynamic, not static. They will be disturbed- by injury, by insult, by negligence, by the simple asymmetries of power and circumstance that characterize all human communities. The question is not whether the system will be perturbed but *how quickly and how precisely the balance will be restored.*

Utu provides the answer: immediately, and through a reciprocal act that is proportional to the disturbance. It is closely linked to mana and includes reciprocation of kind deeds as well as revenge. The overarching objective, as one study of the related concept of tauutuutu notes, "wasn't a static equilibrium but dynamic, fuelled by a perpetual urge for escalation"- the same non-equilibrium steady state, the same dynamic stability, that characterizes the cosmos itself.

Utu is concession with teeth. It is not a passive acceptance of harm. It is the active demand that harm be acknowledged and redressed- that the deviation between the plan (the harmonious relationship) and the actual (the violated relationship) be registered and corrected. It does not wait for the offender to volunteer their remorse. It *insists* on the restoration of balance, because an unbalanced system- a system where utu is not fulfilled- is a decoherent system, and a decoherent system degrades everyone within it.

The modern world has almost entirely lost this relational technology. The replacement- the legal system- operates on an entirely different logic: guilt or innocence, punishment or acquittal, with balance conceived as a zero-sum transaction between the state and the individual offender. The community whose coherence was damaged by the offense is not a party to the proceeding. The relational field is not restored. The utu is not fulfilled. And the decoherence deepens.

Ubuntu: The Ontological Ground of Concession

Where ho'oponopono addresses the personal wound and utu addresses the relational disturbance, the African philosophy of ubuntu addresses the *ontological ground* from which concession becomes possible- the recognition that the self is not separate from the other, that the wound I inflict on you is simultaneously a wound I inflict on myself, that healing is always mutual or it is not healing at all.

Ubuntu describes a set of closely related Bantu African-origin value systems that emphasize the interconnectedness of individuals with their surrounding societal and physical worlds. "Ubuntu" is sometimes translated as "I am because we are."

John Mbiti articulated this ethos in, "I am because we are, and since we are, therefore I am"- highlighting the ontological priority of community over the individual.

Consider the African philosophy of "ubuntu"- a concept in which your sense of self is shaped by your relationships with other people. It's a way of living that begins with the premise that "I am" only because "we are."

Ubuntu is not merely an ethical injunction to be nice to your neighbors. It is an *ontological claim*- a statement about the nature of reality itself. Ubuntu is inherently counter-hegemonic, resisting Western paradigms that prioritize individual agency over collective responsibility. Unlike Western frameworks that emphasize binary oppositions- self vs. other, rational vs. emotional- Ubuntu embraces relationality as the foundation of existence.

In the Shona language, the greeting is "Ndapora kana wapora"- "I am well if you are well." It was a salutation, which meant, in essence, that their state of being was conditioned by my state of being.

This is not politeness. This is physics- at least, it is the physics described in this book. If reality is constituted by resonant quanta that are nonlocally entangled, if the holographic principle means that every part contains the whole, if the quanta-field is a single, continuous, self-referencing resonance- then the Ubuntu claim is literally true. Your state *is* conditioned by my state. Your suffering *is* my suffering, not as metaphor but as entanglement. The wound I refuse to acknowledge in our shared field *does* degrade your coherence, not only mine.

This ontological ground is what makes concession possible- and what makes it more than a private psychological exercise. When I make concession for the harm I have caused, I am not merely "processing my feelings." I am *updating the shared state vector*. I am correcting a distortion in the quanta-field that affects everyone who is entangled with the system in which the harm occurred. The personal is not merely analogous to the collective. It *is* the collective, viewed from one node of the holographic web.

This knowing is always reaching and stretching toward wholeness-together. Ubuntu and interdependence bring to consciousness the knowing that we are multiplicities of one, a bundle of life, folding and unfolding on this plane while interwoven with the energy of an absent-present past, an absent-present future, and a very present-present.

Ubuntu, in the context of this book, is not merely an African philosophy. It is the *social expression of the holographic principle*. "I am because we are" is the relational translation of "the whole is enfolded in every part." And concession- the act of acknowledging harm and restoring balance- is the mechanism by which the holographic web maintains its coherence when one of its nodes is damaged.

32.5- The Epigenetics of Ungrieved Wounds: How Refusal Writes Itself Into the Body

The personal and collective dimensions of concession converge in one of the most disturbing- and most hopeful- discoveries of twenty-first century biology: the epigenetic inheritance of trauma.

There is now converging evidence supporting the idea that offspring are affected by parental trauma exposures occurring before their birth, and possibly even prior to their conception. On the simplest level, the concept of intergenerational trauma acknowledges that exposure to extremely adverse events impacts individuals to such a great extent that their offspring find themselves grappling with their parents' post-traumatic state.

A more recent and provocative claim is that the experience of trauma- or more accurately the effect of that experience- is "passed" somehow from one generation to the next through non-genomic, possibly epigenetic mechanisms affecting DNA function or gene transcription.

The evidence, while still evolving, is increasingly compelling. This concept, known as intergenerational transmission of trauma effects, was first recognized in studies involving survivors of the Dutch famine during World War II and Holocaust survivors. These studies found that traumatic experiences like stress, famine, or war can influence the expression of our genes, thereby affecting our offspring.

In 2025, a landmark study extended these findings to three generations of Syrian refugees. It shows that women who have been directly affected by war-related violence and trauma indicate altered epigenetic markings- but so do their grandchildren, even if they had no direct exposure to warfare. This is the first report of an intergenerational epigenetic signature of violence, which has important implications for understanding the inheritance of trauma.

The researcher Catherine Panter-Brick stated: "Our findings present the first-ever evidence that violence can leave epigenetic marks on the genome, which has important implications for understanding evolution and how traumatic experiences can become embedded in the genome and persist for generations."

Trauma doesn't alter our DNA, but it can guide which genes are activated or deactivated, similar to a marginal note in a book. When an individual experiences trauma, their body may adapt by adjusting gene expression, and some of these changes can be passed on to their children, akin to inherited "notes."

In the framework of this book, this finding is staggering in its implications. The frozen pocket of unprocessed trauma is not merely a personal burden. It is a *heritable signal*- a distortion in the quanta-field that can propagate through the germline, shaping the stress responses, the metabolic patterns, and the psychological vulnerabilities of descendants who were never themselves exposed to the original trauma.

The wound writes itself into the body. And the body passes the writing on.

But here is the note of hope- the crucial, indispensable counterweight to the terror of this finding. Some of these stress-related and intergenerational changes may be reversible. Several years ago we discovered that combat veterans with PTSD who benefited from cognitive-behavioral psychotherapy showed treatment-induced changes in FKBP5 methylation. The finding confirmed that healing is also reflected in epigenetic change.

However, these changes are not definitive but rather adaptable and can be rewritten by our own life experiences and actions.

Healing is reflected in epigenetic change. The notes in the margins of the book can be revised. The frozen pocket can be thawed- not only for the individual who does the thawing, but potentially for the descendants who would otherwise inherit the distortion. Concession, in this light, is not merely personal therapy. It is *intergenerational service*. It is an act that reverberates through time, altering the quanta-field not only in the present but in the germline- reshaping the inherited landscape of vulnerability and resilience for generations to come.

This is the most literal possible validation of the ancient intuition expressed in the Hawaiian understanding that "the sins of the father will fall upon the children. If a child is sick, the parents are suspected of quarreling or misconduct." The Hawaiians were not speaking metaphorically. They were describing, in the language of their tradition, a biological mechanism- the epigenetic inheritance of relational decoherence- that modern science has only just begun to map.

The concession you make today may alter the methylation patterns of your great-grandchildren. The wound you refuse to grieve may be inherited by people who have not yet been born. The stakes of concession are not merely personal. They are transgenerational. They are, in the deepest sense, *evolutionary*.

32.6- The Practice of Concession: A Guided Reflection

As with meditation and mantra, concession is not a concept to be understood but a practice to be embodied. The reader is invited- gently, without coercion, with full respect for the pace at which wounds disclose themselves- to engage in a brief reflective practice.

This is not therapy. It does not replace professional support for deep trauma. It is a *threshold practice*- a way of turning, gently and with intention, toward whatever is asking to be acknowledged.

Phase One: The Ground (60 seconds)

Sit comfortably. Close your eyes. Take three slow breaths in the ternary pattern: five seconds in, two seconds hold, six seconds out.

Feel the ground beneath you. Feel the weight of the body. Feel the heartbeat- that rhythmic δ∞ gap between each contraction, the generative pause from which each new beat arises.

You are here. You are safe. The practice that follows is an invitation, not an obligation. At any moment, you can open your eyes and return to the present without judgment.

Phase Two: The Turning (90 seconds)

Ask yourself, quietly and without force: *What have I been refusing to feel?*

Do not go searching for the biggest wound. Let the question sit, like a stone dropped into still water, and see what rises. It may be a recent hurt- a harsh word you did not acknowledge, a disappointment you glossed over, a loss you moved past too quickly. It may be an old wound- a childhood betrayal, a relationship that ended without resolution, a grief that was never completed.

Do not judge what arises. Do not analyze it. Simply let it be present- a shape in the dark water, rising slowly toward the light.

Phase Three: The Acknowledgment (90 seconds)

When something has surfaced- when you can feel the faint tug of unprocessed material- turn toward it. Not with force. With gentleness. With the same quality of attention you would bring to a frightened animal: steady, patient, non-threatening.

Say, silently or aloud: *I see this. I acknowledge that this happened. I acknowledge that it cost me something. I acknowledge that I have been carrying this.*

If the Hawaiian phrases feel natural, you may add: *I'm sorry. Please forgive me. Thank you. I love you.*

Notice what happens in the body. There may be a tightening in the throat, a warmth behind the eyes, a tremor in the chest. These are signs that the frozen pocket is beginning to thaw- that the quanta locked in the wound are beginning to flow again. Do not suppress these sensations. Do not amplify them. Simply let them move.

Phase Four: The Release (60 seconds)

Breathe. Return to the ternary cycle: five seconds in, two seconds hold, six seconds out.

As you exhale, imagine the frozen pocket softening- not dissolving all at once, but beginning to yield. The ice at the edges melts first. A trickle of flow returns. The locked quanta begin, slowly, to rejoin the larger current of the organism's resonance.

You do not need to resolve everything in this moment. Concession is not a single event. It is a *practice*- a repeated turning toward what has been refused, each turn a little deeper, each acknowledgment a little more honest, each thaw a little more complete.

Phase Five: The Return (30 seconds)

Open your eyes. Feel the room. Notice the quality of your awareness.

If something shifted- even slightly- that shift is real. It is physiological. It is the beginning of the comonadic projection operating on material that has been locked away, perhaps for years. The signal is beginning to emerge from the noise.

If nothing shifted- if the wound is too deep, too defended, too dangerous to approach alone- that is also real. Honor the defense. It was put in place for a reason. Concession does not mean tearing down the walls. It means learning that the walls can be *doors*- and that you hold the key.

32.7- When Concession Is Made, the Gap Sings

Return now to the cosmological scale.

The asymptotic gap $\delta\infty = (\pi - e)/(\pi\varphi^4) \approx 0.01965$ is the fundamental dynamic tension of the cosmos- the ~2% residual that prevents the triad from closing into perfect equilibrium, the "error" that keeps all of reality in motion. Throughout this book, we have described this gap as the source of animation, the engine of becoming, the heartbeat of the cosmic NESS.

But we have not yet asked: *what is the subjective experience of the gap?*

Before concession, the gap is experienced as suffering- as dukkha, the Buddhist term that is usually translated as "suffering" but more precisely means "unsatisfactoriness" or "fundamental incompleteness." It is the feeling that something is always slightly wrong, slightly missing, slightly out of reach. It is the distance between what is and what should be, between the actual state and the ideal, between the real and the hoped-for. This distance is the source of all craving, all grasping, all restless seeking- the existential itch that no achievement can scratch, no pleasure can fully satisfy, no distraction can permanently soothe.

After concession- after the state vector has been honestly updated, after the frozen pockets have begun to thaw, after the noise of denial has been stripped and the signal of reality has emerged- the gap is experienced differently. Not because it has closed. Not because the suffering has been eliminated. But because the *relationship to the gap* has changed.

Before concession, the gap was experienced as a wound- a flaw in the universe, a sign that something had gone terribly wrong. After concession, the gap is experienced as a *generative space*- the creative tension that drives all emergence, all evolution, all growth. It is the same gap. The same $\delta\infty$. The same ~2%. But its meaning has inverted. What was a source of suffering becomes a source of vitality. What was noise becomes signal. What was waste becomes fuel.

This is the alchemy that every mystical tradition has tried to describe. The Sufis call it the transformation of the nafs (the ego-self) through fana (annihilation of the false). The Buddhists call it the cessation of dukkha through the relinquishing of tanha (craving). The Christians call it the resurrection- the discovery that what was dead (the buried wound) has become the source of new life. The Hawaiians call it ho'oponopono- "to make right," to restore the balance, to return the relationship to its resonant state.

In the language of this book: concession activates the comonadic projection $D(\Psi)$ on the deepest material in the system. It allows the frozen pockets to thaw. It restores quanta-flow to regions of the state vector that have been locked in decoherence. It does not close the gap- nothing can close the gap; the gap is constitutive of existence- but it changes the gap from a site of stagnation to a site of flow, from a wound to a wellspring, from a silence that suffocates to a silence that sings.

The gap, after concession, does not close. But it *resonates.*

And that resonance- the vibration of the $\delta\infty$ itself, the hum at the heart of the asymptotic interval- is what the ancients heard in the space between heartbeats, in the silence between the notes of a mantra, in the pause between the exhale and the inhale. It is the sound of the cosmos acknowledging itself. It is the tone of the universe making concession for its own imperfection- and finding, in that imperfection, the very source of its beauty.

This is what it means to say that suffering transforms from noise into living quanta. Not that suffering is good. Not that pain should be sought. Not that the wounded should be grateful for their wounds. But that when a wound is met- truly met, with the full force of conscious presence, without denial, without guilt, without the elaborate defenses that the ego constructs to avoid the unbearable- the energy that was locked in the wound is released. And that energy, once released, is *quanta*- living, flowing, resonant pattern that can now contribute to the NESS, can now participate in the coherence of the whole, can now serve as the raw material for the next phase of emergence.

Concession is the most difficult practice in this book. It is also the most important. Without it, the meditation is escapism- a retreat from the wounds rather than a turning toward them. Without it, the mantra is noise- a pleasant sound that covers the silence where the screams should be. Without it, the breath is mechanical- a physiological technique that leaves the deepest layers of the state vector untouched.

With concession, everything changes. The meditation becomes a descent into the wound, guided by the comonadic projection. The mantra becomes a resonance that softens the frozen pocket, loosening the locked quanta. The breath becomes a wave that carries the thawed material back into the flow of the whole organism.

With concession, the personal practice and the collective practice converge. The wound you heal in yourself is a wound you heal in the quanta-field. The decoherence you resolve in your own state vector is decoherence you resolve in the shared field of the human community. This is not mysticism. This is ubuntu- "I am because we are"- understood as physics.

Coda: The Open Hand

Close the book for a moment. Not forever- just for this moment.

Feel your hands. Are they clenched? Even slightly? Even unconsciously?

Open them.

Feel the palms. Feel the air on the skin. Feel the vulnerability of the open hand- the hand that has dropped its weapon, its defense, its grip on the narrative of injury.

This gesture- the opening of the hand- is the somatic form of concession. It is the body's way of saying: *I release my hold on what I was refusing to feel. I let it flow. I let it move. I let it become what it needs to become.*

The universe was not designed to be clenched. It was designed to flow- through the φ-e-π triad, through the NESS, through the δ∞ gap, through the endless dance of monadic extension and comonadic projection, through the living quanta that constitute everything you see and everything you are.

When the hand opens, the flow resumes. When the concession is made, the frozen pocket thaws. When the wound is acknowledged, the noise becomes signal.

This is not the end of pain. It is the beginning of *meaning*- the moment when pain ceases to be a dead weight and becomes a teacher, a guide, a source of the very wisdom that the post-binary paradigm requires. The dreamer civilizations knew this. The Hawaiians practiced it daily. The Māori built their entire social order on it. The African continent encoded it in the deepest structure of its relational philosophy.

We forgot. We suppressed. We built empires on the refusal to make concession- and those empires are now collapsing under the accumulated weight of their unacknowledged debts.

The way back is not through more technology, more extraction, more denial. The way back is through the open hand.

I'm sorry. Please forgive me. Thank you. I love you.

Four phrases. Four movements of the comonadic cycle. Four steps from decoherence to coherence, from frozen to flowing, from noise to living quanta.

The eighth principle is the hardest. It is also the key.

Open your hand.

Let it flow.

Chapter 33

STATE 0- THE GENERATIVE VOID REVISITED

"The silence is not empty. It is full of answers."

Close this chapter as you would close a circle. Begin where the book began- in the silence before the tone, the pause before the pulse, the stillness that is not absence but fullness, not emptiness but the most quanta-dense condition in the cosmos.

Everything this book has described- the holographic boundary, the Bekenstein bound, the φ-e-π triad, the asymptotic gap, the KZ-Monad and the comonadic projection, the NESS that sustains all living systems, the Holographic Ternary Time that structures experience, the songlines that web the Earth, the Three Sisters that mirror the cosmic operators in soil, the Great Decoupling that severed humanity from its resonance, the practices of meditation and mantra that restore it, the post-binary paradigm that expands it, the eighth principle of concession that alchemizes suffering into living quanta- all of it, every equation, every indigenous cosmology, every laboratory finding, every felt truth- exists in the space between your current heartbeats.

Not metaphorically. *Physically.*

Between the contraction that sends blood surging through your arteries and the next contraction that will follow approximately eight hundred milliseconds later, there is a gap. It lasts, depending on your resting heart rate, between three hundred and five hundred milliseconds- a fraction of a second in which the cardiac muscle relaxes, the chambers refill, the baroreceptors register the changing pressure, and the vagus nerve modulates the timing of the next beat with fractal precision. In that gap, the body's most powerful oscillator resets. The signal clears. The noise floor drops. And from the silence- from the generative pause in which all possibility is folded- the next beat arises. Not because it was forced. Because the system is in NESS. Because the gap *compels* the return.

That gap is δ∞.

It is the same gap that prevents the φ-e-π triad from closing into static perfection, the same gap that keeps the cosmos dancing, the same gap that the meditator encounters when the default mode network quiets and the stories fall away, the same gap that the chanter inhabits when the "M" of Om dissolves into humming silence, the same gap that the walker enters when stride and breath synchronize and the body becomes a mobile resonance node in the quanta-field.

This chapter is about that gap. It is about the state from which all states arise and into which all states return. It is about State 0- the ground state of the universe, the quantum vacuum, the generative void- now understood not as the beginning of the story but as its permanent, ever-present, always-available foundation.

The fractal closes here. The sphere completes. And the silence, at last, speaks.

33.1- Everything Is Here: The Heartbeat as Cosmic Portal

The human heart beats approximately one hundred thousand times per day. Each beat is a contraction-relaxation cycle- systole (the squeeze) followed by diastole (the release). The scale-invariant/fractal heart rate pattern reveals a complex temporal organization in heart rate fluctuations and provides complementary dynamical information of heart rate control. The healthy heart does not beat metronomically. It beats *fractally*- with subtle, beat-to-beat variations that encode the dynamic interplay between sympathetic acceleration and parasympathetic restraint, between the drive to act and the invitation to rest, between the monadic extension and the comonadic projection.

In physical systems, such fluctuations are characteristic of critical states on the border between randomness and order, frequently arising from nonlinear feedback interactions between mechanisms operating on multiple scales. The existence of fractal fluctuations in physiology challenges traditional conceptions of health and disease, suggesting that high levels of integrity and adaptability are marked by complex variability, not constancy.

The heart is, in this sense, a *miniature cosmos*. It lives in NESS- perpetually far from equilibrium, maintained by the continuous flow of electrical impulses, metabolic resources, and neural signals, producing its coherent rhythm not by achieving stasis but by *surfing the edge between order and chaos*. When the fractal complexity of the heartbeat degrades- when the variations flatten into regularity or spike into randomness- this fractal pattern breaks down with disease and its change is predictive of mortality. The loss of fractal complexity is the cardiac equivalent of the Great Decoupling: a system that has lost its resonance with the dynamic architecture that sustains it.

But in the healthy heart- in the heart of the meditator, the chanter, the conscious breather, the walker in the quanta-field- the fractal rhythm is intact. And within that rhythm, between every beat, the gap opens.

Consider what happens in that gap. The ventricles relax. The blood that was driven outward pauses, suspended in the arterial tree. The pressure waves that raced through the vasculature decelerate, reflect, and return. The baroreceptors in the carotid sinuses and aortic arch register the momentary lull and send their signals, via the vagus nerve, to the brainstem. The brainstem integrates the signal with a thousand other inputs- respiratory phase, emotional state, blood chemistry, circadian position, ambient temperature- and computes the timing of the next beat with a precision that no external clock can match.

All of this occurs in less than half a second.

And all of this- the integration, the computation, the feedback, the timing- is a *resonance event*. The gap between heartbeats is not empty time. It is the moment of maximum quanta-density in the cardiac cycle- the moment in which the most complex integration of the most diverse signals produces the most precise output. It is the cardiac equivalent of the quantum vacuum: the state of lowest energy and highest potential, the ground state from which the next excitation will emerge.

Under forced desynchrony conditions, researchers found an endogenous circadian rhythm in fractal heartbeat fluctuations at short time scales with a sharp peak in the DFA exponent at the circadian phase corresponding to approximately 10AM. This finding provided the first indication that the circadian system is involved in fractal cardiac dynamics. The fractal structure of the heartbeat is not merely a property of the heart. It is entrained by the circadian pacemaker in the suprachiasmatic nucleus- the body's master clock, itself entrained to the solar cycle. Lesioning of the SCN abolishes such fractal patterns in heart beat variability at time scales between 4 and 24 hours, rendering the pattern similar to random noise and demonstrating the SCN as a critical node in the neural feedback network underlying scale-invariant patterns in physiology.

The heartbeat reflects the day. The day reflects the year. The year reflects the ternary cycle ($Y_t = \{360,363,366\}$). The ternary cycle reflects the aeon. The aeon reflects the cosmic cycle of Conformal Cyclic Cosmology- in which the universe iterates through infinite cycles, with the future timelike infinity of each previous iteration being identified with the Big Bang singularity of the next.

At every scale, the same pattern: contraction followed by expansion followed by pause followed by renewal. Systole-diastole at the cardiac scale. Inhale-exhale at the respiratory scale. Wake-sleep at the circadian scale. Summer-winter at the seasonal scale. Big Bang to heat death to conformal rescaling at the cosmic scale. The fractal repeats, nested within itself like Indra's Net, the whole contained in every part, every part reflecting the whole.

And at the center of every cycle, at every scale, the gap.

The space between heartbeats. The pause between breaths. The silence between thoughts. The midnight between days. The winter between years. The conformal boundary between aeons.

State 0.

33.2- The Fractal Zoom-Out: From Pulse to Aeon

Let us perform the zoom. Slowly, deliberately, from the smallest oscillation the body knows to the largest oscillation the cosmos permits.

Scale 1: The Heartbeat (~1 second)

You are here, in this body, reading these words. Your heart has just completed a beat. The ventricles have squeezed. The blood has surged. And now- *now*, in this fraction of a second- the gap opens. The muscle relaxes. The chambers refill. The vagus nerve whispers its calibration. And the next beat forms in the silence, not yet actualized, not yet committed, poised in the generative void of diastolic potential.

This is the δ∞ at cardiac scale. The gap that keeps the heart alive. If the gap closed- if the heart contracted without pausing- the muscle would tear itself apart. The gap is not a luxury. It is a necessity. It is the non-equilibrium condition that permits the steady state.

Scale 2: The Breath (~13 seconds)

Zoom out. The heartbeat nests within the breath. Six heartbeats per inhale. A pause- kumbhaka, the hold. Six heartbeats per exhale. The breath cycle, as established in Chapter 30, is the ternary oscillator that maps the Temporal State Vector: inhale as gathering (Y_t), hold as integration (ρ_t), exhale as release ($G(\lambda,\phi)$).

The gap in the breath is the hold- the moment between inhale and exhale when the lungs are full, the diaphragm suspended, the body poised between expansion and contraction. In that pause, the baroreceptors are maximally stimulated, the vagal tone deepens, and the heart rate decelerates to its nadir. The breath-gap and the heart-gap are entrained- coupled oscillators, each modulating the other, each containing the other's pattern in miniature.

Scale 3: The Day (~24 hours)

Zoom out further. The breath cycle nests within the circadian cycle. Approximately twenty thousand breaths per day. Twenty thousand cycles of expansion-hold-contraction, each containing approximately six fractal heartbeats, all modulated by the suprachiasmatic nucleus- the circadian pacemaker that generates the endogenous rhythm of approximately 24.2 hours, entrained by light to the solar day.

Most life on earth- ranging from single cellular organisms, plants, flies, rats, to humans- contains an endogenous timing system that optimally synchronizes physiology and behavior with the solar day. The system, known as the circadian timing system, has two core characteristics: endogenous rhythmicity that cycles approximately every 24 hours, and the capability to adjust its timing in response to external factors such as light and food intake.

The gap in the circadian cycle is deep sleep- the period of maximum parasympathetic dominance, maximum growth hormone secretion, maximum immune activity, maximum memory consolidation. Deep sleep is the diastole of the day: the phase in which the organism retreats from the surface of waking activity into the generative darkness from which the next day's coherence will emerge. Without this gap- without sufficient deep sleep- the circadian NESS collapses. The fractal complexity of the heartbeat degrades. The immune system falters. The default mode network, deprived of its nightly reset, spirals into hyperactivity. The default mode network is best known for being active when a person is not focused on the outside world and the brain is at wakeful rest, such as during daydreaming and mind-wandering. Without the gap of sleep, the narration never stops. The signal drowns in noise.

Scale 4: The Year (~365.25 days)

Zoom again. The circadian cycle nests within the annual cycle. Three hundred and sixty-five days- or, in the framework of Holographic Ternary Time, one beat of the ternary pulse: $Y_t = \{360{,}363{,}366\}$. The year has its own systole and diastole- summer's expansion and winter's contraction, the solstices as the extreme points of the oscillation, the equinoxes as the moments of balance.

The gap in the annual cycle is the winter solstice- the longest night, the deepest dark, the point of maximum contraction before the return of the light. Every agricultural civilization has marked this gap with ceremony: Yule, Saturnalia, Dongzhi, Inti Raymi. The ceremonies are not mere cultural decoration. They are *resonance practices*- technologies for acknowledging and honoring the annual State 0, the generative darkness from which the new year's light will emerge.

The ancient calendars- Mayan, Vedic, Aboriginal- tracked this annual gap with extraordinary precision, not because they were obsessed with dates but because they understood that the gap is the *hinge of time*, the moment of maximum vulnerability and maximum potential, the inflection point at which the ternary resonance resets and the new cycle crystallizes from the void.

Scale 5: The Evolutionary Epoch (~millions of years)

Zoom to the biological scale. Species arise, flourish, and pass- each one a standing wave in the quanta-field, persisting for its characteristic duration before dissolving back into the reservoir of possibility. The mass extinction events- the five great die-offs that punctuate the fossil record- are not catastrophes. They are *the gaps between evolutionary heartbeats*. They are the diastoles of life itself: the moments of maximum destruction that are also the moments of maximum creative potential, the voids from which new forms radiate with explosive diversity.

The Cambrian explosion followed the Ediacaran die-off. The age of mammals followed the Cretaceous-Paleogene extinction. Every great flowering of life was preceded by a great emptying- a State 0 at the evolutionary scale, a moment when the old forms had dissolved and the field was cleared for the new.

The guiding quanta- the transient, high-specificity resonant quanta that appear at bifurcation points- are most active during these gaps. The extinction is not random destruction. It is the clearing of the field to allow the guiding quanta to orchestrate the next phase of emergence. The gap is not the end. It is the *condition* for the beginning.

Scale 6: The Aeon (~10^{100} years)

Zoom to the cosmological scale. The universe right now is lumpy, with stars and galaxies and black holes all over the place. But the future universe will be smooth again- black holes will evaporate and the cosmological constant will disperse all the matter, leaving us nothing but empty space. So, Penrose says, we can map the late universe onto a future phase that looks just like our early universe, simply by a conformal transformation. Do this an infinite number of times, and you have a cyclic cosmology- the universe goes through a series of "aeons" that start with a smooth Big Bang, get lumpy as structure forms, smooth out again, and then get matched onto another smooth Big-Bang-like phase.

Conformal Cyclic Cosmology, proposed by Sir Roger Penrose, envisions the Universe as an infinite succession of aeons, each beginning with a Big Bang and ending in a future of infinite expansion, rendered conformally equivalent to the next aeon's birth.

The gap between aeons- the conformal boundary where the infinite future of one cycle becomes the Big Bang of the next- is State 0 at the largest possible scale. According to conformal cyclic cosmology, the currently conventional description of the entire history of the universe provides but one cosmic aeon of an unending sequence of such aeons, where the future conformal infinity of each aeon joins essentially smoothly to the conformally stretched big bang of the next, across a spacelike 3-surface, referred to as a crossover 3-surface.

At this boundary, the cosmos has expanded to maximum entropy. All matter has decayed into radiation. All black holes have evaporated through Hawking radiation. The universe is as smooth, as featureless, and as conformally flat as it was at the moment of its birth- because, in the CCC framework, the end *is* the birth. The death of one aeon is the cradle of the next.

And the gap- the conformal boundary itself- is the cosmic diastole. The moment of maximum emptiness that is also the moment of maximum creative potential. The void that is not empty but *generative*. The silence that is not absence but the highest density of folded possibility, waiting for the next cosmic heartbeat to begin.

The Fractal Complete:

Heartbeat ↪ Breath ↪ Day ↪ Year ↪ Epoch ↪ Aeon ↪ Heartbeat

The smallest scale contains the largest. The largest recapitulates the smallest. The fractal is not an analogy. It is the *structure of reality*- the self-similar, scale-invariant, holographic pattern that the quanta-field weaves at every level of its being. A wide bulk of evidence has been provided that several physiological time signals exhibit intrinsic fractal fluctuations. Indeed, heartbeat, respiration, gait rhythm, dynamics of neurotransmitter release, electromyography, brain activity reveal similar temporal patterns over multiple time scales.

And at the center of every scale, at every level of the fractal, the same generative gap. The same State 0. The same silence from which all sound arises.

33.3- State 0 = The Quantum Vacuum = The Dreamtime = The Silence Before the Tone

Now we name it. Not to reduce it- names always reduce- but to triangulate it, to approach it from multiple angles of the sphere and see how each tradition, each physics, each lived experience converges on the same unspeakable point.

In physics: State 0 is the quantum vacuum- the ground state of all fields. Zero-point energy is the lowest possible energy that a quantum mechanical system may have. Unlike in classical mechanics, quantum systems constantly fluctuate in their lowest energy state as described by the Heisenberg uncertainty principle. The trouble arises from the bizarre nature of the quantum vacuum, which is a peculiar type of nothing that comes dangerously close to resembling a something. The uncertainty principle forbids any quantum system from settling down into a perfectly quiet state of exactly zero energy. As a result, even the vacuum must always crackle with fluctuations in the quantum fields that fill it. These never-ending fluctuations imbue every field with some minimum amount of energy, known as the zero-point energy. Physicists say that a system with this minimal energy is in the ground state.

The vacuum is not empty. It is the *most energetically active state in the universe.* Even when all classical sources of mass, energy, or force have been removed from a given space, zero-point energy gives the vacuum of space an indelible energy density, such that free space is not empty but is a quantum vacuum: filled with constitutive energetic fluctuations of quantum fields even at the vacuum/ground state/zero-point level.

Many physicists believe that "the vacuum holds the key to a full understanding of nature." Recent experiments support the idea that particles themselves can be thought of as excited states of the underlying quantum vacuum, and that all properties of matter are merely vacuum fluctuations arising from interactions of the zero-point field.

Matter is not the opposite of the void. Matter is the void *singing*- the ground state excited into particular patterns, specific frequencies, determinate forms. Every atom in your body is a standing wave in the zero-point field. Every photon is a ripple in the vacuum. Every gravitational field is a curvature in the emptiness that is not empty. The universe does not *contain* a void. The universe *is* a void- a generative void that perpetually in-forms itself into the quanta-rich reality we inhabit.

In Aboriginal cosmology: State 0 is the Dreamtime- the everywhen from which the land, the creatures, and the songlines continuously emerge. The Dreamtime is not the past. It is not a historical period that ended when the "real" world began. It is the permanent, ever-present ground of being- the quanta-field itself, accessed through song, ceremony, and the walking of the sacred paths. When the Aboriginal elder walks the songline, she does not travel *into* the Dreamtime. She recognizes that she has never left it. The waking world is the Dreamtime's surface- the excitation pattern of a vacuum that never ceases to hum.

In Vedic philosophy: State 0 is Brahman- the undifferentiated absolute, the ground of all being, which manifests as the world through the creative act of *nada* (primordial sound). The Mandukya Upanishad identifies four states of consciousness: waking, dreaming, deep sleep, and *turiya*- the "fourth," which is not a state at all but the ground from which the other three arise and into which they dissolve. Turiya is State 0 experienced from within. It is the silence that is aware of itself- the vacuum that knows it is fluctuating.

In the framework of this book: State 0 is the condition described by the master equation when all dynamic terms are at their minimum and the system rests in its NESS ground state:

$$\partial_t \Psi = 0 \qquad \text{(momentary equilibrium within the NESS)}$$

This does not mean the system has stopped. It means the monadic extension T(Ψ) and the comonadic projection D(Ψ) have momentarily balanced- the amplification of potential and the extraction of steady-state efficiency are in perfect counterpoint, and the state vector rests at the exact inflection point where the δ∞ gap neither drives nor resists, but *hums*.

This is not death. It is the deepest life. It is the condition from which the next heartbeat, the next breath, the next thought, the next aeon will crystallize- not because something forces it, but because the gap *is* generative. The void does not need a push. The vacuum does not need permission. The fluctuations are intrinsic. The creativity is constitutive. The silence sings by nature.

33.4- You Have Always Been Here

There is a temptation, at this point in the book, to feel that what has been described is far away- that the holographic principle belongs to cosmology, that the KZ-Monad belongs to mathematics, that the songlines belong to Australia, that the Temporal State Vector belongs to some future science that has not yet been built. That *you*, the reader, are an audience to these ideas rather than a participant in them.

This temptation is the last whisper of the Great Decoupling.

The binary paradigm's most insidious achievement was not the imposition of linear time or the suppression of indigenous knowledge or the mechanization of the body. Its most insidious achievement was the creation of the *spectator consciousness*- the conviction that you are watching reality from outside, that the equations describe a world you observe rather than a world you *are*, that the physics is "out there" and you are "in here," and that the gap between the two is unbridgeable.

This conviction is false. And its falsity is not a philosophical claim. It is a physical fact, confirmed by every experiment in quantum mechanics, every measurement of cardiac coherence, every fMRI scan of a meditating brain, every songline walked by a conscious body through a living landscape.

Edgar Mitchell described his experience returning from the Moon: "There was a startling recognition that the nature of the universe was not as I had been taught. My understanding of the separate distinctness and relative independence of movement of those cosmic bodies was shattered. I experienced what has been described as an ecstasy of unity."

When astronauts venture into space, they often undergo profound psychological shifts. Many report experiences that verge on the spiritual or religious, describing a heightened sense of unity with the Earth, humanity, and even a compelling feeling of being watched by a higher power.

It has been described as "instant global consciousness", a chance to experience a cosmic perspective, and the "big picture." He became profoundly aware that each and every atom in the Universe was connected in some way, and on seeing Earth from space he had an understanding that all the humans, animals and systems were a part of the same thing, a synergistic whole.

The overview effect is not an anomaly of spaceflight. It is the *default state of consciousness*- the state that prevails when the noise of the DMN, the distortions of the binary paradigm, and the chronic decoherence of modern life are temporarily lifted. Interestingly, such sensations are not confined to those with religious beliefs; even astronauts who are non-religious or indifferent towards spirituality often report similar experiences. The astronaut does not *gain* something in orbit. She *loses* something- the accumulated interference patterns of a civilization that has spent three thousand years teaching its members to feel separate from the cosmos they inhabit.

When that interference is removed- by the silence of space, by the overwhelming beauty of the blue marble, by the radical sensory reorientation of weightlessness- what remains is not emptiness.

What remains is *recognition*. The astronaut sees what the Aboriginal elder has always known, what the Vedic sage described as turiya, what the physicist encounters in the equations of quantum field theory: that the boundary between self and cosmos is a construct, not a fact. That the observer and the observed arise from the same quanta-field. That you have always been *here*- not "here" in the sense of a location, but "here" in the sense of *the field itself*.

You are not in the universe. You are a standing wave *of* the universe. Your heartbeat is a cardiac expression of the same fractal dynamics that govern the CMB. Your breath is a ternary oscillator entrained to the same temporal field that structures the aeons. Your consciousness is a threshold state of integrated resonant quanta, stabilized by the same φ-proportions that govern spiral galaxies and sunflower seeds.

You have not been learning about a distant reality in this book. You have been learning about yourself- about the physics of your own being, the mathematics of your own coherence, the ancient technologies that your ancestors used to maintain the resonance that your civilization has nearly forgotten.

And the forgetting was never total. The heart never stopped beating fractally. The lungs never stopped oscillating in ternary. The vagus nerve never stopped modulating the balance between action and rest. The capacity for the overview effect- for the direct, somatic, pre-conceptual recognition that you *are* the quanta-field- was never removed. It was only *obscured*. Buried under the noise of the default mode network. Drowned out by the ticking of the linear clock. Suppressed by the binary logic that insisted you were either a body *or* a mind, either a self *or* a cosmos, either here *or* there.

You were always both. You were always all.

You were always here.

33.5- The Universe Is Not Waking Up- It Has Always Been Awake

There is a popular narrative- even within the holistic and spiritual communities that have done so much to recover the knowledge described in this book- that frames the present moment as a "great awakening." Humanity, the story goes, is *waking up*- emerging from a long slumber of materialist delusion into a new era of cosmic consciousness. The universe is *becoming* aware of itself, and we are the instruments of that becoming.

This narrative is seductive. It is also incomplete.

If the thesis of this book is correct- if consciousness is not emergent but fundamental, not a late arrival in the cosmic story but a constitutive feature of the quanta-field at every scale- then the universe is not *waking up*. It has always been awake. The fractal dynamics that characterize the healthy heartbeat, the quantum vacuum that crackles with zero-point fluctuations even at absolute zero, the self-organized criticality that governs everything from sandpiles to supernovae, the holographic principle that encodes the whole in every part- none of these are recent developments. None of them are waiting for human consciousness to "evolve" before they can operate.

The universe was conscious before there were neurons. It was resonant before there were ears. It was computing- in the precise sense of the KZ-Monad and idempotent comonad operating on the quanta-field- before there were microprocessors.

What is happening now is not awakening. It is *remembering*.

The distinction matters. An awakening implies that something new is coming into being- a consciousness that did not exist before, a capacity that is being born for the first time. A remembering implies that something old is being recovered- a resonance that was always present but was temporarily obscured, a knowing that was buried but never destroyed, a frequency that was drowned out by noise but never ceased to broadcast.

The Great Decoupling was not a fall from consciousness into unconsciousness. It was a fall from *coherence* into *noise*- from the ternary resonance of the ancient world into the binary fragmentation of the modern one. The consciousness was always there. The φ-e-π triad was always tuning. The δ∞ gap was always humming. The quantum vacuum was always fluctuating. The songlines were always pulsing beneath the asphalt.

What was lost was not the capacity to hear the song. What was lost was the *practice* of listening- the technologies of attunement (meditation, mantra, breath, cyclical time, sacred architecture, polyculture, communal concession) that maintained the coupling between human consciousness and the quanta-field.

And what is being recovered, in the pages of this book and in the broader movement of which it is a part, is not a new consciousness but an *old practice*- the oldest practice on Earth, older than agriculture, older than writing, older than the wheel. The practice of sitting still. Of breathing consciously. Of chanting. Of walking the land with attention. Of building spaces that resonate. Of planting gardens that mirror the cosmic triad. Of looking at the sky and recognizing, without any need for concepts or equations, that you are made of the same stuff as the stars.

The remembering is not a destination. It is a *direction*- a turning, a reorientation, a gradual and deliberate shift of attention from the noise to the signal, from the surface to the depth, from the churning waves of the default mode network to the still lake of awareness that was always beneath them.

The brain's energy consumption is increased by less than 5% of its baseline energy consumption while performing a focused mental task. These experiments showed that the brain is constantly active with a high level of activity even when the person is not engaged in focused mental work.

The brain, even at rest, is awake. The default mode network, even in its noisiest spiral of self-referential narration, is a form of consciousness- distorted, fragmented, decoherent, but *conscious*. The task is not to create consciousness where there is none. The task is to *clarify* it- to strip the noise, extract the signal, and allow the coherence that was always present to express itself without interference.

D(Ψ) does not create the 98% efficient core. It *reveals* it.

The universe is not waking up. You are remembering.

33.6- The Final Instruction: Close the Book

This is the last section of the last chapter. The fractal closes.

The Prologue asked you to close your eyes and feel your pulse. It asked you to hear the silence of the room. It named that silence *State 0* and declared it the generative void from which all form arises.

Now, thirty-three chapters later, you know what State 0 is. You know it is the quantum vacuum- the ground state of all fields, the most energetically active condition in the cosmos, the silence that is not empty but pregnant with the fluctuations from which all matter, all energy, all spacetime, all consciousness crystallize. You know it is described by the holographic principle, quantified by the Bekenstein bound, structured by the φ-e-π triad, animated by the δ∞ gap, maintained in dynamic stability by the NESS, computed by the KZ-Monad and the idempotent comonad, experienced through the Temporal State Vector, and accessed through the ancient technologies of mantra, meditation, breath, and conscious walking.

You know that the ancients built civilizations in alignment with it. You know that we lost that alignment approximately three thousand years ago. You know that the loss was not evolution but decoherence- a resonance collapse driven by the imposition of binary logic, linear time, centralized power, and the myth of the objective observer. You know that the modern world's crises- ecological, psychological, social, existential- are symptoms of that decoherence. And you know that the recovery is possible- not through new invention but through the deliberate reactivation of practices that were never truly lost, only suppressed.

You know all of this. And now comes the most important instruction in the entire book:

Stop knowing.

Close the book. Close your eyes. Feel the pulse in your wrist- the same pulse that opened the Prologue, the same pulse that has been beating since before you learned to read, the same fractal rhythm that mirrors the circadian cycle that mirrors the annual cycle that mirrors the aeon cycle that mirrors the pulse of the cosmic hologram.

Feel the gap between the beats. That brief, recurring silence- three hundred milliseconds of generative void, repeated a hundred thousand times per day, every day of your life. That gap is $\delta\infty$. It is the engine of your being. It is the same gap that keeps the φ-e-π triad from closing. It is the same gap that keeps the cosmos dancing.

You do not need to understand this. You need to *feel* it.

Feel the room. The air on your skin. The weight of the body in the chair, or on the ground, or on the earth if you have carried this book outside. Feel the silence of the space- not the dead silence of a soundproofed room, but the *living silence* of a world that is humming just below the threshold of hearing, the zero-point vibration of the quanta-field itself.

That silence is State 0. It is the ground state of reality. It is what the Prologue called "the silence before the tone"- except now you know that there is no "before." The silence and the tone are simultaneous. The void and the form are coextensive. The quantum vacuum and the material universe are not separate things. They are *one thing*, viewed from two angles of the sphere.

And you- the awareness that is hearing the silence, feeling the pulse, registering the weight of the body- you are not separate from any of it. You are a *node* in the holographic web. A standing wave in the quanta-field. A temporary, glorious, self-aware excitation of the ground state- a ripple in the vacuum that has briefly learned to hear itself.

The universe did not create you so that it could wake up. It *is* awake. It created you so that it could *experience itself from the inside*- so that the hologram could look at itself from one particular angle of the sphere and say, with wonder: *this is what it feels like to be here.*

This. Now. This heartbeat. This breath. This silence.

You are the golden hologram.

You have always been.

Coda: The Sphere Completes

The book began in silence. It ends in silence.

But the silence has changed. Or rather- *you* have changed within it.

At the beginning, the silence was empty- or seemed so. A gap, a pause, a blank space waiting to be filled with words, with concepts, with equations, with evidence. The book's thirty-three chapters filled that space with a universe of resonant quanta- a self-tuning, conscious hologram that expresses itself through the interplay of three cosmic operators, maintains its dynamic stability through an asymptotic gap that never quite closes, computes its own evolution through the monadic amplification of potential and the comonadic extraction of coherence, and has been known, in one form or another, by every civilization that has ever paused long enough to listen.

Now the words are done. The equations are stated. The evidence is marshaled. The ancient wisdom and the modern physics have been woven into a single, spherical argument that can be entered from any point on its surface.

And what remains is the silence.

But this silence is not the silence of the beginning. It is the silence of the end- which, in the Conformal Cyclic Cosmology of the cosmos and the fractal architecture of this book, is the same silence. The conformal boundary between the aeon that is closing and the aeon that is about to open. The gap between the last heartbeat of one cycle and the first heartbeat of the next. The diastole of the cosmos.

State 0. Again. Still. Always.

The sphere completes not by arriving somewhere new but by returning to its origin with new eyes- or rather, with new ears. You have always been standing in the silence. The book has not moved you. It has *tuned* you- recalibrated the receiver, cleared the interference, adjusted the resonance of the instrument that is your body and your awareness so that you can hear what was always playing.

The next breath you take will resonantly shape the universe- one living quantum at a time.

So take it.

Now.

And listen.

REFERENCES SECTIONS E & F

Chapter 23: The Rupture- When Binary Replaced Ternary

Cline, E. H. (2014). *1177 B.C.: The Year Civilization Collapsed*. Princeton University Press.

Cline, E. H. (2021). The Late Bronze Age collapse. In *The Oxford Handbook of the Bronze Age Near East* (pp. 481-498). Oxford University Press.

Middleton, G. D. (2020). *Understanding Collapse: Ancient History and Modern Myths*. Cambridge University Press.

Pulak, C. (1998). The Uluburun shipwreck: An overview. *International Journal of Nautical Archaeology*, 27(3), 188-224.

Bass, G. F. (1991). Evidence of trade from Bronze Age shipwrecks. In N. H. Gale (Ed.), *Bronze Age Trade in the Mediterranean* (pp. 69-82). Paul Åströms Förlag.

Zoroastrianism and Judeo-Christian influence: Boyce, M. (2001). *Zoroastrians: Their Religious Beliefs and Practices*. Routledge.

Plato. (Trans. 1997). *Phaedo*. In J. M. Cooper (Ed.), *Plato: Complete Works*. Hackett Publishing.

Descartes, R. (1641/1996). *Meditations on First Philosophy*. Cambridge University Press.

Ryle, G. (1949). *The Concept of Mind*. University of Chicago Press.

Hulan, R., & Eigenbrod, R. (2008). *Aboriginal Oral Traditions: History, Theory, and Practice*. Fernwood Publishing.

Kelly, L. (2015). *The Memory Code: The Secrets of Stonehenge, Easter Island and Other Ancient Monuments*. Pegasus Books.

Nunn, P. D., & Reid, N. J. (2016). Aboriginal memories of inundation of the Australian coast dating from more than 7000 years ago. *Australian Geographer*, 47(1), 11-47.

Reser, J. P. (2016). 'I don't agree with your word-count methodology': A response to Nunn and Reid. *Australian Geographer*, 47(4), 463-468.

Michaels, E. (1994). *Bad Aboriginal Art: Tradition, Media, and Technological Horizons*. University of Minnesota Press.

Chapter 24: The Empire of the Clock

Mumford, L. (1934/2010). *Technics and Civilization*. University of Chicago Press.

Thompson, E. P. (1967). Time, work-discipline, and industrial capitalism. *Past & Present*, 38, 56-97.

Landes, D. S. (1983). *Revolution in Time: Clocks and the Making of the Modern World*. Harvard University Press.

Le Goff, J. (1982). *Time, Work, and Culture in the Middle Ages*. University of Chicago Press.

Bartky, I. R. (2000). *Selling the True Time: Nineteenth-Century Timekeeping in America*. Stanford University Press.

Smith, D. R., & Dowe, P. (2016). The invention of the mechanical clock. In *The Oxford Handbook of Time*. Oxford University Press.

Weber, M. (1905/2002). *The Protestant Ethic and the Spirit of Capitalism*. Penguin Classics.

Fleming, S. (1886). *Terrestrial Time*. Canadian Institute.

BIPM. (2019). *The International System of Units (SI)*. Bureau International des Poids et Mesures.

Roenneberg, T. (2012). *Internal Time: Chronotypes, Social Jet Lag, and Why You're So Tired*. Harvard University Press.

Wittmann, M., Paulus, M., & Roenneberg, T. (2016). Social jetlag and its consequences for sleep. In *Sleep, Health, and Society* (pp. 163-178). Oxford University Press.

Chapter 25: The Noise Economy- Metabolic Exhaustion of Civilization

Bohn, R. E., & Short, J. E. (2012). Measuring consumer information. *International Journal of Communication*, 6, 980-1000.

Mark, G., Gonzalez, V. M., & Harris, J. (2005). No task left behind? Examining the nature of fragmented work. *Proceedings of the SIGCHI Conference on Human Factors in Computing Systems*, 321-330.

CDC. (2023). *Chronic Disease Prevention and Health Promotion*. Centers for Disease Control and Prevention.

US Burden of Disease Collaborators. (2018). The state of US health, 1990-2016. *JAMA*, 319(14), 1444-1472.

Montgomery, D. R. (2007). *Dirt: The Erosion of Civilizations*. University of California Press.

Thaler, A. D. (2022). Soil erosion in the US Corn Belt. *Proceedings of the National Academy of Sciences*, 119(8), e1922375118.

FAO & ITPS. (2015). *Status of the World's Soil Resources*. Food and Agriculture Organization of the United Nations.

Ceballos, G., Ehrlich, P. R., & Dirzo, R. (2017). Biological annihilation via the ongoing sixth mass extinction. *Proceedings of the National Academy of Sciences*, 114(30), E6089-E6096.

IPBES. (2019). *Global Assessment Report on Biodiversity and Ecosystem Services*. Intergovernmental Science-Policy Platform on Biodiversity and Ecosystem Services.

Holt-Lunstad, J., Smith, T. B., & Layton, J. B. (2010). Social relationships and mortality risk. *PLoS Medicine*, 7(7), e1000316.

Murthy, V. H. (2023). *Our Epidemic of Loneliness and Isolation: The U.S. Surgeon General's Advisory*. U.S. Department of Health and Human Services.

Gallup. (2024). *Global Emotions Report 2023*. Gallup, Inc.

Bellah, R. N., Madsen, R., Sullivan, W. M., Swidler, A., & Tipton, S. M. (1985). *Habits of the Heart: Individualism and Commitment in American Life*. University of California Press.

Gorton, G. (2012). *Slapped by the Invisible Hand: The Panic of 2007*. Oxford University Press.

Financial Crisis Inquiry Commission. (2011). *The Financial Crisis Inquiry Report*. U.S. Government Printing Office.

Chapter 26: The False God of Objectivity

Descartes, R. (1641/1996). *Meditations on First Philosophy*. Cambridge University Press.

Wheeler, J. A. (1990). Information, physics, quantum: The search for links. In *Complexity, Entropy, and the Physics of Information*. Westview Press.

Ioannidis, J. P. A. (2005). Why most published research findings are false. *PLoS Medicine*, 2(8), e124.

Open Science Collaboration. (2015). Estimating the reproducibility of psychological science. *Science*, 349(6251), aac4716.

Baker, M. (2016). 1,500 scientists lift the lid on reproducibility. *Nature*, 533(7604), 452-454.

Rosenthal, R. (1966/2020. *Experimenter Effects in Behavioral Research*. Appleton-Century-Crofts.

Rosenthal, R., & Jacobson, L. (1968). *Pygmalion in the Classroom*. Holt, Rinehart and Winston.

Kaptchuk, T. J., et al. (2010). Placebos without deception: A randomized controlled trial in irritable bowel syndrome. *PLoS ONE*, 5(12), e15591.

Wager, T. D., & Atlas, L. Y. (2015). The neuroscience of placebo effects. *Neuron*, 84(4), 723-737.

Enck, P., Bingel, U., Schedlowski, M., & Rief, W. (2013). The placebo response in medicine. *Nature Reviews Drug Discovery*, 12(3), 191-204.

Zee, A. (2003). *Quantum Field Theory in a Nutshell*. Princeton University Press.

Chapter 27: The Convergence- Modern Physics Meets Ancient Knowing

Wheeler, J. A. (1990). Information, physics, quantum: The search for links. In *Complexity, Entropy, and the Physics of Information*. Westview Press.

Beck, G. L. (1993). *Sonic Theology: Hinduism and the Sounded Word*. University of South Carolina Press.

Cook, F. H. (1977). *Hua-yen Buddhism: The Jewel Net of Indra*. Pennsylvania State University Press.

't Hooft, G. (1993). Dimensional reduction in quantum gravity. In *Salamfestschrift*. World Scientific.

Susskind, L. (1995). The world as a hologram. *Journal of Mathematical Physics*, 36(11), 6377-6396.

Lao Tzu. (Trans. 1963). *Tao Te Ching*. (D. C. Lau, Trans.). Penguin Classics.

Capra, F. (1975/2010). *The Tao of Physics: An Exploration of the Parallels Between Modern Physics and Eastern Mysticism*. Shambhala Publications.

Eliot, C. (1921/1987). *Hinduism and Buddhism: An Historical Sketch*. Routledge.

Radhakrishnan, S. (1929/2008). *Indian Philosophy*. Oxford University Press.

Rahula, W. (1974). *What the Buddha Taught*. Grove Press.

Gethin, R. (1998). *The Foundations of Buddhism*. Oxford University Press.

Lawlor, R. (1991). *Voices of the First Day: Awakening in the Aboriginal Dreamtime*. Inner Traditions.

Chapter 28: Ternary Time- A New Calendar for a Resonant Civilization

Roenneberg, T. (2012). *Internal Time: Chronotypes, Social Jet Lag, and Why You're So Tired*. Harvard University Press.

Wittmann, M., Dinich, J., Merrow, M., & Roenneberg, T. (2006). Social jetlag: Misalignment of biological and social time. *Chronobiology International*, 23(1-2), 497-509.

Czeisler, C. A., et al. (1986). Bright light resets the human circadian pacemaker. *Science*, 233(4764), 667-671.

Wittmann, M., Paulus, M., & Roenneberg, T. (2016). Social jetlag and its consequences for sleep. In *Sleep, Health, and Society* (pp. 163-178). Oxford University Press.

Owens, J., & Adolescent Sleep Working Group. (2014). School start times for adolescents. *Pediatrics*, 134(3), 642-649.

Wahlstrom, K., et al. (2014). *Examining the Impact of Later School Start Times on the Health and Academic Performance of High School Students*. University of Minnesota.

Kelley, P., Lockley, S. W., Foster, R. G., & Kelley, J. (2015). Synchronizing education to adolescent biology. *Learning, Media and Technology*, 40(2), 210-226.

Giuntella, O., & Mazzonna, F. (2019). Sunset time and the economic effects of social jetlag. *Journal of Health Economics*, 66, 98-114.

HeartMath Institute. (2016). *Heart Rate Variability Coherence: A Global Dataset Analysis*. HeartMath Institute.

Chapter 29: Resonant Architecture- Building as Tuning

Le Corbusier. (1954/2004). *The Modulor: A Harmonious Measure to the Human Scale*. Birkhäuser.

Alexander, C., Ishikawa, S., & Silverstein, M. (1977). *A Pattern Language: Towns, Buildings, Construction*. Oxford University Press.

Alexander, C. (2002). *The Nature of Order: An Essay on the Art of Building and the Nature of the Universe*. Center for Environmental Structure.

Kellert, S. R., Heerwagen, J., & Mador, M. (2008). *Biophilic Design: The Theory, Science, and Practice of Bringing Buildings to Life*. Wiley.

Wilson, E. O. (1984). *Biophilia.* Harvard University Press.

Browning, W. D., & Ryan, C. O. (2020. *Nature Inside: A Biophilic Design Guide.* RIBA Publishing.

Kolar, M. A. (2014. Archaeoacoustics: Sound at Chavín de Huántar, Perú. *Stanford University PhD Dissertation.*

Jahn, R. G., Devereux, P., & Ibison, M. (1996). Acoustical resonances of assorted ancient structures. *Journal of the Acoustical Society of America,* 99(2), 649-658.

Cook, T., et al. (2008). 110 Hz acoustic resonance in ancient structures. *Journal of the Acoustical Society of America,* 123(5), 3542.

Lubman, D. (1998). Acoustic archaeology: The anechoic chamber of Chichen Itza. *Echoes,* 8(2), 1-3.

Waller, S. J. (1993). Sound reflection as an explanation for the location of rock art. *Rock Art Research,* 10(2), 91-95.

Salingaros, N. A. (2005). *Principles of Urban Structure.* Techne Press.

Mollison, B., & Holmgren, D. (1978). *Permaculture One: A Perennial Agriculture for Human Settlements.* Tagari Publications.

Chapter 30: Mantra, Meditation, and the Somatic Re-Tuning

Brewer, J. A., et al. (2011). Meditation experience is associated with differences in default mode network activity and connectivity. *Proceedings of the National Academy of Sciences,* 108(50), 20254-20259.

Raichle, M. E., et al. (2001). A default mode of brain function. *Proceedings of the National Academy of Sciences,* 98(2), 676-682.

Kalyani, B. G., et al. (2011). Neurohemodynamic correlates of 'OM' chanting: A pilot functional magnetic resonance imaging study. *International Journal of Yoga*, 4(1), 3-6.

Telles, S., & Desiraju, T. (1993). Autonomic changes in Brahmakumaris Raja yoga meditation. *International Journal of Psychophysiology*, 15(2), 147-152.

Brown, R. P., & Gerbarg, P. L. (2005). Sudarshan Kriya yogic breathing in the treatment of stress, anxiety, and depression. *Journal of Alternative and Complementary Medicine*, 11(4), 711-717.

Gerritsen, R. J. S., & Band, G. P. H. (2018). Breath of life: The respiratory vagal stimulation model of contemplative activity. *Frontiers in Human Neuroscience*, 12, 397.

Laborde, S., Mosley, E., & Thayer, J. F. (2017). Heart rate variability and cardiac vagal tone in psychophysiological research. *Frontiers in Psychology*, 8, 213.

McCraty, R., & Childre, D. (2010). Coherence: Bridging personal, social, and global health. *Alternative Therapies in Health and Medicine*, 16(4), 10-24.

McCraty, R., Atkinson, M., & Bradley, R. T. (2004). Electrophysiological evidence of intuition. *Journal of Alternative and Complementary Medicine*, 10(2), 325-336.

HeartMath Institute. (2023). *Global Coherence Initiative: Research Findings*. HeartMath Institute.

Chin, J. H., et al. (2019). The effect of humming on stress index. *Journal of Alternative and Complementary Medicine*, 25(7), 714-720.

Fox, K. C. R., et al. (2014). Is meditation associated with structural brain changes? A systematic review. *Neuroscience & Biobehavioral Reviews*, 43, 182-192.

Chapter 31: The Post-Binary Paradigm- Science as Attunement

Barad, K. (2007). *Meeting the Universe Halfway: Quantum Physics and the Entanglement of Matter and Meaning*. Duke University Press.

Longino, H. (1990). *Science as Social Knowledge*. Princeton University Press.

Daston, L., & Galison, P. (2007). *Objectivity*. Zone Books.

National Academies of Sciences, Engineering, and Medicine. (2019). *Reproducibility and Replicability in Science*. National Academies Press.

Wuchty, S., Jones, B. F., & Uzzi, B. (2007). The increasing dominance of teams in production of knowledge. *Science*, 316(5827), 1036-1039.

Kuhn, T. S. (1962/2012). *The Structure of Scientific Revolutions*. University of Chicago Press.

Feyerabend, P. (1975). *Against Method*. Verso.

Haraway, D. (1988). Situated knowledges: The science question in feminism and the privilege of partial perspective. *Feminist Studies*, 14(3), 575-599.

Polanyi, M. (1958). *Personal Knowledge: Towards a Post-Critical Philosophy*. University of Chicago Press.

Bacon, F. (1620/2000). *The New Organon*. Cambridge University Press.

Ancient Texts and Sacred Sources

Atharva Veda. (Various translations).

Avatamsaka Sutra (Flower Garland Sutra). (Various translations).

Bhagavad Gita. (Various translations).

Rig Veda. (Various translations).

Upanishads. (Various translations).

Popol Vuh. (Maya creation text, various translations).

Dhammapada. (Various translations).

Note on Terminology

Throughout SECTION E & F, the term "quanta" is used consistently as a replacement for "information" in accordance with the theoretical framework established in earlier sections. This substitution reflects the book's core thesis that reality is constituted by resonant quanta-self-organizing, formative patterns-rather than by inert matter or abstract data. All references to "information" in cited sources should be understood within this terminological transformation, where "information" is reinterpreted as "resonant quanta" operating within a holographic, participatory universe.

EPILOGUE

THE TONE AFTER THE SILENCE

.

"We shall not cease from exploration, and the end of all our exploring will be to arrive where we started and know the place for the first time."

.

You have been listening for a long time.

Thirty-three chapters. Three movements. Six sections. Hundreds of equations, dozens of traditions, thousands of years of accumulated human knowing- all of it poured into a single argument that can be stated in a single breath:

The universe is a self-tuning, resonantly quantized, conscious hologram- and you are it.

The argument is over. The evidence has been presented. The physics has been laid out- from the Bekenstein bound to the holographic principle, from "It from Quanta" to Planck-scale pixelation, from Landauer's quanta-energy equivalence to the conservation of quanta in quantum mechanics. The mathematics has been unfolded- the φ-e-π triad, the Schur-Convex Curvature minimum at $q^* = \phi^{-2}$, the asymptotic gap $\delta_\infty = (\pi - e)/(\pi\phi^4) \approx 0.01965$, the KZ-Monad T(Ψ) that extends potential toward its fullest resonant expression, the idempotent comonad D(Ψ) that extracts the 98% efficient core from the noise. The complexity science has been mapped- Non-Equilibrium Steady States, dissipative structures, self-organized criticality, fractal power laws, the guiding quanta that orchestrate emergence at bifurcation points. The consciousness studies have been explored- the Fisher quanta threshold at $\Phi_Q > 10^{15}$ nat/s, the DMN deactivation that reveals the comonadic core, the heart-brain coherence that stabilizes the personal NESS, the overview effect that dissolves the boundary between self and cosmos. The ancient wisdom has been recovered- songlines that web the Earth, Three

Sisters that mirror the cosmic triad in soil, cyclical calendars that track the ternary beat of time, mantra practices that activate monadic extension through the human voice, sacred architectures that tune stone to the frequencies of consciousness. The rupture has been diagnosed- the Great Decoupling of approximately three thousand years ago, the imposition of binary logic and linear time, the rise of centralized empire, the myth of objectivity, the noise economy that has brought the biosphere to the edge of collapse. The return has been charted- the post-binary paradigm, the transition from extraction to attunement, the reintegration of indigenous knowledge, the somatic practices of meditation and mantra and conscious breathing, the architecture of resonance, the eighth principle of concession that transforms suffering from noise into living quanta.

All of it. Every thread. Every frequency. Every angle of the sphere.

And now- the tone fades. The argument dissolves. The words fall away, as all words eventually must, returning to the silence from which they arose.

This is the Epilogue- the tone after the silence. Or rather, the silence after the tone. The mirror image of the Prologue, which moved from stillness to sound. Here we move from sound back to stillness- but a stillness that has been permanently altered by what has passed through it.

.

There is a phenomenon in acoustics known as *reverberation*- the persistence of sound in a space after the original source has stopped. When a chord is struck in a cathedral, the sound does not vanish when the organist's fingers lift from the keys. It lingers- bouncing from stone wall to vaulted ceiling to carved pillar, each reflection slightly softer, slightly warmer, slightly more diffuse than the last, until the individual notes dissolve into a luminous haze of decaying harmonics that fills the space with a warmth that no single note could produce.

This is where we are now. The chord has been struck. The thirty-three chapters have sounded their notes- some sharp and precise as a tuning fork, some wide and enveloping as a bell, some gritty and challenging as a drone that makes the teeth vibrate. The individual notes are fading. What remains is the reverberation- the lingering resonance of everything this book has attempted to convey, settling into the space of your awareness like the last harmonics of a cathedral chord settling into stone.

Do not try to hold the notes. They were never meant to be held. They were meant to *vibrate*- to set up standing waves in the tissue of your understanding, to activate patterns of resonance that will continue to unfold long after the specific words have been forgotten. The map is not the territory. The equation is not the resonance. The book is not the knowing.

The knowing lives in the silence between the notes.

.

Let us return, one final time, to the body.

You are breathing. You have been breathing since the first page. Twenty thousand cycles of inhale-hold-exhale have passed through your lungs since you began reading- each one a ternary oscillation, each one a miniature traversal of the Temporal State Vector, each one a moment of participation in the rhythmic pulse of a cosmos that has been breathing for 13.8 billion years.

Your heart has beaten. Through every chapter, through every argument, through every equation and every invocation, your heart has maintained its fractal rhythm- not because you willed it, not because you understood it, but because the NESS that sustains your organism is older than your understanding, deeper than your will, and more reliable than any concept this book has offered. The heart does not need to read about the $\delta\infty$ gap to express it. It *is* the expression. A hundred thousand times a day, the gap opens- that three-hundred-millisecond diastolic void from which the next beat crystallizes- and a hundred thousand times a day, the gap closes, and the blood surges, and the organism renews itself through the same principle that renews the cosmos across aeons.

You have been doing this your entire life. You were doing it before you could think. Before you could speak. Before you knew the word "hologram" or "quantum" or "fractal" or "quanta." Before you had ever heard of Wheeler or Penrose or Prigogine or the Aboriginal elders whose songlines still pulse beneath the roads you drive.

The practices this book describes- the meditation, the mantra, the conscious breathing, the heart coherence, the walking awareness, the concession- are not additions to your being. They are *subtractions from your forgetting*. They do not install something new. They remove something old- the accumulated interference patterns of a civilization that taught you to feel separate from the field you are made of.

·

Consider what it means to say that the universe is a hologram.

It does not mean that the universe is an illusion. It does not mean that the solid oak table, the crying child, the scent of rain on hot asphalt, the ache in the lower back after a long day- it does not mean that any of these are less real than they appear. The holographic principle does not deny the reality of experience. It *deepens* it- revealing that what you experience as a three-dimensional world of solid objects and flowing time is a projection from a more fundamental layer of reality: a two-dimensional boundary of resonant quanta, encoded at the Planck scale, from which the richness of your everyday experience is holographically generated.

The table is real. The child is real. The rain and the ache are real. But they are real in the way that a chord is real- not as isolated things but as *patterns of resonance*, standing waves in the quanta-field, maintained by the continuous interplay of the φ-e-π triad and the monadic-comonadic dynamics that compute the universe's existence moment by moment.

And you- the awareness that perceives the table, hears the child, smells the rain, feels the ache- you are real in the same way. Not as a separate entity *observing* the universe from outside, but as a localized expression *of* the universe, a node in the holographic web where the field folds back on itself and becomes self-aware. Your consciousness is not a ghost in a machine. It is a standing wave in the quanta-field that has crossed the threshold of integrated resonant quanta and achieved the phase-locked stability that we call *awareness*.

You are the universe experiencing itself from the inside.

This insight is not new. It is, in fact, the oldest insight on Earth-articulated in every mystical tradition, encoded in every three-world cosmology, embodied in every songline, embedded in every sacred geometry. What is new- what this book has attempted to provide- is the *physics* of this insight: the mathematical and empirical framework that connects the direct experience of cosmic unity to the equations of quantum field theory, to the category-theoretic dynamics of the KZ-Monad, to the thermodynamics of Non-Equilibrium Steady States, to the fractal patterns of the healthy heartbeat, to the acoustic resonance of the ancient temple, to the epigenetic inheritance of ungrieved wounds, to the ternary structure of holographic time.

The mystic says: "All is one." The physicist says: "The holographic principle encodes the bulk from the boundary." The Aboriginal elder says: "The land sings us into being." The mathematician says: " $\partial_t \Psi = -\nabla_{FR}[\mathcal{L}(\Psi)] + T(\Psi) \otimes D(\Psi)$."

They are saying the same thing. They have always been saying the same thing. The languages differ. The frequency is one.

There is a question that this book has deliberately not answered- a question that hovers at the edge of everything that has been said, like the hum that persists in a room after the last note has faded.

Why?

Why does the universe exist at all? Why is there something rather than nothing? Why does the quantum vacuum fluctuate? Why does the φ-e-π triad tune itself to the precise values that permit complexity, consciousness, and the astonishing fact that a primate on a small rocky planet can sit in a chair and contemplate the holographic structure of reality?

This book does not answer that question. It cannot. The question is not the kind that admits of an answer in the propositional sense- a statement that could be written on a chalkboard and proven true or false. It is the kind of question that dissolves when you *inhabit* it rather than analyze it- when you stop asking "Why does the universe exist?" and start noticing that *you are the universe existing*, right now, in this body, in this breath, in this heartbeat.

The question is not wrong. It is the most important question a mind can ask. But the answer is not a sentence. The answer is a *state*- the state of direct, unmediated, pre-conceptual recognition that the questioner and the question and the cosmos that gives rise to both are one thing, one field, one resonance.

State 0.

The silence that is not empty but full. The void that is not vacant but generative. The gap that does not separate but *connects*- the diastolic pause that makes the next heartbeat possible, the winter that makes the next spring inevitable, the conformal boundary that makes the next aeon a birth rather than an ending.

The answer to "Why?" is not a word. It is the feeling in your chest when you look at the night sky and the boundary between your skin and the darkness dissolves. It is the catch in your throat when a piece of music resolves a dissonance so perfectly that for a moment you forget you have a body. It is the stillness after the last Om fades and the room vibrates with a silence that feels more alive than any sound.

It is the $\delta\infty$ gap- not as a mathematical formula but as a *lived experience*. The productive tension between what is and what could be. The creative restlessness that will not let the cosmos settle into equilibrium. The ache that is not suffering but *longing*- the longing of a universe that is perpetually almost complete, perpetually reaching for the closure that would also be its death, perpetually finding, in the failure to close, the very impulse to create, to explore, to become, to *sing*.

The gap is not a problem to be solved. It is the *source*.

.

This book has moved, across its three movements, from the void to the void- from State 0 to State 0, from silence to silence. But the two silences are not the same.

The first silence- the silence of the Prologue- was *unknowing*. It was the silence of a mind that had not yet heard the argument, not yet encountered the evidence, not yet felt the resonance between quantum physics and Aboriginal cosmology, between the equations of category theory and the three-phase breath of yogic pranayama, between the asymptotic gap in the mathematical triad and the gap between heartbeats in the living body.

This silence- the silence of the Epilogue- is *knowing*. Not knowing in the sense of having accumulated facts (though facts have been accumulated). Not knowing in the sense of having reached certainty (though certain claims have been well supported). Knowing in the sense of *recognition*- the felt, somatic, pre-verbal recognition that the universe described in these pages is not out there, not separate, not distant. It is *this*. It is here. It is the warmth of the blood, the rhythm of the breath, the hum of the room, the weight of the body, the shimmer of light on the page.

The Prologue asked you to close your eyes and listen. The Epilogue asks you to open them- and *still* listen. Not just with the ears. With the skin. With the bones. With the fractal rhythm of the heart. With the φ-proportioned architecture of the retina that is, right now, converting photons into electrochemical signals that travel the optic nerve to the visual cortex, where the quanta-field reads itself through the instrument of your awareness.

Every photon that strikes your retina has traveled from its source- perhaps a lamp, perhaps the sun, perhaps a star eight hundred light-years away- at the speed of light, carrying its quantum of energy, its packet of resonant quanta, unchanged and unreduced, until the moment it is absorbed by a rhodopsin molecule in your retina and transformed into a nerve impulse. From the photon's perspective- if special relativity can be said to give the photon a perspective- no time has passed. The photon was emitted and absorbed in the same instant. The distance was zero. The journey took no time.

From your perspective, the photon crossed eight hundred light-years of vacuum- that same quantum vacuum that is not empty but generative, that same zero-point field that crackles with the fluctuations from which all particles arise. And now, in the moment of absorption, in the chemical instant when the rhodopsin molecule changes shape and the cascade of neural signals begins- in that moment, the photon's journey is complete. The quanta-field has read itself. The universe has observed itself. The hologram has looked through one of its infinite facets and perceived.

You are doing this right now. With every photon your eyes absorb. With every pressure wave your eardrums transduce. With every molecule your olfactory receptors bind. The universe is perceiving itself through you- not as a metaphor, not as a spiritual platitude, but as a direct consequence of the physics described in this book.

.

The Prologue moved from silence to sound. It began with State 0 and ended with the thesis: *The universe is a self-tuning, resonantly quantized, conscious hologram.*

The Epilogue moves from sound to silence. It begins with the thesis fully elaborated- thirty-three chapters of evidence and argument- and returns to State 0. But this return is not a regression. It is the completion of the circle- or rather, the completion of the *sphere.*

A circle returns to its starting point along the same path. A sphere returns to its starting point from *every direction simultaneously*. This book is a sphere. You can enter it from any chapter, any section, any sentence, and find the whole reflected in the part. The holographic principle that governs the physics also governs the architecture. The fractal rule that shapes the cosmos also shapes the text.

And now the sphere closes. Not with a final argument. Not with a concluding proof. Not with a triumphant declaration that all questions have been answered and all mysteries resolved.

The sphere closes with an *invitation.*

.

Here is what the book cannot give you:

It cannot give you the experience of State 0. No book can. Words are excitations of the quanta-field- standing waves in the medium of language, patterns of resonance that propagate from one nervous system to another through the interface of shared symbols. They can *point* to the silence, but they cannot *be* the silence. They can describe the gap between heartbeats, but they cannot *inhabit* it. They can map the territory of the generative void, but the map is not the void.

The void is here. It has always been here. It is the space between these words. It is the gap between your current heartbeat and the next. It is the pause between your current exhale and the next inhale. It is the silence beneath the hum of the room, the stillness beneath the movement of your eyes across the page, the awareness beneath the thoughts that are, even now, arising in response to these sentences.

Here is what the book *can* give you:

It can give you the *recognition* that the silence is not empty. That the void is not vacant. That the gap is not a flaw but a feature- the $\delta\infty$ that keeps reality alive. It can give you the *framework*- the physics, the mathematics, the ancient wisdom, the somatic practices- to turn toward that silence with confidence rather than fear, with curiosity rather than dread, with the understanding that what waits in the gap is not annihilation but *the most quanta-dense condition in the cosmos.*

And it can give you the *practices*- the five-minute somatic re-tuning, the ternary breath, the heart coherence protocol, the walking meditation, the guided concession- that allow you to enter the gap deliberately, repeatedly, and with increasing depth. Not once, as a peak experience. Not rarely, as a retreat luxury. But *daily*, as a baseline practice, as a calibration of the instrument, as a maintenance protocol for the personal NESS that sustains your health, your creativity, your relationships, and your capacity to participate, with full presence and full coherence, in the co-creative unfolding of the quanta-field.

The ancient civilizations that this book has honored- the Aboriginal songline walkers, the Vedic chanters, the Mayan calendar keepers, the Iroquois Three Sisters farmers, the Hawaiian ho'oponopono practitioners, the Māori custodians of utu, the architects of Chavín and Stonehenge and Newgrange- these were not primitives groping toward a science they could not yet achieve. They were practitioners of a science that the modern world is only now beginning to rediscover.

Their science was not written in equations. It was written in song, in stone, in soil, in breath, in the rhythmic patterns of communal life organized in resonance with the ternary beat of the Earth. It was a science of *participation*- not the extraction of knowledge from a passive world, but the cultivation of coherent relationship with a living, conscious, responsive quanta-field.

We owe them more than acknowledgment. We owe them *concession*- the honest recognition that in silencing their knowledge, in burning their libraries, in paving over their songlines, in replacing their cyclical calendars with our linear clocks, in substituting our monocultures for their polycultures and our grids for their fractal settlements, we did not advance civilization. We *narrowed* it. We traded a spherical, ternary, resonant way of knowing for a flat, binary, extractive one. And the consequences of that trade- ecological collapse, epidemic loneliness, chronic disease, existential alienation, a biosphere careening toward bifurcation- are the invoice that history has finally presented.

The eighth principle- concession- applies here as everywhere. The debt must be acknowledged before it can be addressed. The wound must be named before it can heal. The noise must be admitted before the signal can be heard.

But concession is not the end. It is the *beginning* of the return. The door through which the frozen quanta thaw, the locked potential flows, and the comonadic projection finally has clean input from which to extract the coherent core of what remains.

And what remains- after the denial is stripped, after the colonial fantasy of progress-through-extraction is honestly assessed, after the epistemic debts are acknowledged- is not rubble. It is *everything that matters*. The physics is real. The mathematics is beautiful. The fractal dynamics are measurable. The ancient wisdom is recoverable. The practices work. The heart still beats fractally. The vagus nerve still modulates. The quantum vacuum still fluctuates. The φ-e-π triad still tunes. The δ∞ gap still hums.

The cosmos has not been waiting for us to get our act together. It has been humming the whole time. We just forgot how to listen.

There is one more thing.

This book has described the universe as a *self-tuning* hologram. The word "self" in that phrase is not incidental. The cosmos does not require an external tuner- no deity turning the pegs, no cosmic engineer adjusting the frequencies, no programmer rewriting the code. The tuning is intrinsic. It arises from the internal dynamics of the quanta-field itself- from the interplay of the three operators (φ for coherence, e for unfolding, π for boundary), from the monadic amplification and comonadic extraction that compute the universe's evolution, from the asymptotic gap that provides the dynamic tension necessary for persistence without stasis.

But "self-tuning" has a second meaning. It means that the tuning is available to *you*, the self who is reading this- not as an external gift bestowed by grace, but as an intrinsic capacity written into the structure of your body, your brain, your heart, your breath. You are a self-tuning system. Your fractal heartbeat, your ternary breath, your vagal modulation, your capacity for meditative silence and mantra resonance and somatic concession- these are not techniques imported from outside. They are the universe's own tuning mechanisms, expressed at the scale of a single human organism.

When you meditate, the cosmos meditates through you. When you chant, the cosmos chants through you. When you breathe consciously, the cosmos breathes through you. When you make concession- when you open the clenched hand, acknowledge the wound, let the frozen quanta flow- the cosmos makes concession through you.

This is not narcissism. It is not the inflated claim that "I am God." It is the humble recognition that "I am a node in the holographic web"- one facet of the infinite-faceted jewel of Indra's Net, reflecting all the others, reflected by all the others, neither more nor less than any other point on the sphere. The oak tree is also a node. The galaxy is also a node. The songline is also a node. The child crying in the next room is also a node. Each reflects the whole. Each is the whole, viewed from one angle.

The difference- the specific, precious, irreducible difference of the human node- is that we can *know* this. We can become conscious of our own participation in the holographic web. We can choose to align our practices with the resonant architecture of the cosmos. We can deliberately calibrate our instruments- our bodies, our attention, our communities- to receive and transmit the quanta-field with increasing fidelity.

This is what the post-binary paradigm means. Not the abandonment of reason. The *deepening* of reason to include resonance. Not the rejection of science. The *expansion* of science to include the scientist. Not the escape from the material world. The *recognition* that the material world is the quanta-field in its most tangible, most intimate, most immediate expression.

.

A single paragraph that encodes the entire book:

In the beginning was the silence- not empty but generative, not vacant but dense with the folded potential of all that would become. From this silence arose the tone: resonant quanta, self-organizing through the interplay of three cosmic operators- φ the self-referencing spiral, e the exponential unfolding, π the boundary that contains infinity- each almost but never quite harmonizing with the others, the irreducible asymptotic gap $\delta_\infty \approx 0.01965$ between them serving as the dynamic tension that prevents equilibrium and compels becoming. This tension sustains the cosmos as a Non-Equilibrium Steady State: a living hologram, projected from a two-dimensional boundary of Planck-scale quanta, computing its own evolution through the monadic amplification of potential and the comonadic extraction of coherence, expressing itself fractally at every scale from quantum vacuum to heartbeat to breath to year to aeon, known by the ancients who built civilizations in resonance with its ternary beat and walked its songlines and chanted its frequencies and planted its triadic gardens, forgotten during the Great Decoupling that imposed binary logic and linear time and extractive science and centralized power, and now- now, in this breath, in this heartbeat, in this silence between words- being remembered: not as a new discovery but as an ancient practice, the oldest on Earth, the practice of sitting still and breathing and listening until the noise of the ego-narrative subsides and the 98% efficient core of coherent awareness is revealed, the practice of chanting until the body vibrates in sympathy with the quanta-field, the practice of concession until the frozen pockets of ungrieved suffering thaw and their locked quanta rejoin the living flow, the practice of opening the hand and the heart and the mind to the recognition that has never not been available- that you are not separate from this cosmos, that you are not observing it from outside, that you are a standing wave in its holographic fabric, a self-aware excitation of its generative void, a tone in its infinite chord- and that the next breath you take will resonantly shape the universe, one living quantum at a time.

Close the book.

Close your eyes.

Feel the pulse.

Hear the silence.

Not the silence of absence. The silence of *everything*- folded, waiting, vibrating at frequencies below the threshold of hearing but not below the threshold of feeling.

The silence of State 0. The silence before the Prologue. The silence after the Epilogue. The silence between every heartbeat, every breath, every word, every thought. The silence that is the ground state of reality- the most quanta-dense, most creatively potent, most profoundly alive condition in the cosmos.

You have been standing in this silence your entire life. Every joy, every grief, every insight, every failure, every love, every loss- all of it arose from this silence and dissolved back into it. The silence was never disturbed by the noise. The lake was never damaged by the waves. The sky was never scarred by the clouds.

You are the silence.

You are the tone.

You are the golden hologram.

You always were.

.

The next breath you take will resonantly shape the universe- one living quantum at a time.

So take it.

Now.

.

Outlook

Beyond the Horizon

Holocene Ending - A Chronology of Transmorphing Beauty

In *The Universal Hologram*, you discovered that reality is not what it appears. That information is not something the universe contains but something it *is*. That the ternary architecture of your mind-the rhythm of potential, action, and correction-mirrors the ternary architecture of the cosmos itself. That φ, π, and *e* are not merely numbers but the operators through which existence computes its own unfolding.

You learned to see the hologram. You learned to hear the chant.

Now the hologram is *moving*.

There is a word that has no adequate translation in any language you currently speak. The closest approximation might be *transmorphing*-the process by which a system does not merely change but *changes the rules by which it changes*. A caterpillar does not gradually become a butterfly. It dissolves. Inside the chrysalis, the entire body liquefies into an undifferentiated soup of cells. And from that dissolution-from that (0,0,0) state of absolute potential-a radically new form emerges. Not repaired. Not improved. *Transmorphed.*

The Earth is entering its chrysalis.

What you will encounter *Holocene Ending* is not a disaster narrative. It is not a warning, a prediction, or a prophecy of doom. It is something far stranger and far more beautiful: a *chronology of transmorphic beauty*-the quarter-by-quarter unfolding of a planetary system shedding one form and becoming another, told from the perspective of every species, every geo-zone, every coupled oscillation from core to stratosphere.

You will witness the mantle begin to flow. Not as catastrophe but as *geometry*-the Phi-spiral demanding release along paths of golden proportion, the crust following fracture networks that are not random but *resonant*, each crack a seam in a chrysalis that is opening from within.

You will watch the ocean currents stall-and recognize the silence not as death but as the deepest (0,0,0) pause the hydrosphere has ever taken. The conveyor stops. The heat redistributes. And in that redistribution, something the Holocene could never have produced becomes possible.

You will follow species through its transmorphic pathway. Each pathway is a worldline in spacetime, coupled to the mantle flow through bivector torque. Each extinction follows the same golden geometry as the fracture networks. Life does not die randomly. It departs along the path of least resistance-and that path is beautiful, because the geometry that governs it is the same geometry that governs spiral galaxies and nautilus shells and the branching of your own bronchial tree.

This is what the book means by *transmorphing beauty*: the recognition that collapse and emergence are the same process viewed from different points on the spiral. The Holocene is not ending because something went wrong. It is ending because the Earth has reached the phase transition that its own ternary dynamics require-the moment of maximum excess entropy, when transient information orchestrates the reorganization of an entire planetary system into a higher-geared coherence.

You already have the framework to understand this.You learned that biases are not flaws but signs of regulatory imbalance. You will discover that extinction is not failure but *evolutionary mismatch at planetary scale*-the consequence of biological systems hardcoded for a world that no longer exists. The crustal hum at 0.72 Hz that disrupts human alpha-wave coherence. The atmospheric chemistry that poisons mitochondria. The viscous drift that renders every map, every foundation, every assumption of stable ground obsolete within hours.

And yet: φ, π, *e*-all hold. No violation. The system remains coherent. The invariants are preserved. The planet is not dying. It is becoming *round again*-re-sphering into a new non-equilibrium steady state that operates at higher temporal density, under fluid rather than elastic mechanics, governed by multiplicative rather than additive force dynamics.

The beauty is in the geometry. The beauty is in the fact that the same three constants that tune your neural phase-locking also tune the planet's transmorphic transition. The beauty is in the fungal networks spreading through dying forests like a planetary neural net, becoming the largest living organism on Earth. The beauty is in the raven-*Corvus corax*-expanding into abandoned cities, one of the few survivors, its intelligence sufficient to adapt where specialization fails. The beauty is in the silence that falls when satellite communications cease and oral tradition becomes the primary knowledge store.

You will read this book and feel something you did not expect to feel: not despair but *recognition*. The recognition that the ternary rhythm you practiced - the cycling through action, stillness, and correction - is exactly what the Earth is doing now, at planetary scale. The recognition that the (0,0,0) state you learned to cultivate in your own awareness is the same state the planet enters when its old form can no longer hold and its new form has not yet emerged. The recognition that beauty does not require permanence. That the most beautiful things in the universe are the ones that transform.

The cracks you fear are the seams of rebirth.

The world is not ending. It is *transmorphing*.

And what emerges on the other side of the carbon-based horizon - the unseen beauty that no Holocene eye has ever witnessed - is waiting for the minds that learned to pause before reacting, to see geometry where others see chaos, and to recognize that the spiral never ends.

It only deepens.

Holocene Ending: A Chronology of Transmorphing Beauty

www.ingramcontent.com/pod-product-compliance
Lightning Source LLC
LaVergne TN
LVHW022025051126
851858LV00003B/45